The Portable

Darwin

Edited, with an Introduction, Notes and Epilogue by

DUNCAN M. PORTER
and
PETER W. GRAHAM

PENGUIN BOOKS

*We thank our colleagues Richard Bambach,
Frederick Burkhardt, and David West for their
helpful comments and Connie Noonkester for
assistance in typing the manuscript.
It was a pleasure working with our editor,
Michael Millman.*

PENGUIN BOOKS
Published by the Penguin Group
Penguin Books USA Inc., 375 Hudson Street, New York, New York 10014, U.S.A.
Penguin Books Ltd, 27 Wrights Lane, London W8 5TZ, England
Penguin Books Australia Ltd, Ringwood, Victoria, Australia
Penguin Books Canada Ltd, 10 Alcorn Avenue,
Toronto, Ontario, Canada M4V 3B2
Penguin Books (N.Z.) Ltd, 182–190 Wairau Road,
Auckland 10, New Zealand

Penguin Books Ltd, Registered Offices:
Harmondsworth, Middlesex, England

First published in Penguin Books 1993

3 5 7 9 10 8 6 4 2

LIBRARY OF CONGRESS CATALOGING IN PUBLICATION DATA
Darwin, Charles, 1809–1882.
[Selections. 1993]
The portable Darwin/ edited, with an introduction, notes, and
epilogue by Duncan M. Porter and Peter W. Graham.
p. cm.
Includes bibliographical references.
ISBN 0 14 015.109 5
1. Darwin, Charles, 1809–1882. 2. Evolution (Biology) 3. Natural
history. 4. Naturalists—England—Biography. I. Porter, Duncan M.
II. Graham, Peter W., 1951– . III. Title.
QH365.Z9P67 1993
575.01'62—dc20 93-17106

Printed in the United States of America
Set in Sabon

CONTENTS

INTRODUCTION

CHARLES DARWIN has been called "the most important person of the nineteenth century." A sweeping assertion of this sort might easily provoke a list of rival contenders, yet the fact remains that Darwin, by mounting a compelling case for the theories of evolution and natural selection, did more than any other person to make the intellectual climate of his age what it was. Despite his acknowledged importance, or perhaps in part because of it, Darwin is notably difficult to characterize. He was as much a cultural product as a shaper of culture. Famous as a theorizer, he was also an empiricist of the classic British sort, a man who tirelessly collected data from others and conducted his own observations and experiments. Though his ideas, and especially their implications, were potentially subversive, his mode of life and personal conduct were impeccably orthodox. He resists categorization as either amateur or professional. Indeed it is worth noting that "darwinise" became an English word before "scientist" did (the former being Samuel Taylor Coleridge's term for theorizing of the sort done by Charles Darwin's paternal grandfather, the latter being an 1830s coinage not widely accepted until William Whewell's *History of the Inductive Sciences* in 1840). If Charles Darwin was a scientist, he was also—and every bit as much—a rich country squire, a liberal Victorian intellectual, and a member of the Wedgwood-Darwin clan that had made its mark on Britain in earlier generations and would continue to do so in ensuing ones.

The second son and fifth of six children born to Dr. Robert Darwin and his wife, Susannah Wedgwood, at The Mount, their substantial house near Shrewsbury, Charles Darwin was from the start of life securely situated in the British ruling class. His grandfathers, both members of the free-thinking Lunar Society of Birmingham, were the potter Josiah Wedgwood and the doctor-poet Erasmus Darwin, who articulated—in such works as *The Loves of the Plants, Zoonomia*, and *The Temple of Nature*—the most prominent eighteenth-century English case for evolutionary development. Charles Darwin disavowed the influence of his grandfather's transformationist speculations upon his own. Nevertheless the younger man recapitulated a family pattern in his choice of problems and projects if not in his particular conclusions, just as his marriage in 1839 to his first cousin Emma Wedgwood conformed to the precedent of his father's and his older sister Caroline's marriages to Wedgwoods.

The young Darwin's early life followed a well-worn path in other more explicit ways. Like his older brother Erasmus, Charles was sent to Shrewsbury School, where a classical curriculum centered on Greek and Latin offered little to interest a boy with a passion for such rural pastimes as hunting, shooting, and collecting rocks and insects. Hoping that his second son could be molded into a professional man of his own sort, Dr. Darwin in 1825 sent Charles with Erasmus, who was already pursuing a medical degree, to study medicine at Edinburgh University. Edinburgh's medical faculty and curriculum were as distinguished as any in Europe, but Charles's interests lay elsewhere: in the gathering, dissecting, and stuffing of new specimens for his collections, and in the papers of the Plinian Society, a club that focused on topics in natural history. A crucial relationship in his Edinburgh days was with the Lamarckian zoologist Robert Grant, who introduced Darwin to the study of marine invertebrates.

Recognizing that Charles did not have the makings of a medical man, Dr. Darwin settled on holy orders and the comfortable, undemanding life of a country clergyman as offering a congenial future for his less-than-promising son. Christ's College, Cambridge, became the next academic institution to accept the young Darwin, who did what was necessary to pass his exams but saved his energy for more favored pastimes: the sporting pursuits of undergraduate gentry and his old love natural history, which now took the form of competitive beetle-hunting in the company of such fellow enthusiasts as his cousin William Darwin Fox. Other new and important friends made at Cambridge were two clergymen with scientific interests. The geologist Adam Sedgwick, who at the time was engaged in work that resulted in his Cambrian time scale, took Darwin on an excursion through North Wales that stimulated the younger man's interest in sedimentary stratification, possibly laid the foundation for his conversion to geologic gradualism, and certainly convinced him, as no mere book had yet done, "that science consists in grouping facts so that general laws or conclusions may be drawn from them." The botanist John Stevens Henslow's influence on Darwin's development was yet more crucial. It was thanks to Henslow's intervention that Darwin was considered for the post of unpaid naturalist and companion to the captain on HMS *Beagle*. The *Beagle* was returning to southern South America to complete its coastal surveys, begun in 1826.

There were some obstacles to Darwin's embarking. The most substantial was his father, whose opinion that the adventure would spoil Charles's chances of settling down to clerical life was overcome by Uncle Josiah Wedgwood, who argued for natural history being suitable to a clergyman. Once family support was secured, there was doubt as to whether Robert FitzRoy, the aristocratic and stoutly Tory captain of the *Beagle*, would make the offer. But in December 1831 Darwin did ship out on the *Beagle*, a small, rebuilt vessel whose surveying expedition took some five years. Despite formidable obstacles—tedium, isolation, perennial seasickness and crowding on shipboard, clashes of temperament with the volatile and politically conservative FitzRoy—Darwin found the voyage in many ways the most rewarding event of his life.

The *Beagle*'s voyage took her past the Canary and Cape Verde Islands to the east coast of South America. While the coastal survey went on, Darwin disembarked at such spots as Bahia, Rio de Janeiro, and Montevideo for long trips into the countryside, where, as on the ocean (when it was calm), he was tireless in gathering botanical, zoological, and geological specimens. The luxuriance of exotic life-forms in the rain forest particularly overwhelmed him and made his mind "a chaos of delight." The ship surveyed the Argentine coast, the Falkland Islands, and the windy, desolate headlands of Tierra del Fuego, where it left an English missionary and three Fuegians who had been indoctrinated in British culture and Christian values. Darwin's first contact with native Fuegians in their "untamed" state filled him with feelings as strong as, but antithetical to, those aroused by the tropical rain forest's flora and fauna. "How entire the difference between savage & civilized man is," he wrote. "It is greater than between a wild & domesticated animal."

Battered and buffeted by winter winds, the *Beagle* succeeded with some difficulty in rounding Cape Horn; and as the Pacific shore was charted, Darwin left the ship to explore Chiloe Island, the Chilean coast, and the Andes. In the volcanic Galapagos archipelago off the South American coast, Darwin encountered populations of tortoises, lizards, and finches that differed slightly but distinctly from island to island. Though he did not realize it at the time, these varying populations offered compelling evidence of evolutionary divergence. The voyage continued across the Pacific to Tahiti and New Zealand, then on to several landfalls on the southern edge of Australia and Tasmania, to Keeling Island, across the Indian Ocean to Mauritius, and around the Cape of Good Hope to St. Helena and Ascension Island. From there, the *Beagle* did not proceed north to England but across the Atlantic to Brazil, to recheck longitude measurements. Heading home after this detour, Darwin and his shipmates returned to England in October 1836. He, at least, would not leave Great Britain again—but like many a British connoisseur back from the Grand Tour or imperial adventurer returned from a tour of duty in the colonies, he carried home exotic collections and recollections that would enrich, and entirely alter, the rest of his life.

Back in England, Darwin found that his expedition had made him, if not famous, at least a rising star among the naturalists. The many crates of specimens (some brought back on the *Beagle*, others shipped home at Dr. Darwin's expense) now needed to be unpacked, arranged, and studied. Over the next few months, he found various specialists eager to help with this vast project: John Gould of the British Museum took on the birds, Thomas Bell of King's College, London, studied the reptiles, and Richard Owen of the College of Surgeons, later to be a hostile critic of Darwin's evolutionary views, examined the fossil mammals from South America. Established in London, increasingly involved with the Geological and Zoological societies, Darwin began to make intellectual order of five years' worth of impressions and observations. At the same time he solidified social and collegial alliances of the sort that would serve his needs. Darwin took up again with his Cambridge mentor Henslow, dined with the liberal intellectuals of his brother Erasmus's circle, and became friends with Charles Lyell, whose *Principles of Geology* had been a key book in Darwin's shipboard library, one that helped him see the brave new world encountered on his voyage through the eyes of a geological gradualist. Most important, he gave up the life of a bachelor. The chosen bride: his domestic, intelligent, rich, and completely familiar Wedgwood cousin Emma. Darwin's health, robust enough during the terrestrial portions of his travels, broke down under the bustle and pressure of metropolitan scientific preoccupations. With his wife (and sometimes nurse), Emma, he soon withdrew from London life to the convenient yet secluded rural village of Down in the chalk hills of

Kent. There, as squire and lady of Down House, they would raise a large family, and Darwin would produce a succession of books and studies, some of which profoundly disturbed the Victorian world in which the Darwin family was so comfortably ensconced. From 1842 on, the outward form of his life was set.

The form his life took warrants some comment, for popular perceptions of Darwin's years as "sage of Down" tend to dwell on certain oversimplifications or exaggerations fostered by Darwin himself. One such oversimplification might be called the myth of Darwin the recluse. In letters Darwin sometimes characterized himself as living a hermit's life; and while it is true that his world-wandering days ended when he disembarked from the *Beagle*, his decision to leave London and settle at Down was far from eccentric—it was even conventional, given his class and financial situation. In Darwin's day, financially secure gentlemen fled jobs or professions rather than seeking them. After spending their early adult years broadening their minds or fattening their purses in the wide world or the metropolis, gentlemen who could afford to do so—particularly those who had married and begun families—were likely to settle on country estates. Here, like Darwin, they could pursue their individual interests with maximum freedom, while at the same time serving their communities as vestrymen, magistrates, or parliamentary representatives. Darwin's scientific work at Down was by no means conducted in solitude, or even in merely domestic company. Such allies and disciples as his tenacious "bulldog" T. H. Huxley, the botanist Joseph Dalton Hooker, and John Lubbock from the local manor were regular visitors; and as the years went by a steady and international stream of the eminent flowed in and out of Darwin's study. Darwin kept in touch with a larger circle of far-flung colleagues and an extensive group of informants through personal correspondence and through letters and queries to journals. Furthermore, as his journal shows, Darwin ventured forth more than is popularly supposed. He occasionally attended scientific meetings, went up to London for lectures, exhibits, and entertainments, took vacations with his growing family, and patronized the fashionable hydropathic spas in his search for improved health. In short, Darwin preferred staying at home in the country; but his tastes and habits were quite usual for his class and time, and they by no means isolated him from the scientific community.

The image of Darwin as invalid is no less vivid—and ultimately not much more helpful to our understanding of the man's achievement—than is that of Darwin as recluse. It is undeniable that once the intrepid explorer of rain forests and volcanic islands had returned to England, his health was intermittently bad for the rest of his life. Darwin's bouts with boils, eczema, dizziness, and especially nausea, vomiting, and headaches are perennial motifs in his and Emma's private writings. For long periods of time he was in daily distress. But it remains true that he got an extraordinary amount of experimenting, data-gathering, theorizing, and writing done, that he fathered ten children, and that he survived to the age of seventy-three. As has been variously argued, Darwin may have suffered from Chagas disease caused by a blood protozoan transmitted by the "great black bug of the pampas," or from hyperventilation, or from arsenic poisoning, or from allergies, whether to organisms he studied (pigeons, for example), chemicals he used, or to food he ate (Emma's cookery book reveals that the Darwins' diet was rich and heavy, typical of their class and day). His "tormented evolutionist's" mind may have punished his body; indeed, most of his health problems were somatic complaints that can have psychic causes. He may have been victim of the self-generated, and partly self-protective, hypochondria common among Victorian intellectuals and virtually endemic among generations of Darwins. Several or all of these causes may have contributed to Darwin's life of ill health. The psycho-medical mystery is fascinating in its own right, important to an understanding of Darwin the man, but not directly relevant to the development and expression of his ideas. Perhaps Darwin gave up geology because ill health made field work impossible, but otherwise his physical condition seems to have had little effect on his scientific work—neither on his choice of problems, procedures, and perspectives nor on the quantity or quality of what he managed to accomplish.

If misunderstandings of Darwin's secluded life at Down and overemphasis on his bad health can obstruct a balanced view of his intellectual achievement, so can the tendency to see Darwin's list of publications, and the scientific problems they address, as intellectually diffuse. According to this view, Darwin becomes a man with one overwhelmingly important achievement (*The Origin of Species*, of course), a few other notable ones (such as *The Voyage of the* Beagle, *The Descent of Man*, and *The Structure and Distribution of Coral Reefs*), and a great many minor pieces, each essentially independent. The facts are quite different. Darwin's first works rose directly out of his voyage, whether they involved scientific observations made on the *Beagle*, travel impressions in the style of Alexander von Humboldt, whose *Personal Narrative* was a lens through which Darwin saw his own journey, or articulation of theories (as in his book on the developmental history of coral islands and reefs or in the various geological works of the 1830s and 1840s) related to his shipboard observations, readings, and meditations. Even when engaged in these projects, though, Darwin was compiling a series of notebooks on "Transmutation of Species." He sketched out his views in an 1842 manuscript unknown until well after his death. This was expanded into an 1844 essay read by Hooker, put aside that same year when the anonymous *Vestiges of the Natural History of Creation* (by the scientific populariser Robert Chambers) provoked deep controversy, and wrestled with for years in a vast project he intended to publish under the title *Natural Selection.* In 1858 the now-legendary arrival of Alfred Russel Wallace's paper on variation and evolution pushed him, at the advice of his scientific friends Hooker and Lyell, into the joint Darwin-Wallace publication "On the Tendency of Species to Form Varieties, and on the Perpetuation of Varieties and Species by Natural Means of Selection" that first put his evolutionary ideas before the public. It is important to recognize that the *Origin* itself, appearing in 1859, was very far from being the book Darwin had envisioned his work on species as being. He saw the *Origin* as only a summary of ideas and data that had developed slowly and through an exhaustive intellectual apprenticeship of observation, information gathering, reading, and experimenting.

Thus the projects preceding *The Origin of Species* bear upon it in various crucial ways. For instance, through his meticulous work on barnacles, Darwin established beyond doubt his credentials as a taxonomist. Similarly, the books subsequent to the *Origin* expand, illustrate, or modify Darwin's evolutionary ideas. *The Variation of Animals and Plants under Domestication*, for example, presents the minute details of how artificial and natural selection have operated in different plant and animal groups—material essential to his theory but too extensive and particular to fit into the abstract that is the *Origin.* *Variation* was the first volume of a planned three-volume expansion of the data that the *Origin* summarized. *The Descent of Man* and *The Expression of the Emotions* extend evolutionary theory to include the human race, thereby fulfilling Darwin's promise in the *Origin*: "Light will be thrown on the origin of man and his history." Books on such subjects as the power of movement in plants, the ways insects pollinate orchids, and the role earthworms play in forming vegetable mold show Darwin, decades after the *Origin*, still concerned with how adaptive change operates in the plant and animal world. The post-*Origin* books address, in a pioneering way, subjects that have in our time become specialized disciplines; but all of these interests branch out in different ways from one root: his preoccupation with evolution. Indeed, it might be said that adaptive radiation is as much an operating principle in Darwin's scientific career as it is in the development of new life-forms.

Although Darwin died over a century ago, his name remains as familiar to the general public as it was in his own time. One reason is that the essential ideas of evolution remain compelling. No one has yet advanced a scientific theory that successfully challenges Darwin's assertions that the earth slowly but constantly changes, that forms of life therefore must change in order to survive, that random variation among the individual members of a species occurs naturally, and that in the unceasing struggle for existence the best-adapted individuals survive to pass on their characteristics. But Darwin's name is also known to many people who lack firsthand knowledge of his scientific writings or the theories they advance. The religious and ideological implications of Darwin's theory engage diverse audiences by raising a variety of questions, some that Darwin himself addressed and some that he refrained from handling. What is the nature of humanity in an evolutionary world? How are we related to other organisms? What are our responsibilities to one another and to other forms of life? Do the implications of natural selection extend into human societies? Is there a place for God in a naturally evolving world? If so, what kind of God? All these questions are of more than academic interest—and it seems worth emphasizing that over the years religious, philosophical, social, political, and racial attitudes have played important roles in how different individuals or groups accept, reject, or modify Darwin's thesis. But whatever our particular orientations, we must take Darwin's ideas into account. In the long conversation that is natural history, Darwin spoke neither the first word nor the last one; nevertheless, the questions he asked and the answers he offered changed the subject in a profound and enduring way.

We hope to have made *The Portable Darwin* both as concise as a reasonably comprehensive anthology can be and as comprehensive as a relatively concise one can be. The balancing act has been difficult. Within the constraints of a single volume of some six hundred pages, we have tried to give the fullest and most faithful possible representation of Charles Darwin's complex intellectual life-project—and to do so by allowing him to speak for himself in the diverse range of tones found in his writings. We have aimed at inclusiveness, if not exhaustiveness, with regard to Darwin's titles. Selections from the best-known books (*Origin of Species, Voyage of the* Beagle, *Descent of Man*) appear, as would be expected. So do passages drawn from less familiar volumes, such as *The Variation of Animals and Plants*, which Darwin considered among his most important books, and his charming studies of orchids and earthworms. Scientific papers, letters to journals, and other writings are also represented. Darwin was truly a "man of enlarged curiosity"; and the best way to appreciate the scope and coherence of his achievement is to see how minute observations and particular queries find their way, years later, into massive, ambitiously theoretical projects. Darwin's scientific writings, experimental and theoretical, loom largest in this collection; but we have also taken care to represent other, less familiar aspects of his intellect. Darwin the country squire, Darwin the archaeologist, Darwin the family memorialist, and Darwin the moralizing explorer all appear on these pages, which contain, among other things, his thoughts on dogs, carnivorous plants, atoll islands, missionaries, infant psychology, topsoil, blushing, and the deity.

A consistently edifying writer and sometimes a delightful one, Darwin laid out his lines of argument with an eye toward clarity and accessibility. He is one of the few important scientific theorists who aims his rhetoric squarely and successfully at the general educated reader. Because Darwin took great care in articulating his ideas and arranging his points, the way to keep faith with his work is to present self-contained extracts with little, if any, internal cutting. For example, rather than assembling bits and pieces from all or most of the chapters of *The Origin of Species*, this volume offers the book's first four chapters, and its final one, in full. But if Darwin was a careful writer, he was also a deferential one. His editorial efforts to make his published ideas palatable or plausible to various readers, individual and collective, caused him to revise various works (most notably the *Origin*) in ways that sometimes obscured his insights and original intentions. Thus we have chosen to reprint selections from the first editions of Darwin's books, where his prose is likeliest to do the best job of saying what he actually wishes to say. In cases where a later edition's amplifying materials or newly available data illuminate or significantly alter a passage, an editorial note to that effect is included. With the exception of the *Origin*, Darwin's works are peppered with footnotes, mostly references to the literature he is citing. In order to make his prose more easily readable, we have removed most of these footnotes. Our own commentary has been deliberately brief. There are many scholarly and general discussions to which a reader in search of more extensive interpretation can turn (the list of further reading includes a good number of these resources), and it seemed best to reserve as much of this volume as possible for Darwin's own words.

CHRONOLOGY

This chronology includes the original titles and dates of publication of all of Darwin's books. In addition, he published about 170 papers in journals, some of which were quite substantial.

1809 Charles Robert Darwin, second son and fifth of six children of Robert Waring Darwin, successful physician, and Susannah Wedgwood, daughter of Josiah Wedgwood, founder of the Wedgwood Pottery Works, is born in Shrewsbury, February 12.

1817 Goes to Rev. George Case's day school, spring. Mother dies, July 15.

1818 Enters Shrewsbury School, midsummer.

1825 Leaves Shrewsbury School, June 17. Charles and his brother Erasmus enter Edinburgh University to study medicine, October 22.

1826 Elected to Plinian Society, an undergraduate scientific club, November 10. Examines marine animals and algae on shores of Firth of Forth with zoologist Dr. Robert Grant, winter.

1827 Reads papers on original research on marine animals to Plinian Society, March 27. Visits Belfast and Dublin, spring. Visits Paris, May. Enters Christ's College, Cambridge University, intending to study for the ministry, October 15.

1829 Insect-collecting expedition to Wales with entomologist Frederick William Hope, summer.

1830 Insect collecting in North Wales with Hope, August.

1831 Plans a trip to the Canary Islands, spring. Receives B.A. degree, tenth out of 178 students who chose not to take honors, April 26. On geological tour of North Wales with Prof. Adam Sedgwick, August. Receives letter from Cambridge mentor Prof. John Stevens Henslow with invitation to accompany Capt. Robert FitzRoy on surveying voyage to South America, August 24. Spends September through December preparing for voyage. Sails from England, December 27.

1831–36 Naturalist and companion to Captain FitzRoy on survey ship HMS *Beagle* on voyage around the world.

1836 First published paper, "A Letter, Containing Remarks on the Moral State of Tahiti, New Zealand, &c." in the *South African Christian Recorder*, by FitzRoy and Darwin, September. Returns to England, October 2. Elected Fellow of the Geological Society, November. One other paper published this year.

1836–37 Lives in Cambridge while working on *Beagle* collections and journal of the voyage, December 16 to March 6.

1837 Reads first paper on *Beagle* research to the Geological Society, January 4. Moves to London and continues working on journal, March 6. Begins first notebook on transmutation of species, July. Awarded a grant of £1000 from the Treasury for publication of a *Zoology of the* Beagle, August 16. Five papers published.

1838 Elected Fellow of the Royal Geographical Society. First part of *The Zoology of the Voyage of HMS* Beagle, *under the Command of Captain FitzRoy, during the Years 1832 to 1836* (*Fossil Mammalia*, by Richard Owen) published, February. Elected vice president of the Entomological Society, February 5. Elected secretary of the Geological Society, February 16. Geological fieldtrip to Scotland, June 23 to July 13. Reads Thomas Malthus's *An Essay on the Principle of Population*, September. Four papers published.

1839 Elected Fellow of the Royal Society, January 24. Marries his first cousin Emma Wedgwood, January 29. Elected Fellow of the Zoological Society, March. *Narrative of the Surveying Voyages of His Majesty's Ships* Adventure *and* Beagle. Vol. III. *Journal and Remarks, 1832–1836* published, June 1; issued separately as *Journal of Researches into the Geology and Natural History of the Various Countries Visited by HMS* Beagle, August 15. Attends British Association for the Advancement of Science meeting in Birmingham, August. William Erasmus Darwin is born, December 27. Two papers published.

1840 Elected to Council of the Royal Geographical Society, May 25. Two papers published.

1841 Resigns from Council of the Royal Geographical Society. Resigns as secretary of the Geological Society, February 19. Anne Elizabeth Darwin is born, March 2. Three papers published.

1842 Member of the British Association Committee on Zoological Nomenclature, February through August. *The Structure and Distribution of Coral Reefs. Being the First Part of the Geology of the Voyage of the* Beagle, *under the Command of Capt. FitzRoy, R.N. during the Years 1832 to 1836* published, May. Geological fieldtrip to North Wales, June. Completes thirty-five-page sketch on species theory, June. Moves to Down, Kent, September 17. Mary Eleanor Darwin is born, September 23, dies October 16. Two papers published.

1843 Henrietta Emma Darwin is born, September 25. Last part of *Zoology of the* Beagle (*Reptiles*, by Thomas Bell) published, October. Two papers published.

1844 Made a vice president of the Geological Society, January. Writes to Joseph Dalton Hooker that he believes in the transmutation of species, January 11. Completes 231-page essay on evolution by natural selection, February 13. *Geological Observations on the Volcanic Islands Visited during the Voyage of HMS* Beagle, *Together with Some Brief Notices of the Geology of Australia and the Cape of Good Hope. Being the Second Part of the Geology of the Voyage of the* Beagle, *under the Command of Capt. FitzRoy, R.N. during the Years 1832 to 1836* published, November. Five papers published.

1845 George Howard Darwin is born, July 9. Second edition of *Journal of Researches* published, August 30. One paper published.

1846 Attends British Association meeting in Southampton, September. Begins study of barnacles, October. *Geological Observations on South America. Being the Third Part of the Geology of the Voyage of the* Beagle, *under the Command of Capt. FitzRoy, R.N. during the Years 1832 to 1836* published, late in the year. Three papers published.

1847 Elizabeth Darwin is born, July 8. Attends British Association meeting at Oxford. Two papers published.

1848 Francis Darwin is born, August 16. Father dies, November 13. One paper published.

1849 Attends British Association meeting at Birmingham, September. Elected to Council of Royal Society. Two papers published.

1850 Leonard Darwin is born, January 15. One paper published.

1851 Anne Elizabeth Darwin dies, April 23, age 10. Horace Darwin is born, May 13. *A Monograph of Subclass Cirripedia, with Figures of All the Species. The Lepadidae; or Pedunculated Cirripedes* published, after November 12. *A Monograph of the Fossil Lepadidae, or Pedunculated Cirripedes of Great Britain* published, after November 12.

1853 Awarded the Royal Society Medal for geological and barnacle research, November 30.

1854 Elected a Fellow of the Linnean Society, March 7. *A Monograph of the Subclass Cirripedia, with Figures of all the Species. The Balanidae, (or Sessile Cirripedes); the Verrucidae, etc., etc., etc.*, published, after July 15. Begins work on species, September 9. *A Monograph of the Fossil Balanidae and Verrucidae of Great Britain* published, after mid-September. Elected to council of Royal Society, November 2.

1855 Attends British Association meeting at Glasgow, September. Ten papers published.

1856 Begins writing *Natural Selection*, May 14. Charles Waring Darwin is born, December 6. Three papers published.

1857 Five papers published.

1858 Arrival of letter and manuscript on evolution from the East Indies by Alfred Russel Wallace, June 18. Charles Waring Darwin dies, June 28. "On the tendency of species to form varieties, and on the perpetuation of varieties and species by natural means of selection," by Darwin and Wallace, read at a meeting of the Linnean Society, July 1, published August 30. Begins an abstract of *Natural Selection*, July 20. Two other papers published.

1859 Awarded the Wollaston Medal by the Geological Society, February 18. *On the Origin of Species by Means of Natural Selection, or the Preservation of Favoured Races in the Struggle for Life* published, November 24.

1860 Second edition of *Origin* published, January 7. Begins work on *Variation*, January 9. Four papers published.

1861 Elected Fellow of the Ethnological Society. Third edition of *Origin* published, April. Eight papers published.

1862 *On the Various Contrivances by which British and Foreign Orchids are Fertilised by Insects, and on the Good Effects of Intercrossing* published, May 15. Nine papers published.

1863 Ten papers published.

1864 Awarded the Copley Medal of the Royal Society, November 30. Two papers published.

1865 *On the Movements and Habits of Climbing Plants* published, June 12.

1866 Fourth edition of *Origin* published, after May 10. Three papers published.

1867 Two papers published.

1868 Elected Honorary Member of the Medico-Chirugical Society. *The Variation of Animals and Plants under Domestication* published, January 30, revised printing February 10. Begins work on *Descent of Man*, February 4. Five papers published.

1869 Fifth edition of *Origin* published, August 7 (?). Six papers published.

1870 Two papers published.

1871 *The Descent of Man, and Selection in Relation to Sex* published, February 24, second printing in March. Five papers published.

1872 Sixth edition of *Origin* published, February 19. *The Expression of the Emotions in Man and Animals* published, November 26. One paper published.

1873 Eight papers published.

1874 Second edition of *Descent of Man* published. Second edition of *Coral Reefs* published. Five papers published.

1875 Second edition of *Variation under Domestication* published. *Insectivorous Plants* published, July 2. Second edition of *Climbing Plants* published, September. Gives evidence before the Royal Commission on Vivisection, November 8.

1876 Begins recollections, May 3, published in 1887. *The Effects of Cross and Self-Fertilisation in the Vegetable Kingdom* published, November 10. Two papers published.

1877 Makes geological excursion to Stonehenge to observe the action of worms, June. *The Different Forms of Flowers on Plants of the Same Species* published, July 9. Second edition of *Orchids* published, November 17. Receives honorary LL.D. from Cambridge University, November 18. Nine papers published.

1878 Second edition of *Cross and Self-Fertilisation* published. Two papers published.

1879 Awarded the Baly Medal of the Royal College of Physicians. Preliminary notice for *Erasmus Darwin* published, November. Initiates a proposal to grant Wallace a government pension, December. Two papers published.

1880 Second edition of *Forms of Flowers* published, July. *The Power of Movement in Plants* published, November 6. Five papers published.

1881 *The Formation of Vegetable Mould, through the Action of Worms, with Observations on Their Habits* published, October 10. Ten papers published.

1882 Five papers published. Dies at Down, April 19. Buried in Westminster Abbey, April 25.

The Portable

Darwin

Remarks upon the Habits of the Genera *Geospiza, Camarynchus, Cactornis* and *Certhidea* of Gould. [1837]

Soon after he returned from the Beagle *voyage, Darwin began to report on his collections. One of the first such reports, presented on May 10, 1837, to the Zoological Society of London, concerned the finches he had collected in the Galapagos Islands of Ecuador. John Gould, the ornithologist describing the birds from the voyage, had informed Darwin that the Galapagos finches were closely related to one another and were not to be found elsewhere.*

In the 1930s and 1940s the Galapagos finches, now called "Darwin's finches," were found by the English ornithologist David Lack to be an excellent example of adaptive radiation—that is, the evolution of a number of closely related species occupying different ecological niches from a single ancestral species. Scientists assumed that Darwin had recognized the significance of the finches he collected, and thus that the Galapagos finches had played a major role in the development of his evolutionary ideas. But in spite of its presence in many biology texts, this account is largely inaccurate. Unfortunately, Darwin had not paid enough attention to where he had collected the individual finches, and in most cases he was never able to determine the island of origin. The Galapagos Islands were important to Darwin's understanding of how evolution takes place, but in fact it was not the island finches but mockingbirds and tortoises that provided him with examples.

The group of groundfinches, characterized, at a previous meeting, by Mr. Gould, under the generic appellations of *Geospiza, Camarhyn-*

chus, *Certhidea*, and *Cactornis*, were upon the table; and Mr. Darwin being present, remarked that these birds were exclusively confined to the Gallapagos Islands; but their general resemblance in character, and the circumstance of their indiscriminately associating in large flocks, rendered it almost impossible to study the habits of particular species. In common with nearly all the birds of these islands, they were so tame that the use of the fowling-piece in procuring specimens was quite unnecessary. They appeared to subsist on seeds, deposited on the ground in great abundance by a rich annual crop of herbage.

On Certain Areas of Elevation and Subsidence in the Pacific and Indian Oceans, as Deduced from the Study of Coral Formations. [1838]

POSTERITY HAS COME TO THINK of Darwin as a biologist because of his many publications on the evolution of plants and animals. The only time, however, that Darwin referred to himself in print as something other than a naturalist, he called himself a geologist. His early interest in geology, stimulated by his Cambridge friendship with Prof. Adam Sedgwick, was clear on the voyage of the Beagle: *he took 1383 pages of notes on geology as compared to 368 pages on zoology, and he kept no separate notebook for botany. Likewise, most of the publications that directly resulted from the voyage are geological.*

Darwin read the following paper at a meeting of the Geological Society of London on May 31, 1837. The published account, which appeared in the Society's Proceedings *later that year, is a summary of Darwin's talk. The paper is important for several reasons. First, it is a fine early example of Darwin's reasoning from evidence. Second, the publication reports Darwin's first scientific breakthrough, his explanation of how coral reefs are formed. The previously accepted theory, proposed by Britain's leading geologist Sir Charles Lyell, had been that reefs formed atop submarine craters rising from the ocean depths. Lyell, however, was persuaded by Darwin's counterargument that reefs were formed by the upward growth of coral as the ocean floor slowly subsided; and Darwin's reputation soared. Finally, this paper offers the first published hint that Darwin might harbor evolutionist tenden-*

cies. Examination of his early notebook on geology and other subjects (published in 1980 as the Red Notebook*) shows that Darwin began to write about evolution in March 1837.*

The author commenced by observing on some of the most remarkable points in the structure of Lagoon islands. He then proceeded to show that the lamelliform corals, the only efficient agents in forming a reef, do not grow at any great depths; and that beyond twelve fathoms the bottom generally consists of calcareous sand, or of masses of dead coral rock. As long as Lagoon islands were considered the only difficulty to be solved, the belief that corals constructed their habitations (or speaking more correctly, their skeletons), on the crests of submarine craters, was both plausible and very ingenious; although the immense size, sinuous outline and great number, must have startled any one who adopted this theory. Mr. Darwin remarked that a class of reefs which he calls "encircling" are quite, if not more, extraordinary. These form a ring round mountainous islands, at the distance of two and three miles from the shore; rising on the outside from a profoundly deep ocean, and separated from the land by a channel, frequently about 200 and sometimes 300 feet deep. This structure as observed by Balbi resembles a lagoon, or an atoll, surrounding another island. In this case it is impossible, on account of the nature of the central mass, to consider the reef as based on an external crater, or on any accumulation of sediment; for such reefs encircle the submarine prolongation of islands, as well as the islands themselves. Of this case New Caledonia presents an extraordinary instance, the double line of reef extending 140 miles beyond the island. Again the Barrier reef, running for nearly 1000 miles parallel to the North-East coast of Australia, and including a wide and deep arm of the sea, forms a third class, and is the grandest and most extraordinary coral formation in the world.

The reef itself in the three classes, encircling, barrier and lagoon, is most closely similar; the difference entirely lying in the absence or presence of neighbouring land, and the relative position which the reefs bear to it. The author particularly points out one difficulty in understanding the structure in the barrier and encircling classes, namely, that the reef extends so far from the shore, that a line drawn perpendicularly from its outer edge down to the solid rock on which the reef must be based, very far exceeds that small limit at which corals can grow. A distinct class of reefs however exists, which the author calls "fringing reefs" which extend only so far from the shore, that there is no difficulty in understanding their growth. The theory which Mr. Darwin then offered, so as to include every kind of structure, is simply that as the land with the attached reefs subsides very gradually from the action of subterranean causes, the coral building polypi soon again raise their solid masses to the level of the water; but not so with the land; each inch lost is irreclaimably gone:—as the whole gradually sinks, the water gains foot by foot on the shore, till the last and highest peak is finally submerged. Before explaining this view in detail, the author offered some considerations on the probability of general subsidences,—such as the small portion of land in the Pacific, where many causes tend to its production, an argument first suggested by Mr. Lyell, and the extreme difficulty (with the knowledge that corals grow at but limited depths) in explaining the existence of a vast number of reefs on one level, without we grant subsidence, so that one mountain top should be submerged after another; the zoophytes always bringing up their stony masses to the surface of the water. Subsidence being thus rendered almost necessary, it was shown by the aid of sections, that a simple fringing reef would thus necessarily be converted by the upward growth of the coral into one of the encircling order, and this finally, by the disappearance through the agency of the same movement of the central land, into a lagoon island. In the same manner a reef skirting a shore would be changed into a barrier extending parallel to, but at some distance from, the mainland.

Mr. Darwin then showed, that there existed every intermediate form between a simple well characterized encircling reef, and a lagoon island; that new Caledonia supplied a link between encircling and barrier reefs; that the different reefs produced by the same order of movement were always in juxtaposition, of which the Australian barrier associated with encircled islets and true lagoons, affords a good example. He then proceeded to show that within the lagoon of Keeling Island, proofs of subsidence might be deduced from many falling trees and a ruined storehouse; these movements appearing to take place at the period of bad earthquakes, which likewise affect Sumatra, 600 miles distant. It was thence inferred as probable, that as Sumatra rises, (of which proofs are well known to exist,) the other end of the lever sinks down; Keeling Island thus acting as an index of the movement of the bottom of the Indian Ocean. Again at Vanikoro, where the structure indicates according to the theory recent subsidence, violent earthquakes are known lately to have occurred.

The author then removed an apparent objection to the theory, namely, that subsidence would form a disc of coral but not a cup-shaped mass or lagoon, by showing that the corals which grow in tranquil water are very different from those on the outside, and less effective; and that as the basin becomes shallower they are subject to various causes of injury. The lagoon nevertheless is constantly filling up to the height of lowest water spring tides, (the utmost possible limit of living coral,) and in that state it long remains, for no means exist to complete the work. Mr. Darwin then proceeded to the main object of the paper, in showing that as continental elevations act over wide areas, so might we suppose continental subsidences would do, and in conformity to these views, that the Pacific and Indian seas could be divided into symmetrical areas of the two kinds; the one sinking, as deduced from the presence of encircling and barrier reefs, and lagoon islands, and the other rising, as known from uplifted shells and corals, and skirting reefs. The absence of lagoon islands in certain wide tracts, such as in both the West and East Indies, Red Sea, &c., was thus easily explained, for proofs of recent elevation are there abundant. In a like manner, in very many cases where islands are only fringed with reefs, which according to the theory had not been subsiding, actual proofs of elevation were adduced. Mr. Darwin remarked that, excepting on the theory of the configuration of reefs being determined by the order of movement, the circumstance that certain classes which are characteristic and universal in some parts of the sea, being never found in others, is quite anomalous, and has never been attempted to be explained.

Mr. Darwin then pointed out the above areas both in the Pacific and Indian Oceans, and deduced the following as the principal results. 1st. That linear spaces of great extent are undergoing movements of an astonishing uniformity, and that the bands of elevation and subsidence alternate. 2. From an extended examination, that the points of eruption all fall on the areas of elevation. The author insisted on the importance of this law, as thus affording some means of speculating, wherever volcanic rocks occur, on the changes of level even during ancient geological periods. 3. That certain coral formations acting as monuments over subsided land, the geographical distribution of organic beings (as consequent on geological changes as laid down by Mr. Lyell) is elucidated, by the discovery of former centres whence the germs could be disseminated. 4. That some degree of light might thus be thrown on the questions, whether certain groups of living beings peculiar to small spots are the remnants of a former large population, or a new one springing into existence. Lastly, when beholding more than a hemisphere, divided into symmetrical areas, which within a limited period of time have undergone certain known movements, we obtain some insight into the system by which the crust of the globe is modified during the endless cycle of changes.

Narrative of the Surveying Voyages of His Majesty's Ships *Adventure* and *Beagle*, between the years 1826 and 1836, describing Their Examination of the Southern Shores of South America, and the *Beagle*'s Circumnavigation of the Globe. Vol. III. Journal and Remarks, 1832–1836. [1839]

In 1831 Capt. Robert FitzRoy was seeking a companion to accompany him on a surveying voyage to South America. FitzRoy had joined the earlier South American expedition of HMS Beagle *following the suicide of the ship's previous captain. Warned by his predecessor's fate, understanding the isolated status of a captain, and recognizing his own propensity to suffer from depression, FitzRoy undoubtedly concluded that having another gentleman on board to dine and converse with him would be effective preventive medicine. If that gentleman were also interested in natural history, so much the better: he could observe and collect the geological, zoological, and botanical wonders FitzRoy had noticed but to which he, as captain, could devote little systematic attention.*

FitzRoy wrote to Capt. Francis Beaufort, hydrographer of the Admiralty, requesting such a companion. Beaufort wrote to his friend the Reverend George Peacock, a fellow of Trinity College and lecturer in mathematics, for recommendations. Peacock turned for suggestions to the Reverend John Stevens Henslow, Cambridge's professor of botany. Henslow briefly considered going himself, but the look on his wife's face when he broached the subject convinced him not to pursue the idea. Henslow then asked his brother-in-law, the Reverend Leonard Jenyns, an Anglican priest and naturalist, if he were interested. Jenyns considered the possibility overnight but decided that he could not leave his parish. Then Jenyns and Henslow thought of young Charles Darwin, a divinity student and protegé of Henslow's.

On August 24, Henslow wrote to Darwin, who was then geologizing in North Wales with the Reverend Adam Sedgwick, Woodwardian professor of geology at Cambridge: "I have stated that I consider you to be the best qualified person I know of who is likely to undertake such a situation. I state this not in the supposition of your being a finished naturalist, but as amply qualified for collecting, observing, and noting, anything worthy to be noted in Natural History. Peacock has the appointment at his disposal, and if he cannot find a man willing to take the office, the opportunity will probably be lost." Darwin, who had earlier entertained visions of voyaging to Tenerife in the wake of Alexander von Humboldt, wanted to jump at the chance; but his father disapproved of the scheme and offered several objections, including his sense that once again the wayward Charles was contemplating a change of professional course. Despite his personal opposition, however, Dr. Darwin said, "If you can find any man of common sense who advises you to go I will give my consent." Luckily for Charles, his uncle Josiah Wedgwood refuted the paternal objections; and the trip was on.

After several false starts and some anxiety on the part of Darwin, HMS Beagle sailed from England on December 27, 1831, bound for South America and a place in history. During the voyage of almost five years, the officers and men surveyed coasts and islands around the world, while Darwin observed and collected. He was on board the Beagle only about half the time. While the ship was sailing along a coast, Darwin would often be on brief or prolonged excursions inland. The large numbers of rocks, animals, and plants he collected (now mostly housed at Cambridge University and The Natural History Museum in London) show how busy he was at land and on sea.

Darwin also amassed extensive notes on the voyage. He filled fifteen field notebooks, three specimen notebooks for dried plants and stuffed animals, and three specimens in spirits of wine notebooks for the preserved materials. Longer comments on the collections and general observations went into the zoological diary, which despite its title *gave about 20 percent of its pages to plants. The zoological collections and diary formed the basis for the five-volume* Zoology of the *Beagle, published in nineteen installments between 1838 and 1843 under Darwin's editorship.*

Besides Darwin's geological diary and 1383 pages of geological notes, there were four geological specimen notebooks for rocks and fossils. From these sources came the three volumes on coral reefs, volcanic islands, and South America collected as The Geology of the *Beagle (1842, 1844, 1846). Darwin also kept a journal, now known as the* Diary, *that offers his running account of the voyage, though some entries may have been made days or weeks after the happenings discussed. Darwin periodically sent home portions of the* Diary *with his letters to let family and friends know what he was doing. These documents became the book Darwin called his* Journal of Researches *(1839), which since 1905 has been known as* The Voyage of the *Beagle. The* Diary *proper was published in 1933 and republished in 1988. On the other hand,* The Voyage of the *Beagle, which went into a revised second edition in 1845, has been reprinted over 250 times and translated into at least twenty-two languages. In perhaps the worst financial decision he ever made, Darwin sold the copyright in 1845 to the well-known publisher John Murray, for £150. The book has remained in print ever since.*

The passages that follow come from its first edition, which stood as the third part of the three-volume Narrative of the Surveying Voyages of His Majesty's Ships *Adventure* and *Beagle, edited by Captain FitzRoy and published by Henry Colburn in 1839. The first volume of the trilogy is written by Capt. Philip Parker King, commander of the 1826–30 voyage. The second, chronicling the 1831–36 expedition, is by FitzRoy himself. Neither captain comes close to rivaling Darwin's prose.*

Included here are Darwin's preface and conclusion to his volume along with several chapters or parts of chapters concerned with subjects to which he returns again and again in his later publications: his first experience of the tropics (on Atlantic islands and in Brazil), his discovery of Patagonian fossils resembling larger versions of existing South American mammals, and his encounters with aboriginal humans in Tierra del Fuego and with remarkable birds and reptiles in the Galapagos. Darwin appended to his book a section of valuable firsthand advice to collectors. Later he would amplify the advice concerning geology for a chapter in Sir John Frederick Herschel's A Manual of Scientific Enquiry: Prepared for Use of Her Majesty's Navy: and

Adapted for Travellers in General. *First published in 1849, Herschel's book went through six editions and stayed in print until 1906.*

PREFACE.

I have stated in the preface to the Zoology of the Voyage of the Beagle, that it was in consequence of a wish expressed by Captain FitzRoy, of having some scientific person on board, accompanied by an offer from him, of giving up part of his own accommodations, that I volunteered my services, which received, through the kindness of the hydrographer, Captain Beaufort, the sanction of the Lords of the Admiralty. As I feel that the opportunities, which I enjoyed of studying the Natural History of the different countries we visited, have been wholly due to Captain FitzRoy, I hope I may here be permitted to express my gratitude to him; and to add that, during the five years we were together, I received from him the most cordial friendship and steady assistance. Both to Captain FitzRoy and to all the Officers of the Beagle, I shall ever feel most thankful for the undeviating kindness with which I was treated, during our long voyage.

The present volume contains in the form of a journal, a sketch of those observations in Geology and Natural History, which I thought would possess some general interest. As it was originally intended to have preceded any more detailed account, and as its publication has been unavoidably delayed, the briefness and imperfection of several parts, I hope, will be excused. I have given a list of those errata (partly caused by my absence from town when some of the sheets were in the press) which affect the sense; and have added an Appendix, containing some additional facts (especially on the theory of the transportation of erratic blocks) which I have accidentally met with during the past year. I hope shortly to publish my geological observations; the first Part of which will be on the Volcanic Islands of the Atlantic and Pacific Oceans, and on Coral Formations; and the second Part will treat of South America. Several numbers of the Zoology of the Voyage of the Beagle, due to the disinterested zeal of several of our first naturalists, have already appeared. These works could not have been undertaken, had it not been for the liberality of the Lords Commissioners of Her Majesty's Treasury, who, through the representation of the Right Honourable the Chancellor of the Exchequer, have been pleased to grant a sum of one thousand pounds towards defraying part of the expenses of publication. I have repeated in this volume my account of the habits of some of the birds and quadrupeds of South America, as I thought such observations might interest those readers who would not, probably, consult the larger work. But I trust that naturalists will remember that mere sketches are here given on several subjects, which will hereafter be more fully entered on, or have already been so: for instance, the notices of the strange fossil quadrupeds of the eastern plains of South America are exceedingly imperfect, whilst an admirable account of them by Mr. Owen now forms the first part of the Zoology of the Voyage of the Beagle.

I shall have the pleasure of acknowledging the great assistance I have received from several naturalists, in the course of this and the succeeding works; but I must be here allowed to return my most sincere thanks to the Reverend Professor Henslow, who, when I was an under-graduate at Cambridge, was one chief means of giving me a taste for Natural History,—who, during my absence, took charge of the collections I sent home, and by his correspondence directed my endeavours,—and who, since my return, has constantly rendered me every assistance which the kindest friend could offer.

ST. JAGO—CAPE DE VERD ISLANDS.

JAN. 16TH, 1832.—The neighbourhood of Porto Praya, viewed from the sea, wears a desolate aspect. The volcanic fire of past ages, and the scorching heat of a tropical sun, have in most places rendered the soil sterile and unfit for vegetation. The country rises in successive steps of tableland, interspersed with some truncate conical hills, and the horizon is bounded by an irregular chain of more lofty mountains. The scene, as beheld through the hazy atmosphere of this climate, is one of great interest; if, indeed, a person, fresh from the sea, and who has just walked, for the first time, in a grove of cocoa-nut trees, can be a judge of any thing but his own happiness. The island would generally be considered as very uninteresting; but to any one accustomed only to an English landscape, the novel prospect of an utterly sterile land possesses a grandeur which more vegetation might spoil. A single green leaf can scarcely be discovered over wide tracts of the lava plains; yet flocks of goats, together with a few cows, contrive to exist. It rains very seldom, but during a short portion of the year heavy torrents fall, and immediately afterwards a light vegetation springs out of every crevice. This soon withers; and upon such naturally-formed hay the animals live. At the present time it has not rained for an entire year. The broad, flat-bottomed, valleys, many of which serve during a few days only in the season as a water-course, are clothed with thickets of leafless bushes. Few living creatures inhabit these valleys. The commonest bird is a kingfisher (*Dacelo jagoensis*), which tamely sits on the branches of the castor-oil plant, and thence darts on the grasshoppers and lizards. It is brightly coloured, but not so beautiful as the European species: in its flight, manners, and place of habitation, which is generally in the driest valleys, there is also a wide difference.

One day, two of the officers and myself rode to Ribeira Grande, a village a few miles to the eastward of Porto Praya. Until we reached the valley of St. Martin, the country presented its usual dull brown appearance; but there, a very small rill of water produces a most refreshing margin of luxuriant vegetation. In the course of an hour we arrived at Ribeira Grande, and were surprised at the sight of a large ruined fort and cathedral. The little town, before its harbour was filled up, was the principal place in the island: it now presents a melancholy, but very picturesque appearance. Having procured a black Padre for a guide, and a Spaniard, who had served in the Peninsular war, as an interpreter, we visited a collection of buildings, of which an ancient church formed the principal part. It is here the governors and captain-generals of the islands have been buried. Some of the tombstones recorded dates of the sixteenth century. The heraldic ornaments were the only things in this retired place that reminded us of Europe. The church or chapel formed one side of a quadrangle, in the middle of which a large clump of bananas were growing. On another side was a hospital, containing about a dozen miserable-looking inmates.

We returned to the "Vênda" to eat our dinners. A considerable number of men, women, and children, all as black as jet, were collected to watch us. Our companions were extremely merry; and every thing we said or did was followed by their hearty laughter. Before leaving the town we visited the cathedral. It does not appear so rich as the smaller church, but boasts of a little organ, which sent forth most singularly inharmonious cries. We presented the black priest with a few shillings, and the Spaniard, patting him on the head, said, with much candour, he thought his colour made no great difference. We then returned, as fast as the ponies would go, to Porto Praya.

Another day we rode to the village of St. Domingo, situated near the centre of the island. On a small plain which we crossed, a few stunted acacias were growing; their tops, by the action of the steady trade-wind, were bent in a singular manner—some of them even at a right angle to the trunk. The direction of the branches was exactly N.E. by N., and S.W. by S. These natural vanes must indicate the prevailing direction of the force of the trade wind. The travelling had made so little impression on the barren soil, that we here missed our track, and took that to Fuentes. This we did not find out till we arrived there; and we were afterwards very glad of our mistake. Fuentes is a pretty village, with a small stream; and every thing appeared to prosper well, excepting, indeed, that which ought to do so most—its inhabitants. The black children, completely naked, and looking very wretched, were carrying bundles of firewood half as big as their own bodies.

Near Fuentes we saw a large flock of guinea-fowl—probably fifty or sixty in number. They were extremely wary, and could not be approached. They avoided us, like partridges on a rainy day in September, running with their heads cocked up; and if pursued, they readily took to the wing.

The scenery of St. Domingo possesses a beauty totally unexpected, from the prevalent gloomy character of the rest of the island. The village is situated at the bottom of a valley, bounded by lofty and jagged walls of stratified lava. The black rocks afford a most striking contrast with the bright green vegetation, which follows the banks of a little stream of clear water. It happened to be a grand feast-day, and the village was full of people. On our return we overtook a party of about twenty young black girls, dressed in most excellent taste; their black skins and snow-white linen being set off by their coloured turbans and large shawls. As soon as we approached near, they suddenly all turned round, and covering the path with their shawls, sung with great energy a wild song, beating time with their hands upon their legs. We threw them some vintéms, which were received with screams of laughter, and we left them redoubling the noise of their song.

It has already been remarked, that the atmosphere is generally very hazy; this appears chiefly due to an impalpable dust, which is constantly falling, even on vessels far out at sea. The dust is of a brown colour, and under the blowpipe easily fuses into a black enamel. It is produced, as I believe, from the wear and tear of volcanic rocks, and must come from the coast of Africa. One morning the view was singularly clear; the distant mountains being projected with the sharpest outline, on a heavy bank of dark blue clouds. Judging from the appearance, and from similar cases in England, I supposed that the air was saturated with moisture. The fact, however, turned out quite the contrary. The hygrometer gave a difference of 29.6 degrees, between the temperature of the air, and the point at which dew was precipitated. This difference was nearly double that which I had observed on the previous mornings. This unusual degree of atmospheric dryness was accompanied by continual flashes of lightning. Is it not an uncommon case, thus to find a remarkable degree of aerial transparency with such a state of weather?

The geology of this island is the most interesting part of its natural history. On entering the harbour, a perfectly horizontal white band, in the face of the sea cliff, may be seen running for some miles along the coast, and at the height of about forty-five feet above the water. Upon examination, this white stratum is found to consist of calcareous matter, with numerous shells embedded, such as now exist on the neighbouring coast. It rests on ancient volcanic rocks, and has been covered by a stream of basalt, which must have entered the sea, when the white shelly bed was lying at the bottom. It is interesting to trace the changes, produced by the heat of the overlying lava, on the friable mass. For a thickness of several inches it is converted, in some parts, into a firm stone, as hard as the best freestone; and the earthy matter, originally mingled with the calcareous, has been separated into little spots, thus leaving the limestone white and pure. In other parts a highly crystalline marble has been formed, and so perfect are the crystals of carbonate of lime, that they can easily be measured by the reflecting goniometer. The change is even more extraordinary, where the lime has been caught up by the scoriaceous fragments of the lower surface of the stream; for it is there converted into groups of beautifully radiated fibres resembling arragonite. The beds of lava rise in successive gently-sloping plains, towards the interior, whence the deluges of melted stone originally proceeded. Within historical times, no signs of volcanic activity have, I believe, been manifested in any part of St. Jago. This state of quiescence is, probably, owing to the neighbouring island of Fogo being frequently in eruption. Even the form of a crater can but rarely be discovered on the summits of any of the red cindery hills; yet the more recent streams can be distinguished on the coast, forming a line of cliffs of less height, but stretching out in advance of those belonging to an older series: the height of the cliff thus affording a rude measure of the age.

During our stay, I observed the habits of some marine animals. A large Aplysia is very common. This sea-slug is about five inches long;

and is of a dirty yellowish colour, veined with purple. At the anterior extremity, it has two pair of feelers; the upper ones of which resemble in shape the ears of a quadruped. On each side of the lower surface, or foot, there is a broad membrane, which appears sometimes to act as a ventilator, in causing a current of water to flow over the dorsal branchiæ. It feeds on delicate sea-weeds, which grow among the stones in muddy and shallow water; and I found in its stomach several small pebbles, as in the gizzards of birds. This slug, when disturbed, emits a very fine purplish-red fluid, which stains the water for the space of a foot around. Besides this means of defence, an acrid secretion, which is spread over its body, causes a sharp, stinging sensation, similar to that produced by the Physalia, or Portuguese man-of-war.

I was much interested, on several occasions, by watching the habits of an Octopus or cuttle-fish. Although common in the pools of water left by the retiring tide, these animals were not easily caught. By means of their long arms and suckers, they could drag their bodies into very narrow crevices; and when thus fixed, it required great force to remove them. At other times they darted tail first, with the rapidity of an arrow, from one side of the pool to the other, at the same instant discolouring the water with a dark chestnut-brown ink. These animals also escape detection by a very extraordinary, chameleon-like, power of changing their colour. They appear to vary the tints, according to the nature of the ground over which they pass: when in deep water, their general shade was brownish purple, but when placed on the land, or in shallow water, this dark tint changed into one of a yellowish green. The colour, examined more carefully, was a French gray, with numerous minute spots of bright yellow: the former of these varied in intensity; the latter entirely disappeared and appeared again by turns. These changes were effected in such a manner, that clouds, varying in tint between a hyacinth red and a chestnut brown, were continually passing over the body. Any part being subjected to a slight shock of galvanism, became almost black: a similar effect, but in a less degree, was produced by scratching the skin with a needle. These clouds, or blushes, as they may be called, when examined under a glass, are described as being produced by the alternate expansions and contractions of minute vesicles, containing variously-coloured fluids.

This cuttle-fish displayed its chameleon-like power both during the act of swimming and whilst remaining stationary at the bottom. I was much amused by the various arts to escape detection used by one individual, which seemed fully aware that I was watching it. Remaining for a time motionless, it would then stealthily advance an inch or two, like a cat after a mouse; sometimes changing its colour: it thus proceeded, till having gained a deeper part, it darted away, leaving a dusky train of ink to hide the hole into which it had crawled.

While looking for marine animals, with my head about two feet above the rocky shore, I was more than once saluted by a jet of water, accompanied by a slight grating noise. At first I did not know what it was, but afterwards I found out that it was the cuttle-fish, which, though concealed in a hole, thus often led me to its discovery. That it possesses the power of ejecting water there is no doubt, and it appeared to me certain that it could, moreover, take good aim by directing the tube or siphon on the under side of its body. From the difficulty which these animals have in carrying their heads, they cannot crawl with ease when placed on the ground. I observed that one which I kept in the cabin was slightly phosphorescent in the dark.

ST. PAUL'S ROCKS.—In crossing the Atlantic we hove to, during the morning of February 16th, close to the island of St. Paul. This cluster of rocks is situated in 0° 58′ north latitude, and 29° 15′ west longitude. It is 540 miles distant from the coast of America, and 350 from the island of Fernando Noronha. The highest point is only fifty feet above the level of the sea, and the entire circumference is under three-quarters of a mile. This small point rises abruptly out of the depths of the ocean. Its mineralogical constitution is not simple; in some parts, the rock is of a cherty, in others, of a felspathic nature; and in the latter case it contains thin veins of serpentine, mingled with calcareous matter.

The circumstance of these rocks not being of volcanic origin is of interest, because, with very few exceptions, the islands situated in the midst of the great oceans are thus constituted. As the highest pinnacles of the great mountain ranges probably once existed as islands distant from any continent, we are led to expect that they would frequently consist of volcanic rocks. It becomes, therefore, a curious point to speculate on the changes which many of the present islands would undergo, during the lapse of the countless ages, which would be required to elevate them into snow-clad summits. If we take the case of Ascension, or St. Helena, both of which have long existed in an extinct condition, we may feel assured, before so vast a period could elapse, during the whole of which the surface would be exposed to constant wear and tear, that the mere nucleus or core of the island would remain; perhaps, every fragment of cellular rock having been decomposed, a mass of some compact stone, as phonolite or greenstone, would crown our new Chimborazo.

The rocks of St. Paul appear from a distance of a brilliantly white colour. This is partly owing to the dung of a vast multitude of seafowl, and partly to a coating of a glossy white substance, which is intimately united to the surface of the rocks. This, when examined with a lens, is found to consist of numerous exceedingly thin layers, its total thickness being about the tenth of an inch. The surface is smooth and glossy, and has a pearly lustre; it is considerably harder than calcareous spar, although it can be scratched by a knife: under the blowpipe it decrepitates, slightly blackens, and emits a fetid odour. It consists of phosphate of lime, mingled with some impurities; and its origin without doubt is due to the action of the rain or spray on the bird's dung. I may here mention, that I found in some hollows in the lava rocks of Ascension considerable masses of the substance called *guano*, which on the west coast of the intertropical parts of South America occurs in great beds, some yards thick, on the islets frequented by seafowl. According to the analysis of Fourcroy and Vauquelin, it consists of the urates, phosphates, and oxalates of lime, ammonia, and potash, together with some other salts, and some fatty and earthy matter. I believe there is no doubt of its being the richest manure which has ever been discovered. At Ascension, close to the *guano*, stalactitic or botryoidal masses of impure phosphate of lime adhered to the basalt. The basal part of these had an earthy texture, but the extremities were smooth and glossy, and sufficiently hard to scratch common glass. These stalactites appeared to have shrunk, perhaps from the removal of some soluble matter, in the act of consolidation; and hence they had an irregular form. Similar stalactitic masses, though I am not aware that they have ever been noticed, are, I believe, by no means of uncommon occurrence.

We only observed two kinds of birds—the booby and the noddy. The former is a species of gannet, and the latter a tern. Both are of a tame and stupid disposition, and are so unaccustomed to visiters, that I could have killed any number of them with my geological hammer. The booby lays her eggs on the bare rock; but the tern makes a very simple nest with sea-weed. By the side of many of these nests a small flying-fish was placed; which, I suppose, had been brought by the male bird for its partner. It was amusing to watch how quickly a large and active crab (*Graspus*), which inhabits the crevices of the rock, stole the fish from the side of the nest, as soon as we had disturbed the birds. Not a single plant, not even a lichen, grows on this island; yet it is inhabited by several insects and spiders. The following list completes, I believe, the terrestrial fauna: a species of Feronia and an acarus, which must have come here as parasites on the birds; a small brown moth, belonging to a genus that feeds on feathers; a staphylinus (*Quedius*) and a woodlouse from beneath the dung; and lastly, numerous spiders, which I suppose prey on these small attendants on, and scavengers of, the waterfowl. The often-repeated description of the first colonists of the coral islets in the South Sea, is not, probably, quite correct: I fear it destroys the poetry of the story to find, that these little vile insects should thus take possession before the cocoa-nut tree and other noble plants have appeared.

The smallest rock in the tropical seas, by giving a foundation, for the growth of innumerable kinds of sea-weed and compound animals, supports likewise a large number of fish. The sharks, and the seamen in the boats maintained a constant struggle, who should secure the greater share of the prey caught by the lines. I have heard, that a rock near the Bermudas, lying many miles out at sea, and covered by a considerable depth of water, was first discovered by the circumstance of fish having been observed in the neighbourhood.

FERNANDO NORONHA, FEB. 20TH.—As far as I was enabled to observe, during the few hours we staid at this place, the constitution of the island is volcanic, but probably not of a recent date. The most remarkable feature is a conical hill, about one thousand feet high, the upper part of which is exceedingly steep, and on one side overhangs its base. The rock is phonolite, and is divided into irregular columns. From the first impression, on viewing one of these isolated masses, one is inclined to believe, that the whole has been suddenly pushed up in a semi-fluid state. At St. Helena, however, I ascertained that some pinnacles, of a nearly similar figure and constitution, had been formed by the injection of melted rock among the yielding strata; which thus formed the model for these gigantic obelisks. The whole island is covered with wood; but from the dryness of the climate there is no appearance of luxuriance. At some elevation great masses of the columnar rock, shaded by laurels, and ornamented by a tree covered by fine pink flowers like those of a foxglove, but without a single leaf, gave a pleasing effect to the nearer parts of the scenery.

BAHIA, OR SAN SALVADOR. BRAZIL, FEB. 29TH.—The day has past delightfully. Delight itself, however, is a weak term to express the feelings of a naturalist who, for the first time, has been wandering by himself in a Brazilian forest. Among the multitude of striking objects, the general luxuriance of the vegetation bears away the victory. The elegance of the grasses, the novelty of the parasitical plants, the beauty of the flowers, the glossy green of the foliage, all tend to this end. A most paradoxical mixture of sound and silence pervades the shady parts of the wood. The noise from the insects is so loud, that it may be heard even in a vessel anchored several hundred yards from the shore; yet within the recesses of the forest a universal silence appears to reign. To a person fond of natural history, such a day as this, brings with it a deeper pleasure than he ever can hope again to experience. After wandering about for some hours, I returned to the landing-place; but, before reaching it, I was overtaken by a tropical storm. I tried to find shelter under a tree which was so thick, that it would never have been penetrated by common English rain; but here, in a couple of minutes, a little torrent flowed down the trunk. It is to this violence of the rain we must attribute the verdure at the bottom of the thickest woods: if the showers were like those of a colder clime, the greater part would be absorbed or evaporated before it reached the ground. I will not at present attempt to describe the gaudy scenery of this noble bay, because, in our homeward voyage, we called here a second time, and I shall then have occasion to remark on it.

The geology of the surrounding country possesses little interest. Throughout the coast of Brazil, and certainly for a considerable space inland, from the Rio Plata to Cape St. Roque, lat. 5° S., a distance of more than 2000 geographical miles, wherever solid rock occurs, it belongs to a granitic formation. The circumstance of this enormous area being thus constituted of materials, which almost every geologist believes to have been crystallized by the action of heat under pressure, gives rise to many curious reflections. Was this effect produced beneath the depths of a profound ocean? or did a covering of strata formerly extend over it, which has since been removed? Can we believe that any power, acting for a time short of infinity, could have denuded the granite over so many thousand square leagues?

On a point not far from the city, where a rivulet entered the sea, I observed a fact connected with a subject discussed by Humboldt. At the cataracts of the great rivers Orinoco, Nile, and Congo, the syenitic rocks are coated by a black substance, appearing as if they had been polished with plumbago. The layer is of extreme thinness; and on analysis by Berzelius it was found to consist of the oxides of manganese and iron. In the Orinoco it occurs on the rocks periodically washed by the floods, and in those parts alone, where the stream is rapid; or, as the Indians say, "the rocks are black, where the waters are white." The coating is here of a rich brown instead of a black colour, and seems to be composed of ferrugineous matter alone. Hand specimens fail to give a just idea of these brown, burnished, stones which glitter in the sun's rays. They occur only within the limits of tidal action; and as the rivulet slowly trickles down, the surf must supply the polishing power of the cataracts in the great rivers. In the same manner, the rise and fall of the tide probably answers to the periodical inundations; and thus the same causes are present under apparently very different circumstances. The real origin, however, of these coatings of metallic oxides, which seem as if cemented to the rocks, is not understood; and no reason, I believe, can be assigned for their thickness remaining constant.

One day I was amused by watching the habits of a Diodon, which was caught swimming near the shore. This fish is well known to possess the singular power of distending itself into a nearly spherical form. After having been taken out of water for a short time, and then again immersed in it, a considerable quantity both of water and air was absorbed by the mouth, and perhaps likewise by the branchial apertures. This process is effected by two methods; the air is swallowed, and is then forced into the cavity of the body, its return being prevented by a muscular contraction which is externally visible; but the water, I observed, entered in a stream through the mouth, which was wide open and motionless: this latter action must, therefore, depend on suction. The skin about the abdomen is much looser than that of the back; hence, during the inflation, the lower surface becomes far more distended than the upper; and the fish, in consequence, floats with its back downwards. Cuvier doubts whether the Diodon, in this position, is able to swim; but not only can it thus move forward in a straight line, but likewise it can turn round to either side. This latter movement is effected solely by the aid of the pectoral fins; the tail being collapsed, and not used. From the body being buoyed up with so much air, the branchial openings were out of water; but a stream drawn in by the mouth, constantly flowed through them.

The fish, having remained in this distended state for a short time, generally expelled the air and water with considerable force from the branchial apertures and mouth. It could emit, at will, a certain portion of the water; and it appears, therefore, probable, that this fluid is taken in partly for the sake of regulating its specific gravity. This diodon possessed several means of defence. It could give a severe bite, and could eject water from its mouth to some distance, at the same time it made a curious noise by the movement of its jaws. By the inflation of its body, the papillæ, with which the skin is covered, became erect and pointed. But the most curious circumstance was, that it emitted from the skin of its belly, when handled, a most beautiful carmine red and fibrous secretion, which stained ivory and paper in so permanent a manner, that the tint is retained with all its brightness to the present day. I am quite ignorant of the nature and use of this secretion.

MARCH 18TH.—We sailed from Bahia. A few days afterwards, when not far distant from the Abrolhos islets, my attention was called to a discoloured appearance in the sea. The whole surface of the water, as it appeared under a weak lens, seemed as if covered by chopped bits of hay, with their ends jagged. One of the larger particles measured .03 of an inch in length, and .009 in breadth. Examined more carefully, each is seen to consist of from twenty to sixty cylindrical filaments, which have perfectly rounded extremities, and are divided at regular intervals by transverse septa, containing a brownish-green flocculent matter. The filaments must be enveloped in some viscid fluid, for the bundles adhered together without actual contact. I do not know to what family these bodies properly belong, but they have a close general resemblance in structure with the confervæ which grow in every ditch. These simple vegetables, thus constituted for floating in the open ocean, must in certain places exist in countless numbers. The ship passed through several bands of them, one of which was about ten yards wide, and, judging from the mud-like colour of the water, at least two and a half miles long. In almost every long voyage some account is given of these confervæ. They appear especially common in the sea near Australia. Off Cape Leeuwin, I found some very similar to those above described; they differed chiefly in the bundles being rather smaller, and being composed of fewer filaments. Captain Cook, in his third voyage, remarks, that the sailors gave to this appearance the name of sea-sawdust.

I may here mention that during two days preceding our arrival at the Keeling Islands in the Indian Ocean, I saw in many parts masses of flocculent matter, of a brownish-green colour, floating in the sea. They varied in size, from half to to three or four inches square; and were quite irregular in figure. In an opake vessel they could barely be distinguished, but in a glass one they were clearly visible. Under the microscope the flocculent matter was seen to consist of two kinds of confervæ, between which I am quite ignorant whether there exists any connection. Minute cylindrical bodies, conical at each extremity, are involved in vast numbers, in a mass of fine threads. These threads have a diameter of about 2⁄3000 of an inch; they possess an internal lining, and are divided at irregular and very wide intervals by transverse septa. Their length is so great, that I could never with certainty ascertain the form of the uninjured extremity; they are all curvilinear,

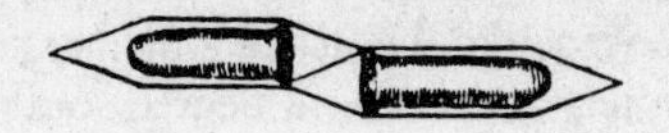

and resemble in mass a handful of hair, coiled up and squeezed together. In the midst of these threads, and probably connected by some viscid fluid, the other kind, or the cylindrical transparent bodies, float in great numbers. These have their two extremities terminated by cones, produced into the finest points: their diameter is tolerably constant between .006 and .008 of an inch; but their length varies considerably from .04 to .06, and even sometimes to .08. Near one extremity of the cylindrical part, a green septum, formed of granular matter, and thickest in the middle, may generally be seen. This, I believe, is the bottom of a most delicate, colourless sack, composed of a pulpy substance, which lines the exterior case, but does not extend within the extreme conical points. In some, small but perfect spheres of brownish granular matter supplied the place of the septa; and I observed the curious process by which they were produced. The pulpy matter of the internal coating suddenly grouped itself into lines, some of which assumed a form radiating from a common centre; it then continued, with an irregular and rapid movement, to contract itself, so that, in the course of a second, the whole was united into a perfect little sphere, which occupied the position of the septum at one end of the now quite hollow case. The appearance was as if an elastic membrane, for instance, a thin Indian-rubber ball, had been distended with air, and then burst, in which case the edges would instantly shrink up and contract towards a point. The formation of the granular sphere was hastened by any accidental injury. I may add, that frequently a pair of these bodies were attached to each other, as represented in the accompanying rude drawing, cone beside cone, at that end where the septum occurs. When floating uninjured in the sea, the formation of the spherical gemmules perhaps only takes place, when two of the plants (or rather animals, according to Bory St. Vincent) thus become attached, and married to each other. Nevertheless, I certainly witnessed this curious process in several individuals, when separate, and where there was no apparent cause of disturbance. In any case it does not seem probable, from the fixed structure of the septum, that the whole of the granular matter is transferred from one to the other body, as with the true Conjugatæ.

I will here add a few other observations connected with the discoloration of the sea from organic causes. On the coast of Chile, a few leagues north of Concepcion, the Beagle one day passed through great bands of muddy water; and again, a degree south of Valparaiso, the same appearance was still more extensive. Although we were nearly fifty miles from the coast, I at first attributed this circumstance to real streams of muddy water brought down by the river Maypo. Mr. Sulivan, however, having drawn up some in a glass, thought he distinguished, by the aid of a lens, moving points. The water was slightly stained as if by red dust; and after leaving it for some time quiet, a cloud collected at the bottom. With a lens, of one-fourth of an inch focal distance, small hyaline points could be seen darting about with great rapidity, and frequently exploding. Examined with a much higher power, their shape was found to be oval, and contracted by a ring round the middle, from which line curved little setæ proceeded on all sides; and these were the organs of motion. One end of the body was narrower and more pointed than the other. According to the arrangement of Bory St. Vincent, they are animalcula, belonging to the family of Trichodes: it was, however, very difficult to examine them with care, for almost the instant motion ceased, even while crossing the field of vision, their bodies burst. Sometimes both ends burst at once, sometimes only one, and a quantity of coarse brownish granular matter was ejected, which cohered very slightly. The ring with the setæ sometimes retained its irritability for a little while after the contents of the body had been emptied, and continued a riggling, uneven motion. The animal an instant before bursting expanded to half again its natural size; and the explosion took place about fifteen seconds after the rapid progressive motion had ceased: in a few cases it was preceded for a short interval by a rotatory movement on the longer axis. About two minutes after any number were isolated in a drop of water, they thus perished. The animals move with the narrow apex forwards, by the aid of their vibratory ciliæ, and generally by rapid starts. They are exceedingly minute, and quite invisible to the naked eye, only covering a space equal to the square of the thousandth of an inch. Their numbers were infinite; for the smallest drop of water which I could remove contained very many. In one day we passed through two spaces of water thus stained, one of which alone must have extended over several square miles. What incalculable numbers of these microscopical animals! The colour of the water, as seen at some distance, was like that of a river which has flowed through a red clay district; but under the shade of the vessel's side, it was quite as dark as chocolate. The line where the red and blue water joined was distinctly defined. The weather for some days previously had been calm, and the ocean abounded, to an unusual degree, with living creatures. In Ulloa's voyage an account is given of crossing, in nearly the same latitude, some discoloured water, which was mistaken for a shoal: no soundings were obtained, and I have no doubt, from the description, that this little animalcule was the cause of the alarm.

In the sea around Tierra del Fuego, and at no great distance from the land, I have seen narrow lines of water of a bright red colour, from the number of crustacea, which somewhat resemble in form large prawns. The sealers call them whale-food. Whether whales feed on them I do not know; but terns, cormorants, and immense herds of great unwieldly seals, on some parts of the coast, derive their chief sustenance from these swimming crabs. Seamen invariably attribute the discoloration of the water to spawn; but I found this to be the case only on one occasion. At the distance of several leagues from the Archipelago of the Galapagos, the ship sailed through three strips of a dark yellowish, or mud-like, water; these strips were some miles long; but only a few yards wide, and they were separated from the surrounding surface by a sinuous yet distinct margin. The colour was caused by little gelatinous balls, about the fifth of an inch in diameter, in which numerous minute spherical ovules were embedded: they were of two distinct kinds, one being of a reddish colour and of a different shape from the other. I cannot form a conjecture as to what two kinds of animals these belonged. Captain Colnett remarks, that this appearance is very common among the Galapagos Islands, and that the direction of the bands indicates that of the currents; in the described case, however, the line was caused by the wind. The only other appearance which I have to notice, is a thin oily coat on the surface which displays iridescent colours. I saw a considerable tract of the ocean thus covered on the coast of Brazil; the seamen attributed it to the putrefying carcass of some whale, which probably was floating at no great distance. I do not here mention the minute gelatinous particles which are frequently dispersed throughout the water, for they are not sufficiently abundant to create any change of colour.

There are two circumstances in the above accounts which appear very remarkable: first, how do the various bodies which form the bands with defined edges keep together? In the case of the prawn-like crabs, their movements were as coinstantaneous as in a regiment of soldiers; but this cannot happen from any thing like voluntary action with the ovules, or the confervæ, nor is it probable among the infusoria. Secondly, what causes the length and narrowness of the bands? The appearance so much resembles that which may be seen in every torrent, where the stream uncoils into long streaks, the froth collected in the eddies, that I must attribute the effect to a similar action either of the currents of the air, or sea. Under this supposition we must believe that the various organized bodies are produced in certain favourable places, and are thence removed by the set of either wind or water. I confess, however, there is a very great difficulty in imagining any one spot to be the birthplace of the millions of millions of animalcula and confervæ: for whence come the germs at such points?—the parent bodies having been distributed by the winds and waves over the immense ocean. But on no other hypothesis can I understand their linear grouping. I may add that Scoresby remarks, that green water abounding with pelagic animals, is invariably found in a certain part of the Arctic Sea.

BAHIA BLANCA.

The Beagle arrived on the 24th of August, and a week afterwards sailed for the Plata. With Captain Fitzroy's consent I was left behind, to travel by land to Buenos Ayres. I will here add some observations, which were made during this visit, and on a previous occasion, when the Beagle was employed in surveying the harbour. Not much can be made out respecting the geology. At the distance of some miles inland, an escarpment of a great argillaceo-calcareous formation of rock extends. The space near the coast consists of plains of hardened mud, and broad bands of sand-dunes, which present appearances, that can easily be accounted for by a rise of the land; and of this phenomenon, although to a trifling amount, we have other proofs.

At Punta Alta, a low cliff, about twenty feet high, exposes a mass of partly consolidated shingle, irregularly interstratified with a reddish muddy clay, and containing numerous recent shells. We may believe a similar accumulation would now take place, on any point, where tides and waves were opposed. In the gravel a considerable number of bones were embedded. Mr. Owen, who has undertaken the description of these remains, has not yet examined them with care; but the following list may give some idea of their nature: 1st, a tolerably perfect head of a megàtherium, and a fragment and teeth of two others; 2d, an animal of the order Edentata, as large as a pony, and with great scratching claws; 3d and 4th, two great Edentata related to the megatherium, and both fully as large as an ox or horse; 5th, another equally large animal, closely allied or perhaps identical with the Toxodon (hereafter to be described), which had very flat grinding teeth, somewhat resembling those of a rodent; 6th, a large piece of the tesselated covering like that of the armadillo, but of gigantic size; 7th, a tusk which in its prismatic form, and in the disposition of the enamel, closely resembles that of the African boar; it is probable that it belonged to the same animal with the singular flat grinders. Lastly, a tooth in the same state of decay with the others: its broken condition does not allow Mr. Owen, without further comparison, to come to any definite conclusion; but the part that is perfect, resembles in every respect the tooth of the common horse.* All these remains were found embedded in a beach which is covered at spring tides; and the space in which they were collected could not have exceeded one hundred and fifty yards square. It is a remarkable circumstance that so many different species should be found together; and it proves how numerous in kind the ancient inhabitants of this country must have been.

At the distance of about thirty miles, in another cliff of red earth, I found several fragments of bones. Among them were the teeth of a rodent, much narrower, but even larger than those of the *Hydrochærus capybara*; the animal which has been mentioned as exceeding in dimensions every existing member of its order. There was also part of the head of a Ctenomys; the species being different from the Tucutuco, but with a close general resemblance.

The remains at Punta Alta were associated, as before remarked, with shells of existing species. These have not as yet been examined with scrupulous care, but it may be safely asserted, that they are most closely similar to the species now living in the same bay: it is also very remarkable, that not only the species, but the proportional numbers of each kind, are nearly the same with those now cast up on the pebble beaches. There are eleven marine species (some in an imperfect state), and one terrestrial. If I had not collected living specimens from the same bay, some of the fossils would have been thought extinct; for Mr. Sowerby, who was kind enough to look at my collection, had not

* With respect to the remains of the last animal, as some doubt may be entertained by others, respecting its origin; it must be remarked, that it was fairly embedded in the gravel with the other bones; and that its state of decay was equal. To this circumstance it may be added, that the surrounding country is without fresh water, and is uninhabited, and that the settlement, itself only of five years standing, is twenty-five miles distant.

previously seen them. We may feel certain that the bones have not been washed out of an older formation, and embedded in a more recent one, because the remains of one of the Edentata were lying in their proper relative position (and partly so in a second case); which could not have happened, without the carcass had been washed to the spot where the skeleton is now entombed.

We here have a strong confirmation of the remarkable law so often insisted on by Mr. Lyell, namely, that the "longevity of the species in the mammalia, is upon the whole inferior to that of the testacea." When we proceed to the southern part of Patagonia, I shall have occasion to describe the case of an extinct camel, from which the same result may be deduced.

From the shells being littoral species (including one terrestrial), and from the character of the deposit, we may feel absolutely certain that the remains were embedded in a shallow sea, not far from the coast. From the position of the skeleton being undisturbed, and likewise from the fact that full-grown serpulæ were attached to some of the bones, we know that the mass could not have been accumulated on the beach itself. At the present time, part of the bed is daily washed by the tide, while another part has been raised a few feet above the level of the sea. Hence we may infer, that the elevation has here been trifling, since the period when the mammalia, now extinct, were living. This conclusion is in harmony with several other considerations (such as the recent character of the beds underlying the Pampas deposit), but which I have not space in this work to enter on.

From the general structure of the coast of this part of South America, we are compelled to believe, that the changes of level have all (at least of late) been in one direction, and that they have been very gradual. If, then, we look back to the period when these quadrupeds lived, the land probably stood at a level, less elevated only by a few fathoms than at present. Therefore, its general configuration since that epoch cannot have been greatly modified; a conclusion which certainly would be drawn from the close similarity in every respect, between the shells now living in the bay (as well as in the case of the one terrestrial species) with those which formerly lived there.

The surrounding country, as may have been gathered from this journal, is of a very desert character. Trees nowhere occur, and only a few bushes, which are chiefly confined to depressions among the sand-hillocks, or to the borders of the saline marshes. Here, then, is an apparent difficulty: we have the strongest evidence that there has occurred no great physical change to modify the features of the country, yet in former days, numerous large animals were supported on the plains now covered by a thin and scanty vegetation.

That large animals require a luxuriant vegetation, has been a general assumption, which has passed from one work to another. I do not hesitate, however, to say that it is completely false; and that it has vitiated the reasoning of geologists, on some points of great interest in the ancient history of the world. The prejudice has probably been derived from India, and the Indian islands, where troops of elephants, noble forests, and impenetrable jungles, are associated together in every account. If, on the other hand, we refer to any work of travels through the southern parts of Africa, we shall find allusions in almost every page either to the desert character of the country, or to the numbers of large animals inhabiting it. The same thing is rendered evident by the many sketches which have been published of various parts of the interior. When the Beagle was at Cape Town, I rode a few leagues into the country, which at least was sufficient to render that which I had read more fully intelligible.

Dr. Andrew Smith, who, at the head of his adventurous party, has so lately succeeded in passing the Tropic of Capricorn, informs me that, taking into consideration the whole of the southern part of Africa, there can be no doubt of its being sterile country. On the southern and south-eastern coasts there are some fine forests, but with these exceptions, the traveller may pass, for days together, through open plains, covered by a poor and scanty vegetation. It is difficult to convey any accurate idea of degrees of comparative fertility; but it may be safely said, that the amount of vegetation supported at any one time* by Great Britain, exceeds, perhaps even tenfold, the quantity on an equal area, in the interior parts of Southern Africa. The fact that bullock-waggons can travel in any direction, excepting near the coast, without more than occasionally half an hour's delay, gives, perhaps, a more definite notion of its scantiness. Now if we look to the animals inhabiting these wide plains, we shall find their numbers extraordinarily great, and their bulk immense. We must enumerate the elephant, three species of rhinoceros, and as Dr. Smith is convinced two others also, the hippopotamus, giraffe, the bos caffer—as large as a full-grown bull, and the elan—but little less, two zebras, and the quaccha, two gnus, and several antelopes even larger than these latter animals. It may be supposed that although the species are numerous, the indi-

* I mean by this to exclude the total amount, which may have been successively produced and consumed during a given period.

viduals of each kind are few. By the kindness of Dr. Smith, I am enabled to show that the case is very different. He informs me, that in lat. 24°, in one day's march with the bullock-waggons, he saw, without wandering to any great distance on either side, between one hundred and one hundred and fifty rhinoceroses, which belonged to three species. That the same day he saw several herds of giraffes, amounting together to nearly a hundred; and that, although no elephant was observed, yet they are found in this district. At the distance of a little more than one hour's march from their place of encampment on the previous night, his party actually killed at one spot eight hippopotamoses, and saw many more. In this same river there were likewise crocodiles. Of course it was a case quite extraordinary, to see so many great animals crowded together, but it evidently proves that they must exist in great numbers. Dr. Smith describes the country passed through that day, as "being thinly covered with grass, and bushes about four feet high, and still more thinly with mimosa-trees." The waggons were not prevented travelling in a nearly direct line.

Besides these large animals, every one the least acquainted with the natural history of the Cape, has read of the herds of antelopes, which can be compared only to flocks of migratory birds. The numbers indeed of the lion,* panther, and hyæna, and the multitude of birds of prey, plainly tell of the abundance of the smaller quadrupeds. As Dr. Smith remarked to me, the carnage each day in Southern Africa must indeed be terrific! I confess it is truly surprising, how such a number of animals can find support in a country producing so little food. The larger quadrupeds no doubt roam over wide extents in search of it; and their food chiefly consists of underwood, which probably contains much nutriment in a small bulk. Dr. Smith also informs me that the vegetation has a rapid growth; no sooner is a part consumed, than its place is supplied by a fresh stock. I apprehend, however, that our ideas respecting the quantity necessary for the support of large quadrupeds are exaggerated. It should have been remembered that the camel, an animal of no mean bulk, has always been considered as the emblem of the desert.

The belief that where large quadrupeds exist, the vegetation must necessarily be luxuriant, is the more remarkable, because the converse is far from true. Mr. Burchell observed to me that when entering Brazil, nothing struck him more forcibly than the splendour of the South

* Dr. Smith mentioned to me, that one evening seven lions were counted at one time walking on the plain, round the encampment.

American vegetation, contrasted with that of South Africa, together with the absence of all large quadrupeds. In his Travels, he has suggested that the comparison of the respective weights (if there were sufficient data) of an equal number of the largest herbivorous quadrupeds of each country would be extremely curious. If we take on the one side the elephant, hippopotamus, giraffe, bos caffer, elan, certainly three species of rhinoceros, and probably five; and on the other side, two tapirs, the guanaco, three deer, the vicuna, peccari, capybara (after which we must choose from the monkeys to complete the number), and then place these two groups alongside each other, it is not easy to conceive ranks more disproportionate. After the above facts, we are compelled to conclude, against anterior probability, that among the mammalia there exists no close relation between the *bulk* of the species, and the *quantity* of the vegetation, in the countries they inhabit.

With regard to the number of large quadrupeds, there certainly exists no quarter of the globe which will bear comparison with Southern Africa. After the different statements which have been given, the extremely desert character of that region will not be disputed. In the European division of the world, we must look back to the tertiary epochs, to find a condition of things among the mammalia resembling that which is now found at the Cape of Good Hope. That tertiary epoch, which we are apt to consider as abounding to an astonishing degree with large animals, because we find the remains of many ages accumulated at certain spots, could boast of but few more of the large quadrupeds, than Southern Africa does at present. If we speculate on the condition of the vegetation during that epoch, we are at least bound so far to consider existing analogies, as not to urge as absolutely necessary a luxuriant vegetation, when we see a state of things so totally different in the region to which we refer.

We know that the extreme regions of North America, many degrees beyond the limit where the ground at the depth of a few feet remains perpetually congealed, are covered by forests of large and tall trees. In a like manner, in Siberia, we have woods of birch, fir, aspen, and larch, growing in a latitude (64°), where the mean temperature of the air falls below the freezing point, and where the earth is so completely frozen, that the carcass of an animal embedded in it is perfectly preserved. With these facts we must grant, as far as *quantity alone* of vegetation is concerned, that the great quadrupeds of the later tertiary epochs might, in most parts of Northern Europe and Asia, have lived on the spots where their remains are now found. I do not here speak of the *kind* of vegetation necessary for their support; because, as there is evidence of physical changes, and as the animals have become extinct, so may we suppose that the species of plants have likewise been changed.

These remarks directly bear on the case of the Siberian animals preserved in ice. The firm conviction of the necessity of a vegetation, possessing a character of tropical luxuriance, to support such large animals, and the impossibility of reconciling this with the proximity of perpetual congelation, was one chief cause of the several theories of sudden revolutions of climate, and of overwhelming catastrophes, which were invented to account for their entombment. I am far from supposing that the climate has not changed since the period when those animals lived, which now lie buried in the ice. At present I only wish to show, that as far as *quantity* of food *alone* is concerned, the ancient rhinoceroses might have roamed over the *steppes* of central Siberia (the northern parts probably being under water) even in their present condition, as well as the living rhinoceroses and elephants over the *Karros* of Southern Africa.

After our long digression, if we return to the case of the fossil animals at Bahia Blanca, there is a difficulty from our not knowing on what food the great Edentata probably lived. If on insects and larvæ, like their nearest representatives the armadilloes and anteaters, there is an end to all conjecture. But as vegetation is the first source of life in every part of the world, I think we may safely conclude that the country around Bahia Blanca, with a very little increase of fertility, would support large animals. The plains of the Rio Negro, thickly scattered over with thorny bushes, I do not doubt would supply sufficient food equally well with the *Karros* of Africa. As there is evidence of a physical change to a small amount, so may we allow it to be probable that the productiveness of the soil has decreased in an equally small degree. With this concession I apprehend every difficulty is removed. On the other hand, if we imagine a luxuriant vegetation to be necessary for the support of these animals, we become involved in a series of contradictions and improbabilities.

As the notices of the remains of several quadrupeds, which I discovered in South America, are scattered in different parts of this volume, I will here give a catalogue of them. After having enlarged on the diminutive size of the present races, it may be of interest to see that formerly a very different order of things prevailed. First, the megatherium, and the four or five other large Edentata, already alluded to; 6th, an immense mastodon, which must have abounded over the whole country; 7th, the horse (I do not now refer to the broken tooth at Bahia Blanca, but to more certain evidence); 8th, the toxodon, an extraordinary animal as large as a hippopotamus; 9th, a fragment of the head of an animal larger than a horse, and of a very singular character; 10th, 11th, and 12th, parts of rodents—one of considerable size; lastly, a llama or *guanaco*, fully as large as the camel. All these animals coexisted during an epoch which, geologically speaking is so recent, that it may be considered as only just gone by. These remains have been presented to the College of Surgeons, where they are now in the hands of those best qualified to appreciate whatever value they may possess.

• • •

TIERRA DEL FUEGO.

DECEMBER 17TH, 1832.—Having now finished with Patagonia, I will describe our first arrival in Tierra del Fuego. A little after noon we doubled Cape St. Diego, and entered the famous strait of Le Maire. We kept close to the Fuegian shore, but the outline of the rugged, inhospitable Staten land was visible amidst the clouds. In the afternoon we anchored in the bay of Good Success. While entering we were saluted in a manner becoming the inhabitants of this savage land. A group of Fuegians partly concealed by the entangled forest, were perched on a wild point overhanging the sea; and as we passed by, they sprang up, and waving their tattered cloaks sent forth a loud and sonorous shout. The savages followed the ship, and just before dark we saw their fire, and again heard their wild cry. The harbour consists of a fine piece of water half surrounded by low rounded mountains of clay-slate, which are covered to the water's edge by one dense gloomy forest. A single glance at the landscape was sufficient to show me, how widely different it was from any thing I had ever beheld. At night it blew a gale of wind, and heavy squalls from the mountains swept past us. It would have been a bad time out at sea, and we, as well as others, may call this Good Success Bay.

In the morning, the Captain sent a party to communicate with the Fuegians. When we came within hail, one of the four natives who were present advanced to receive us, and began to shout most vehemently, wishing to direct us where to land. When we were on shore the party looked rather alarmed, but continued talking and making gestures with great rapidity. It was without exception the most curious and interesting spectacle I had ever beheld. I could not have believed how wide was the difference, between savage and civilized man. It is greater than between a wild and domesticated animal, in as much as in man there is a greater power of improvement. The chief spokesman was old, and appeared to be the head of the family; the three others were powerful young men, about six feet high. The women and children had been sent away. These Fuegians are a very different race from the stunted miserable wretches further to the westward. They are much superior in person, and seem closely allied to the famous Patagonians of the Strait of Magellan. Their only garment consists of a mantle made of guanaco skin, with the wool outside; this they wear just thrown over their shoulders, as often leaving their persons exposed as covered. Their skin is of a dirty coppery red colour.

The old man had a fillet of white feathers tied round his head, which partly confined his black, coarse, and entangled hair. His face was crossed by two broad transverse bars; one painted bright red reached from ear to ear, and included the upper lip; the other, white like chalk, extended parallel and above the first, so that even his eyelids were thus coloured. Some of the other men were ornamented by streaks of black powder, made of charcoal. The party altogether closely resembled the devils which come on the stage in such plays as Der Freischutz.

Their very attitudes were abject, and the expression of their countenances distrustful, surprised, and startled. After we had presented them with some scarlet cloth, which they immediately tied round their necks, they became good friends. This was shown by the old man patting our breasts, and making a chuckling kind of noise, as people do when feeding chickens. I walked with the old man, and this demonstration of friendship was repeated several times; it was concluded by three hard slaps, which were given me on the breast and back at the same time. He then bared his bosom for me to return the compliment, which being done, he seemed highly pleased. The language of these people, according to our notions, scarcely deserves to be called articulate. Captain Cook has compared it to a man clearing his throat, but certainly no European ever cleared his throat with so many hoarse, guttural, and clicking sounds.

They are excellent mimics: as often as we coughed or yawned, or made any odd motion, they immediately imitated us. Some of our party began to squint and look awry; but one of the young Fuegians (whose whole face was painted black, excepting a white band across his eyes) succeeded in making far more hideous grimaces. They could repeat with perfect correctness, each word in any sentence we addressed them, and they remembered such words for some time. Yet we Europeans all know how difficult it is to distinguish apart the sounds in a foreign language. Which of us, for instance, could follow an American Indian through a sentence of more than three words? All savages appear to possess, to an uncommon degree, this power of mimicry. I was told almost in the same words, of the same ludicrous habits among the Caffres: the Australians, likewise, have long been notorious for being able to imitate and describe the gait of any man, so that he may be recognised. How can this faculty be explained? is it a consequence of the more practised habits of perception and keener senses, common to all men in a savage state, as compared to those long civilized?

When a song was struck up by our party, I thought the Fuegians would have fallen down with astonishment. With equal surprise they viewed our dancing; but one of the young men, when asked, had no objection to a little waltzing. Little accustomed to Europeans as they appeared to be, yet they knew, and dreaded our fire-arms; nothing would tempt them to take a gun in their hands. They begged for knives, calling them by the Spanish word "cuchilla." They explained also what they wanted, by acting as if they had a piece of blubber in their mouth, and then pretending to cut instead of tear it.

It was interesting to watch the conduct of these people towards Jemmy Button (one of the Fuegians* who had been taken, during the former voyage, to England): they immediately perceived the difference between him and the rest, and held much conversation between themselves on the subject. The old man addressed a long harangue to Jemmy, which it seems was to invite him to stay with them. But Jemmy understood very little of their language, and was, moreover, thoroughly ashamed of his countrymen. When York Minster (another of these men) came on shore, they noticed him in the same way, and told him he ought to shave; yet he had not twenty dwarf hairs on his face, whilst we all wore our untrimmed beards. They examined the colour of his skin, and compared it with ours. One of our arms being bared, they expressed the liveliest surprise and admiration at its whiteness. We thought that they mistook two or three of the officers, who were

* Captain FitzRoy has given a history of these people. Four were taken to England; one died there, and the three others (two men and one woman) were now brought back and settled in their own country.

rather shorter and fairer (though adorned with large beards), for the ladies of our party. The tallest amongst the Fuegians was evidently much pleased at his height being noticed. When placed back to back with the tallest of the boat's crew, he tried his best to edge on higher ground, and to stand on tiptoe. He opened his mouth to show his teeth, and turned his face for a side view; and all this was done with such alacrity, that I dare say he thought himself the handsomest man in Tierra del Fuego. After the first feeling on our part of grave astonishment was over, nothing could be more ludicrous or interesting than the odd mixture of surprise and imitation which these savages every moment exhibited.

• • •

December 21st.—The Beagle got under way: and on the succeeding day, favoured to an uncommon degree by a fine easterly breeze, we closed in with the Barnevelts, and, running past Cape Deceit with its stony peaks, about three o'clock doubled the weatherbeaten Cape Horn. The evening was calm and bright, and we enjoyed a fine view of the surrounding isles. Cape Horn, however, demanded his tribute, and before night sent us a gale of wind directly in our teeth. We stood out to sea, and on the second day again made the land, when we saw on our weather-bow this notorious promontory in its proper form—veiled in a mist, and its dim outline surrounded by a storm of wind and water. Great black clouds were rolling across the heavens, and squalls of rain, with hail, swept by us with extreme violence; so that the captain determined to run into Wigwam Cove. This is a snug little harbour, not far from Cape Horn; and here, at Christmas-eve, we anchored in smooth water. The only thing which reminded us of the gale outside, was every now and then a puff from the mountains, which seemed to wish to blow us out of the water.

December 25th.—Close by the cove, a pointed hill, called Kater's Peak, rises to the height of 1700 feet. The surrounding islands all consist of conical masses of greenstone, associated sometimes with less regular hills of baked and altered clay-slate. This part of Tierra del Fuego may be considered as the extremity of the submerged chain of mountains already alluded to. The cove takes its name of "Wigwam" from some of the Fuegian habitations; but every bay in the neighbourhood might be so called with equal propriety. The inhabitants living chiefly upon shell-fish, are obliged constantly to change their place of residence; but they return at intervals to the same spots, as is evident from the pile of old shells, which must often amount to some tons in weight. These heaps can be distinguished at a long distance by the bright green colour of certain plants, which invariably grow on them. Among these may be enumerated the wild celery and scurvy grass, two very serviceable plants, the use of which has not been discovered by the natives.

The Fuegian wigwam resembles, in size and dimensions, a haycock. It merely consists of a few broken branches stuck in the ground, and very imperfectly thatched on one side with a few tufts of grass and rushes. The whole cannot be so much as the work of an hour, and it is only used for a few days. At Goeree Roads I saw a place where one of these naked men had slept, which absolutely offered no more cover than the form of a hare. The man was evidently living by himself, and York Minster said he was "very bad man," and that probably he had stolen something. On the west coast, however, the wigwams are rather better, for they are covered with seal-skins. We were detained here several days by the bad weather. The climate is certainly wretched; the summer solstice was now passed, yet every day snow fell on the hills, and in the valleys there was rain, accompanied by sleet. The thermometer generally stood about 45°, but in the night fell to 38° or 40°. From the damp and boisterous state of the atmosphere, not cheered by a gleam of sunshine, one fancied the climate even worse than it really was.

At a subsequent period the Beagle anchored for a couple of days under Wollaston Island, which is a short way to the northward. While going on shore we pulled alongside a canoe with six Fuegians. These were the most abject and miserable creatures I any where beheld.* On the east coast the natives, as we have seen, have guanaco cloaks, and on the west, they possess seal-skins. Amongst these central tribes the men generally possess an otter-skin, or some small scrap about as large

* I believe, in this extreme part of South America, man exists in a lower state of improvement than in any other part of the world. The South Sea islander of either race is comparatively civilized. The Esquimaux, in his subterranean hut, enjoys some of the comforts of life, and in his canoe, when fully equipped, manifests much skill. Some of the tribes of Southern Africa, prowling about in search of roots, and living concealed on the wild and arid plains, are sufficiently wretched. But the Australian, in the simplicity of the arts of life, comes nearest the Fuegian. He can, however, boast of his boomerang, his spear and throwing-stick, his method of climbing trees, tracking animals, and scheme of hunting. Although thus superior in acquirements, it by no means follows that he should likewise be so in capabilities. Indeed, from what we saw of the Fuegians, who were taken to England, I should think the case was the reverse.

as a pocket-handkerchief, which is barely sufficient to cover their backs as low down as their loins. It is laced across the breast by strings, and according as the wind blows, it is shifted from side to side. But these Fuegians in the canoe were quite naked, and even one full-grown woman was absolutely so. It was raining heavily, and the fresh water, together with the spray, trickled down her body. In another harbour not far distant, a woman, who was suckling a recently-born child, came one day alongside the vessel, and remained there whilst the sleet fell and thawed on her naked bosom, and on the skin of her naked child. These poor wretches were stunted in their growth, their hideous faces bedaubed with white paint, their skins filthy and greasy, their hair entangled, their voices discordant, their gestures violent and without dignity. Viewing such men, one can hardly make oneself believe they are fellow-creatures, and inhabitants of the same world. It is a common subject of conjecture what pleasure in life some of the less gifted animals can enjoy: how much more reasonably the same question may be asked with respect to these barbarians. At night, five or six human beings, naked and scarcely protected from the wind and rain of this tempestuous climate, sleep on the wet ground coiled up like animals. Whenever it is low water, they must rise to pick shellfish from the rocks; and the women, winter and summer, either dive to collect sea eggs, or sit patiently in their canoes, and, with a baited hair-line, jerk out small fish. If a seal is killed, or the floating carcass of a putrid whale discovered, it is a feast: such miserable food is assisted by a few tasteless berries and fungi. Nor are they exempt from famine, and, as a consequence, cannibalism accompanied by parricide.

The tribes have no government or head, yet each is surrounded by other hostile ones, speaking different dialects; and the cause of their warfare would appear to be the means of subsistence. Their country is a broken mass of wild rock, lofty hills, and useless forests: and these are viewed through mists and endless storms. The habitable land is reduced to the stones which form the beach; in search of food they are compelled to wander from spot to spot, and so steep is the coast, that they can only move about in their wretched canoes. They cannot know the feeling of having a home, and still less that of domestic affection; unless indeed the treatment of a master to a laborious slave can be considered as such. How little can the higher powers of the mind be brought into play! What is there for imagination to picture, for reason to compare, for judgment to decide upon? to knock a limpet from the rock does not even require cunning, that lowest power of the mind. Their skill in some respects may be compared to the instinct of animals; for it is not improved by experience: the canoe, their most ingenious work, poor as it is, has remained the same, for the last two hundred and fifty years.

Whilst beholding these savages, one asks, whence have they come? What could have tempted, or what change compelled a tribe of men to leave the fine regions of the north, to travel down the Cordillera or backbone of America, to invent and build canoes, and then to enter on one of the most inhospitable countries within the limits of the globe? Although such reflections must at first occupy one's mind, yet we may feel sure that many of them are quite erroneous. There is no reason to believe that the Fuegians decrease in number; therefore we must suppose that they enjoy a sufficient share of happiness (of whatever kind it may be) to render life worth having. Nature by making habit omnipotent, and its effects hereditary, has fitted the Fuegian to the climate and the productions of his country.

• • •

GALAPAGOS ARCHIPELAGO.

SEPTEMBER 15TH.—The Beagle arrived at the southernmost of the Galapagos Islands. This archipelago consists of ten principal islands, of which five much exceed the others in size. They are situated under the equatorial line, and between five and six hundred miles to the westward of the coast of America. The constitution of the whole is volcanic. With the exception of some ejected fragments of granite, which have been most curiously glazed and altered by the heat, every part consists of lava, or of sandstone resulting from the attrition of such materials. The higher islands, (which attain an elevation of three, and even four thousand feet) generally have one or more principal craters towards their centre, and on their flanks smaller orifices. I have no exact data from which to calculate, but I do not hesitate to affirm, that there must be, in all the islands of the archipelago, at least two thousand craters. These are of two kinds; one, as in ordinary cases, consisting of scoriæ and lava, the other of finely-stratified volcanic sandstone. The latter in most instances have a form beautifully symmetrical: their origin is due to the ejection of mud,—that is, fine volcanic ashes and water,—without any lava.

Considering that these islands are placed directly under the equator, the climate is far from being excessively hot; a circumstance which, perhaps, is chiefly owing to the singularly low temperature of the surrounding sea. Excepting during one short season, very little rain falls, and even then it is not regular: but the clouds generally hang low. From these circumstances the lower parts of the islands are extremely arid, whilst the summits, at an elevation of a thousand feet or more, possess a tolerably luxuriant vegetation. This is especially the case on the windward side, which first receives and condenses the moisture from the atmosphere.

In the morning (17th,) we landed on Chatham Island, which, like the others, rises with a tame and rounded outline, interrupted only here and there by scattered hillocks—the remains of former craters. Nothing could be less inviting than the first appearance. A broken field of black basaltic lava is every where covered by astunted brushwood, which shows little signs of life. The dry and parched surface, having been heated by the noonday sun, gave the air a close and sultry feeling, like that from a stove: we fancied even the bushes smelt unpleasantly. Although I diligently tried to collect as many plants as possible, I succeeded in getting only ten kinds; and such wretched-looking little weeds would have better become an arctic, than an equatorial Flora.

The thin woods, which cover the lower parts of all the islands, excepting where the lava has recently flowed, appear from a short distance quite leafless, like the deciduous trees of the northern hemisphere in winter. It was some time before I discovered, that not only almost every plant was in full leaf, but that the greater number were now in flower. After the period of heavy rains, the islands are said to appear for a short time partially green. The only other country, in which I have seen a vegetation with a character at all approaching to this, is at the volcanic island of Fernando Noronha, placed in many respects under similar conditions.

The natural history of this archipelago is very remarkable: it seems to be a little world within itself; the greater number of its inhabitants, both vegetable and animal, being found nowhere else. As I shall refer to this subject again, I will only here remark, as forming a striking character on first landing, that the birds are strangers to man. So tame and unsuspecting were they, that they did not even understand what was meant by stones being thrown at them; and quite regardless of us, they approached so close that any number might have been killed with a stick.

The Beagle sailed round Chatham Island, and anchored in several bays. One night I slept on shore, on a part of the island where some black cones—the former chimneys of the subterranean heated fluids —were extraordinarily numerous. From one small eminence, I counted sixty of these truncated hillocks, which were all surmounted by a more or less perfect crater. The greater number consisted merely of a ring of red scoriæ, or slags, cemented together: and their height above the plain of lava, was not more than from fifty to a hundred feet. From their regular form, they gave the country a *workshop* appearance, which strongly reminded me of those parts of Staffordshire where the great iron-foundries are most numerous.

The age of the various beds of lava was distinctly marked by the comparative growth, or entire absence, of vegetation. Nothing can be imagined more rough and horrid than the surface of the more modern streams. These have been aptly compared to the sea petrified in its most boisterous moments: no sea, however, would present such irregular undulations, or would be traversed by such deep chasms. All the craters are in an extinct condition; and although the age of the different streams of lava could be so clearly distinguished, it is probable they have remained so for many centuries. There is no account in any of the old voyagers of any volcano on this island having been seen in activity; yet since the time of Dampier (1684), there must have been some increase in the quantity of vegetation, otherwise so accurate a person would not have expressed himself thus:—"Four or five of the easternmost islands are rocky, barren, and hilly, producing neither tree, herb, nor grass, but a few dildoe (cactus) trees, except by the seaside." This description is at present applicable only to the western islands, where the volcanic forces are in frequent activity.

The day, on which I visited the little craters, was glowing hot, and the scrambling over the rough surface, and through the intricate thickets, was very fatiguing; but I was well repaid by the Cyclopian scene. In my walk I met two large tortoises, each of which must have weighed at least two hundred pounds. One was eating a piece of cactus, and when I approached, it looked at me, and then quietly walked away: the other gave a deep hiss and drew in its head. These huge reptiles, surrounded by the black lava, the leafless shrubs, and large cacti, appeared to my fancy like some antediluvian animals.

SEPTEMBER 23D.—The Beagle proceeded to Charles Island. This archipelago has long been frequented, first by the Bucaniers, and latterly by whalers, but it is only within the last six years, that a small colony has been established on it. The inhabitants are between two and three hundred in number: they nearly all consist of people of colour, who have been banished for political crimes from the Republic of the Equator (Quito is the capital of this state) to which these islands belong. The settlement is placed about four and a half miles inland, and at an elevation probably of a thousand feet. In the first part of the road we passed through leafless thickets, as in Chatham Island. Higher up, the wood gradually became greener; and immediately we had crossed the ridge of the island, our bodies were cooled by the fine southerly trade-wind, and our senses refreshed by the sight of a green and thriving vegetation. The houses are irregularly scattered over a flat space of ground, which is cultivated with sweet potatoes and bananas. It will not easily be imagined how pleasant the sight of black mud was to us, after having been so long accustomed to the parched soil of Peru and Chile.

The inhabitants, although complaining of poverty, gain, without much trouble, the means of subsistence from the fertile soil. In the woods there are many wild pigs and goats, but the main article of animal food is derived from the tortoise. Their numbers in this island have of course been greatly reduced, but the people yet reckon on two days' hunting supplying food for the rest of the week. It is said that formerly single vessels have taken away as many as seven hundred of these animals, and that the ship's company of a frigate some years since brought down two hundred to the beach in one day.

We staid at this island four days, during which time I collected many plants and birds. One morning I ascended the highest hill, which has an altitude of nearly 1800 feet. The summit consists of a broken-down crater, thickly clothed with coarse grass and brushwood. Even in this one island, I counted thirty-nine hills, each of which was terminated by a more or less perfect circular depression.

SEPTEMBER 29TH.—We doubled the south-west extremity of Albermarle Island, and the next day were nearly becalmed between it and Narborough Island. Both are covered with immense streams of black naked lava; which, having either flowed over the rims of the great caldrons, or having burst forth from the smaller orifices on the flanks, have in their descent spread over miles of the sea-coast. On both of these islands eruptions are known occasionally to take place; and in Albermarle we saw a small jet of smoke curling from the summit of one of the more lofty craters. In the evening we anchored in Bank's Cove, in Albermarle Island.

When morning came, we found that the harbour in which we were at anchor was formed by a broken-down crater, composed of volcanic sandstone. After breakfast I went out walking. To the southward of this first crater, there was another of similar composition, and beautifully symmetrical. It was elliptic in form; the longer axis being less than a mile, and its depth about 500 feet. The bottom was occupied by a shallow lake, and in its centre a tiny crater formed an islet. The day was overpoweringly hot, and the lake looked clear and blue. I hurried down the cindery slope, and choked with dust eagerly tasted the water—but to my sorrow I found it salt as brine.

The rocks on the coast abounded with great black lizards, between three and four feet long; and on the hills, another species was equally common. We saw several of the latter, some clumsily running out of our way, and others shuffling into their burrows. I shall presently describe in more detail the habits of both these reptiles.

OCTOBER 3D.—We sailed round the northern end of Albermarle Island. Nearly the whole of this side is covered with recent streams of dark-coloured lavas, and is studded with craters. I should think it would be difficult to find in any other part of the world, an island situated within the tropics, and of such considerable size (namely 75 miles long), so sterile and incapable of supporting life.

On the 8th we reached James Island. Captain FitzRoy put Mr. Bynoe, myself, and three others on shore, leaving with us a tent and provisions, to wait there till the vessel returned from watering. This was an admirable plan for the collections, as we had an entire week of hard work. We found here a party of Spaniards, who had been sent from Charles Island to dry fish, and to salt tortoise-meat.

At the distance of about six miles, and at the height of nearly 2000 feet, the Spaniards had erected a hovel in which two men lived, who were employed in catching tortoises, whilst the others were fishing on the coast. I paid this party two visits, and slept there one night. In the same manner as in the other islands, the lower region is covered by nearly leafless bushes: but here many of them grow to the size of trees. I measured several which were two feet in diameter, and some even two feet nine inches. The upper region being kept damp, from the moisture of the condensed clouds, supports a green and flourishing vegetation. So damp was the ground, that there were large beds of a coarse carex, in which great numbers of a very small water-rail lived and bred. While staying in this upper region, we lived entirely upon tortoise-meat. The breastplate roasted (as the Gauchos do *carne con cuero*), with the flesh attached to it, is very good; and the young tortoises make excellent soup; but otherwise the meat to my taste is very indifferent.

During another day we accompanied a party of the Spaniards in their whale-boat to a salina, or lake from which salt is procured. After landing, we had a very rough walk over a rugged field of recent lava, which has almost surrounded a sandstone crater, at the bottom of which the salt-lake is situated. The water was only three or four inches deep, and rested on a layer of beautifully crystallized white salt. The lake was quite circular, and fringed with a border of brightly green succulent plants: the precipitous walls of the crater were also clothed with wood, so that the scene was both picturesque and curious. A few years since, the sailors belonging to a sealing-vessel murdered their captain in this quiet spot; and we saw his skull lying among the bushes.

During the greater part of our week on shore, the sky was cloudless, and if the trade-wind failed for an hour, the heat became very oppressive. On two days, the thermometer within the tent stood for some hours at 93°; but in the open air, in the wind and sun, at only 85°. The sand was extremely hot; the thermometer placed in some of a brown colour immediately rose to 137°, and how much higher it would have risen, I do not know, for it was not graduated above that number. The *black* sand felt much hotter, so that even in thick boots it was disagreeable, on this account, to walk over it.

I will now offer a few general observations on the natural history of these islands. I endeavoured to make as nearly a perfect collection in every branch as time permitted. The plants have not yet been examined, but Professor Henslow, who has kindly undertaken the description of them, informs me that there are probably many new species, and perhaps even some new genera. They all have an extremely weedy character, and it would scarcely have been supposed, that they had grown at an inconsiderable elevation directly under the equator. In the lower and sterile parts, the bush, which from its minute brown leaves chiefly gives the leafless appearance to the brushwood, is one of the Euphorbiacæ. In the same region an acacia and a cactus (*Opuntia Galapageia*), with large oval compressed articulations, springing from a cylindrical stem, are in some parts common. These are the only trees which in that part afford any shade. Near the summits of the different islands, the vegetation has a very different character; ferns and coarse grasses are abundant; and the commonest tree is one of the Compositæ. Tree-ferns are not present. One of the most singular characters of the Flora, considering the position of this archipelago, is the absence of every member of the palm family. Cocos Island, on the other hand, which is the nearest point of land, takes its name from the great number of cocoa-nut trees on it. From the presence of the Opuntias and some other plants, the vegetation partakes more of the character of that of America than of any other country.

Of mammalia a large kind of mouse forms a well-marked species. From its large thin ears, and other characters, it approaches in form a section of the genus, which is confined to the sterile regions of South America. There is also a rat which Mr. Waterhouse believes is probably distinct from the English kind; but I cannot help suspecting that it is only the same altered by the peculiar conditions of its new country.

In my collections from these islands, Mr. Gould considers that there are twenty-six different species of land birds. With the exception of one, all probably are undescribed kinds, which inhabit this archipelago, and no other part of the world. Among the waders and waterfowl it is more difficult, without detailed comparison, to say what are new. But a water-sail which lives near the summits of the mountains, is undescribed, as perhaps is a Totanus and a heron. The only kind of gull which is found among these islands, is also new; when the wandering habits of this genus are considered, this is a very remarkable circumstance. The species most closely allied to it, comes from the Strait of Magellan. Of the other aquatic birds, the species appear the same with well-known American birds.

The general character of the plumage of these birds is extremely plain, and like the Flora possesses little beauty. Although the species are thus peculiar to the archipelago, yet nearly all in their general structure, habits, colour of feathers, and even tone of voice, are strictly American. The following brief list will give an idea of their kinds. 1st. A buzzard, having many of the characters of Polyborus or Caracara; and in its habits not to be distinguished from that peculiar South American genus; 2d. Two owls; 3d. Three species of tyrant-flycatchers—a form strictly American. One of these appears identical with a common kind (*Muscicapa coronata*? Lath.), which has a very wide range, from La Plata throughout Brazil to Mexico; 4th. A sylvicola, an American form, and especially common in the northern division of the continent; 5th. Three species of mocking-birds, a genus common to both Americas; 6th. A finch, with a stiff tail and a long claw to its hinder toe, closely allied to a North American genus; 7th. A swallow belonging to the American division of that genus; 8th. A dove, like, but distinct from, the Chilian species; 9th. A group of finches, of which Mr. Gould considers there are thirteen species; and these he has distributed into four new sub-genera. These birds are the most singular of any in the archipelago. They all agree in many points; namely, in a peculiar structure of their bill, short tails, general form, and in their plumage. The females are gray or brown, but the old cocks jet-black. All the species, excepting two, feed in flocks on the ground, and have very similar habits. It is very remarkable that a nearly perfect gradation of structure in this one group can be traced in the form of the beak, from one exceeding in dimensions that of the largest grosbeak, to another differing but little from that of a warbler. Of the aquatic birds I have already remarked that some are peculiar to these islands, and some common to North and South America.

We will now turn to the order of reptiles, which forms, perhaps, the most striking feature in the zoology of these islands. The species are not numerous, but the number of individuals of each kind, is extraordinarily great. There is one kind both of the turtle and tortoise; of lizards four; and of snakes about the same number.

I will first describe the habits of the tortoise (*Testudo Indicus*) which has been so frequently alluded to. These animals are found, I believe, in all the islands of the Archipelago; certainly in the greater number. They frequent in preference the high damp parts, but likewise inhabit the lower and arid districts. I have already mentioned* proofs, from the numbers which have been taken in a single day, how very numerous they must be. Some individuals grow to an immense size: Mr. Lawson, an Englishman, who had at the time of our visit charge of the colony, told us that he had seen several so large, that it required six or eight men to lift them from the ground; and that some had afforded as much as two hundred pounds of meat. The old males are the largest, the females rarely growing to so great a size. The male can readily be distinguished from the female by the greater length of its tail. The tortoises which live on those islands where there is no water, or in the lower and arid parts of the others, chiefly feed on the succulent cactus. Those which frequent the higher and damp regions, eat the leaves of various trees, a kind of berry (called guayavita) which is acid and austere, and likewise a pale green filamentous lichen, that hangs in tresses from the boughs of the trees.

The tortoise is very fond of water, drinking large quantities, and wallowing in the mud. The larger islands alone possess springs, and these are always situated towards the central parts, and at a considerable elevation. The tortoises, therefore, which frequent the lower districts, when thirsty, are obliged to travel from a long distance. Hence broad and well-beaten paths radiate off in every direction from

* Dampier says, "The land-turtles are here so numerous, that five or six hundred men might subsist on them for several months without any other sort of provisions. They are so extraordinarily large and fat, and so sweet, that no pullet eats more pleasantly."—Vol. i, p. 110.

the wells even down to the sea-coast; and the Spaniards by following them up, first discovered the watering-places. When I landed at Chatham Island, I could not imagine what animal travelled so methodically along the well-chosen tracks. Near the springs it was a curious spectacle to behold many of these great monsters; one set eagerly travelling onwards with outstretched necks, and another set returning, after having drunk their fill. When the tortoise arrives at the spring, quite regardless of any spectator, it buries its head in the water above its eyes, and greedily swallows great mouthfuls, at the rate of about ten in a minute. The inhabitants say each animal stays three or four days in the neighbourhood of the water, and then returns to the lower country; but they differed in their accounts respecting the frequency of these visits. The animal probably regulates them according to the nature of the food which it has consumed. It is, however, certain, that tortoises can subsist even on those islands where there is no other water, than what falls during a few rainy days in the year.

I believe it is well ascertained, that the bladder of the frog acts as a reservoir for the moisture necessary to its existence: such seems to be the case with the tortoise. For some time after a visit to the springs, the urinary bladder of these animals is distended with fluid, which is said gradually to decrease in volume, and to become less pure. The inhabitants, when walking in the lower district, and overcome with thirst, often take advantage of this circumstance, by killing a tortoise, and if the bladder is full, drinking its contents. In one I saw killed, the fluid was quite limpid, and had only a *very slightly* bitter taste. The inhabitants, however, always drink first the water in the pericardium, which is described as being best.

The tortoises, when moving towards any definite point, travel by night and day, and arrive at their journey's end much sooner than would be expected. The inhabitants, from observations on marked individuals, consider that they can move a distance of about eight miles in two or three days. One large tortoise, which I watched, I found walked at the rate of sixty yards in ten minutes, that is 360 in the hour, or four miles a day,—allowing also a little time for it to eat on the road.

During the breeding season, when the male and female are together, the male utters a hoarse roar or bellowing, which it is said, can be heard at the distance of more than a hundred yards. The female never uses her voice, and the male only at such times; so that when the people hear this noise, they know the two are together. They were at this time (October) laying their eggs. The female, where the soil is sandy, deposits them together, and covers them up with sand; but wherc the ground is rocky she drops them indiscriminately in any hollow. Mr. Bynoe found seven placed in a line in a fissure. The egg is white and spherical; one which I measured was seven inches and three-eighths in circumference. The young animals, as soon as they are hatched, fall a prey in great numbers to the buzzard, with the habits of the Caracara. The old ones seem generally to die from accidents, as from falling down precipices. At least several of the inhabitants told me, they had never found one dead without some such apparent cause.

The inhabitants believe that these animals are absolutely deaf; certainly they do not overhear a person walking close behind them. I was always amused, when overtaking one of these great monsters as it was quietly pacing along, to see how suddenly, the instant I passed, it would draw in its head and legs, and uttering a deep hiss fall to the ground with a heavy sound, as if struck dead. I frequently got on their backs, and then, upon giving a few raps on the hinder part of the shell, they would rise up and walk away;—but I found it very difficult to keep my balance.

The flesh of this animal is largely employed, both fresh and salted; and a beautifully clear oil is prepared from the fat. When a tortoise is caught, the man makes a slit in the skin near its tail, so as to see inside its body, whether the fat under the dorsal plate is thick. If it is not, the animal is liberated; and it is said to recover soon from this strange operation. In order to secure the tortoises, it is not sufficient to turn them like turtle, for they are often able to regain their upright position.

It was confidently asserted, that the tortoises coming from different islands in the archipelago were slightly different in form; and that in certain islands they attained a larger average size than in others. Mr. Lawson maintained that he could at once tell from which island any one was brought. Unfortunately, the specimens which came home in the Beagle were too small to institute any certain comparison. This tortoise, which goes by the name of *Testudo Indicus*, is at present found in many parts of the world. It is the opinion of Mr. Bell, and some others who have studied reptiles, that it is not improbable that they all originally came from this archipelago. When it is known how long these islands have been frequented by the bucaniers, and that they constantly took away numbers of these animals alive, it seems very probable that they should have distributed them in different parts of the world. If this tortoise does not originally come from these islands, it is a remarkable anomaly; inasmuch as nearly all the other land inhabitants seem to have had their birthplace here.

Of lizards there are four or five species; two probably belong to the South American genus Leiocephalus, and two to Amblyrhyncus. This remarkable genus was characterized by Mr. Bell, from a stuffed specimen sent from Mexico, but which I conceive there can be little doubt originally came through some whaling ship from these islands. The two species agree pretty closely in general appearance; but one is aquatic and the other terrestrial in its habits. Mr. Bell thus concludes his description of *Amb. cristatus*: "On a comparison of this animal with the true Iguanas, the most striking and important discrepancy is in the form of the head. Instead of the long, pointed, narrow muzzle of those species, we have here a short, obtusely truncated head, not so long as it is broad, the mouth consequently only capable of being opened to a very short space. These circumstances, with the shortness and equality of the toes, and the strength and curvature of the claws, evidently indicate some striking peculiarity in its food and general habits, on which, however, in the absence of all certain information, I shall abstain from offering any conjecture." The following account of these two lizards, will, I think, show with what judgment Mr. Bell foresaw a variation in habit, accompanying change in structure.

First for the aquatic kind (*Amb. cristatus*). This lizard is extremely common on all the islands throughout the Archipelago. It lives exclusively on the rocky sea-beaches, and is never found, at least I never saw one, even ten yards inshore. It is a hideous-looking creature, of a dirty black colour, stupid and sluggish in its movements. The usual length of a full-grown one is about a yard, but there are some even four feet long: I have seen a large one which weighed twenty pounds. On the island of Albemarle they seem to grow to a greater size than on any other. These lizards were occasionally seen some hundred yards from the shore swimming about; and Captain Collnett, in his Voyage, says, "they go out to sea in shoals to fish." With respect to the object, I believe he is mistaken; but the fact stated on such good authority cannot be doubted. When in the water the animal swims with perfect ease and quickness, by a serpentine movement of its body and flattened tail,—the legs, during this time, being motionless and closely collapsed on its sides. A seaman on board sank one, with a heavy weight attached to it, thinking thus to kill it directly; but when an hour afterwards he drew up the line, the lizard was quite active. Their limbs and strong claws are admirably adapted for crawling over the rugged and fissured masses of lava, which every where form the coast. In such situations, a group of six or seven of these hideous reptiles may oftentimes be seen on the black rocks, a few feet above the surf, basking in the sun with outstretched legs.

I opened the stomach of several, and in each case found it largely distended with minced sea-weed, of that kind which grows in thin foliaceous expansions of a bright green or dull red colour. I do not recollect having observed this sea-weed in any quantity on the tidal rocks; and I have reason to believe it grows at the bottom of the sea, at some little distance from the coast. If such is the case, the object of these animals occasionally going out to sea is explained. The stomach contained nothing but the sea-weed. Mr. Bynoe, however, found a piece of a crab in one; but this might have got in accidentally, in the same manner as I have seen a caterpillar, in the midst of some lichen, in the paunch of a tortoise. The intestines were large, as in other herbivorous animals.

The nature of this lizard's food, as well as the structure of its tail, and the certain fact of its having been seen voluntarily swimming out at sea, absolutely prove its aquatic habits; yet there is in this respect one strange anomaly; namely, that when frightened it will not enter the water. From this cause, it is easy to drive these lizards down to any little point overhanging the sea, where they will sooner allow a person to catch hold of their tail than jump into the water. They do not seem to have any notion of biting; but when much frightened they squirt a drop of fluid from each nostril. One day I carried one to a deep pool left by the retiring tide, and threw it in several times as far as I was able. It invariably returned in a direct line to the spot where I stood. It swam near the bottom, with a very graceful and rapid movement, and occasionally aided itself over the uneven ground with its feet. As soon as it arrived near the margin, but still being under water, it either tried to conceal itself in the tufts of sea-weed, or it entered some crevice. As soon as it thought the danger was past, it crawled out on the dry rocks, and shuffled away as quickly as it could. I several times caught this same lizard, by driving it down to a point, and though possessed of such perfect powers of diving and swimming, nothing would induce it to enter the water; and as often as I threw it in, it returned in the manner above described. Perhaps this singular piece of apparent stupidity may be accounted for by the circumstance, that this reptile has no enemy whatever on shore, whereas at sea it must often fall a prey to the numerous sharks. Hence, probably urged by a fixed and hereditary instinct that the shore is its place of safety, whatever the emergency may be, it there takes refuge.

During our visit (in October) I saw extremely few small individuals of this species, and none I should think under a year old. From this circumstance it seems probable that the breeding season had not commenced. I asked several of the inhabitants if they knew where it laid its eggs: they said, that although well acquainted with the eggs of the other kind, they had not the least knowledge of the manner in which this species is propagated;—a fact, considering how common an animal this lizard is, not a little extraordinary.

We will now turn to the terrestrial species (*Amb. subcristatus* of Gray). This species, differently from the last, is confined to the central islands of the Archipelago, namely to Albemarle, James, Barrington, and Indefatigable. To the southward, in Charles, Hood, and Chatham islands, and to the northward, in Towers, Bindloes, and Abington, I neither saw nor heard of any. It would appear as if this species had been created in the centre of the Archipelago, and thence had been dispersed only to a certain distance.

In the central islands they inhabit both the higher and damp, as well as the lower and sterile parts; but in the latter they are much the most numerous. I cannot give a more forcible proof of their numbers, than by stating, that when we were left at James Island, we could not for some time find a spot free from their burrows, on which to pitch our tent. These lizards, like their brothers the sea-kind, are ugly animals; and from their low facial angle have a singularly stupid appearance. In size perhaps they are a little inferior to the latter, but several of them weighed between ten and fifteen pounds each. The colour of their belly, front legs, and head (excepting the crown which is nearly white), is a dirty yellowish-orange: the back is a brownish-red, which in the younger specimens is darker. In their movements they are lazy and half torpid. When not frightened, they slowly crawl along with their tails and bellies dragging on the ground. They often stop, and doze for a minute with closed eyes, and hind legs spread out on the parched soil.

They inhabit burrows; which they sometimes excavate between fragments of lava, but more generally on level patches of the soft volcanic sandstone. The holes do not appear to be very deep, and they enter the ground at a small angle; so that when walking over these lizard *warrens*, the soil is constantly giving way, much to the annoyance of the tired walker. This animal when excavating its burrow, alternately works the opposite sides of its body. One front leg for a short time scratches up the soil, and throws it towards the hind foot, which is well placed so as to heave it beyond the mouth of the hole. This side of the body being tired, the other takes up the task, and so on alternately. I watched one for a long time, till half its body was buried; I then walked up and pulled it by the tail; at this it was greatly astonished, and soon shuffled up to see what was the matter; and then stared me in the face, as much as to say, "What made you pull my tail?"

They feed by day, and do not wander far from their burrows; and if frightened they rush to them with a most awkward gait. Except when running down hill, they cannot move very fast; which appears chiefly owing to the lateral position of their legs.

They are not at all timorous: when attentively watching any one, they curl their tails, and raising themselves on their front legs, nod their heads vertically, with a quick movement, and try to look very fierce: but in reality they are not at all so; if one just stamps the ground, down go their tails, and off they shuffle as quickly as they can. I have frequently observed small muscivorous lizards, when watching any thing, nod their heads in precisely the same manner; but I do not at all know for what purpose. If this Amblyrhyncus is held, and plagued with a stick, it will bite it very severely; but I caught many by the tail, and they never tried to bite me. If two are placed on the ground and held together, they will fight and bite each other till blood is drawn.

The individuals (and they are the greater number) which inhabit the lower country, can scarcely taste a drop of water throughout the year; but they consume much of the succulent cactus, the branches of which are occasionally broken off by the wind. I have sometimes thrown a piece to two or three when together; and it was amusing enough to see each trying to sieze and carry it away in its mouth, like so many hungry dogs with a bone. They eat very deliberately, but do not chew their food. The little birds are aware how harmless these creatures are: I have seen one of the thick-billed finches picking at one end of a piece of cactus (which is in request among all the animals of the lower region), whilst a lizard was eating at the other; and afterwards the little bird with the utmost indifference hopped on the back of the reptile.

I opened the stomachs of several, and found them full of vegetable fibres, and leaves of different trees, especially of a species of acacia. In the upper region they live chiefly on the acid and astringent berries of the guayavita, under which trees I have seen these lizards and the huge tortoises feeding together. To obtain the acacia-leaves, they crawl up the low stunted trees; and it is not uncommon to see one or a pair quietly browsing, whilst seated on a branch several feet above the ground.

The meat of these animals when cooked is white, and by those whose stomachs rise above all prejudices, it is relished as very good food. Humboldt has remarked that in intertropical South America, all lizards which inhabit *dry* regions are esteemed delicacies for the table. The inhabitants say, that those inhabiting the damp region drink water, but that the others do not travel up for it from the sterile country like the tortoises. At the time of our visit, the females had within their bodies numerous large elongated eggs. These they lay in their burrows, and the inhabitants seek them for food.

These two species of Amblyrhyncus agree, as I have already stated, in general structure, and in many of their habits. Neither have that rapid movement, so characteristic of true Lacerta and Iguana. They are both herbivorous, although the kind of vegetation consumed in each case is so very different. Mr. Bell has given the name to the genus from the shortness of the snout: indeed, the form of the mouth may almost be compared to that of the tortoise. One is tempted to suppose this is an adaptation to their herbivorous appetites. It is very interesting thus to find a well-characterized genus, having its aquatic and terrestrial species, belonging to so confined a portion of the world. The former species is by far the most remarkable, because it is the only existing Saurian, which can properly be said to be a maritime animal. I should perhaps have mentioned earlier, that in the whole archipelago, there is only one rill of fresh water that reaches the coast; yet these reptiles frequent the sea-beaches, and no other parts in all the islands. Moreover, there is no existing lizard, as far as I am aware, excepting this Amblyrhyncus, that feeds exclusively on aquatic productions. If, however, we refer to epochs long past, we shall find such habits common to several gigantic animals of the Saurian race.

To conclude with the order of reptiles. Of snakes there are several species, but all harmless. Of toads and frogs there are none. I was surprised at this, considering how well the temperate and damp woods in the elevated parts appeared adapted for their habits. It recalled to my mind the singular statement made by Bory St. Vincent, namely, that none of this family are to be found on the volcanic islands in the great oceans. There certainly appears to be some foundation for this observation; which is the more remarkable, when compared with the case of lizards, which are generally among the earliest colonists of the smallest islet. It may be asked, whether this is not owing to the different facilities of transport through salt-water, of the eggs of the latter protected by a calcareous coat, and of the slimy spawn of the former?

As I at first observed, these islands are not so remarkable for the number of species of reptiles, as for that of individuals; when we remember the well-beaten paths made by the many hundred great tortoises—the warrens of the terrestrial Amblyrhyncus—and the groups of the aquatic species basking on the coast-rocks—we must admit that there is no other quarter of the world, where this order replaces the herbivorous mammalia in so extraordinary a manner. It is worthy of observation by the geologist (who will probably refer back in his mind to the secondary periods, when the Saurians were developed with dimensions, which at the present day can be compared only to the cetaceous mammalia), that this archipelago, instead of possessing a humid climate and rank vegetation, cannot be considered otherwise than extremely arid, and for an equatorial region, remarkably temperate.

To finish with the zoology: I took great pains in collecting the insects, but I was surprised to find, even in the high and damp region, how exceedingly few they were in number. The forests of Tierra del Fuego are certainly much more barren; but with that exception I never collected in so poor a country. In the lower and sterile land I took seven species of Heteromera, and a few other insects; but in the fine thriving woods towards the centre of the islands, although I perseveringly swept under the bushes during all kinds of weather, I obtained only a few minute Diptera and Hymenoptera. Owing to this scarcity of insects, nearly all the birds live in the lower country; and the part which any one would have thought much the most favourable for them, is frequented only by a few of the small tyrant-flycatchers. I do not believe a single bird, excepting the water-rail, is confined to the damp region. Mr. Waterhouse informs me that nearly all the insects belong to European forms, and that they do not by any means possess an equatorial character. I did not take a single one of large size, or of bright colours. This last observation applies equally to the birds and flowers. It is worthy of remark, that the only land-bird with bright colours, is that species of tyrant-flycatcher, which seems to be a wanderer from the continent. Of shells, there are a considerable number of land kinds, all of which, I believe are confined to this archipelago. Even of marine species, a large proportion were not known, before the collection made by Mr. Cuming on these islands was brought to England.

I will not here attempt to come to any definite conclusions, as the species have not been accurately examined; but we may infer, that, with the exception of a few wanderers, the organic beings found on this archipelago are peculiar to it; and yet that their general form strongly partakes of an American character. It would be impossible for any one accustomed to the birds of Chile and La Plata to be placed on these islands, and not to feel convinced that he was, as far as the organic world was concerned, on American ground. This similarity in type, between distant islands and continents, while the species are distinct, has scarcely been sufficiently noticed. The circumstance would be explained, according to the views of some authors, by saying that the creative power had acted according to the same law over a wide area.

It has been mentioned, that the inhabitants can distinguish the tortoises, according to the islands whence they are brought. I was also informed that many of the islands possess trees and plants which do not occur on the others. For instance the berry-bearing tree, called Guyavita, which is common on James Island, certainly is not found on Charles Island, though appearing equally well fitted for it. Unfortunately, I was not aware of these facts till my collection was nearly completed: it never occurred to me, that the productions of islands only a few miles apart, and placed under the same physical conditions, would be dissimilar. I therefore did not attempt to make a series of specimens from the separate islands. It is the fate of every voyager, when he has just discovered what object in any place is more particularly worthy of his attention, to be hurried from it. In the case of the mocking-bird, I ascertained (and have brought home the specimens) that one species (*Orpheus trifasciatus*, Gould) is exclusively found in Charles Island: a second (*O. parvulus*) on Albemarle Island; and a third (*O. melanotus*) common to James and Chatham Islands. The two last species are closely allied, but the first would be considered by every naturalist as quite distinct. I examined many specimens in the different islands, and in each the respective kind was *alone* present. These birds agree in general plumage, structure, and habits; so that the different species replace each other in the economy of the different islands. These species are not characterized by the markings on the plumage alone, but likewise by the size and form of the bill, and other differences. I have stated, that in the thirteen species of ground-finches, a nearly perfect gradation may be traced, from a beak extraordinarily thick, to one so fine, that it may be compared to that of a warbler. I very much suspect, that certain members of the series are confined to different islands; therefore, if the collection had been made on any *one* island, it would not have presented so perfect a gradation. It is clear, that if several islands have each their peculiar species of the same genera, when these are placed together, they will have a wide range of character. But there is not space in this work, to enter on this curious subject.

Before concluding my account of the zoology of these islands, I must describe more in detail the tameness of the birds. This disposition is common to all the terrestrial species; namely, to the mocking-birds, the finches, sylvicolæ, tyrant-flycatchers, doves, and hawks. There is not one which will not approach sufficiently near to be killed with a switch, and sometimes, as I have myself tried, with a cap or hat. A gun is here almost superfluous; for with the muzzle of one I pushed a hawk off the branch of a tree. One day a mocking-bird alighted on the edge of a pitcher (made of the shell of a tortoise), which I held in my hand whilst lying down. It began very quietly to sip the water, and allowed me to lift it with the vessel from the ground. I often tried, and very nearly succeeded, in catching these birds by their legs. Formerly the birds appear to have been even tamer than at present. Cowley (in the year 1684) says that the "Turtle-doves were so tame that they would often alight upon our hats and arms, so as that we could take them alive: they not fearing man, until such time as some of our company did fire at them, whereby they were rendered more shy." Dampier (in the same year) also says that a man in a morning's walk might kill six or seven dozen of these birds. At present, although certainly very tame, they do not alight on people's arms; nor do they suffer themselves to be killed in such numbers. It is surprising that the change has not been greater; for these islands during the last hundred and fifty years, have been frequently visited by bucaniers and whalers; and the sailors, wandering through the woods in search of tortoises, always take delight in knocking down the little birds.

These birds, although much persecuted, do not become wild in a short time: in Charles Island, which had then been colonized about six years, I saw a boy sitting by a well with a switch in his hand, with which he killed the doves and finches as they came to drink. He had already procured a little heap of them for his dinner; and he said he had constantly been in the habit of waiting there for the same purpose. We must conclude that the birds, not having as yet learnt that man is a more dangerous animal than the tortoise, or the Amblyrhyncus, disregard us, in the same manner as magpies in England do the cows and horses grazing in the fields.

The Falkland Islands offer a second instance of this disposition among its birds. The extraordinary tameness of the dark-coloured Furnarius has been remarked by Pernety, Lesson, and other voyagers. It is not, however, peculiar to that bird: the Caracara, snipe, upland and lowland goose, thrush, Emberiza, and even some true hawks, are all more or less tame. Both hawks and foxes are present; and as the birds are so tame, we may infer that the absence of all rapacious animals at the Galapagos, is not the cause of their tameness there. The geese at the Falklands, by the precaution they take in building on the islets, show that they are aware of their danger from the foxes; but they are not by this rendered wild towards man. This tameness of the birds, especially the waterfowl, is strongly contrasted with the habits of the same species in Tierra del Fuego, where for ages past they have been persecuted by the wild inhabitants. In the Falklands, the sportsman may sometimes kill more of the upland geese in one day, than he is able to carry home; whereas in Tierra del Fuego, it is nearly as difficult to kill one, as it is in England of the common wild species.

In the time of Pernety (1763), all the birds appear to have been much tamer than at present. Pernety states that the Furnarius would almost perch on his finger; and that with a wand he killed ten in half an hour. At that period, the birds must have been about as tame as they now are at the Galapagos. They appear to have learnt caution more quickly at the Falklands than at the latter place, and they have had proportionate means of experience; for besides frequent visits from vessels, the islands have been at intervals colonized during the whole period.

Even formerly, when all the birds were so tame, by Pernety's account it was impossible to kill the black-necked swan. It is rather an interesting fact, that this is a bird of passage, and therefore brings with it the wisdom learnt in foreign countries.

I have not met with any account of the *land* birds being so tame, in any other quarter of the world, as at the Galapagos and Falkland Islands. And it may be observed that of the few archipelagoes of any size, which when discovered were uninhabited by man, these two are among the most important. From the foregoing statements we may, I think, conclude; first, that the wildness of birds with regard to man, is a particular instinct directed against *him*, and not dependent on any general degree of caution arising from other sources of danger; secondly, that it is not acquired by them in a short time, even when much persecuted; but that in the course of successive generations it becomes hereditary. With domesticated animals we are accustomed to see instincts becoming hereditary; but with those in a state of nature, it is more rare to discover instances of such acquired knowledge. In regard to the wildness of birds towards men, there is no other way of accounting for it. Few young birds in England have been injured by man, yet all are afraid of him: many individuals, on the other hand, both at the Galapagos and at the Falklands, have been injured, but yet have not learned that salutary dread. We may infer from these facts, what havoc the introduction of any new beast of prey must cause in a country, before the instincts of the aborigines become adapted to the stranger's craft or power.

• • •

ADVICE TO COLLECTORS.

As this volume may possibly fall into the hands of some one about to undertake a similar expedition, I will offer a few pieces of advice, some of which I observed with much advantage, but others, to my cost, neglected. Let the collector's motto be, "Trust nothing to the memory;" for the memory becomes a fickle guardian when one interesting object is succeeded by another still more interesting. Keep a list with the date of the ships by which *every* box of specimens, or even a letter, is transmitted to England; let the receiver do the same: it will afterwards save much anxiety. Put a number on every specimen, and every fragment of a specimen; and during the very same minute let it be entered in the catalogue, so that if hereafter its locality be doubted, the collector may say in good truth, "Every specimen of mine was ticketed on the spot." Any thing which is folded up in paper, or put into a separate box, ought to have a number on the outside (with the exception perhaps of geological specimens), but more *especially* a duplicate number on the inside attached to the specimen itself. A series of small numbers should be printed from 0 to 5000; a stop must be added to those numbers which can be read upside down (as 699. or 86.). It is likewise convenient to have the different thousands printed on differently coloured paper, so that when unpacking, a single glance tells the approximate number.

For specimens in spirits of wine, I found the following plan answered admirably: Get a set of steel dies from 0 to 9, a small punch, and some sheets of *trebly*-thick tinfoil. The numbers may at any time be stamped in a line, with a hole punched in front of each, and then cut off with a pair of scissors as wanted. These tickets cost little trouble in making, and *do not corrode.* Each specimen in spirits should be loosely folded up in *very open* gauze, or some such stuff; the string which ties up the corners may likewise secure the number. Use nothing but glass jars; but these are difficult to be obtained of any size out of Europe. Jars of earthenware, and wooden casks, either leak, or allow of evaporation; and when such are used, it is not easy to know whether the specimens are too much crowded (a very common fault), or in what state the spirit is in, which through glass can be judged of by its colour. Bear in mind, that in nine out of ten specimens which are spoiled, it is owing to the spirit being too weak. The jars should be closed with a bung covered by bladder, twice by common tinfoil, and by bladder again; let the bladder soak till half putrid. I found this plan quite worth the trouble it cost.

Few, excepting those who have travelled in ships, know the extreme inconvenience of want of room; and on this much depends: but if it be practicable, keep three or four sets of bottles open at the same time, so that one may serve for crustacea, another for animals for dissection, another for minute specimens, another for fish, always putting the latter into the strongest spirit. Any how it is absolutely necessary to keep a couple of receiving bottles in which every thing can at first be put, and afterwards transferred to the permanent bottles with *fresh spirits.* Without assistance from government, and plenty of room, it is most disheartening work to attempt to bring home many specimens in spirits, although without doubt in such a state they are very far the most valuable. I should recommend any one circumstanced as I was to preserve the skins only of large fish and reptiles. But with room and means at command, let the collector place no limit to the number of his glass jars.

With respect to the catalogues it is inconvenient to have many; but there must at least be two, one for the tin labels or specimens in spirits, and another for the paper numbers, which should be applied indiscriminately to every kind of specimen. If the observer has any particular branch to which he devotes much attention, a third catalogue exclusively for such specimens is desirable: I kept a third for geological specimens and fossils. In a like manner notes should be as simple as possible: I kept one set for geology, and another for zoological and all other observations. It is well to endeavour to write upon separate pages remarks on different specimens; for much copying will thus be saved. My journal was likewise kept distinct from the other subjects. I found an arrangement carried thus far very useful: a traveller by land would, I suppose, be obliged to adopt a still more simple plan.

Use arsenical soap* for all skins, but do not neglect to brush the

* Seeds must not be sent home in the same case with skins prepared with poison, camphor, or essential oils; scarcely any of mine germinated, and Professor Henslow thinks they were thus killed.

legs and beak with a solution of corrosive sublimate. Likewise slightly brush over all dried plants with the solution. For collecting insects use a plain strong sweeping-net, and pack the specimens of all orders, excepting lepidoptera, between layers of rag in pill-boxes, placing at the bottom a bit of camphor; this costs *scarcely any* trouble, and the insects, especially thousands of unknown minute ones, arrive in an excellent state. Take a good stock of chip pill-boxes—a simple plain strong microscope, such as that long ago described by Ellis—a good stock of lace-needles, with glass tubes and sealingwax, for the purpose of making dissecting instruments. I need not mention small collecting bottles covered with leather, tin boxes, dissecting scissors, blowpipe case, compasses, mountain barometer, &c. I should recommend a sort of work-box fitted up to hold watch-glasses, glass micrometers, pins, string, printed numbers, &c; and I found a small cabinet with drawers, some lined with cork, and others with cross partitions, most useful as a temporary storehouse.

Pack up for shipment *every* specimen of *every* kind in boxes lined with tinned plates, and *soldered* together: if the case be large the specimens should further be packed into light pasteboard or other boxes, for by long pressure even skins of quadrupeds are injured. On no account whatever put bottles with spirits of wine, though ever so well packed, in the same case with other specimens, for if one should break every thing near it will be spoiled, as I found to my cost in one instance.

When limited either in time, funds, or space, let not the collector crowd too many specimens either into one bottle, or into one case. For he should *constantly* bear in mind as his second motto, that "It is better to send home a few things well preserved, than a multitude in a bad condition." As long as due steps are taken that the harvest may not be spoiled, let him not be disheartened, because he may for a long time be labouring by himself; let him work hard from morning to night, for every day and every hour is precious in a foreign clime; and then most assuredly his own satisfaction will one day well repay him.

CONCLUSION.

Our voyage having come to an end, I will take a short retrospect of the advantages and disadvantages, the pains and pleasures, of our five years' wandering. If a person should ask my advice, before undertaking a long voyage, my answer would depend upon his possessing a decided taste for some branch of knowledge, which could by such means be improved. No doubt it is a high satisfaction to behold various countries, and the many races of mankind, but the pleasures gained at the time do not counterbalance the evils. It is necessary to look forward to a harvest, however distant it may be, when some fruit will be reaped, some good effected.

Many of the losses which must be experienced are obvious; such as that of the society of all old friends, and of the sight of those places, with which every dearest remembrance is so intimately connected. These losses, however, are at the time partly relieved by the exhaustless delight of anticipating the long wished-for day of return. If, as poets say, life is a dream, I am sure in a voyage these are the visions which serve best to pass away the long night. Other losses, although not at first felt, tell heavily after a period; these are, the want of room, of seclusion, of rest;—the jading feeling of constant hurry;—the privation of small luxuries, the comforts of civilization and domestic society, and, lastly, even of music and the other pleasures of imagination. When such trifles are mentioned, it is evident that the real grievances (excepting from accidents) of a sea life are at an end. The short space of sixty years has made an astonishing difference in the facility of distant navigation. Even in the time of Cook, a man who left his comfortable fireside for such expeditions, underwent severe privations. A yacht now with every luxury of life might circumnavigate the globe. Besides the vast improvements in ships and naval resources, the whole western shores of America are thrown open, and Australia has become the metropolis of a rising continent. How different are the circumstances to a man shipwrecked at the present day in the Pacific, to what they were in the time of Cook! since his voyage a hemisphere has been added to the civilized world.

If a person suffer much from sea-sickness, let him weigh it heavily in the balance. I speak from experience: it is no trifling evil which may be cured in a week. If, on the other hand, he takes pleasure in naval tactics, he will assuredly have full scope for his taste. But it must be borne in mind, how large a proportion of the time, during a long voyage, is spent on the water, as compared with the days in harbour. And what are the boasted glories of the illimitable ocean? A tedious waste, a desert of water, as the Arabian calls it. No doubt there are some delightful scenes. A moonlight night, with the clear heavens and the dark glittering sea, and the white sails filled by the soft air of a gently-blowing trade-wind;—a dead calm, with the heaving surface polished like a mirror, and all still, except the occasional flapping of the sails. It is well once to behold a squall with its rising arch and coming fury, or the heavy gale of wind and mountainous waves. I confess, however, my imagination had painted something more grand, more terrific in the full-grown storm. It is an incomparably finer spectacle when beheld on shore, where the waving trees, the wild flight of the birds, the dark shadows and bright lights, the rushing of the torrents, all proclaim the strife of the unloosed elements. At sea the albatross and petrel fly as if the storm were their proper sphere, the water rises and sinks as if fulfilling its usual task, the ship alone and its inhabitants seem the objects of wrath. On a forlorn and weather-beaten coast, the scene is indeed different, but the feelings partake more of horror than of wild delight.

Let us now look at the brighter side of the past time. The pleasure derived from beholding the scenery and the general aspect of the various countries we have visited, has decidedly been the most constant and highest source of enjoyment. It is probable that the picturesque beauty of many parts of Europe exceeds any thing we have beheld. But there is a growing pleasure in comparing the character of scenery in different countries, which to a certain degree is distinct from merely admiring its beauty. It depends more on an acquaintance with the individual parts of each view. I am strongly induced to believe that, as in music, the person who understands every note will, if he also possesses a proper taste, more thoroughly enjoy the whole, so he who examines each part of a fine view, may also thoroughly comprehend the full and combined effect. Hence, a traveller should be a botanist, for in all views plants form the chief embellishment. Group masses of naked rock even in the wildest forms, and they may for a time afford a sublime spectacle, but they will soon grow monotonous. Paint them with bright and varied colours, they will become fantastic; clothe them with vegetation, they must form at least a decent, if not a most beautiful picture.

When I said that the scenery of Europe was probably superior to any thing which we have beheld, I excepted, as a class by itself, that of the intertropical regions. The two classes cannot be compared together; but I have already often enlarged on the grandeur of these climates. As the force of impressions generally depends on preconceived ideas, I may add, that all mine were taken from the vivid descriptions in the Personal Narrative of Humboldt, which far exceed in merit any thing I have read on the subject. Yet with these high-wrought ideas, my feelings were far from partaking of any tinge of disappointment on first landing on the shores of Brazil.

Among the scenes which are deeply impressed on my mind, none exceed in sublimity the primeval forests undefaced by the hand of man; whether those of Brazil, where the powers of Life are predominant, or those of Tierra del Fuego, where Death and Decay prevail. Both are temples filled with the varied productions of the God of Nature:—no one can stand in these solitudes unmoved, and not feel that there is more in man than the mere breath of his body. In calling up images of the past, I find the plains of Patagonia frequently cross before my eyes: yet these plains are pronounced by all most wretched and useless. They are characterized only by negative possessions; without habitations, without water, without trees, without mountains, they support merely a few dwarf plants. Why then, and the case is not peculiar to myself, have these arid wastes taken so firm possession of the memory? Why have not the still more level, the greener and more fertile Pampas, which are serviceable to mankind, produced an equal impression? I can scarcely analyze these feelings: but it must be partly owing to the free scope given to the imagination. The plains of Patagonia are boundless, for they are scarcely practicable, and hence unknown: they bear the stamp of having thus lasted for ages, and there appears no limit to their duration through future time. If, as the ancients supposed, the flat earth was surrounded by an impassable breadth of water, or by deserts heated to an intolerable excess, who would not look at these last boundaries to man's knowledge with deep but ill-defined sensations?

Lastly, of natural scenery, the views from lofty mountains, though certainly in one sense not beautiful, are very memorable. When looking down from the crest of the highest Cordillera, the mind undisturbed by minute details, was filled with the stupendous dimensions of the surrounding masses.

Of individual objects, perhaps no one is more certain to create astonishment than the first sight in his native haunt of a real barbarian,—of man in his lowest and most savage state. One's mind hurries back over past centuries, and then asks, could our progenitors have been such as these? Men, whose very signs and expressions are less intelligible to us than those of the domesticated animals; men, who do not possess the instinct of those animals, nor yet appear to boast of human reason, or at least of arts consequent on that reason. I do not believe it is possible to describe or paint the difference between savage and civilized man. It is the difference between a wild and tame animal: and part of the interest in beholding a savage, is the same which would lead every one to desire to see the lion in his desert, the tiger tearing his prey in the jungle, the rhinoceros on the wide plain, or the hippopotamus wallowing in the mud of some African river.

Among the other most remarkable spectacles which we have beheld, may be ranked the stars of the southern hemisphere—the water-spout—the glacier leading its blue stream of ice in a bold precipice overhanging the sea—a lagoon island raised by the coral-forming polypi—an active volcano—and the overwhelming effects of a violent earthquake. The three latter phenomena, perhaps, possess for me a peculiar interest, from their intimate connection with the geological structure of the world. The earthquake must however, be to every one a most impressive event: the earth, considered from our earliest childhood as the type of solidity, has oscillated like a thin crust beneath our feet; and in seeing the most beautiful and laboured works of man in a moment overthrown, we feel the insignificance of his boasted power.

It has been said, that the love of the chase is an inherent delight in man—a relic of an instinctive passion. If so, I am sure the pleasure of living in the open air, with the sky for a roof, and the ground for a table, is part of the same feeling: it is the savage returning to his wild and native habits. I always look back to our boat cruises, and my land journeys, when through unfrequented countries, with a kind of extreme delight, which no scenes of civilization could have created. I do not doubt that every traveller must remember the glowing sense of happiness he experienced, from the simple consciousness of breathing in a foreign clime, where the civilized man has seldom or never trod.

There are several other sources of enjoyment in a long voyage, which are, perhaps, of a more reasonable nature. The map of the world ceases to be a blank; it becomes a picture full of the most varied and animated figures. Each part assumes its true dimensions: continents are not looked at in the light of islands, or those islands considered as mere specks, which are, in truth, larger than many kingdoms of Europe. Africa, or North and South America, are well-sounding names, and easily pronounced; but it is not till having sailed for some weeks along small portions of their coasts that one is thoroughly convinced how large a portion of our immense world these names imply.

From seeing the present state, it is impossible not to look forward with high expectation to the future progress of nearly an entire hemisphere. The march of improvement, consequent on the introduction of Christianity throughout the South Sea, probably stands by itself on the records of history. It is the more striking when we remember that only sixty years since, Cook, whose most excellent judgment none will dispute, could foresee no prospect of such change. Yet these changes have now been effected by the philanthropic spirit of the British nation.

In the same quarter of the globe Australia is rising, or indeed may be said to have risen, into a grand centre of civilization, which at some not very remote period, will rule as empress over the southern hemisphere. It is impossible for an Englishman to behold these distant colonies, without a high pride and satisfaction. To hoist the British flag, seems to draw with it as a certain consequence, wealth, prosperity, and civilization.

In conclusion, it appears to me that nothing can be more improving to a young naturalist, than a journey in distant countries. It both sharpens, and partly likewise allays that want and craving, which, as Sir J. Herschel remarks, a man experiences although every corporeal sense is fully satisfied. The excitement from the novelty of objects, and the chance of success, stimulate him to increased activity. Moreover as a number of isolated facts soon become uninteresting, the habit of comparison leads to generalization. On the other hand, as the traveller stays but a short space of time in each place, his descriptions must generally consist of mere sketches, instead of detailed observation. Hence arises, as I have found to my cost, a constant tendency to fill up the wide gaps of knowledge, by inaccurate and superficial hypotheses.

But I have too deeply enjoyed the voyage, not to recommend any naturalist, although he must not expect to be so fortunate in his companions as I have been, to take all chances, and to start, on travels by land if possible, if otherwise on a long voyage. He may feel assured, he will meet with no difficulties or dangers (excepting in rare cases) nearly so bad as he beforehand anticipated. In a moral point of view, the effect ought to be, to teach him good-humoured patience, freedom from selfishness, the habit of acting for himself, and of making the best of every thing, or in other words contentment. In short he should partake of the characteristic qualities of the greater number of sailors. Travelling ought also to teach him distrust; but at the same time he will discover, how many truly good-natured people there are, with whom he never before had, or ever again will have any further communication, who yet are ready to offer him the most disinterested assistance.

Humble-Bees. [1841]

THIS IS THE FIRST of a thirty-six-year series of letters on various topics that Darwin published in the weekly Gardeners' Chronicle and Agricultural Gazette, *from which he gleaned information on species. He generally wrote to the magazine to gather additional information or to clarify points made in the articles of others. The following letter conveys, in an accessible and even light-hearted way, some of his first published observations on animal behavior, pioneering work in what would become the science of ethology. The letter is in part a response to four previous correspondents' queries and comments in the 1841* Gardeners' Chronicle. *In his text Darwin cites the German botanist Christian Konrad Sprengel's 1793 book* Das entdeckte Geheimniss der Natur im Bau und in der Befruchtung der Blumen (Nature's Secret Revealed in the Structure and Fertilization of Flowers), *a source to which he often turns in his later botanical works.*

Perhaps some of your readers may like to hear a few more particulars about the humble-bees which bore holes in flowers, and thus extract the nectar. This operation has been performed on a large scale in the Zoological Gardens:—Near the refection-house there is a fine bed of Stachys coccinea, every flower in which has one, and sometimes two, small irregular slits, or orifices, on the upper side of the corolla near its base. I observed some plants of Marvel of Peru, and of Salvia coccinea, with holes in similar positions; but in Salvia Grahami they were without exception cut through the calyx, which is in this species elongated. The tubular corolla of Pentstemon argutus is rather broader than in the above flowers, and two holes are always bored in it by the side of each other, and just above the calyx. All these orifices are so small that they might easily be overlooked; I first noticed them a week since, when, from the brown colour of their edges, they appeared to have been made some time before. The beds of Stachys and Pentstemon are frequented by numerous humble-bees of many very different kinds; at one moment I saw between twenty and thirty round a bed of the latter flower; they fly very quickly from flower to flower, and always alight with their heads just over the little orifices, into which they most dexterously insert their proboscis, and in the case of the Pentstemon, first into the orifice on one side and then into the other, so that they thus extract the nectar on both sides of the german. Besides the humble-bees, I saw some hive-bees on the Pentstemon; they were, however, much less dexterous, and generally alighted across the flower, or on the calyx, and thus lost time. The orifices in all the abovementioned flowers are made on the upper side of the corolla: I was, therefore, surprised to find, close by, a large bed of the common Antirrhinum in which all the flowers had one or two irregular slits, or holes, on the under side of the corolla at its base, close to the small protuberance which represents the spur in Linaria, and therefore directly in front of the nectary at the foot of the german. From the position of these orifices they cannot be seen without turning up the flower; but the humble-bees seemed to understand this method of picking pockets full as well as the other, and never hesitated where to go, but quickly flew from the under side of one flower to that of another. Now I can speak positively, as far as the experience of part of two summers goes, that country humble-bees are not so cunning, and invariably crawl into the flower by forcing open the elastic lower lip; and a very pretty spectacle it is to watch them. All the flowers of Salvia Grahami and the Antirrhinum which I looked at in different parts of the garden, were bored; and out of the many hundreds in bloom in the two large beds of Stachys and Pentstemon, I could not find one without its little orifice, nor did I see one bee crawl in at the mouth. Nevertheless I found, and the fact appears to me very curious, two separate plants of the Stachys coccinea, and one large one of the Penstemon argutus, with all their flowers unbored; from the scratches on the lower lip of the flowers of two former plants I have no doubt that many bees had entered in the usual way, and I actually saw one bee crawling into the flowers of the Penstemon. One is tempted to conjecture that in these plants each humble-bee as it came, not finding a hole ready cut, thought it less trouble to extract the nectar by the mouth than to make one; but that on the beds of the same flowers, where very many bees were rivalling each other in getting honey, some few set to work boring holes, and others copied the example. From the comparative fewness of the hive-bees on the Pentstemon, their evident awkwardness in finding the orifices, and the smallness of their mandibles, I can hardly doubt they were profiting by the workmanship and the example of the humble-bees: should this be verified, it will, I think, be a very instructive case of acquired knowledge in insects. We should be astonished did one genus of monkeys adopt from another a particular manner of opening hard-shelled fruit; how much more so ought we to be in a tribe of insects so preeminent for their instinctive faculties, which are generally supposed to be in inverse ratio to the intellectual! Moreover, from what I have above stated regarding the Antirrhinum, I much suspect that the practice of boring holes in its flowers is likewise a piece of acquired knowledge, whether the humble-bees do it instinctively or not in other cases. Although I have said that country humble-bees appear to be less cunning than London ones, yet I confess I saw this June, in Staffordshire, some in the act of cutting holes at the base of the corolla of the Rhododendron azaleoides; the greater number entered the mouth of the corolla, as indeed was evident from the quantity of pollen on the stigma, brought by the bees from neighboring Azaleas—this hybrid not having a single grain of pollen of its own. One bee was seen which entered the mouth of some of the flowers and cut holes in others; this shows that the orifices are made simply to save trouble, and not because the bee cannot extract the nectar from the long tube. In the Stachys and Pentstemon it is also evident that the bees cut the holes because they can fly much quicker from the upper surface of one flower to that of another, than clamber in on the fringed edge of the lower lip. I have no doubt by this means they are able to visit twice the number of flowers in the same time. Your correspondent says that the Honeysuckle is sometimes bored; I never happened to notice this, but I have seen pollen-gathering humble-bees show much skill in forcing open the yet-closed mouth of the young flowers and extracting the pollen; flowers which had been open, apparently even for a day, they at once passed over, whereas the nectar-seeking humble-bees stopped at them. If the mouth of the flower was absolutely closed, without any one segment having started, the bees from the difficulty of the attempt immediately gave it up. Your correspondent attributes the failure of his Bean-crop to the apertures made by the bees; but when we remember how the petals of many flowers may be manipulated in hybridising them, without preventing their fructification, we may well doubt this view. But I conceive they may be indirectly the cause of the crop failing, not by their making the orifice, but by their not extracting the nectar in the manner nature intended them; for I have observed that when papilionaceous flowers are mature, (and actually in the case of the Bean,) bees alighting on the wing-petal, as they always do to reach the nectar at the base of the standard-petal, depress the wing-petals together with the keel, by which movement the grains of pollen together with the stigma are forced out, and both rubbed against one side of the bee's body, already generally well dusted with the pollen of other flowers of the same species. If all those flowers, even hermaphrodite ones, which are attractive to insects, almost necessarily require their intervention, as is supposed with much probability by Christian Sprengel (Entd. Geheim.), to remove the pollen from the anthers to the stigma, what unworthy members of society are these humble-bees, thus to cheat, by boring a hole into the flower instead of brushing over the stamens and pistils, the, so imagined, final cause of their existence! Although I can believe that such wicked bees may be injurious to the seedsman, one would lament to see these industrious, happy-looking creatures punished with the severity proposed by your correspondent. Moreover, the florist, I believe, ought rather to praise them for this ingenious method of obtaining the nectar, instead of by the old-fashioned natural one; for let him look how torn and scratched the lower petals of some flowers are—for instance, those of the Mimulus roseus, and the wing-petals of some Everlasting-peas. The little orifice which the bees make, in order to avoid clambering in at the mouth, is hardly visible; whereas all the flowers in some beds of the Mimulus, at the Zoological Gardens, are sadly defaced. Let any one who doubts the use of bees in the fructification of hermaphrodite flowers, watch and admire the manner in which the flat surface of the divided stigma of this Mimulus licks the back of the entering bees, which is generally well-dusted already with pollen; and then how admirably the two divisions of the stigma, endowed with a sensitive faculty, close like a forceps on the included granules of pollen! I will only farther remark, that after the facts here noticed, one may well doubt C. K. Sprengel's view, that the streaks and spots of colour (saft-maal) on the corolla of most nectariferous flowers, serve as guides to insects, they may readily find out where the nectar-vessel lies. I think the bees which flew so quickly from flower to flower on the under sides of the Antirrhinum, or those which bored the pair of holes on the Pentstemon, or those which bored through calyx and corolla in the Salvia, would tell Mr. Sprengel that although he might want such aids, they did not. I know hardly any flower which bees open and insert their proboscis into, more rapidly, than the common tall Linaria, which has a little purplish well-closed flower; I have watched one humble-bee suck twenty-four flowers in one minute; yet on this flower there are no streaks of colour to guide these quick and clever workmen.

The Structure and Distribution of Coral Reefs. [1842]

On the west coast of South America, probably in April 1835, Darwin worked out his theory of coral reef development—some seven months before seeing his first coral reef in Tahiti that November. While aboard the Beagle, *Darwin had read widely in the surprisingly extensive library of books on geographical exploration and biology. He constructed his theory concerning reefs as a thought experiment that he later verified in the field. Articulating this theory greatly satisfied Darwin at the time and encouraged him in future speculative work. Like his later theory of natural selection, his theory of coral reef formation remains accepted to this day as the best explanation for the phenomenon. When Darwin writes of the movements of the earth's crusts, however, he is referring to vertical movements, not the horizontal movements now known to occur through plate tectonics and continental drift. These forces were not discovered until the twentieth century.*

We have chosen to excerpt the introduction and the recapitulation from Darwin's book on coral reefs. Most of the book consists of descriptions of the various types of reefs, with information on where and how they grow. The book, which contains a good colored map showing the distribution of reefs and plates providing typical examples of atolls, barrier reefs, and fringing reefs, first came out in 1842. A second, extensively revised, edition appeared in 1874.

INTRODUCTION.

The object of this volume is to describe from my own observation and the works of others, the principal kinds of coral reefs, more especially

those occurring in the open ocean, and to explain the origin of their peculiar forms. I do not here treat of the polypifers, which construct these vast works, except so far as relates to their distribution, and to the conditions favourable to their vigorous growth. Without any distinct intention to classify coral-reefs, most voyagers have spoken of them under the following heads: 'lagoon-islands,' or 'atolls,' 'barrier,' or 'encircling reefs' and 'fringing,' or 'shore reefs.' The lagoon-islands have received much the most attention; and it is not surprising, for every one must be struck with astonishment, when he first beholds one of these vast rings of coral-rock, often many leagues in diameter, here and there surmounted by a low verdant island with dazzling white shores, bathed on the outside by the foaming breakers of the ocean, and on the inside surrounding a calm expanse of water, which, from reflection, is of a bright but pale green colour. The naturalist will feel this astonishment more deeply after having examined the soft and almost gelatinous bodies of these apparently insignificant creatures, and when he knows that the solid reef increases only on the outer edge, which day and night is lashed by the breakers of an ocean never at rest. Well did François Pyrard de Laval, in the year 1605, exclaim, "C'est une merueille de voir chacun de ces atollons, enuironné d'un grand banc de pierre tout autour, n'y ayant point d'artifice humain." ["It is a marvel to see each of these atolls, completely encircled by a great stone bank, having nothing at all to do with human construction."—editors] The accompanying sketch of Whitsunday Island, in the S. Pacific, taken from Capt. Beechey's admirable Voyage, although excellent of its kind, gives but a faint idea of the singular aspect of one of these lagoon-islands. Whitsunday Island is of small size, and the whole circle has been converted into land, which is a comparatively

rare circumstance. As the reef of a lagoon-island generally supports many separate small islands, the word 'island,' applied to the whole, is often the cause of confusion; hence I have invariably used in this volume the term 'atoll,' which is the name given to these circular groups of coral islets by their inhabitants in the Indian Ocean, and is synonymous with 'lagoon-island.'

Barrier reefs, when encircling small islands, have been comparatively little noticed by voyagers; but they well deserve attention. In their structure they are little less marvellous than atolls, and they give a singular and most picturesque character to the scenery of the islands they surround. In the accompanying sketch, taken from the voyage of the Coquille, the reef is seen from within, from one of the high peaks of the island of Bolabola.* Here, as in Whitsunday island, the whole of that part of the reef which is visible is converted into land. This is a circumstance of rare occurrence; more usually a snow-white line of great breakers, with here and there an islet crowned by cocoa-nut trees, separates the smooth waters of the lagoon-like channel from the waves of the open sea. The barrier reefs of Australia and of New Caledonia, owing to their enormous dimensions, have excited much attention: in structure and form they resemble those encircling many of the smaller islands in the Pacific Ocean.

With respect to fringing, or shore reefs, there is little in their structure which needs explanation; and their name expresses their comparatively small extension. They differ from barrier-reefs in not lying so far from the shore, and in not having within a broad channel of deep

* I have taken the liberty of simplifying the foreground, and leaving out a mountainous island in the far distance.

water. Reefs also occur around submerged banks of sediment and of worn-down rock; and others are scattered quite irregularly where the sea is very shallow: these in most respects are allied to those of the fringing class, but they are of comparatively little interest.

I have given a separate chapter to each of the above classes, and have described some one reef or island, on which I possessed most information, as typical; and have afterwards compared it with others of a like kind. Although this classification is useful from being obvious, and from including most of the coral reefs existing in the open sea, it admits of a more fundamental division into barrier and atoll-formed reefs on the one hand, where there is a great apparent difficulty with respect to the foundation on which they must first have grown; and into fringing reefs on the other, where, owing to the nature of the slope of the adjoining land, there is no such difficulty. The two blue tints and the red colour on the map represent this main division, as explained in the beginning of the last chapter. In the Appendix, every existing coral reef, except some on the coast of Brazil not included in the map, is briefly described in geographical order, as far as I possessed information; and any particular spot may be found by consulting the Index.

Several theories have been advanced to explain the origin of atolls or lagoon-islands, but scarcely one to account for barrier-reefs. From the limited depths at which reef-building polypifers can flourish, taken into consideration with certain other circumstances, we are compelled to conclude, as it will be seen, that both in atolls and barrier-reefs, the foundation on which the coral was primarily attached, has subsided; and that during this downward movement, the reefs have grown upwards. This conclusion, it will be further seen, explains most satisfactorily the outline and general form of atolls and barrier-reefs, and likewise certain peculiarities in their structure. The distribution, also, of the different kinds of coral-reefs, and their position with relation to the areas of recent elevation, and to the points subject to volcanic eruptions, fully accord with this theory of their origin.

• • •

Recapitulation.—In the three first chapters, the principal kinds of coral-reefs were described in detail, and they were found to differ little, as far as relates to the actual surface of the reef. An atoll differs from an encircling barrier-reef only in the absence of land within its central expanse; and a barrier-reef differs from a fringing-reef, in being placed at a much greater distance from the land with reference to the probable inclination of its submarine foundation, and in the presence of a deep-water lagoon-like space or moat within the reef. In the fourth chapter the growing powers of the reef-constructing polypifers were discussed; and it was shown, that they cannot flourish beneath a very limited depth. In accordance with this limit, there is no difficulty respecting the foundations, on which fringing-reefs are based; where-as, with barrier-reefs and atolls, there is a great apparent difficulty on this head; —in barrier-reefs from the improbability of the rock of the coast or of banks of sediment extending, in every instance, so far seaward within the required depth;—and in atolls, from the immensity of the spaces over which they are interspersed, and the apparent necessity for believing that they are all supported on mountain-summits, which, although rising very near to the surface-level of the sea, in no one instance emerge above it. To escape this latter most improbable admission, which implies the existence of submarine chains of mountains of almost the same height, extending over areas of many thousand square miles, there is but one alternative; namely, the prolonged subsidence of the foundations, on which the atolls were primarily based, together with the upward growth of the reef-constructing corals. On this view every difficulty vanishes: fringing-reefs are thus converted into barrier-reefs; and barrier-reefs, when encircling islands, are thus converted into atolls, the instant the last pinnacle of land sinks beneath the surface of the ocean.

Thus the ordinary forms and certain peculiarities in the structure of atolls and barrier-reefs can be explained;—namely, the wall-like structure on their inner sides—the basin or ring-like shape both of the marginal and central reefs in the Maldiva atolls—the union of some atolls as if by a ribbon—the apparent disseverment of others—and the occurrence, in atolls as well as in barrier-reefs, of portions of reef, and of the whole of some reefs, in a dead and submerged state, but retaining the outline of living reefs. Thus can be explained the existence of breaches through barrier-reefs in front of valleys, though separated from them by a wide space of deep water; thus, also, the ordinary outline of groups of atolls, and the relative forms of the separate atolls one to another; thus can be explained the proximity of the two kinds of reefs formed during subsidence, and their separation from the spaces where fringing-reefs abound. On searching for other evidence of the movements supposed by our theory, we find marks of change in atolls and in barrier-reefs, and of subterranean disturbances under them; but from the nature of things, it is scarcely possible to detect any direct proofs of subsidence, although some appearances are strongly in favour of it. On the fringed coasts, however, the presence of upraised marine bodies of a recent epoch, plainly show, that these coasts, instead of having remained stationary, which is all that can be directly inferred from our theory, have generally been elevated.

Finally, when the two great types of structure, namely barrier-reefs and atolls on the one hand, and fringing-reefs on the other, were laid down in colours on our map, a magnificent and harmonious picture of the movements, which the crust of the earth has within a late period undergone, is presented to us. We there see vast areas rising, with volcanic matter every now and then bursting forth through the vents or fissures with which they are traversed. We see other wide spaces slowly sinking without any volcanic outbursts; and we may feel sure, that this sinking must have been immense in amount as well as in area, thus to have buried over the broad face of the ocean every one of those mountains, above which atolls now stand like monuments, marking the place of their former existence. Reflecting how powerful an agent with respect to denudation, and consequently to the nature and thickness of the deposits in accumulation, the sea must ever be, when acting for prolonged periods on the land, during either its slow emergence or subsidence; reflecting, also, on the final effects of these movements in the inter-change of land and ocean-water, on the climate of the earth, and on the distribution of organic beings, I may be permitted to hope, that the conclusions derived from the study of coral-formations, originally attempted merely to explain their peculiar forms, may be thought worthy of the attention of geologists.

Geological Observations on South America. [1846]

GEOLOGICAL OBSERVATIONS ON SOUTH AMERICA *followed* Coral Reefs *and* Volcanic Islands *as the third and last book on geology spawned by the* Beagle *voyage. Published in 1846 and little known today to those outside the earth sciences, it was reprinted twice during Darwin's lifetime but never revised. Like many of his works, it is mainly descriptive. One of the fascinating findings it reports is Darwin's observation that the west coast of South America has been, in a relatively short space of geological time, uplifted above the sea from several hundred to several thousand feet. We now know that this elevation resulted from the Pacific plate comprising the coastal seafloor being forced against and beneath the continent, with other results being the rise of the Andes and the formation of a chain of active volcanoes from Colombia in the north to Tierra del Fuego in the south. Always on the lookout for evidence pertinent to his theories, Darwin here describes some Peruvian human artifacts that have been uplifted, a sign that large elevations still take place.*

ON THE ELEVATION OF THE EASTERN COAST OF SOUTH AMERICA.

In the following volume, which treats of the geology of South America, and almost exclusively of the parts southward of the Tropic of Capricorn, I have arranged the chapters according to the age of the deposits, occasionally departing from this order, for the sake of geographical simplicity.

The elevation of the land within the recent period, and the modifications of its surface through the action of the sea (to which subjects I paid particular attention) will be first discussed; I will then pass on to the tertiary deposits, and afterwards to the older rocks. Only those districts and sections will be described in detail which appear to me to deserve some particular attention; and I will, at the end of each chapter, give a summary of the results. We will commence with the proofs of the upheaval of the eastern coast of the continent, from the Rio Plata southward; and, in the next chapter, follow up the same subject along the shores of Chile and Peru.

• • •

Fossil Remains of Human Art. In the midst of these shells on San Lorenzo, I found light corallines, the horny ovule-cases of Mollusca, roots of sea-weed, bones of birds, the heads of Indian corn and other vegetable matter, a piece of woven rushes, and another of nearly decayed *cotton* string. I extracted these remains by digging a hole, on a level spot; and they had all indisputably been embedded with the shells. I compared the plaited rush, the *cotton* string, and Indian corn, at the house of an antiquary, with similar objects, taken from the Huacas or burial-grounds of the ancient Peruvians, and they were undistinguishable; it should be observed that the Peruvians used string only of cotton. The small quantity of sand or gravel with the shells, the absence of large stones, the width and thickness of the bed, and the time requisite for a ledge to be cut into the sandstone, all show that these remains were not thrown high up by an earthquake-wave: on the other hand, these facts, together with the number of dead shells, and of floating objects, both marine and terrestrial, both natural and human, render it almost certain that they were accumulated on a true beach, since upraised eighty-five feet, and upraised this much since *Indian man inhabited Peru.* The elevation may have been, either by several small sudden starts, or quite gradual; in this latter case the unrolled shells having been thrown up during gales beyond the reach of the waves which afterwards broke on the slowly emerging land. I have made these remarks, chiefly because I was at first surprised at the complete difference in nature, between this broad, smooth, upraised bed of shells, and the present shingle-beach at the foot of the low sandstone-cliffs; but a beach formed, when the sea is cutting into the land, as is shown now to be the case by the low bare sandstone-cliffs, ought not to be compared with a beach accumulated on a gently inclined rocky surface, at a period when the sea (probably owing to the elevatory movement in process) was not able to eat into the land. With respect to the mass of nearly angular, salt-cemented fragments of sandstone, which lie under the shells, and which are so unlike the materials of an ordinary sea beach; I think it probable after having seen the remarkable effects of the earthquake of 1835, in absolutely shattering as if by gunpowder the *surface* of the primary rocks near Concepcion, that a smooth bare surface of stone was left by the sea covered by the shelly mass, and that afterwards when upraised, it was superficially shattered by the severe shocks so often experienced here.

The very low land surrounding the town of Callao, is to the south joined by an obscure escarpment to a higher plain (south of Bella Vista), which stretches along the coast for a length of about eight miles. This plain appears to the eye quite level; but the sea cliffs show that its height varies (as far as I could estimate) from 70 to 120 feet. It is composed of thin, sometimes waving, beds of clay, often of bright red and yellow colours, of layers of impure sand, and in one part with a great stratified mass of granitic pebbles. These beds are capped by a remarkable mass, varying from two to six feet in thickness, of reddish loam or mud, containing many scattered and broken fragments of recent marine shells, sometimes though rarely single large round pebbles, more frequently short irregular layers of fine gravel, and very many pieces of red coarse earthenware, which from their curvatures must once have formed parts of large vessels. The earthenware is of Indian manufacture; and I found exactly similar pieces accidentally included within the bricks, of which the neighboring ancient Peruvian burial-mounds are built. These fragments abounded in such numbers in certain spots, that it appeared as if waggon-loads of earthenware had been smashed to pieces. The broken sea-shells and pottery are strewed both on the surface, and throughout the whole thickness of this upper loamy mass. I found them wherever I examined the cliffs, for a space of between two and three miles, and for half a mile inland; and there can be little doubt that this same bed extends with a smooth surface several miles further over the entire plain. Besides the little included irregular layers of small pebbles, there are occasionally very obscure traces of stratification.

At one of the highest parts of the cliff, estimated 120 feet above the sea, where a little ravine came down, there were two sections, at right angles to each other, of the floor of a shed or building. In both sections or faces, two rows, one over the other, of large round stones could be distinctly seen; they were packed close together on an artificial layer of sand two inches thick, which had been placed on the natural clay-beds; the round stones were covered by three feet in thickness of the loam with broken sea-shells and pottery. Hence, before this widely spread-out bed of loam was deposited, it is certain that the plain was inhabited; and it is probable, from the broken vessels being so much more abundant in certain spots than in others, and from the underlying clay being fitted for their manufacture, that the kilns stood here.

The smoothness and wide extent of the plain, the bulk of matter deposited, and the obscure traces of stratification seem to indicate that the loam was deposited under water; on the other hand, the presence of sea-shells, their broken state, the pebbles of various sizes, and the artificial floor of round stones, almost prove that it must have originated in a rush of water from the sea over the land. The height of the plain, namely, 120 feet, renders it improbable that an earthquake-wave, vast as some have here been, could have broken over the surface at its present level; but when the land stood eighty-five feet lower, at the period when the shells were thrown up on the ledge at S. Lorenzo, and when as we know man inhabited this district, such an event might well have occurred; and if we may further suppose, that the plain was at that time converted into a temporary lake, as actually occurred, during the earthquakes of 1713 and 1746, in the case of the low land round Callao owing to its being encircled by a high shingle-beach, all the appearances above described will be perfectly explained. I must add, that at a lower level near the point where the present low land round Callao joins the higher plain, there are appearances of two distinct deposits both apparently formed by debacles: in the upper one, a horse's tooth and a dog's jaw were embedded; so that both must have been formed after the settlement of the Spaniards: according to Acosta, the earthquake-wave of 1586 rose eighty-four feet.

The inhabitants of Callao do not believe, as far as I could ascertain, that any change in level is now in progress. The great fragments of brickwork, which it is asserted can be seen at the bottom of the sea, and which have been adduced as a proof of a late subsidence, are, as I am informed by Mr. Gill, a resident engineer, loose fragments; this is probable, for I found on the beach, and not near the remains of any building, masses of brickwork, three and four feet square, which had been washed into their present places, and smoothed over with shingle during the earthquake of 1746. The spit of land, on which the ruins of *Old* Callao stand, is so extremely low and narrow, that it is improbable in the highest degree that a town should have been founded on it in its present state; and I have lately heard that M. Tschudi has come to the conclusion, from a comparison of old with modern charts, that the coast both south and north of Callao has subsided. I have shown that the island of San Lorenzo has been upraised eighty-five feet since the Peruvians inhabited this country; and whatever may have been the amount of recent subsidence, by so much more must the elevation have exceeded the eighty-five feet. In several places in this neighbourhood, marks of sea-action have been observed: Ulloa gives a detailed account of such appearances at a point five leagues northward of Callao: Mr. Cruikshank found near Lima successive lines of sea-cliffs, with rounded blocks at their bases, at a height of 700 feet above the present level of the sea.

Does Salt-water Kill Seeds?
[1855]

From 1846 through 1854, Darwin spent much of his time working out the taxonomy of living and fossil barnacles. As it turned out, his barnacle research, published in two volumes in 1851 and two in 1854, successfully demonstrated the extent of variation in a good-sized group of organisms. Darwin began the project, however, with a different and much more modest objective: to solve questions of identity and relationship associated with an odd barnacle he had collected in Chile. As was true with so many of his projects, great things came from small beginnings.

Following the completion of the barnacle monographs, Darwin returned to the question of the origin of species, a subject that had filled his notebooks from 1837 to 1844. Seeking information on the variation of species and sometimes even mentioning that he was amassing evidence on their origins, he wrote to a number of people around the world. He also began to plan a series of experiments to test his hypothesis of evolutionary change. One major project involved looking at how plants and animals were dispersed. This article, the first of four on the subject published in the Gardeners' Chronicle *in 1855, describes an experiment whose results were later presented to the Linnean Society in 1856, published in its journal in 1857, and cited in* The Origin of Species.

I have begun making some few experiments on the effects of immersion in sea-water on the germinating powers of seeds, in the hope of being able to throw a very little light on the distribution of plants, more especially in regard to the same species being found in many cases in far outlying islands and on the mainland. Will any of your readers be so kind as to inform me whether such experiments have already been tried? And, secondly, what class of seeds, or particular species, they have any reason to suppose would be eminently liable to be killed by sea-water? The results at which I have already arrived are too few and unimportant to be worth mentioning.

Productiveness of Foreign Seed. [1857]

THE FOLLOWING PIECE from the Gardeners' Chronicle *responds to an article by a Mr. Wood, who had written on the spread of agricultural weeds. It raises the issue of the "struggle for existence" made famous in the* Origin.

Will the writer of the highly remarkable article on weeds in your last Number have the kindness to state why he supposes that "there is too much reason to believe that foreign seed of an indigenous species is often more prolific than that grown at home?" Is it meant that the plant produced from the foreign seed actually produces more seed, or merely that the introduced stock is more vigorous than the native stock? I have no doubt that so acute an observer has some good reason for his belief. The point seems to me of considerable interest in regard to the great battle for life which is perpetually going on all around us. The great American botanist, Dr. Asa Gray, believes that in the United States there are several plants now naturalised in abundance from imported seed, which are likewise indigenous; and my impression is (but writing from home I cannot refer to his letter to me) that the imported stock prevails over the aboriginal. So again, Dr. Hooker in his admirable Flora of New Zealand has told us that the common Sonchus has spread extensively from imported seed, whilst the same species is likewise an aboriginal; the natives in this instance being able from trifling differences to distinguish the two stocks. Might I further ask whether it is now some years since the seed of Sinapis nigra was accidentally introduced on the farm described; and if so, whether the common Charlock still remains in lessened numbers owing to the presence of the invader, and without, as far as known, fresh seed of the invading S. nigra having been introduced?—whether, in short, it was a fair fight between the two species, ending in the victory of the Black Mustard? Would it be trespassing too much on the kindness of the writer of the article to ask whether he knows of any other analogous cases of a weed introduced from other land beating out, to a greater or lesser extent, a weed previously common in any particular field or farm?

On the Tendency of Species to Form Varieties; and on the Perpetuation of Varieties and Species by Natural Means of Selection. [1858]

EARLY IN 1856 Darwin's geological mentor, Sir Charles Lyell, told him that he had just read an interesting paper by Alfred Russel Wallace, an English naturalist who had, like Darwin, collected in Brazil and was currently collecting specimens in the East Indies. Lyell discerned that Wallace's paper, "On the law which has regulated the introduction of new species," seemed to anticipate some of Darwin's ideas regarding evolution, ideas Darwin had been working out since early 1837 and Lyell had known about since 1844. At Lyell's urging, Darwin gathered his piles of notes, manuscripts, and references on evolution. On May 14, 1856, he began to write what he called his "big book" on evolution, published only in 1975 as Natural Selection.

While engaged on the "big book," Darwin wrote to Wallace, as he did to a number of others in 1855–57, with the objective of gathering further information on the variation of plant and animal species. After the exchange of several letters, Darwin revealed to Wallace on May 1, 1857: "This summer will make the 20th year (!) since I opened my first note-book, on the question of how and in what way do species and varieties differ from each other. I am now preparing my work for publication, but I find the subject so very large, that though I have written many chapters, I do not suppose I shall go to press for two years."

Thus it is not surprising that on June 18, 1858, Darwin received a letter from Wallace containing a manuscript presenting Wallace's ideas as to how species originated. Wallace asked Darwin, if he found the manuscript worthwhile, to forward it to Sir Charles Lyell for publication. Tormented by the ethical, personal, and practical implications of this request, Darwin wrote to Lyell the same day: "I never saw a more striking coincidence: if Wallace had my MS sketch written out in 1842, he could not have made a better short abstract! Even his terms now stand as heads of my chapters." Lyell shared Darwin's dilemma with the botanist Joseph Dalton Hooker, who had been the first of Darwin's friends to be told of his belief in the evolution of species and the only one to have read drafts of the chapters for Natural Selection.

Lyell and Hooker advised Darwin that Wallace's manuscript should be published immediately, along with proof that Darwin had formulated these ideas first. The first part of this proof consisted of extracts from Darwin's 1844 essay first sketched out in 1842 (not 1839, as Lyell and Hooker's prefatory letter states) and which Hooker had read in 1847 (not in 1844, as specified in the letter). The second part of the proof is an abstract of a letter sent in September (rather than October) 1857 from Darwin to the Harvard botanist Asa Gray, who had become one of his most trusted scientific confidants. Here, Darwin had explained natural selection to Gray, as he had previously done only to Lyell and Hooker.

Lyell and Hooker presented the so-called "joint paper" with two parts by Darwin and one by Wallace to the Linnean Society, which was and still is today the leading British natural history organization, at a late-night meeting on July 1, 1858. Publication in the society's journal followed on August 30. Darwin was not present at the London meeting. His youngest child, the infant Charles Waring Darwin, had died of scarlet fever on June 28, and several other members of the household were suffering with the disease. Nor was Wallace in attendance. Still in the East Indies, from which he did not return until 1862, he did not even know that his manuscript had been published until he received a copy in October 1858.

No one knows how long Darwin, who was meticulous about marshaling evidence and reluctant to provoke the storms of controversy he knew his hypothesis would engender, might have waited before publishing on natural selection had Wallace not sent his manuscript. The large-minded Wallace, for his part, displayed no resentment at the means taken to publish that manuscript—instead, he was honored to be linked with Darwin, and he became one of Darwin's stout allies in the evolutionary debates. Lyell and Hooker had seen to it that neither Darwin's priority nor Wallace's originality was eclipsed and that the important but hastily contrived "joint paper" was presented in the most appropriate and prestigious context. We include the entire document below: reading Wallace's portion shows just why Darwin was so concerned about being scooped. It is also interesting to note Wallace's claim that artificial selection is no aid to understanding what happens in the natural world, which directly contradicts Darwin's use of artificial selection to explain the action of natural selection.

London, June 30th, 1858.

My Dear Sir,—The accompanying papers, which we have the honour of communicating to the Linnean Society, and which all relate to the same subject, viz. the Laws which affect the Production of Varieties, Races, and Species, contain the results of the investigations of two indefatigable naturalists, Mr. Charles Darwin and Mr. Alfred Wallace.

These gentlemen having, independently and unknown to one another, conceived the same very ingenious theory to account for the appearance and perpetuation of varieties and of specific forms on our planet, may both fairly claim the merit of being original thinkers in this important line of inquiry; but neither of them having published his views, though Mr. Darwin has for many years past been repeatedly urged by us to do so, and both authors having now unreservedly placed their papers in our hands, we think it would best promote the interests of science that a selection from them should be laid before the Linnean Society.

Taken in the order of their dates, they consist of:—

1. Extracts from a MS. work on Species, by Mr. Darwin, which was sketched in 1839, and copied in 1844, when the copy was read by Dr. Hooker, and its contents afterwards communicated to Sir Charles Lyell. The first Part is devoted to "The Variation of Organic Beings under Domestication and in their Natural State;" and the second chapter of that Part, from which we propose to read to the Society the extracts referred to, is headed, "On the Variation of Organic Beings in a state of Nature; on the Natural Means of Selection; on the Comparison of Domestic Races and true Species."

2. An abstract of a private letter addressed to Professor Asa Gray, of Boston, U.S., in October 1857, by Mr. Darwin, in which he repeats his views, and which shows that these remained unaltered from 1839 to 1857.

3. An Essay by Mr. Wallace, entitled "On the Tendency of Va-

rieties to depart indefinitely from the Original Type." This was written at Ternate in February 1858, for the perusal of his friend and correspondent Mr. Darwin, and sent to him with the expressed wish that it should be forwarded to Sir Charles Lyell, if Mr. Darwin thought it sufficiently novel and interesting. So highly did Mr. Darwin appreciate the value of the views therein set forth, that he proposed, in a letter to Sir Charles Lyell, to obtain Mr. Wallace's consent to allow the Essay to be published as soon as possible. Of this step we highly approved, provided Mr. Darwin did not withhold from the public, as he was strongly inclined to do (in favour of Mr. Wallace), the memoir which he had himself written on the same subject, and which, as before stated, one of us had perused in 1844, and the contents of which we had both of us been privy to for many years. On representing this to Mr. Darwin, he gave us permission to make what use we thought proper of his memoir, &c.; and in adopting our present course, of presenting it to the Linnean Society, we have explained to him that we are not solely considering the relative claims to priority of himself and his friend, but the interests of science generally; for we feel it to be desirable that views founded on a wide deduction from facts, and matured by years of reflection, should constitute at once a goal from which others may start, and that, while the scientific world is waiting for the appearance of Mr. Darwin's complete work, some of the leading results of his labours, as well as those of his able correspondent, should together be laid before the public.

We have the honour to be yours very obediently,

Charles Lyell
Jos. D. Hooker

J. J. Bennett, Esq.,
Secretary of the Linnean Society

I. Extract from an Unpublished Work on Species, by C. Darwin, Esq., Consisting of a Portion of a Chapter Entitled, "On the Variation of Organic Beings in a State of Nature; on the Natural Means of Selection; on the Comparison of Domestic Races and True Species."

De Candolle, in an eloquent passage, has declared that all nature is at war, one organism with another, or with external nature. Seeing the contented face of nature, this may at first well be doubted; but reflection will inevitably prove it to be true. The war, however, is not constant, but recurrent in a slight degree at short periods, and more severely at occasional more distant periods; and hence its effects are easily overlooked. It is the doctrine of Malthus applied in most cases with tenfold force. As in every climate there are seasons, for each of its inhabitants, of greater and less abundance, so all annually breed; and the moral restraint which in some small degree checks the increase of mankind is entirely lost. Even slow-breeding mankind has doubled in twenty-five years; and if he could increase his food with greater ease, he would double in less time. But for animals without artificial means, the amount of food for each species must, on an average, be constant, whereas the increase of all organisms tends to be geometrical, and in a vast majority of cases at an enormous ratio. Suppose in a certain spot there are eight pairs of birds, and that *only* four pairs of them annually (including double hatches) rear only four young, and that these go on rearing their young at the same rate, then at the end of seven years (a short life, excluding violent deaths, for any bird) there will be 2048 birds, instead of the original sixteen. As this increase is quite impossible, we must conclude either that birds do not rear nearly half their young, or that the average life of a bird is, from accident, not nearly seven years. Both checks probably concur. The same kind of calculation applied to all plants and animals affords results more or less striking, but in very few instances more striking than in man.

Many practical illustrations of this rapid tendency to increase are on record, among which, during peculiar seasons, are the extraordinary numbers of certain animals; for instance, during the years 1826 to 1828, in La Plata, when from drought some millions of cattle perished, the whole country actually *swarmed* with mice. Now I think it cannot be doubted that during the breeding-season all the mice (with the exception of a few males or females in excess) ordinarily pair, and therefore that this astounding increase during three years must be attributed to a greater number than usual surviving the first year, and then breeding, and so on till the third year, when their numbers were brought down to their usual limits on the return of wet weather. Where man has introduced plants and animals into a new and favourable country, there are many accounts in how surprisingly few years the whole country has become stocked with them. This increase would necessarily stop as soon as the country was fully stocked; and yet we have every reason to believe, from what is known of wild animals, that *all* would pair in the spring. In the majority of cases it is most difficult to imagine where the checks fall—though generally, no doubt, on the seeds, eggs, and young; but when we remember how impossible, even in mankind (so much better known than any other animal), it is to infer from repeated casual observations what the average duration of life is, or to discover the different percentage of deaths to births in different countries, we ought to feel no surprise at our being unable to discover where the check falls in any animal or plant. It should always be remembered, that in most cases the checks are recurrent yearly in a small, regular degree, and in an extreme degree during unusually cold, hot, dry, or wet years, according to the constitution of the being in question. Lighten any check in the least degree, and the geometrical powers of increase in every organism will almost instantly increase the average number of the favoured species. Nature may be compared to a surface on which rest ten thousand sharp wedges touching each other and driven inwards by incessant blows. Fully to realize these views much reflection is requisite. Malthus on man should be studied; and all such cases as those of the mice in La Plata, of the cattle and horses when first turned out in South America, of the birds by our calculation, &c., should be well considered. Reflect on the enormous multiplying power *inherent and annually in action* in all animals; reflect on the countless seeds scattered by a hundred ingenious contrivances, year after year, over the whole face of the land; and yet we have every reason to suppose that the average percentage of each of the inhabitants of a country usually remains constant. Finally, let it be borne in mind that this average number of individuals (the external conditions remaining the same) in each country is kept up by recurrent struggles against other species or against external nature (as on the borders of the Arctic regions, where the cold checks life), and that ordinarily each individual of every species holds its place, either by its own struggle and capacity of acquiring nourishment in some period of its life, from the egg upwards; or by the struggle of its parents (in short-lived organisms, when the main check occurs at longer intervals) with other individuals of the *same* or *different* species.

But let the external conditions of a country alter. If in a small degree, the relative proportions of the inhabitants will in most cases simply be slightly changed; but let the number of inhabitants be small, as on an island, and free access to it from other countries be circumscribed, and let the change of conditions continue progressing (forming new stations), in such a case the original inhabitants must cease to be as perfectly adapted to the changed conditions as they were originally. It has been shown in a former part of this work, that such changes of external conditions would, from their acting on the reproductive system, probably cause the organization of those beings which were most affected to become, as under domestication, plastic. Now, can it be doubted, from the struggle each individual has to obtain subsistence, that any minute variation in structure, habits, or instincts, adapting that individual better to the new conditions, would tell upon its vigour and health? In the struggle it would have a better *chance* of surviving; and those of its offspring which inherited the variation, be it ever so slight, would also have a better *chance*. Yearly more are bred than can survive; the smallest grain in the balance, in the long run, must tell on which death shall fall, and which shall survive. Let this work of selection on the one hand, and death on the other, go on for a thousand generations, who will pretend to affirm that it would produce no effect, when we remember what, in a few years, Bakewell effected in cattle, and Western in sheep, by this identical principle of selection?

To give an imaginary example from changes in progress on an island:—let the organization of a canine animal which preyed chiefly on rabbits, but sometimes on hares, become slightly plastic; let these same changes cause the number of rabbits very slowly to decrease, and the number of hares to increase; the effect of this would be that the fox or dog would be driven to try to catch more hares: his organization, however, being slightly plastic, those individuals with the lightest forms, longest limbs, and best eyesight, let the difference be ever so small, would be slightly favoured, and would tend to live longer, and to survive during that time of the year when food was scarcest; they would also rear more young, which would tend to inherit these slight peculiarities. The less fleet ones would be rigidly destroyed. I can see no more reason to doubt that these causes in a thousand generations would produce a marked effect, and adapt the form of the fox or dog to the catching of hares instead of rabbits, than that greyhounds can be improved by selection and careful breeding. So would it be with plants under similar circumstances. If the number of individuals of a species with plumed seeds could be increased by greater powers of dissemination within its own area (that is, if the check to increase fell chiefly on the seeds), those seeds which were provided with ever so little more down, would in the long run be most disseminated; hence a greater number of seeds thus formed would germinate, and would tend to produce plants inheriting the slightly better-adapted down.

Besides this natural means of selection, by which those individuals are preserved, whether in their egg, or larval, or mature state, which are best adapted to the place they fill in nature, there is a second agency at work in most unisexual [misprint for "bisexual"—editors] animals, tending to produce the same effect, namely, the struggle of the males for the females. These struggles are generally decided by the law of battle, but in the case of birds, apparently, by the charms of their song, by their beauty or their power of courtship, as in the dancing rock-thrush of Guiana. The most vigorous and healthy males, implying perfect adaptation, must generally gain the victory in their contests. This kind of selection, however, is less rigorous than the other; it does not require the death of the less successful, but gives to them fewer descendants. The struggle falls, moreover, at a time of year when food is generally abundant, and perhaps the effect chiefly produced would be the modification of the secondary sexual characters, which are not related to the power of obtaining food, or to defence from enemies, but to fighting with or rivalling other males. The result of this struggle amongst the males may be compared in some respects to that produced by those agriculturists who pay less attention to the careful selection of all their young animals, and more to the occasional use of a choice mate [misprint for "male"—editors].

II. Abstract of a Letter from C. Darwin, Esq., to Prof. Asa Gray, Boston, U.S., dated Down, September 5th, 1857.

1. It is wonderful what the principle of selection by man, that is the picking out of individuals with any desired quality, and breeding from them, and again picking out, can do. Even breeders have been astounded at their own results. They can act on differences inappreciable to an uneducated eye. Selection has been *methodically* followed in *Europe* for only the last half century; but it was occasionally, and even in some degree methodically, followed in most ancient times. There must have been also a kind of unconscious selection from a remote period, namely in the preservation of the individual animals (without any thought of their offspring) most useful to each race of man in his particular circumstances. The "roguing," as nurserymen call the destroying of varieties which depart from their type, is a kind of selection. I am convinced that intentional and occasional selection has been the main agent in the production of our domestic races; but however this may be, its great power of modification has been indisputably shown in later times. Selection acts only by the accumulation of slight or greater variations, caused by external conditions, or by the mere fact that in generation the child is not absolutely similar to its parent. Man, by this power of accumulating variations, adapts living beings to his wants—may be said to make the wool of one sheep good for carpets, of another for cloth, &c.

2. Now suppose there were a being who did not judge by mere

external appearances, but who could study the whole internal organization, who was never capricious, and should go on selecting for one object during millions of generations; who will say what he might not effect? In nature we have some *slight* variation occasionally in all parts; and I think it can be shown that changed conditions of existence is the main cause of the child not exactly resembling its parents; and in nature geology shows us what changes have taken place, and are taking place. We have almost unlimited time; no one but a practical geologist can fully appreciate this. Think of the Glacial period, during the whole of which the same species at least of shells have existed; there must have been during this period millions on millions of generations.

3. I think it can be shown that there is such an unerring power at work in *Natural Selection* (the title of my book), which selects exclusively for the good of each organic being. The elder De Candolle, W. Herbert, and Lyell have written excellently on the struggle for life; but even they have not written strongly enough. Reflect that every being (even the elephant) breeds at such a rate, that in a few years, or at most a few centuries, the surface of the earth would not hold the progeny of one pair. I have found it hard constantly to bear in mind that the increase of every single species is checked during some part of its life, or during some shortly recurrent generation. Only a few of those annually born can live to propagate their kind. What a trifling difference must often determine which shall survive, and which perish!

4. Now take the case of a country undergoing some change. This will tend to cause some of its inhabitants to vary slightly—not but that I believe most beings vary at all times enough for selection to act on them. Some of its inhabitants will be exterminated; and the remainder will be exposed to the mutual action of a different set of inhabitants, which I believe to be far more important to the life of each being than mere climate. Considering the infinitely various methods which living beings follow to obtain food by struggling with other organisms, to escape danger at various times of life, to have their eggs or seeds disseminated, &c. &c., I cannot doubt that during millions of generations individuals of a species will be occasionally born with some slight variation, profitable to some part of their economy. Such individuals will have a better chance of surviving, and of propagating their new and slightly different structure; and the modification may be slowly increased by the accumulative action of natural selection to any profitable extent. The variety thus formed will either coexist with, or, more commonly, will exterminate its parent form. An organic being,

like the woodpecker or misseltoe, may thus come to be adapted to a score of contingencies—natural selection accumulating those slight variations in all parts of its structure, which are in any way useful to it during any part of its life.

5. Multiform difficulties will occur to every one, with respect to this theory. Many can, I think, be satisfactorily answered. *Natura non facit saltum* ["Nature does not move by jumps."—editors] answers some of the most obvious. The slowness of the change, and only a very few individuals undergoing change at any one time, answers others. The extreme imperfection of our geological records answers others.

6. Another principle, which may be called the principle of divergence, plays, I believe, an important part in the origin of species. The same spot will support more life if occupied by very diverse forms. We see this in the many generic forms in a square yard of turf, and in the plants or insects on any little uniform islet, belonging almost invariably to as many genera and families as species. We can understand the meaning of this fact amongst the higher animals, whose habits we understand. We know that it has been experimentally shown that a plot of land will yield a greater weight if sown with several species and genera of grasses, than if sown with only two or three species. Now, every organic being, by propagating so rapidly, may be said to be striving its utmost to increase in numbers. So it will be with the offspring of any species after it has become diversified into varieties, or sub-species, or true species. And it follows, I think, from the foregoing facts, that the varying offspring of each species will try (only few will succeed) to seize on as many and as diverse places in the economy of nature as possible. Each new variety or species, when formed, will generally take the place of, and thus exterminate its less well-fitted parent. This I believe to be the origin of the classification and affinities of organic beings at all times; for organic beings always *seem* to branch and sub-branch like the limbs of a tree from a common trunk, the flourishing and diverging twigs destroying the less vigorous—the dead and lost branches rudely representing extinct genera and families.

This sketch is *most* imperfect; but in so short a space I cannot make it better. Your imagination must fill up very wide blanks.

III. On the Tendency of Varieties to depart indefinitely from the Original Type, by Alfred Russel Wallace.

One of the strongest arguments which have been adduced to

prove the original and permanent distinctness of species is, that *varieties* produced in a state of domesticity are more or less unstable, and often have a tendency, if left to themselves, to return to the normal form of the parent species; and this instability is considered to be a distinctive peculiarity of all varieties, even of those occurring among wild animals in a state of nature, and to constitute a provision for preserving unchanged the originally created distinct species.

In the absence or scarcity of facts and observations as to *varieties* occurring among wild animals, this argument has had great weight with naturalists, and has led to a very general and somewhat prejudiced belief in the stability of species. Equally general, however, is the belief in what are called "permanent or true varieties,"—races of animals which continually propagate their like, but which differ so slightly (although constantly) from some other race, that the one is considered to be a *variety* of the other. Which is the *variety* and which the original *species*, there is generally no means of determining, except in those rare cases in which the one race has been known to produce an offspring unlike itself and resembling the other. This, however, would seem quite incompatible with the "permanent invariability of species," but the difficulty is overcome by assuming that such varieties have strict limits, and can never again vary further from the original type, although they may return to it, which, from the analogy of the domesticated animals, is considered to be highly probable, if not certainly proved.

It will be observed that this argument rests entirely on the assumption, that *varieties* occurring in a state of nature are in all respects analogous to or even identical with those of domestic animals, and are governed by the same laws as regards their permanence or further variation. But it is the object of the present paper to show that this assumption is altogether false, that there is a general principle in nature which will cause many *varieties* to survive the parent species, and to give rise to successive variations departing further and further from the original type, and which also produces, in domesticated animals, the tendency of varieties to return to the parent form.

The life of wild animals is a struggle for existence. The full exertion of all their faculties and all their energies is required to preserve their own existence and provide for that of their infant offspring. The possibility of procuring food during the least favourable seasons, and of escaping the attacks of their most dangerous enemies, are the primary conditions which determine the existence both of individuals and of entire species. These conditions will also determine the population of a species; and by a careful consideration of all the circumstances we may be enabled to comprehend, and in some degree to explain, what at first sight appears so inexplicable—the excessive abundance of some species, while others closely allied to them are very rare.

The general proportion that must obtain between certain groups of animals is readily seen. Large animals cannot be so abundant as small ones; the carnivora must be less numerous than the herbivora; eagles and lions can never be so plentiful as pigeons and antelopes; the wild asses of the Tartarian deserts cannot equal in numbers the horses of the more luxuriant prairies and pampas of America. The greater or less fecundity of an animal is often considered to be one of the chief causes of its abundance or scarcity; but a consideration of the facts will show us that it really has little or nothing to do with the matter. Even the least prolific of animals would increase rapidly if unchecked, whereas it is evident that the animal population of the globe must be stationary, or perhaps, through the influence of man, decreasing. Fluctuations there may be; but permanent increase, except in restricted localities, is almost impossible. For example, our own observation must convince us that birds do not go on increasing every year in a geometrical ratio, as they would do, were there not some powerful check to their natural increase. Very few birds produce less than two young ones each year, while many have six, eight, or ten; four will certainly be below the average; and if we suppose that each pair produce young only four times in their life, that will also be below the average, supposing them not to die either by violence or want of food. Yet at this rate how tremendous would be the increase in a few years from a single pair! A simple calculation will show that in fifteen years each pair of birds would have increased to nearly ten millions! whereas we have no reason to believe that the number of the birds of any country increases at all in fifteen or in one hundred and fifty years. With such powers of increase the population must have reached its limits, and have become stationary, in a very few years after the origin of each species. It is evident, therefore, that each year an immense number of birds must perish—as many in fact as are born; and as on the lowest calculation the progeny are each year twice as numerous as their parents, it follows that, whatever be the average number of individuals existing in any given country, *twice that number must perish annually,*—a striking result, but one which seems at least highly probable, and is perhaps under rather than over the truth. It would therefore appear that, as far as the continuance of the species and the keeping up the average number of individuals are concerned, large broods are superfluous. On the average all above *one* become food for hawks and kites, wild cats and weasels, or perish of cold and hunger as winter comes on. This is strikingly proved by the case of particular species; for we find that their abundance in individuals bears no relation whatever to their fertility in producing offspring. Perhaps the most remarkable instance of an immense bird population is that of the passenger pigeon of the United States, which lays only one, or at the most two eggs, and is said to rear generally but one young one. Why is this bird so extraordinarily abundant, while others producing two or three times as many young are much less plentiful? The explanation is not difficult. The food most congenial to this species, and on which it thrives best, is abundantly distributed over a very extensive region, offering such differences of soil and climate, that in one part or another of the area the supply never fails. The bird is capable of a very rapid and long-continued flight, so that it can pass without fatigue over the whole of the district it inhabits, and as soon as the supply of food begins to fail in one place is able to discover a fresh feeding-ground. This example strikingly shows us that the procuring a constant supply of wholesome food is almost the sole condition requisite for ensuring the rapid increase of a given species, since neither the limited fecundity, nor the unrestrained attacks of birds of prey and of man are here sufficient to check it. In no other birds are these peculiar circumstances so strikingly combined. Either their food is more liable to failure, or they have not sufficient power of wing to search for it over an extensive area, or during some season of the year it becomes very scarce, and less wholesome substitutes have to be found; and thus, though more fertile in offspring, they can never increase beyond the supply of food in the least favourable seasons. Many birds can only exist by migrating, when their food becomes scarce, to regions possessing a milder, or at least a different climate, though, as these migrating birds are seldom excessively abundant, it is evident that the countries they visit are still deficient in a constant and abundant supply of wholesome food. Those whose organization does not permit them to migrate when their food becomes periodically scarce, can never attain a large population. This is probably the reason why woodpeckers are scarce with us, while in the tropics they are among the most abundant of solitary birds. Thus the house sparrow is more abundant than the redbreast, because its food is more constant and plentiful,—seeds of grasses being preserved during the winter, and our farm-yards and stubble-fields furnishing an almost inexhaustible supply. Why, as a general rule, are aquatic, and especially sea birds, very numerous in individuals? Not because they are more prolific than others, generally the contrary; but because their food never fails, the sea-shores and river-banks daily swarming with a fresh supply of small mollusca and crustacea. Exactly the same laws will apply to mammals. Wild cats are prolific and have few enemies; why then are they never as abundant as rabbits? The only intelligible answer is, that their supply of food is more precarious. It appears evident, therefore, that so long as a country remains physically unchanged, the numbers of its animal population cannot materially increase. If one species does so, some others requiring the same kind of food must diminish in proportion. The numbers that die annually must be immense; and as the individual existence of each animal depends upon itself, those that die must be the weakest—the very young, the aged, and the diseased,—while those that prolong their existence can only be the most perfect in health and vigour—those who are best able to obtain food regularly, and avoid their numerous enemies. It is, as we commenced by remarking, "a struggle for existence," in which the weakest and least perfectly organized must always succumb.

Now it is clear that what takes place among the individuals of a species must also occur among the several allied species of a group,—viz. that those which are best adapted to obtain a regular supply of food, and to defend themselves against the attacks of their enemies and the vicissitudes of the seasons, must necessarily obtain and preserve a superiority in population; while those species which from some defect of power or organization are the least capable of counteracting the vicissitudes of food, supply, &c., must diminish in numbers, and, in extreme cases, become altogether extinct. Between these extremes the species will present various degrees of capacity for ensuring the means of preserving life; and it is thus we account for the abundance or rarity of species. Our ignorance will generally prevent us from accurately tracing the effects to their causes; but could we become perfectly acquainted with the organization and habits of the various species of animals, and could we measure the capacity of each for performing the different acts necessary to its safety and existence under all the varying circumstances by which it is surrounded, we might be able even to calculate the proportionate abundance of individuals which is the necessary result.

If now we have succeeded in establishing these two points—1st, *that the animal population of a country is generally stationary, being kept down by a periodical deficiency of food, and other checks*; and, 2nd, *that the comparative abundance or scarcity of the individuals of the several species is entirely due to their organization and resulting habits, which, rendering it more difficult to procure a regular supply of food and to provide for their personal safety in some cases than in others, can only be balanced by a difference in the population which have to exist in a given area*—we shall be in a condition to proceed to the consideration of *varieties*, to which the preceding remarks have a direct and very important application.

Most or perhaps all the variations from the typical form of a species must have some definite effect, however slight, on the habits or capacities of the individuals. Even a change of colour might, by rendering them more or less distinguishable, affect their safety; a greater or less development of hair might modify their habits. More important changes, such as an increase in the power or dimensions of the limbs or any of the external organs, would more or less affect their mode of procuring food or the range of country which they inhabit. It is also evident that most changes would affect, either favourably or adversely, the powers of prolonging existence. An antelope with shorter or weaker legs must necessarily suffer more from the attacks of the feline carnivora; the passenger pigeon with less powerful wings would sooner or later be affected in its powers of procuring a regular supply of food; and in both cases the result must necessarily be a diminution of the population of the modified species. If, on the other hand, any species should produce a variety having slightly increased powers of preserving existence, that variety must inevitably in time acquire a superiority in numbers. These results must follow as surely as old age, intemperance, or scarcity of food produce an increased mortality. In both cases there may be many individual exceptions; but on the average the rule will invariably be found to hold good. All varieties will therefore fall into two classes—those which under the same conditions would never reach the population of the parent species, and those which would in time obtain and keep a numerical superiority. Now, let some alternation of physical conditions occur in the district—a long period of drought, a destruction of vegetation by locusts, the irruption of some new carnivorous animal seeking "pastures new"—any change in fact tending to render existence more difficult to the species in question, and tasking its utmost powers to avoid complete extermination; it is evident that, of all the individuals composing the species, those forming the least numerous and most feebly organized variety would suffer first, and, were the pressure severe, must soon become extinct. The same causes continuing in action, the parent species would next suffer, would gradually diminish in numbers, and with a recurrence of similar unfavourable conditions might also become extinct. The superior variety would then alone remain, and on a return to favourable circumstances would rapidly increase in numbers and occupy the place of the extinct species and variety.

The *variety* would now have replaced the *species*, of which it would be a more perfectly developed and more highly organized form. It would be in all respects better adapted to secure its safety, and to prolong its individual existence and that of the race. Such a variety *could not* return to the original form; for that form is an inferior one, and could never compete with it for existence. Granted, therefore, a "tendency" to reproduce the original type of the species, still the variety must ever remain preponderant in numbers, and under adverse physical conditions *again alone survive*. But this new, improved, and populous race might itself, in course of time, give rise to new varieties, exhibiting several diverging modifications of form, any of which, tending to increase the facilities for preserving existence, must, by the same general law, in their turn become predominant. Here, then, we have *progression and continued divergence* deduced from the general laws which regulate the existence of animals in a state of nature, and from the undisputed fact that varieties do frequently occur. It is not, however, contended that this result would be invariable; a change of physical conditions in the district might at times materially modify it, rendering the race which had been the most capable of supporting existence under the former conditions now the least so, and even causing the extinction of the newer and, for a time, superior race, while the old or parent species and its first inferior varieties continued to flourish. Variations in unimportant parts might also occur, having no perceptible effect on the life-preserving powers; and the varieties so furnished might run a course parallel with the parent species, either giving rise to further variations or returning to the former type. All we argue for is, that certain varieties have a tendency to maintain their existence longer than the original species, and this tendency must make itself felt; for though the doctrine of chances or averages can never be trusted to on a limited scale, yet, if applied to high numbers, the results come nearer to what theory demands, and, as we approach to an infinity of examples, become strictly accurate. Now the scale on which nature works is so vast—the numbers of individuals and periods of time with which she deals approach so near to infinity, that any cause, however slight, and however liable to be veiled and counteracted by accidental circumstances, must in the end produce its full legitimate results.

Let us now turn to domesticated animals, and inquire how varieties produced among them are affected by the principles here enunciated. The essential difference in the condition of wild and domestic animals is this,—that among the former, their well-being and very existence depend upon the full exercise and healthy condition of all their senses and physical powers, whereas, among the latter, these are only partially exercised, and in some cases are absolutely unused. A wild animal has to search, and often to labour, for every mouthful of food—to exercise sight, hearing, and smell in seeking it, and in avoiding dangers, in procuring shelter from the inclemency of the seasons, and in providing for the subsistence and safety of its offspring. There is no muscle of its body that is not called into daily and hourly activity; there is not sense or faculty that is not strengthened by continual exercise. The domestic animal, on the other hand, has food provided for it, is sheltered, and often confined, to guard it against the vicissitudes of the seasons, is carefully secured from the attacks of its natural enemies, and seldom even rears its young without human assistance. Half of its senses and faculties are quite useless; and the other half are but occasionally called into feeble exercise, while even its muscular system is only irregularly called into action.

Now when a variety of such an animal occurs, having increased power or capacity in any organ or sense, such increase is totally useless, is never called into action, and may even exist without the animal ever becoming aware of it. In the wild animal, on the contrary, all its faculties and powers being brought into full action for the necessities of existence, any increase becomes immediately available, is strengthened by exercise, and must even slightly modify the food, the habits, and the whole economy of the race. It creates as it were a new animal, one of the superior powers, and which will necessarily increase in numbers and outlive those inferior to it.

Again, in the domesticated animal all variations have an equal chance of continuance; and those which would decidedly render a wild animal unable to compete with its fellows and continue its existence are no disadvantage whatever in a state of domesticity. Our quickly fattening pigs, short-legged sheep, pouter pigeons, and poodle dogs could never have come into existence in a state of nature, because the very first step towards such inferior forms would have led to the rapid extinction of the race; still less could they now exist in competition with their wild allies. The great speed but slight endurance of the race horse, the unwieldy strength of the ploughman's team, would both be useless in a state of nature. If turned wild [in] the pampas, such animals would probably soon become extinct, or under favourable circumstances might each lose those extreme qualities which would never be called into action, and in a few generations would revert to a common type, which must be that in which the various powers and faculties are so proportioned to each other as to be best adapted to procure food and secure safety,—that in which by the full exercise of every part of his organization the animal can alone continue to live. Domestic varieties, when turned wild, *must* return to something near the type of the original wild stock, *or become altogether extinct.*

We see, then, that no inferences as to varieties in a state of nature can be deduced from the observation of those occurring among domestic animals. The two are so much opposed to each other in every circumstance of their existence, that what applies to the one is almost sure not to apply to the other. Domestic animals are abnormal, irregular, artificial; they are subject to varieties which never occur and never can occur in a state of nature: their very existence depends altogether on human care; so far are many of them removed from that just proportion of faculties, that true balance of organization, by means of which alone an animal left to its own resources can preserve its existence and continue its race.

The hypothesis of Lamarck—that progressive changes in species have been produced by the attempts of animals to increase the development of their own organs, and thus modify their structure and habits—has been repeatedly and easily refuted by all writers on the subject of varieties and species, and it seems to have been considered that when this was done the whole question has been finally settled; but the view here developed renders such an hypothesis quite unnecessary, by showing that similar results must be produced by the action of principles constantly at work in nature. The powerful retractile talons of the falcon- and the cat-tribes have not been produced or increased by the volition of those animals; but among the different varieties which occurred in the earlier and less highly organized forms of these groups, *those always survived longest which had the greatest facilities for seizing their prey.* Neither did the giraffe acquire its long neck by desiring to reach the foliage of the more lofty shrubs, and constantly stretching its neck for the purpose, but because any varieties which occurred among its antitypes with a longer neck than usual *at once secured a fresh range of pasture over the same ground as their shorter-necked companions, and on the first scarcity of food were thereby enabled to outlive them.* Even the peculiar colours of many animals, especially insects, so closely resembling the soil or the leaves or the trunks on which they habitually reside, are explained on the same principle; for though in the course of ages varieties of many tints may have occurred, *yet those races having colours best adapted to concealment from their enemies would inevitably survive the longest.* We have also here an acting cause to account for that balance so often observed in nature,—a deficiency in one set of organs always being compensated by an increased development of some others—powerful wings accompanying weak feet, or great velocity making up for the absence of defensive weapons; for it has been shown that all varieties in which an unbalanced deficiency occurred could not long continue their existence. The action of this principle is exactly like that of the centrifugal governor of the steam engine, which checks and corrects any irregularities almost before they become evident; and in like manner no unbalanced deficiency in the animal kingdom can ever reach any conspicuous magnitude, because it would make itself felt at the very first step, by rendering existence difficult and extinction almost sure soon to follow. An origin such as is here advocated will also agree with the peculiar character of the modifications of form and structure which obtain in organized beings—the many lines of divergence from a central type, the increasing efficiency and power of a particular organ through a succession of allied species, and the remarkable persistence of unimportant parts such as colour, texture of plumage and hair, form of horns or crests, through a series of species differing considerably in more essential characters. It also furnishes us with a reason for that "more specialized structure" which Professor Owen states to be a characteristic of recent compared with extinct forms, and which would evidently be the result of the progressive modification of any organ applied to a special purpose in the animal economy.

We believe we have now shown that there is a tendency in nature to the continued progression of certain classes of *varieties* further and further from the original type—a progression to which there appears no reason to assign any definite limits—and that the same principle which produces this result in a state of nature will also explain why domestic varieties have a tendency to refer to the original type. This progression, by minute steps, in various directions, but always checked and balanced by the necessary conditions, subject to which alone existence can be preserved, may, it is believed, be followed out so as to agree with all the phenomena presented by organized beings, their extinction and succession in past ages, and all the extraordinary modifications of form, instinct, and habits which they exhibit.

On the Origin of Species by Means of Natural Selection, or the Preservation of Favoured Races in the Struggle for Life. [1859]

THE ORIGIN OF SPECIES *was a direct consequence of the letter and manuscript from Wallace discussed in the headnote to the last reading. Following the publication of the "joint paper" on natural selection, Darwin tried to write a thirty-page paper distilling his ideas for the* Journal of the Proceedings of the Linnean Society. *He failed. Darwin then decided to write an abstract of his unfinished manuscript* Natural Selection. *He began on July 20, 1858. When he finished, in March 1859, the abstract had run to what would be almost five hundred printed pages.*

The well-known publishing house of John Murray brought out Darwin's abstract in book form on November 24, 1859. All 1250 copies were immediately taken by booksellers, who in fact ordered 250 more copies than were available. Accordingly a second edition, to which Darwin made editorial corrections and a few additions, quickly succeeded the first on January 7, 1860. The book continued to sell well, with six editions, thirty-five printings, and translation into eleven other languages in Darwin's lifetime. At latest count, the Origin *has been reprinted over four hundred times and translated into twenty-nine languages.*

Some contemporary critics of the book complained that Darwin had been lax in not citing references to the work of others discussed in the book. As mentioned above, however, Darwin had conceived of the Origin *as merely an abstract of the larger work on which he had been working diligently since 1856,* Natural Selection. *He reasoned that citing one reference would oblige him to cite all—and because* Natural Selection *contained a large number of references to be verified, citation would have proven a long and tedious task. The title Darwin had proposed,* An Abstract of an Essay on the Origin of Species and Varieties through Natural Selection, *highlights its status as abstract. The title under which Murray published the book,* On the Origin of Species by Means of Natural Selection, or the Preservation of Favoured Races in the Struggle for Life, *more precisely reflects the major topics but sacrifices the generic distinction made clear in Darwin's original title.*

Subsequent editions of the Origin *appeared in 1861, 1866, 1869, and 1872. In each new edition, Darwin attempted to answer the objections of his critics, supplied new data on adaptations, and revised his prose, so that the last edition he saw into print differs markedly in some places from his original abstract. What resulted might be called an instance of evolution in the textual sphere—but survival of the fittest is not part of the analogy. Although the sixth and most recent edition is the one generally available in reprints today, its repeatedly bolstered, qualified, and reconsidered argument has lost much of the freshness, flow, and force of the first edition, from which we reprint.*

Any reader of early Victorian science or literature knows that evolution was widely discussed, by scientific and general audiences alike, in the days before the Origin. *Darwin was not unique in being an evolutionist. Where he stood alone, however, was in supplying a rational explanation of how evolution takes place—namely, that animals and plants produce far more offspring than can survive, that these offspring vary in their traits, and that natural selection acts in such a way that the offspring with the most useful traits are likeliest to survive and perpetuate their advantages. Although he relied primarily on natural selection to explain evolution, Darwin did not rule out such factors as the forces of use and disuse of structures, the influence of the environment, the inheritance of acquired characteristics, and sexual selection. In fact, these supplementary explanations played progressively larger roles in succeeding editions of the* Origin. *Today, however, only natural selection and sexual selection are accepted as accounting for evolutionary change.*

Evolution of humans comes up only twice in the Origin, *but the second and more provocative allusion led to much speculation about the possible relationships of humans to the great apes, that group of animals with which we share the most characteristics: "In the distant future I see far more important researches," predicts Darwin. "Psy-*

chology will be based on a new foundation, that of necessary acquirement of each mental power and capacity by gradation. Light will be thrown on the origin of man and his history." Darwin fulfilled this forecast himself in 1871, with the publication of The Descent of Man.

The eminent Darwin scholar Richard Freeman has called The Origin of Species *"certainly the most important biological book ever written." It also may have been the most important book of any kind published in the nineteenth century. We have chosen to include the first four chapters of the* Origin, *where Darwin outlines his original argument, along with the summarizing final chapter. The chapters not included—those covering laws of variation, difficulties associated with the theory, instinct, hybridism, the geological record, geographical distribution, classification, morphology, embryology, and rudimentary organs—are by no means unimportant, but they build upon and support the theoretical portion presented here. Readers interested in going further should find copies of the first edition and explore these chapters as well.*

INTRODUCTION.

When on board H.M.S. 'Beagle,' as naturalist, I was much struck with certain facts in the distribution of the inhabitants of South America, and in the geological relations of the present to the past inhabitants of that continent. These facts seemed to me to throw some light on the origin of species—that mystery of mysteries, as it has been called by one of our greatest philosophers. On my return home, it occurred to me, in 1837, that something might perhaps be made out on this question by patiently accumulating and reflecting on all sorts of facts which could possibly have any bearing on it. After five years' work I allowed myself to speculate on the subject, and drew up some short notes; these I enlarged in 1844 into a sketch of the conclusions, which then seemed to me probable: from that period to the present day I have steadily pursued the same object. I hope that I may be excused for entering on these personal details, as I give them to show that I have not been hasty in coming to a decision.

My work is now nearly finished; but as it will take me two or three more years to complete it, and as my health is far from strong, I have been urged to publish this Abstract. I have more especially been induced to do this, as Mr. Wallace, who is now studying the natural history of the Malay archipelago, has arrived at almost exactly the same general conclusions that I have on the origin of species. Last year he sent to me a memoir on this subject, with a request that I would forward it to Sir Charles Lyell, who sent it to the Linnean Society, and it is published in the third volume of the Journal of that Society. Sir C. Lyell and Dr. Hooker, who both knew of my work—the latter having read my sketch of 1844—honoured me by thinking it advisable to publish, with Mr. Wallace's excellent memoir, some brief extracts from my manuscripts.

This Abstract, which I now publish, must necessarily be imperfect. I cannot here give references and authorities for my several statements; and I must trust to the reader reposing some confidence in my accuracy. No doubt errors will have crept in, though I hope I have always been cautious in trusting to good authorities alone. I can here give only the general conclusions at which I have arrived, with a few facts in illustration, but which, I hope, in most cases will suffice. No one can feel more sensible than I do of the necessity of hereafter publishing in detail all the facts, with references, on which my conclusions have been grounded; and I hope in a future work to do this. For I am well aware that scarcely a single point is discussed in this volume on which facts cannot be adduced, often apparently leading to conclusions directly opposite to those at which I have arrived. A fair result can be obtained only by fully stating and balancing the facts and arguments on both sides of each question; and this cannot possibly be here done.

I much regret that want of space prevents my having the satisfaction of acknowledging the generous assistance which I have received from very many naturalists, some of them personally unknown to me. I cannot, however, let this opportunity pass without expressing my deep obligations to Dr. Hooker, who for the last fifteen years has aided me in every possible way by his large stores of knowledge and his excellent judgment.

In considering the Origin of Species, it is quite conceivable that a naturalist, reflecting on the mutual affinities of organic beings, on their embryological relations, their geographical distribution, geological succession, and other such facts, might come to the conclusion that each species had not been independently created, but had descended, like varieties, from other species. Nevertheless, such a conclusion, even if well founded, would be unsatisfactory, until it could be shown how the innumerable species inhabiting this world have been modified, so as to acquire that perfection of structure and coadaptation which most justly excites our admiration. Naturalists continually refer to external conditions, such as climate, food, &c., as the only possible cause of variation. In one very limited sense, as we shall hereafter see, this may be true; but it is preposterous to attribute to mere external conditions, the structure, for instance, of the woodpecker, with its feet, tail, beak, and tongue, so admirably adapted to catch insects under the bark of trees. In the case of the misseltoe, which draws its nourishment from certain trees, which has seeds that must be transported by certain birds, and which has flowers with separate sexes absolutely requiring the agency of certain insects to bring pollen from one flower to the other, it is equally preposterous to account for the structure of this parasite, with its relations to several distinct organic beings, by the effects of external conditions, or of habit, or of the volition of the plant itself.

The author of the 'Vestiges of Creation' would, I presume, say that, after a certain unknown number of generations, some bird had given birth to a woodpecker, and some plant to the misseltoe, and that these had been produced perfect as we now see them; but this assumption seems to me to be no explanation, for it leaves the case of the coadaptations of organic beings to each other and to their physical conditions of life, untouched and unexplained.

It is, therefore, of the highest importance to gain a clear insight into the means of modification and coadaptation. At the commencement of my observations it seemed to me probable that a careful study of domesticated animals and of cultivated plants would offer the best chance of making out this obscure problem. Nor have I been disappointed; in this and in all other perplexing cases I have invariably found that our knowledge, imperfect though it be, of variation under domestication, afforded the best and safest clue. I may venture to express my conviction of the high value of such studies, although they have been very commonly neglected by naturalists.

From these considerations, I shall devote the first chapter of this Abstract to Variation under Domestication. We shall thus see that a large amount of hereditary modification is at least possible; and, what is equally or more important, we shall see how great is the power of man in accumulating by his Selection successive slight variations. I will then pass on to the variability of species in a state of nature; but I shall, unfortunately, be compelled to treat this subject far too briefly, as it can be treated properly only by giving long catalogues of facts. We shall, however, be enabled to discuss what circumstances are most favourable to variation. In the next chapter the Struggle for Existence amongst all organic beings throughout the world, which inevitably follows from their high geometrical powers of increase, will be treated of. This is the doctrine of Malthus, applied to the whole animal and vegetable kingdoms. As many more individuals of each species are born than can possibly survive; and as, consequently, there is a frequently recurring struggle for existence, it follows that any being, if it vary however slightly in any manner profitable to itself, under the complex and sometimes varying conditions of life, will have a better chance of surviving, and thus be *naturally selected*. From the strong principle of inheritance, any selected variety will tend to propagate its new and modified form.

This fundamental subject of Natural Selection will be treated at some length in the fourth chapter; and we shall then see how Natural Selection almost inevitably causes much Extinction of the less improved forms of life, and induces what I have called Divergence of Character. In the next chapter I shall discuss the complex and little known laws of variation and of correlation of growth. In the four succeeding chapters, the most apparent and gravest difficulties on the theory will be given: namely, first, the difficulties of transitions, or in understanding how a simple being or a simple organ can be changed and perfected into a highly developed being or elaborately constructed organ; secondly, the subject of Instinct, or the mental powers of animals; thirdly, Hybridism, or the infertility of species and the fertility of varieties when intercrossed; and fourthly, the imperfection of the Geological Record. In the next chapter I shall consider the geological succession of organic beings throughout time; in the eleventh and twelfth, their geographical distribution throughout space; in the thirteenth, their classification or mutual affinities, both when mature and in an embryonic condition. In the last chapter I shall give a brief recapitulation of the whole work, and a few concluding remarks.

No one ought to feel surprise at much remaining as yet unexplained in regard to the origin of species and varieties, if he makes due allowance for our profound ignorance in regard to the mutual relations of all the beings which live around us. Who can explain why one species ranges widely and is very numerous, and why another allied species has a narrow range and is rare? Yet these relations are of the highest importance, for they determine the present welfare, and, as I believe, the future success and modification of every inhabitant of this world. Still less do we know of the mutual relations of the innumerable inhabitants of the world during the many past geological epochs in its history. Although much remains obscure, and will long remain obscure, I can entertain no doubt, after the most deliberate study and dispassionate judgment of which I am capable, that the view which most naturalists entertain, and which I formerly entertained—namely, that each species has been independently created—is erroneous. I am fully convinced that species are not immutable; but that those belonging to what are called the same genera are lineal descendants of some other and generally extinct species, in the same manner as the acknowledged varieties of any one species are the descendants of that species. Furthermore, I am convinced that Natural Selection has been the main but not exclusive means of modification.

VARIATION UNDER DOMESTICATION.

When we look to the individuals of the same variety or sub-variety of our older cultivated plants and animals, one of the first points which strikes us, is, that they generally differ much more from each other, than do the individuals of any one species or variety in a state of nature. When we reflect on the vast diversity of the plants and animals which have been cultivated, and which have varied during all ages under the most different climates and treatment, I think we are driven to conclude that this greater variability is simply due to our domestic productions having been raised under conditions of life not so uniform as, and somewhat different from, those to which the parent-species have been exposed under nature. There is, also, I think, some probability in the view propounded by Andrew Knight, that this variability may be partly connected with excess of food. It seems pretty clear that organic beings must be exposed during several generations to the new conditions of life to cause any appreciable amount of variation; and that when the organisation has once begun to vary, it generally continues to vary for many generations. No case is on record of a variable being ceasing to be variable under cultivation. Our oldest cultivated plants, such as wheat, still often yield new varieties: our oldest domesticated animals are still capable of rapid improvement or modification.

It has been disputed at what period of life the causes of variability, whatever they may be, generally act; whether during the early or late period of development of the embryo, or at the instant of conception. Geoffroy St. Hilaire's experiments show that unnatural treatment of the embryo causes monstrosities; and monstrosities cannot be separated by any clear line of distinction from mere variations. But I am strongly inclined to suspect that the most frequent cause of variability may be attributed to the male and female reproductive elements having been affected prior to the act of conception. Several reasons make me believe in this; but the chief one is the remarkable effect which confinement or cultivation has on the functions of the reproductive system; this system appearing to be far more susceptible than any other part of the organisation, to the action of any change in the conditions of life. Nothing is more easy than to tame an animal, and few things more difficult than to get it to breed freely under confinement, even in the many cases when the male and female unite. How many animals there are which will not breed, though living long under not very close confinement in their native country! This is generally attributed to vitiated instincts; but how many cultivated plants display the utmost vigour, and yet rarely or never seed! In some few such cases it has been found out that very trifling changes, such as a little more or less water at some particular period of growth, will determine whether or not the plant sets a seed. I cannot here enter on the copious details which I have collected on this curious subject; but to show how singular the laws are which determine the reproduction of animals under confinement, I may just mention that carnivorous animals, even from the tropics, breed in this country pretty freely under confinement, with the exception of the plantigrades or bear family; whereas, carnivorous birds, with the rarest exceptions, hardly ever lay fertile eggs. Many exotic plants have pollen utterly worthless, in the same exact condition as in the most sterile hybrids. When, on the one hand, we see domesticated animals and plants, though often weak and sickly, yet breeding quite freely under confinement; and when, on the other hand, we see individuals, though taken young from a state of nature, perfectly tamed, long-lived, and healthy (of which I could give numerous instances), yet having their reproductive system so seriously affected by unperceived causes as to fail in acting, we need not be surprised at this system, when it does act under confinement, acting not quite regularly, and producing offspring not perfectly like their parents or variable.

Sterility has been said to be the bane of horticulture; but on this view we owe variability to the same cause which produces sterility; and variability is the source of all the choicest productions of the garden. I may add, that as some organisms will breed most freely under the most unnatural conditions (for instance, the rabbit and ferret kept in hutches), showing that their reproductive system has not been thus affected; so will some animals and plants withstand domestication or cultivation, and vary very slightly—perhaps hardly more than in a state of nature.

A long list could easily be given of "sporting plants"; by this term gardeners mean a single bud or offset, which suddenly assumes a new and sometimes very different character from that of the rest of the plant. Such buds can be propagated by grafting, &c., and sometimes by seed. These "sports" are extremely rare under nature, but far from rare under cultivation; and in this case we see that the treatment of the parent has affected a bud or offset, and not the ovules or pollen. But it is the opinion of most physiologists that there is no essential difference between a bud and an ovule in their earliest stages of formation; so that, in fact, "sports" support my view, that variability may be largely attributed to the ovules or pollen, or to both, having been affected by the treatment of the parent prior to the act of conception. These cases anyhow show that variation is not necessarily connected, as some authors have supposed, with the act of generation.

Seedlings from the same fruit, and the young of the same litter, sometimes differ considerably from each other, though both the young and the parents, as Müller has remarked, have apparently been exposed to exactly the same conditions of life; and this shows how unimportant the direct effects of the conditions of life are in comparison with the laws of reproduction, and of growth, and of inheritance; for had the action of the conditions been direct, if any of the young had varied, all would probably have varied in the same manner. To judge how much, in the case of any variation, we should attribute to the direct action of heat, moisture, light, food, &c., is most difficult: my impression is, that with animals such agencies have produced very little direct effect, though apparently more in the case of plants. Under this point of view, Mr. Buckman's recent experiments on plants seem extremely valuable. When all or nearly all the individuals exposed to certain conditions are affected in the same way, the change at first appears to be directly due to such conditions; but in some cases it can be shown that quite opposite conditions produce similar changes of structure. Nevertheless some slight amount of change may, I think, be attributed to the direct action of the conditions of life—as, in some cases, increased size from amount of food, colour from particular kinds of food and from light, and perhaps the thickness of fur from climate.

Habit also has a decided influence, as in the period of flowering with plants when transported from one climate to another. In animals it has a more marked effect; for instance, I find in the domestic duck that the bones of the wing weigh less and the bones of the leg more, in proportion to the whole skeleton, than do the same bones in the wild-duck; and I presume that this change may be safely attributed to the domestic duck flying much less, and walking more, than its wild parent. The great and inherited development of the udders in cows and goats in countries where they are habitually milked, in comparison with the state of these organs in other countries, is another instance of the effect of use. Not a single domestic animal can be named which has not in some country drooping ears; and the view suggested by some authors, that the drooping is due to the disuse of the muscles of the ear, from the animals not being much alarmed by danger, seems probable.

There are many laws regulating variation, some few of which can be dimly seen, and will be hereafter briefly mentioned. I will here only allude to what may be called correlation of growth. Any change in the embryo or larva will almost certainly entail changes in the mature animal. In monstrosities, the correlations between quite distinct parts are very curious; and many instances are given in Isidore Geoffroy St. Hilaire's great work on this subject. Breeders believe that long limbs are almost always accompanied by an elongated head. Some instances of correlation are quite whimsical: thus cats with blue eyes are invariably deaf; colour and constitutional peculiarities go together, of which many remarkable cases could be given amongst animals and plants. From the facts collected by Heusinger, it appears that white sheep and pigs are differently affected from coloured individuals by certain vegetable poisons. Hairless dogs have imperfect teeth; long-haired and coarse-haired animals are apt to have, as is asserted, long or many horns; pigeons with feathered feet have skin between their outer toes; pigeons with short beaks have small feet, and those with long beaks large feet. Hence, if man goes on selecting, and thus augmenting, any peculiarity, he will almost certainly unconsciously modify other parts of the structure, owing to the mysterious laws of the correlation of growth.

The result of the various, quite unknown, or dimly seen laws of variation is infinitely complex and diversified. It is well worth while carefully to study the several treatises published on some of our old cultivated plants, as on the hyacinth, potato, even the dahlia, &c.; and it is really surprising to note the endless points in structure and constitution in which the varieties and subvarieties differ slightly from each other. The whole organisation seems to have become plastic, and tends to depart in some small degree from that of the parental type.

Any variation which is not inherited is unimportant for us. But the number and diversity of inheritable deviations of structure, both those of slight and those of considerable physiological importance, is endless. Dr. Prosper Lucas's treatise, in two large volumes, is the fullest and the best on this subject. No breeder doubts how strong is the tendency to inheritance: like produces like is his fundamental belief: doubts have been thrown on this principle by theoretical writers alone. When a deviation appears not unfrequently, and we see it in the father and child, we cannot tell whether it may not be due to the same original cause acting on both; but when amongst individuals, apparently exposed to the same conditions, any very rare deviation, due to some extraordinary combination of circumstances, appears in the parent—say, once amongst several million individuals—and it reappears in the child, the mere doctrine of chances almost compels us to attribute its reappearance to inheritance. Every one must have heard of cases of albinism, prickly skin, hairy bodies, &c., appearing in several members of the same family. If strange and rare deviations of structure are truly inherited, less strange and commoner deviations may be freely admitted to be inheritable. Perhaps the correct way of viewing the whole subject, would be, to look at the inheritance of every character whatever as the rule, and noninheritance as the anomaly.

The laws governing inheritance are quite unknown; no one can say why the same peculiarity in different individuals of the same species, and in individuals of different species, is sometimes inherited and sometimes not so; why the child often reverts in certain characters to its grandfather or grandmother or other much more remote ancestor; why a peculiarity is often transmitted from one sex to both sexes, or to one sex alone, more commonly but not exclusively to the like sex. It is a fact of some little importance to us, that peculiarities appearing in the males of our domestic breeds are often transmitted either exclusively, or in a much greater degree, to males alone. A much more important rule, which I think may be trusted, is that, at whatever period of life a peculiarity first appears, it tends to appear in the offspring at a corresponding age, though sometimes earlier. In many cases this could not be otherwise: thus the inherited peculiarities in the horns of cattle could appear only in the offspring when nearly mature; peculiarities in the silkworm are known to appear at the corresponding caterpillar or cocoon stage. But hereditary diseases and some other facts make me believe that the rule has a wider extension, and that when there is no apparent reason why a peculiarity should appear at any particular age, yet that it does tend to appear in the offspring at the same period at which it first appeared in the parent. I believe this rule to be of the highest importance in explaining the laws of embryology. These remarks are of course confined to the first *appearance* of the peculiarity, and not to its primary cause, which may have acted on the ovules or male element; in nearly the same manner as in the crossed offspring from a short-horned cow by a long-horned bull, the greater length of horn, though appearing late in life, is clearly due to the male element.

Having alluded to the subject of reversion, I may here refer to a statement often made by naturalists—namely, that our domestic varieties, when run wild, gradually but certainly revert in character to their aboriginal stocks. Hence it has been argued that no deductions can be drawn from domestic races to species in a state of nature. I have in vain endeavoured to discover on what decisive facts the above statement has so often and so boldly been made. There would be great difficulty in proving its truth: we may safely conclude that very many of the most strongly-marked domestic varieties could not possibly live in a wild state. In many cases we do not know what the aboriginal stock was, and so could not tell whether or not nearly perfect reversion had ensued. It would be quite necessary, in order to prevent the effects of intercrossing, that only a single variety should be turned loose in its new home. Nevertheless, as our varieties certainly do occasionally revert in some of their characters to ancestral forms, it seems to me not improbable, that if we could succeed in naturalising, or were to cultivate, during many generations, the several races, for instance, of the cabbage, in very poor soil (in which case, however, some effect would have to be attributed to the direct action of the poor soil), that they would to a large extent, or even wholly, revert to the wild aboriginal stock. Whether or not the experiment would succeed, is not of great importance for our line of argument; for by the experiment itself the conditions of life are changed. If it could be shown that our domestic varieties manifested a strong tendency to reversion—that is, to lose their acquired characters, whilst kept under unchanged conditions, and whilst kept in a considerable body, so that free intercrossing might check, by blending together, any slight deviations of structure, in such case, I grant that we could deduce nothing from domestic varieties in regard to species. But there is not a shadow of evidence in favour of this view: to assert that we could not breed our cart and race-horses, long- and short-horned cattle, and poultry of various breeds, and esculent vegetables, for an almost infinite number of generations, would be opposed to all experience. I may add, that when under nature the conditions of life do change, variations and reversions of character probably do occur; but natural selection, as will hereafter be explained, will determine how far the new characters thus arising shall be preserved.

When we look to the hereditary varieties or races of our domestic animals and plants, and compare them with species closely allied together, we generally perceive in each domestic race, as already remarked, less uniformity of character than in true species. Domestic races of the same species, also, often have a somewhat monstrous character; by which I mean, that, although differing from each other, and from the other species of the same genus, in several trifling respects, they often differ in an extreme degree in some one part, both when compared one with another, and more especially when compared with all the species in nature to which they are nearest allied. With these exceptions (and with that of the perfect fertility of varieties when crossed,—a subject hereafter to be discussed), domestic races of the same species differ from each other in the same manner as, only in most cases in a lesser degree than, do closely-allied species of the same genus in a state of nature. I think this must be admitted, when we find that there are hardly any domestic races, either amongst animals or plants, which have not been ranked by some competent judges as mere varieties, and by other competent judges as the descendants of aboriginally distinct species. If any marked distinction existed between domestic races and species, this source of doubt could not so perpetually recur. It has often been stated that domestic races do not differ from each other in characters of generic value. I think it could be shown that this statement is hardly correct; but naturalists differ most widely in determining what characters are of generic value; all such valuations being at present empirical. Moreover, on the view of the origin of genera which I shall presently give, we have no right to expect often to meet with generic differences in our domesticated productions.

When we attempt to estimate the amount of structural difference between the domestic races of the same species, we are soon involved in doubt, from not knowing whether they have descended from one or several parent-species. This point, if it could be cleared up, would be interesting; if, for instance, it could be shown that the grey-hound, bloodhound, terrier, spaniel, and bull-dog, which we all know propagate their kind so truly, were the offspring of any single species, then such facts would have great weight in making us doubt about the immutability of the many very closely allied and natural species—for instance, of the many foxes—inhabiting different quarters of the world. I do not believe, as we shall presently see, that all our dogs have descended from any one wild species; but, in the case of some other domestic races, there is presumptive, or even strong, evidence in favour of this view.

It has often been assumed that man has chosen for domestication animals and plants having an extraordinary inherent tendency to vary, and likewise to withstand diverse climates. I do not dispute that these capacities have added largely to the value of most of our domesticated productions; but how could a savage possibly know, when he first tamed an animal, whether it would vary in succeeding generations, and whether it would endure other climates? Has the little variability of the ass or guinea-fowl, or the small power of endurance of warmth by the rein-deer, or of cold by the common camel, prevented their domestication? I cannot doubt that if other animals and plants, equal in number to our domesticated productions, and belonging to equally diverse classes and countries, were taken from a state of nature, and could be made to breed for an equal number of generations under domestication, they would vary on an average as largely as the parent species of our existing domesticated productions have varied.

In the case of most of our anciently domesticated animals and plants, I do not think it is possible to come to any definite conclusion, whether they have descended from one or several species. The argument mainly relied on by those who believe in the multiple origin of our domestic animals is, that we find in the most ancient records, more especially on the monuments of Egypt, much diversity in the breeds; and that some of the breeds closely resemble, perhaps are identical with, those still existing. Even if this latter fact were found more strictly and generally true than seems to me to be the case, what does it show, but that some of our breeds originated there, four or five thousand years ago? But Mr. Horner's researches have rendered it in some degree probable that man sufficiently civilized to have manufactured pottery existed in the valley of the Nile thirteen or fourteen thousand years ago; and who will pretend to say how long before these ancient periods, savages, like those of Tierra del Fuego or Australia, who possess a semi-domestic dog, may not have existed in Egypt?

The whole subject must, I think, remain vague; neverthelsss, I may, without here entering on any details, state that, from geographical and other considerations, I think it highly probable that our domestic dogs have descended from several wild species. In regard to sheep and goats I can form no opinion. I should think, from facts communicated to me by Mr. Blyth, on the habits, voice, and constitution, &c., of the humped Indian cattle, that these had descended from a different aboriginal stock from our European cattle; and several competent judges believe that these latter have had more than one wild parent. With respect to horses, from reasons which I cannot give here, I am doubtfully inclined to believe, in opposition to several authors, that all the races have descended from one wild stock. Mr. Blyth, whose opinion, from his large and varied stores of knowledge, I should value more than that of almost any one, thinks that all the breeds of poultry have proceeded from the common wild Indian fowl (Gallus bankiva). In regard to ducks and rabbits, the breeds of which differ considerably from each other in structure, I do not doubt that they all have descended from the common wild duck and rabbit.

The doctrine of the origin of our several domestic races from several aboriginal stocks, has been carried to an absurd extreme by some authors. They believe that every race which breeds true, let the distinctive characters be ever so slight, has had its wild prototype. At this rate there must have existed at least a score of species of wild cattle, as many sheep, and several goats in Europe alone, and several even within Great Britain. One author believes that there formerly existed in Great Britain eleven wild species of sheep peculiar to it! When we bear in mind that Britain has now hardly one peculiar mammal, and France but few distinct from those of Germany and conversely, and so with Hungary, Spain, &c., but that each of these kingdoms possesses several peculiar breeds of cattle, sheep, &c., we must admit that many domestic breeds have originated in Europe; for whence could they have been derived, as these several countries do not possess a number of peculiar species as distinct parent-stocks? So it is in India. Even in the case of the domestic dogs of the whole world, which I fully admit have probably descended from several wild species, I cannot doubt that there has been an immense amount of inherited variation. Who can believe that animals closely resembling the Italian greyhound, the bloodhound, the bull-dog, or Blenheim spaniel, &c.—so unlike all wild Canidæ—ever existed freely in a state of nature? It has often been loosely said that all our races of dogs have been produced by the crossing of a few aboriginal species; but by crossing we can get only forms in some degree intermediate between their parents; and if we account for our several domestic races by this process, we must admit the former existence of the most extreme forms, as the Italian greyhound, bloodhound, bull-dog, &c., in the wild state. Moreover, the possibility of making distinct races by crossing has been greatly exaggerated. There can be no doubt that a race may be modified by occasional crosses, if aided by the careful selection of those individual mongrels, which present any desired character; but that a race could be obtained nearly intermediate between two extremely different races or speceies, I can hardly believe. Sir J. Sebright expressly experimentised for this object, and failed. The offspring from the first cross between two pure breeds is tolerably and sometimes (as I have found with pigeons) extremely uniform, and everything seems simple enough; but when these mongrels are crossed one with another for several generations, hardly two of them will be alike, and then the extreme difficulty, or rather utter hopelessness, of the task becomes apparent. Certainly, a breed intermediate between *two very distinct* breeds could not be got without extreme care and long-continued selection; nor can I find a single case on record of a permanent race having been thus formed.

On the Breeds of the Domestic Pigeon. Believing that it is always best to study some special group, I have, after deliberation, taken up domestic pigeons. I have kept every breed which I could purchase or obtain, and have been most kindly favoured with skins from several quarters of the world, more especially by the Hon. W. Elliot from India, and by the Hon. C. Murray from Persia. Many treatises in different languages have been published on pigeons, and some of them are very important, as being of considerable antiquity. I have associated with several eminent fanciers, and have been permitted to join two of the London Pigeon Clubs. The diversity of the breeds is something astonishing. Compare the English carrier and the short-faced tumbler, and see the wonderful difference in their beaks, entailing corresponding differences in their skulls. The carrier, more especially the male bird, is also remarkable from the wonderful development of the carunculated skin about the head, and this is accompanied by greatly elongated eyelids, very large external orifices to the nostrils, and a wide gape of mouth. The short faced tumbler has a beak in outline almost like that of a finch; and the common tumbler has the singular and strictly inherited habit of flying at a great height in a compact flock, and tumbling in the air head over heels. The runt is a bird of great size, with long, massive beak and large feet; some of the sub-breeds of runts have very long necks, others very long wings and tails, others singularly short tails. The barb is allied to the carrier, but, instead of a very long beak, has a very short and very broad one. The pouter has a much elongated body, wings, and legs; and its enormously developed crop, which it glories in inflating, may well excite astonishment and even laughter. The turbit has a very short and conical beak, with a line of reversed feathers down the breast; and it has the habit of continually expanding slightly the upper part of the œsophagus. The Jacobin has the feathers so much reversed along the back of the neck that they form a hood, and it has, proportionally to its size, much elongated wing and tail feathers. The trumpeter and laugher, as their names express, utter a very different coo from the other breeds. The fantail has thirty or even forty tail-feathers, instead of twelve or fourteen, the normal number in all members of the great pigeon family; and these feathers are kept expanded, and are carried so erect that in good birds the head and tail touch; the oil-gland is quite aborted. Several other less distinct breeds might have been specified.

In the skeletons of the several breeds, the development of the bones of the face in length and breadth and curvature differs enormously. The shape, as well as the breadth and length of the ramus of the lower jaw, varies in a highly remarkable manner. The number of the caudal and sacral vertebræ vary; as does the number of the ribs, together with their relative breadth and the presence of processes. The size and shape of the apertures in the sternum are highly variable; so is the degree of divergence and relative size of the two arms of the furcula. The proportional width of the gape of mouth, the proportional length of the eyelids, of the orifice of the nostrils, of the tongue (not always in strict correlation with the length of beak), the size of the crop and of the upper part of the œsophagus; the development and abortion of the oil-gland; the number of the primary wing and caudal feathers; the relative length of wing and tail to each other and to the body; the relative length of leg and of the feet; the number of scutellæ on the toes, the development of skin between the toes, are all points of structure which are variable. The period at which the perfect plumage is acquired varies, as does the state of the down with which the nestling birds are clothed when hatched. The shape and size of the eggs vary. The manner of flight differs remarkably; as does in some breeds the voice and disposition. Lastly, in certain breeds, the males and females have come to differ to a slight degree from each other.

Altogether at least a score of pigeons might be chosen, which if shown to an ornithologist, and he were told that they were wild birds, would certainly, I think, be ranked by him as well-defined species. Moreover, I do not believe that any ornithologist would place the English carrier, the short-faced tumbler, the runt, the barb, pouter, and fantail in the same genus; more especially as in each of these breeds several truly-inherited sub-breeds, or species as he might have called them, could be shown him.

Great as the differences are between the breeds of pigeons, I am fully convinced that the common opinion of naturalists is correct, namely, that all have descended from the rock-pigeon (Columba livia), including under this term several geographical races or sub-species, which differ from each other in the most trifling respects. As several of the reasons which have led me to this belief are in some degree applicable in other cases, I will here briefly give them. If the several breeds are not varieties, and have not proceeded from the rock-pigeon, they must have descended from at least seven or eight aboriginal stocks; for it is impossible to make the present domestic breeds by the crossing of any lesser number: how, for instance, could a pouter be produced by crossing two breeds unless one of the parent-stocks possessed the characteristic enormous crop? The supposed aboriginal stocks must all have been rock-pigeons, that is, not breeding or willingly perching on trees. But besides C. livia, with its geographical sub-species, only two or three other species of rock-pigeons are known; and these have not any of the characters of the domestic breeds. Hence the supposed aboriginal stocks must either still exist in the countries where they were originally domesticated, and yet be unknown to ornithologists; and this, considering their size, habits, and remarkable characters, seems very improbable; or they must have become extinct in the wild state. But birds breeding on precipices, and good fliers, are unlikely to be exterminated; and the common rock-pigeon, which has the same habits with the domestic breeds, has not been exterminated even on several of the smaller British islets, or on the shores of the Mediterranean. Hence the supposed extermination of so many species having similar habits with the rock-pigeon seems to me a very rash assumption. Moreover, the several above-named domesticated breeds have been transported to all parts of the world, and, therefore, some of them must have been carried back again into their native country; but not one has ever become wild or feral, though the dovecot-pigeon, which is the rock-pigeon in a very slightly altered state, has become feral in several places. Again, all recent experience shows that it is most difficult to get any wild animal to breed freely under domestication; yet on the hypothesis of the multiple origin of our pigeons, it must be assumed that at least seven or eight species were so thoroughly domesticated in ancient times by half-civilized man, as to be quite prolific under confinement.

An argument, as it seems to me, of great weight, and applicable in several other cases, is, that the above-specified breeds, though agreeing generally in constitution, habits, voice, colouring, and in most parts of their structure, with the wild rock-pigeon, yet are certainly highly abnormal in other parts of their structure: we may look in vain throughout the whole great family of Columbidæ for a beak like that of the English carrier, or that of the short faced tumbler, or barb; for reversed feathers like those of the jacobin; for a crop like that of the pouter; for tail-feathers like those of the fantail. Hence it must be assumed not only that half-civilized man succeeded in thoroughly domesticating several species, but that he intentionally or by chance picked out extraordinarily abnormal species; and further, that these very species have since all become extinct or unknown. So many strange contingencies seem to me improbable in the highest degree.

Some facts in regard to the colouring of pigeons well deserve consideration. The rock-pigeon is of a slaty-blue, and has a white rump (the Indian subspecies, C. intermedia of Strickland, having it bluish); the tail has a terminal dark bar, with the bases of the outer feathers externally edged with white; the wings have two black bars; some semi-domestic breeds and some apparently truly wild breeds have, besides the two black bars, the wings chequered with black. These several marks do not occur together in any other species of the whole family. Now, in every one of the domestic breeds, taking thoroughly well-bred birds, all the above marks, even to the white edging of the outer tail-feathers, sometimes concur perfectly developed. Moreover, when two birds belonging to two distinct breeds are crossed, neither of which is blue or has any of the above-specified marks, the mongrel offspring are very apt suddenly to acquire these characters; for instance, I crossed some uniformly white fantails with some uniformly black barbs, and they produced mottled brown and black birds; these I again crossed together, and one grandchild of the pure white fantail and pure black barb was of as beautiful a blue colour, with the white rump, double black wing-bar, and barred and white-edged tail-feathers, as any wild rock-pigeon! We can understand these facts, on the well-known principle of reversion to ancestral characters, if all the domestic breeds have descended from the rock-pigeon. But if we deny this, we must make one of the two following highly improbable suppositions. Either, firstly, that all the several imagined aboriginal stocks were coloured and marked like the rock-pigeon, although no other existing species is thus coloured and marked, so that in each separate breed there might be a tendency to revert to the very same colours and markings. Or, secondly, that each breed, even the purest, has within a dozen or, at most, within a score of generations, been crossed by the rock-pigeon: I say within a dozen or twenty generations, for we know of no fact countenancing the belief that the child ever reverts to some one ancestor, removed by a greater number of generations. In a breed which has been crossed only once with some distinct breed, the tendency to reversion to any character derived from such cross will naturally become less and less, as in each succeeding generation there will be less of the foreign blood; but when there has been no cross with a distinct breed, and there is a tendency in both parents to revert to a character, which has been lost during some former generation, this tendency, for all that we can see to the contrary, may be transmitted undiminished for an indefinite number of generations. These two distinct cases are often confounded in treatises on inheritance.

Lastly, the hybrids or mongrels from between all the domestic breeds of pigeons are perfectly fertile. I can state this from my own observations, purposely made on the most distinct breeds. Now, it is difficult, perhaps impossible, to bring forward one case of the hybrid offspring of two animals *clearly distinct* being themselves perfectly fertile. Some authors believe that long-continued domestication eliminates this strong tendency to sterility: from the history of the dog I think there is some probability in this hypothesis, if applied to species closely related together, though it is unsupported by a single experiment. But to extend the hypothesis so far as to suppose that species, aboriginally as distinct as carriers, tumblers, pouters, and fantails now are, should yield offspring perfectly fertile, *inter se*, seems to me rash in the extreme.

From these several reasons, namely, the improbability of man having formerly got seven or eight supposed species of pigeons to breed freely under domestication; these supposed species being quite unknown in a wild state, and their becoming nowhere feral; these species having very abnormal characters in certain respects, as compared with all other Columbidae, though so like in most other respects to the rock-pigeon; the blue colour and various marks occasionally appearing in all the breeds, both when kept pure and when crossed; the mongrel offspring being perfectly fertile,—from these several reasons, taken together, I can feel no doubt that all our domestic breeds have descended from the Columba livia with its geographical sub-species.

In favour of this view, I may add, firstly, that C. livia, or the rock-pigeon, has been found capable of domestication in Europe and in India; and that it agrees in habits and in a great number of points of structure with all the domestic breeds. Secondly, although an English carrier or short-faced tumbler differs immensely in certain characters from the rock-pigeon, yet by comparing the several sub-breeds of these breeds, more especially those brought from distant countries, we can make an almost perfect series between the extremes of structure. Thirdly, those characters which are mainly distinctive of each breed, for instance the wattle and length of beak of the carrier, the shortness of that of the tumbler, and the number of tail-feathers in the fantail, are in each breed eminently variable; and the explanation of this fact will be obvious when we come to treat of selection. Fourthly, pigeons have been watched, and tended with the utmost care, and loved by many people. They have been domesticated for thousands of years in several quarters of the world; the earliest known record of pigeons is in the fifth Ægyptian dynasty, about 3000 B.C., as was pointed out to me by Professor Lepsius; but Mr. Birch informs me that pigeons are given in a bill of fare in the previous dynasty. In the time of the Romans, as we hear from Pliny, immense prices were given for pigeons; "nay, they are come to this pass, that they can reckon up their pedigree and race." Pigeons were much valued by Akber Khan in India, about the year 1600; never less than 20,000 pigeons were taken with the court. "The monarchs of Iran and Turan sent him some very rare birds;" and, continues the courtly historian, "His Majesty by crossing the breeds, which method was never practised before, has improved them astonishingly." About this same period the Dutch were as eager about pigeons as were the old Romans. The paramount importance of these considerations in explaining the immense amount of variation which pigeons have undergone, will be obvious when we treat of Selection. We shall then, also, see how it is that the breeds so often have a somewhat monstrous character. It is also a most favourable circumstance for the production of distinct breeds, that male and female pigeons can be easily mated for life; and thus different breeds can be kept together in the same aviary.

I have discussed the probable origin of domestic pigeons at some, yet quite insufficient, length; because when I first kept pigeons and watched the several kinds, knowing well how true they bred, I felt fully as much difficulty in believing that they could ever have descended from a common parent, as any naturalist could in coming to a similar conclusion in regard to the many species of finches, or other large groups of birds, in nature. One circumstance has struck me much; namely, that all the breeders of the various domestic animals and the cultivators of plants, with whom I have ever conversed, or whose treatises I have read, are firmly convinced that the several breeds to which each has attended, are descended from so many aboriginally distinct species. Ask, as I have asked, a celebrated raiser of Hereford cattle, whether his cattle might not have descended from long-horns, and he will laugh you to scorn. I have never met a pigeon, or poultry, or duck, or rabbit fancier, who was not fully convinced that each main breed was descended from a distinct species. Van Mons, in his treatise on pears and apples, shows how utterly he disbelieves that the several sorts, for instance a Ribston-pippin or Codlin-apple, could ever have proceeded from the seeds of the same tree. Innumerable other examples could be given. The explanation, I think, is simple: from long-continued study they are strongly impressed with the differences between the several races; and though they well know that each race varies slightly, for they win their prizes by selecting such slight differences, yet they ignore all general arguments, and refuse to sum up in their minds slight differences accumulated during many successive generations. May not those naturalists who, knowing far less of the laws of inheritance than does the breeder, and knowing no more than he does of the intermediate links in the long lines of descent, yet admit that many of our domestic races have descended from the same parents—may they not learn a lesson of caution, when they deride the idea of species in a state of nature being lineal descendants of other species?

Selection.—Let us now briefly consider the steps by which domestic races have been produced, either from one or from several allied species. Some little effect may, perhaps, be attributed to the direct action of the external conditions of life, and some little to habit; but he would be a bold man who would account by such agencies for the differences of a dray and race horse, a greyhound and bloodhound, a carrier and tumbler pigeon. One of the most remarkable features in our domesticated races is that we see in them adaptation, not indeed to the animal's or plant's own good, but to man's use or fancy. Some variations useful to him have probably arisen suddenly, or by one step; many botanists, for instance, believe that the fuller's teazle, with its hooks, which cannot be rivalled by any mechanical contrivance, is only a variety of the wild Dipsacus; and this amount of change may have suddenly arisen in a seedling. So it has probably been with the turnspit dog; and this is known to have been the case with the ancon sheep. But when we compare the dray-horse and race-horse, the dromedary and camel, the various breeds of sheep fitted either for cultivated land or mountain pasture, with the wool of one breed good for one purpose, and that of another breed for another purpose; when we compare the many breeds of dogs, each good for man in very different ways; when we compare the game-cock, so pertinacious in battle, with other breeds so little quarrelsome, with "everlasting layers" which never desire to sit, and with the bantam so small and elegant; when we compare the host of agricultural, culinary, orchard, and flower-garden races of plants, most useful to man at different seasons and for different purposes, or so beautiful in his eyes, we must, I think, look further than to mere variability. We cannot suppose that all the breeds were suddenly produced as perfect and as useful as we now see them; indeed, in several cases, we know that this has not been their history. The key is man's power of accumulative selection: nature gives successive variations; man adds them up in certain directions useful to him. In this sense he may be said to make for himself useful breeds.

The great power of this principle of selection is not hypothetical. It is certain that several of our eminent breeders have, even within a single lifetime, modified to a large extent some breeds of cattle and sheep. In order fully to realise what they have done, it is almost necessary to read several of the many treatises devoted to this subject, and to inspect the animals. Breeders habitually speak of an animal's organisation as something quite plastic, which they can model almost as they please. If I had space I could quote numerous passages to this effect from highly competent authorities. Youatt, who was probably better acquainted with the works of agriculturists than almost any other individual, and who was himself a very good judge of an animal, speaks of the principle of selection as "that which enables the agriculturist, not only to modify the character of his flock, but to change it altogether. It is the magician's wand, by means of which he may summon into life whatever form and mould he pleases." Lord Somerville, speaking of what breeders have done for sheep, says:—"It would seem as if they had chalked out upon a wall a form perfect in itself, and then had given it existence." That most skilful breeder, Sir John Sebright, used to say, with respect to pigeons, that "he would produce any given feather in three years, but it would take him six years to obtain head and beak." In Saxony the importance of the principle of selection in regard to merino sheep is so fully recognised, that men follow it as a trade: the sheep are placed on a table and are studied, like a picture by a connoisseur; this is done three times at intervals of months, and the sheep are each time marked and classed, so that the very best may ultimately be selected for breeding.

What English breeders have actually effected is proved by the enormous prices given for animals with a good pedigree; and these have now been exported to almost every quarter of the world. The improvement is by no means generally due to crossing different breeds; all the best breeders are strongly opposed to this practice, except sometimes amongst closely allied sub-breeds. And when a cross has been made, the closest selection is far more indispensable even than in ordinary cases. If selection consisted merely in separating some very distinct variety, and breeding from it, the principle would be so obvious as hardly to be worth notice; but its importance consists in the great effect produced by the accumulation in one direction, during successive generations, of differences absolutely inappreciable by an uneducated eye—differences which I for one have vainly attempted to appreciate. Not one man in a thousand has accuracy of eye and judgment sufficient to become an eminent breeder. If gifted with these qualities, and he studies his subject for years, and devotes his lifetime to it with indomitable perseverance, he will succeed, and may make great improvements; if he wants any of these qualities, he will assuredly fail. Few would readily believe in the natural capacity and years of practice requisite to become even a skilful pigeon-fancier.

The same principles are followed by horticulturists; but the variations are here often more abrupt. No one supposes that our choicest productions have been produced by a single variation from the aboriginal stock. We have proofs that this is not so in some cases, in which exact records have been kept; thus, to give a very trifling instance, the steadily-increasing size of the common gooseberry may be quoted. We see an astonishing improvement in many florists' flowers, when the flowers of the present day are compared with drawings made only twenty or thirty years ago. When a race of plants is once pretty well established, the seed-raisers do not pick out the best plants, but merely go over their seed-beds, and pull up the "rogues," as they call the plants that deviate from the proper standard. With animals this kind of selection is, in fact, also followed; for hardly any one is so careless as to allow his worst animals to breed.

In regard to plants, there is another means of observing the accumulated effects of selection—namely, by comparing the diversity of flowers in the different varieties of the same species in the flower-garden; the diversity of leaves, pods, or tubers, or whatever part is valued, in the kitchen-garden, in comparison with the flowers of the same varieties; and the diversity of fruit of the same species in the orchard, in comparison with the leaves and flowers of the same set of varieties. See how different the leaves of the cabbage are, and how extremely alike the flowers; how unlike the flowers of the heartsease are, and how alike the leaves; how much the fruit of the different kinds of gooseberries differ in size, colour, shape, and hairiness, and yet the flowers present very slight differences. It is not that the varieties which differ largely in some one point do not differ at all in other points; this is hardly ever, perhaps never, the case. The laws of correlation of growth, the importance of which should never be overlooked, will ensure some differences; but, as a general rule, I cannot doubt that the continued selection of slight variations, either in the leaves, the flowers, or the fruit, will produce races differing from each other chiefly in these characters.

It may be objected that the principle of selection has been reduced to methodical practice for scarcely more than three-quarters of a century; it has certainly been more attended to of late years, and many treatises have been published on the subject; and the result, I may add, has been, in a corresponding degree, rapid and important. But it is very far from true that the principle is a modern discovery. I could give several references to the full acknowledgment of the importance of the principle in works of high antiquity. In rude and barbarous periods of English history choice animals were often imported, and laws were passed to prevent their exportation: the destruction of horses under a certain size was ordered, and this may be compared to the "roguing" of plants by nurserymen. The principle of selection I find distinctly given in an ancient Chinese encyclopædia. Explicit rules are laid down by some of the Roman classical writers. From passages in Genesis, it is clear that the colour of domestic animals was at that early period attended to. Savages now sometimes cross their dogs with wild canine animals, to improve the breed, and they formerly did so, as is attested by passages in Pliny. The savages in South Africa match their draught cattle by colour, as do some of the Esquimaux their teams of dogs. Livingstone shows how much good domestic breeds are valued by the negroes of the interior of Africa who have not associated with Europeans. Some of these facts do not show actual selection, but they show that the breeding of domestic animals was carefully attended to in ancient times, and is now attended to by the lowest savages. It would, indeed, have been a strange fact, had attention not been paid to breeding, for the inheritance of good and bad qualities is so obvious.

At the present time, eminent breeders try by methodical selection, with a distinct object in view, to make a new strain or sub-breed, superior to anything existing in the country. But, for our purpose, a kind of Selection, which may be called Unconscious, and which results from every one trying to possess and breed from the best individual animals, is more important. Thus, a man who intends keeping pointers naturally tries to get as good dogs as he can, and afterwards breeds from his own best dogs, but he has no wish or expectation of permanently altering the breed. Nevertheless I cannot doubt that this process, continued during centuries, would improve and modify any breed, in the same way as Bakewell, Collins, &c., by this very same process, only carried on more methodically, did greatly modify, even during their own lifetimes, the forms and qualities of their cattle. Slow and insensible changes of this kind could never be recognised unless actual measurements or careful drawings of the breeds in question had been made long ago, which might serve for comparison. In some cases, however, unchanged or but little changed individuals of the same breed may be found in less civilised districts, where the breed has been less improved. There is reason to believe that King Charles's spaniel has been unconsciously modified to a large extent since the time of that monarch. Some highly competent authorities are convinced that the setter is directly derived from the spaniel, and has probably been slowly altered from it. It is known that the English pointer has been greatly changed within the last century, and in this case the change has, it is believed, been chiefly effected by crosses with the fox-hound; but what concerns us is, that the change has been effected unconsciously and gradually, and yet so effectually, that, though the old Spanish pointer certainly came from Spain, Mr. Borrow has not seen, as I am informed by him, any native dog in Spain like our pointer.

By a similar process of selection, and by careful training, the whole body of English racehorses have come to surpass in fleetness and size the parent Arab stock, so that the latter, by the regulations for the Goodwood Races, are favoured in the weights they carry. Lord Spencer and others have shown how the cattle of England have increased in weight and in early maturity, compared with the stock formerly kept in this country. By comparing the accounts given in old pigeon treatises of carriers and tumblers with these breeds as now existing in Britain, India, and Persia, we can, I think, clearly trace the stages through which they have insensibly passed, and come to differ so greatly from the rock-pigeon.

Youatt gives an excellent illustration of the effects of a course of selection, which may be considered as unconsciously followed, in so far that the breeders could never have expected or even have wished to have produced the result which ensued—namely, the production of two distinct strains. The two flocks of Leicester sheep kept by Mr.

Buckley and Mr. Burgess, as Mr. Youatt remarks, "have been purely bred from the original stock of Mr. Bakewell for upwards of fifty years. There is not a suspicion existing in the mind of any one at all acquainted with the subject that the owner of either of them has deviated in any one instance from the pure blood of Mr. Bakewell's flock, and yet the difference between the sheep possessed by these two gentlemen is so great that they have the appearance of being quite different varieties."

If there exist savages so barbarous as never to think of the inherited character of the offspring of their domestic animals, yet any one animal particularly useful to them, for any special purpose, would be carefully preserved during famines and other accidents, to which savages are so liable, and such choice animals would thus generally leave more offspring than the inferior ones; so that in this case there would be a kind of unconscious selection going on. We see the value set on animals even by the barbarians of Tierra del Fuego, by their killing and devouring their old women, in times of dearth, as of less value than their dogs.

In plants the same gradual process of improvement, through the occasional preservation of the best individuals, whether or not sufficiently distinct to be ranked at their first appearance as distinct varieties, and whether or not two or more species or races have become blended together by crossing, may plainly be recognised in the increased size and beauty which we now see in the varieties of the heartsease, rose, pelargonium, dahlia, and other plants, when compared with the older varieties or with their parent-stocks. No one would ever expect to get a first-rate heartsease or dahlia from the seed of a wild plant. No one would expect to raise a first-rate melting pear from the seed of the wild pear, though he might succeed from a poor seedling growing wild, if it had come from a garden-stock. The pear, though cultivated in classical times, appears, from Pliny's description, to have been a fruit of very inferior quality. I have seen great surprise expressed in horticultural works at the wonderful skill of gardeners, in having produced such splendid results from such poor materials; but the art, I cannot doubt, has been simple, and, as far as the final result is concerned, has been followed almost unconsciously. It has consisted in always cultivating the best known variety, sowing its seeds, and, when a slightly better variety has chanced to appear, selecting it, and so onwards. But the gardeners of the classical period, who cultivated the best pear they could procure, never thought what splendid fruit we should eat; though we owe our excellent fruit, in some small degree, to their having naturally chosen and preserved the best varieties they could anywhere find.

A large amount of change in our cultivated plants, thus slowly and unconsciously accumulated, explains, as I believe, the well-known fact, that in a vast number of cases we cannot recognise, and therefore do not know, the wild parent-stocks of the plants which have been longest cultivated in our flower and kitchen gardens. If it has taken centuries or thousands of years to improve or modify most of our plants up to their present standard of usefulness to man, we can understand how it is that neither Australia, the Cape of Good Hope, nor any other region inhabited by quite uncivilised man, has afforded us a single plant worth culture. It is not that these countries, so rich in species, do not by a strange chance possess the aboriginal stocks of any useful plants, but that the native plants have not been improved by continued selection up to a standard of perfection comparable with that given to the plants in countries anciently civilised.

In regard to the domestic animals kept by uncivilised man, it should not be overlooked that they almost always have to struggle for their own food, at least during certain seasons. And in two countries very differently circumstanced, individuals of the same species, having slightly different constitutions or structure, would often succeed better in the one country than in the other, and thus by a process of "natural selection," as will hereafter be more fully explained, two sub-breeds might be formed. This, perhaps, partly explains what has been remarked by some authors, namely, that the varieties kept by savages have more of the character of species than the varieties kept in civilised countries.

On the view here given of the all-important part which selection by man has played, it becomes at once obvious, how it is that our domestic races show adaptation in their structure or in their habits to man's wants or fancies. We can, I think, further understand the frequently abnormal character of our domestic races, and likewise their differences being so great in external characters and relatively so slight in internal parts or organs. Man can hardly select, or only with much difficulty, any deviation of structure excepting such as is externally visible; and indeed he rarely cares for what is internal. He can never act by selection, excepting on variations which are first given to him in some slight degree by nature. No man would ever try to make a fantail, till he saw a pigeon with a tail developed in some slight degree in an unusual manner, or a pouter till he saw a pigeon with a crop of somewhat unusual size; and the more abnormal or unusual any character was when it first appeared, the more likely it would be to catch his attention. But to use such an expression as trying to make a fantail, is, I have no doubt, in most cases, utterly incorrect. The man who first selected a pigeon with a slightly larger tail, never dreamed what the descendants of that pigeon would become through long-continued, partly unconscious and partly methodical selection. Perhaps the parent bird of all fantails had only fourteen tail-feathers somewhat expanded, like the present Java fantail, or like individuals of other and distinct breeds, in which as many as seventeen tail-feathers have been counted. Perhaps the first pouter-pigeon did not inflate its crop much more than the turbit now does the upper part of its œsophagus,—a habit which is disregarded by all fanciers, as it is not one of the points of the breed.

Nor let it be thought that some great deviation of structure would be necessary to catch the fancier's eye: he perceives extremely small differences, and it is in human nature to value any novelty, however slight, in one's own possession. Nor must the value which would formerly be set on any slight differences in the individuals of the same species, be judged of by the value which would now be set on them, after several breeds have once fairly been established. Many slight differences might, and indeed do now, arise amongst pigeons, which are rejected as faults or deviations from the standard of perfection of each breed. The common goose has not given rise to any marked varieties; hence the Thoulouse and the common breed, which differ only in colour, that most fleeting of characters, have lately been exhibited as distinct at our poultry-shows.

I think these views further explain what has sometimes been noticed—namely, that we know nothing about the origin or history of any of our domestic breeds. But, in fact, a breed, like a dialect of a language, can hardly be said to have had a definite origin. A man preserves and breeds from an individual with some slight deviation of structure, or takes more care than usual in matching his best animals and thus improves them, and the improved individuals slowly spread in the immediate neighbourhood. But as yet they will hardly have a distinct name, and from being only slightly valued, their history will be disregarded. When further improved by the same slow and gradual process, they will spread more widely, and will get recognised as something distinct and valuable, and will then probably first receive a provincial name. In semi-civilised countries, with little free communication, the spreading and knowledge of any new sub-breed will be a slow process. As soon as the points of value of the new sub-breed are once fully acknowledged, the principle, as I have called it, of unconscious selection will always tend,—perhaps more at one period than at another, as the breed rises or falls in fashion,—perhaps more in one district than in another, according to the state of civilisation of the inhabitants,—slowly to add to the characteristic features of the breed, whatever they may be. But the chance will be infinitely small of any record having been preserved of such slow, varying, and insensible changes.

I must now say a few words on the circumstances, favourable, or the reverse, to man's power of selection. A high degree of variability is obviously favourable, as freely giving the materials for selection to work on; not that mere individual differences are not amply sufficient, with extreme care, to allow of the accumulation of a large amount of modification in almost any desired direction. But as variations manifestly useful or pleasing to man appear only occasionally, the chance of their appearance will be much increased by a large number of individuals being kept; and hence this comes to be of the highest importance to success. On this principle Marshall has remarked, with respect to the sheep of parts of Yorkshire, that "as they generally belong to poor people, and are mostly *in small lots*, they never can be improved." On the other hand, nurserymen, from raising large stocks of the same plants, are generally far more successful than amateurs in getting new and valuable varieties. The keeping of a large number of individuals of a species in any country requires that the species should be placed under favourable conditions of life, so as to breed freely in that country. When the individuals of any species are scanty, all the individuals, whatever their quality may be, will generally be allowed to breed, and this will effectually prevent selection. But probably the most important point of all, is, that the animal or plant should be so highly useful to man, or so much valued by him, that the closest attention should be paid to even the slightest deviation in the qualities or structure of each individual. Unless such attention be paid nothing can be effected. I have seen it gravely remarked, that it was most fortunate that the strawberry began to vary just when gardeners began to attend closely to this plant. No doubt the strawberry had always varied since it was cultivated, but the slight varieties had been neglected. As soon, however, as gardeners picked out individual plants with slightly larger, earlier, or better fruit, and raised seedlings from them, and again picked out the best seedlings and bred from them, then, there appeared (aided by some crossing with distinct species) those many admirable varieties of the strawberry which have been raised during the last thirty or forty years.

In the case of animals with separate sexes, facility in preventing crosses is an important element of success in the formation of new races,—at least, in a country which is already stocked with other races. In this respect enclosure of the land plays a part. Wandering savages or the inhabitants of open plains rarely possess more than one breed of the same species. Pigeons can be mated for life, and this is a great convenience to the fancier, for thus many races may be kept true, though mingled in the same aviary; and this circumstance must have largely favoured the improvement and formation of new breeds. Pigeons, I may add, can be propagated in great numbers and at a very quick rate, and inferior birds may be freely rejected, as when killed they serve for food. On the other hand, cats, from their nocturnal rambling habits, cannot be matched, and, although so much valued by women and children, we hardly ever see a distinct breed kept up; such breeds as we do sometimes see are almost always imported from some other country, often from islands. Although I do not doubt that some domestic animals vary less than others, yet the rarity or absence of distinct breeds of the cat, the donkey, peacock, goose, &c., may be attributed in main part to selection not having been brought into play: in cats, from the difficulty in pairing them; in donkeys, from only a few being kept by poor people, and little attention paid to their breeding; in peacocks, from not being very easily reared and a large stock not kept; in geese, from being valuable only for two purposes, food and feathers, and more especially from no pleasure having been felt in the display of distinct breeds.

To sum up on the origin of our Domestic Races of animals and plants. I believe that the conditions of life, from their action on the reproductive system, are so far of the highest importance as causing variability. I do not believe that variability is an inherent and necessary contingency, under all circumstances, with all organic beings, as some authors have thought. The effects of variability are modified by various degrees of inheritance and of reversion. Variability is governed by many unknown laws, more especially by that of correlation of growth. Something may be attributed to the direct action of the conditions of life. Something must be attributed to use and disuse. The final result is thus rendered infinitely complex. In some cases, I do not doubt that the intercrossing of species, aboriginally distinct, has played an important part in the origin of our domestic productions. When in any country several domestic breeds have once been established, their occasional intercrossing, with the aid of selection, has, no doubt, largely aided in the formation of new sub-breeds; but the importance of the crossing of varieties has, I believe, been greatly exaggerated, both in regard to animals and to those plants which are propagated by seed. In plants which are temporarily propagated by cuttings, buds, &c., the importance of the crossing both of distinct species and of varieties is immense; for the cultivator here quite disregards the extreme variability both of hybrids and mongrels, and the frequent sterility of hybrids; but the cases of plants not propagated by seed are of little importance to us, for their endurance is only temporary. Over all these causes of Change I am convinced that the accumulative action of Selection, whether applied methodically and more quickly, or unconsciously and more slowly, but more efficiently, is by far the predominant Power.

VARIATION UNDER NATURE.

Before applying the principles arrived at in the last chapter to organic beings in a state of nature, we must briefly discuss whether these latter are subject to any variation. To treat this subject at all properly, a long catalogue of dry facts should be given; but these I shall reserve for my future work. Nor shall I here discuss the various definitions which have been given of the term species. No one definition has as yet satisfied all naturalists; yet every naturalist knows vaguely what he means when he speaks of a species. Generally the term includes the unknown element of a distinct act of creation. The term "variety" is almost equally difficult to define; but here community of descent is almost universally implied, though it can rarely be proved. We have also what are called monstrosities; but they graduate into varieties. By a monstrosity I presume is meant some considerable deviation of structure in one part, either injurious to or not useful to the species, and not generally propagated. Some authors use the term "variation" in a technical sense, as implying a modification directly due to the physical conditions of life; and "variations" in this sense are supposed not to be inherited: but who can say that the dwarfed condition of shells in the brackish waters of the Baltic, or dwarfed plants on Alpine summits, or the thicker fur of an animal from far northwards, would not in some cases be inherited for at least some few generations? and in this case I presume that the form would be called a variety.

Again, we have many slight differences which may be called individual differences, such as are known frequently to appear in the offspring from the same parents, or which may be presumed to have thus arisen, from being frequently observed in the individuals of the same species inhabiting the same confined locality. No one supposes that all the individuals of the same species are cast in the very same mould. These individual differences are highly important for us, as they afford materials for natural selection to accumulate, in the same manner as man can accumulate in any given direction individual differences in his domesticated productions. These individual differences generally affect what naturalists consider unimportant parts; but I could show by a long catalogue of facts, that parts which must be called important, whether viewed under a physiological or classificatory point of view, sometimes vary in the individuals of the same species. I am convinced that the most experienced naturalist would be surprised at the number of the cases of variability, even in important parts of structure, which he could collect on good authority, as I have collected, during a course of years. It should be remembered that systematists are far from pleased at finding variability in important characters, and that there are not many men who will laboriously examine internal and important organs, and compare them in many specimens of the same species. I should never have expected that the branching of the main nerves close to the great central ganglion of an insect would have been variable in the same species; I should have expected that changes of this nature could have been effected only by slow degrees: yet quite recently Mr. Lubbock has shown a degree of variability in these main nerves in Coccus, which may almost be compared to the irregular branching of the stem of a tree. This philosophical naturalist, I may add, has also quite recently shown that the muscles in the larvæ of certain insects are very far from uniform. Authors sometimes argue in a circle when they state that important organs never vary; for these same authors practically rank that character as important (as some few naturalists have honestly confessed) which does not vary; and, under this point of view, no instance of an important part varying will ever be found: but under any other point of view many instances assuredly can be given.

There is one point connected with individual differences, which seems to me extremely perplexing: I refer to those genera which have sometimes been called "protean" or "polymorphic," in which the species present an inordinate amount of variation; and hardly two naturalists can agree which forms to rank as species and which as varieties. We may instance Rubus, Rosa, and Hieracium amongst plants, several genera of insects, and several genera of Brachiopod shells. In most polymorphic genera some of the species have fixed and definite characters. Genera which are polymorphic in one country seem to be, with some few exceptions, polymorphic in other countries, and likewise, judging from Brachiopod shells, at former periods of time. These facts seem to be very perplexing, for they seem to show that this kind of variability is independent of the conditions of life. I am inclined to suspect that we see in these polymorphic genera variations in points of structure which are of no service or disservice to the species, and which consequently have not been seized on and rendered definite by natural selection, as hereafter will be explained.

Those forms which possess in some considerable degree the character of species, but which are so closely similar to some other forms, or are so closely linked to them by intermediate gradations, that naturalists do not like to rank them as distinct species, are in several respects the most important for us. We have every reason to believe that many of these doubtful and closely-allied forms have permanently retained their characters in their own country for a long time; for as long, as far as we know, as have good and true species. Practically, when a naturalist can unite two forms together by others having intermediate characters, he treats the one as a variety of the other, ranking the most common, but sometimes the one first described, as the species, and the other as the variety. But cases of great difficulty, which I will not here enumerate, sometimes occur in deciding whether or not to rank one form as a variety of another, even when they are closely connected by intermediate links; nor will the commonly-assumed hybrid nature of the intermediate links always remove the difficulty. In very many cases, however, one form is ranked as a variety of another, not because the intermediate links have actually been found, but because analogy leads the observer to suppose either that they do now somewhere exist, or may formerly have existed; and here a wide door for the entry of doubt and conjecture is opened.

Hence, in determining whether a form should be ranked as a species or a variety, the opinion of naturalists having sound judgment and wide experience seems the only guide to follow. We must, however, in many cases, decide by a majority of naturalists, for few well-marked and well-known varieties can be named which have not been ranked as species by at least some competent judges.

That varieties of this doubtful nature are far from uncommon cannot be disputed. Compare the several floras of Great Britain, of France or of the United States, drawn up by different botanists, and see what a surprising number of forms have been ranked by one botanist as good species, and by another as mere varieties. Mr. H. C.

Watson, to whom I lie under deep obligation for assistance of all kinds, has marked for me 182 British plants, which are generally considered as varieties, but which have all been ranked by botanists as species; and in making this list he has omitted many trifling varieties, but which nevertheless have been ranked by some botanists as species, and he has entirely omitted several highly polymorphic genera. Under genera, including the most polymorphic forms, Mr. Babington gives 251 species, whereas Mr. Bentham gives only 112,—a difference of 139 doubtful forms! Amongst animals which unite for each birth, and which are highly locomotive, doubtful forms, ranked by one zoologist as a species and by another as a variety, can rarely be found within the same country, but are common in separated areas. How many of those birds and insects in North America and Europe, which differ very slightly from each other, have been ranked by one eminent naturalist as undoubted species, and by another as varieties, or, as they are often called, as geographical races! Many years ago, when comparing, and seeing others compare, the birds from the separate islands of the Galapagos Archipelago, both one with another, and with those from the American mainland, I was much struck how entirely vague and arbitrary is the distinction between species and varieties. On the islets of the little Madeira group there are many insects which are characterized as varieties in Mr. Wollaston's admirable work, but which it cannot be doubted would be ranked as distinct species by many entomologists. Even Ireland has a few animals, now generally regarded as varieties, but which have been ranked as species by some zoologists. Several most experienced ornithologists consider our British red grouse as only a strongly-marked race of a Norwegian species, whereas the greater number rank it as an undoubted species peculiar to Great Britain. A wide distance between the homes of two doubtful forms leads many naturalists to rank both as distinct species; but what distance, it has been well asked, will suffice? if that between America and Europe is ample, will that between the Continent and the Azores, or Madeira, or the Canaries, or Ireland, be sufficient? It must be admitted that many forms, considered by highly-competent judges as varieties, have so perfectly the character of species that they are ranked by other highly competent judges as good and true species. But to discuss whether they are rightly called species or varieties, before any definition of these terms has been generally accepted, is vainly to beat the air.

Many of the cases of strongly-marked varieties or doubtful species well deserve consideration; for several interesting lines of argument, from geographical distribution, analogical variation, hybridism, &c., have been brought to bear on the attempt to determine their rank. I will here give only a single instance,—the well-known one of the primrose and cowslip, or Primula veris and elatior. These plants differ considerably in appearance; they have a different flavour and emit a different odour; they flower at slightly different periods; they grow in somewhat different stations; they ascend mountains to different heights; they have different geographical ranges; and lastly, according to very numerous experiments made during several years by that most careful observer Gärtner, they can be crossed only with much difficulty. We could hardly wish for better evidence of the two forms being specifically distinct. On the other hand, they are united by many intermediate links, and it is very doubtful whether these links are hybrids; and there is, as it seems to me, an overwhelming amount of experimental evidence, showing that they descend from common parents, and consequently must be ranked as varieties.

Close investigation, in most cases, will bring naturalists to an agreement how to rank doubtful forms. Yet it must be confessed, that it is in the best-known countries that we find the greatest number of forms of doubtful value. I have been struck with the fact, that if any animal or plant in a state of nature be highly useful to man, or from any cause closely attract his attention, varieties of it will almost universally be found recorded. These varieties, moreover, will be often ranked by some authors as species. Look at the common oak, how closely it has been studied; yet a German author makes more than a dozen species out of forms, which are very generally considered as varieties; and in this country the highest botanical authorities and practical men can be quoted to show that the sessile and pedunculated oaks are either good and distinct species or mere varieties.

When a young naturalist commences the study of a group of organisms quite unknown to him, he is at first much perplexed to determine what differences to consider as specific, and what as varieties; for he knows nothing of the amount and kind of variation to which the group is subject; and this shows, at least, how very generally there is some variation. But if he confine his attention to one class within one country, he will soon make up his mind how to rank most of the doubtful forms. His general tendency will be to make many species, for he will become impressed, just like the pigeon or poultry-fancier before alluded to, with the amount of difference in the forms which he is continually studying; and he has little general knowledge of analogical variation in other groups and in other countries, by which to correct his first impressions. As he extends the range of his observations, he will meet with more cases of difficulty; for he will encounter a greater number of closely-allied forms. But if his observations be widely extended, he will in the end generally be enabled to make up his own mind which to call varieties and which species; but he will succeed in this at the expense of admitting much variation,—and the truth of this admission will often be disputed by other naturalists. When, moreover, he comes to study allied forms brought from countries not now continuous, in which case he can hardly hope to find the intermediate links between his doubtful forms, he will have to trust almost entirely to analogy, and his difficulties will rise to a climax.

Certainly no clear line of demarcation has as yet been drawn between species and sub-species,—that is, the forms which in the opinion of some naturalists come very near to, but do not quite arrive at the rank of species; or, again, between sub-species and well-marked varieties, or between lesser varieties and individual differences. These differences blend into each other in an insensible series; and a series impresses the mind with the idea of an actual passage.

Hence I look at individual differences, though of small interest to the systematist, as of high importance for us, as being the first step towards such slight varieties as are barely thought worth recording in works on natural history. And I look at varieties which are in any degree more distinct and permanent, as steps leading to more strongly marked and more permanent varieties; and at these latter, as leading to sub-species, and to species. The passage from one stage of difference to another and higher stage may be, in some cases, due merely to the long-continued action of different physical conditions in two different regions; but I have not much faith in this view; and I attribute the passage of a variety, from a state in which it differs very slightly from its parent to one in which it differs more, to the action of natural selection in accumulating (as will hereafter be more fully explained) differences of structure in certain definite directions. Hence I believe a well-marked variety may be justly called an incipient species; but whether this belief be justifiable must be judged of by the general weight of the several facts and views given throughout this work.

It need not be supposed that all varieties or incipient species necessarily attain the rank of species. They may whilst in this incipient state become extinct, or they may endure as varieties for very long periods, as has been shown to be the case by Mr. Wollaston with the varieties of certain fossil land-shells in Madeira. If a variety were to flourish so as to exceed in numbers the parent species, it would then rank as the species, and the species as the variety; or it might come to supplant and exterminate the parent species; or both might co-exist, and both rank as independent species. But we shall hereafter have to return to this subject.

From these remarks it will be seen that I look at the term species, as one arbitrarily given for the sake of convenience to a set of individuals closely resembling each other, and that it does not essentially differ from the term variety, which is given to less distinct and more fluctuating forms. The term variety, again, in comparison with mere individual differences, is also applied arbitrarily, and for mere convenience sake.

Guided by theoretical considerations, I thought that some interesting results might be obtained in regard to the nature and relations of the species which vary most, by tabulating all the varieties in several well-worked floras. At first this seemed a simple task; but Mr. H. C. Watson, to whom I am much indebted for valuable advice and assistance on this subject, soon convinced me that there were many difficulties, as did subsequently Dr. Hooker, even in stronger terms. I shall reserve for my future work the discussion of these difficulties, and the tables themselves of the proportional numbers of the varying species. Dr. Hooker permits me to add, that after having carefully read my manuscript, and examined the tables, he thinks that the following statements are fairly well established. The whole subject, however, treated as it necessarily here is with much brevity, is rather perplexing, and allusions cannot be avoided to the "struggle for existence," "divergence of character," and other questions, hereafter to be discussed.

Alph. De Candolle and others have shown that plants which have very wide ranges generally present varieties; and this might have been expected, as they become exposed to diverse physical conditions, and as they come into competition (which, as we shall hereafter see, is a far more important circumstance) with different sets of organic beings. But my tables further show that, in any limited country, the species which are most common, that is abound most in individuals, and the species which are most widely diffused within their own country (and this is a different consideration from wide range, and to a certain extent from commonness), often give rise to varieties sufficiently well-marked to have been recorded in botanical works. Hence it is the most flourishing, or, as they may be called, the dominant species,—those which range widely over the world, are the most diffused in their own country, and are the most numerous in individuals,—which oftenest produce well-marked varieties, or, as I consider them, incipient species. And this, perhaps, might have been anticipated; for, as varieties, in order to become in any degree permanent, necessarily have to struggle with the other inhabitants of the country, the species which are already dominant will be the most likely to yield offspring which, though in some slight degree modified, will still inherit those advantages that enabled their parents to become dominant over their compatriots.

If the plants inhabiting a country and described in any Flora be divided into two equal masses, all those in the larger genera being placed on one side, and all those in the smaller genera on the other side, a somewhat larger number of the very common and much diffused or dominant species will be found on the side of the larger genera. This, again, might have been anticipated; for the mere fact of many species of the same genus inhabiting any country, shows that there is something in the organic or inorganic conditions of that country favourable to the genus; and, consequently, we might have expected to have found in the larger genera, or those including many species, a large proportional number of dominant species. But so many causes tend to obscure this result, that I am surprised that my tables show even a small majority on the side of the larger genera. I will here allude to only two causes of obscurity. Fresh-water and salt-loving plants have generally very wide ranges and are much diffused, but this seems to be connected with the nature of the stations inhabited by them, and has little or no relation to the size of the genera to which the species belong. Again, plants low in the scale of organisation are generally much more widely diffused than plants higher in the scale; and here again there is no close relation to the size of the genera. The cause of lowly-organised plants ranging widely will be discussed in our chapter on geographical distribution.

From looking at species as only strongly-marked and well-defined varieties, I was led to anticipate that the species of the larger genera in each country would oftener present varieties, than the species of the smaller genera; for wherever many closely related species (*i.e.*, species of the same genus) have been formed, many varieties or incipient species ought, as a general rule, to be now forming. Where many large trees grow, we expect to find saplings. Where many species of a genus have been formed through variation, circumstances have been favourable for variation; and hence we might expect that the circumstances would generally be still favourable to variation. On the other hand, if we look at each species as a special act of creation, there is no apparent reason why more varieties should occur in a group having many species, than in one having few.

To test the truth of this anticipation I have arranged the plants of twelve countries, and the coleopterous insects of two districts, into two nearly equal masses, the species of the larger genera on one side, and those of the smaller genera on the other side, and it has invariably proved to be the case that a larger proportion of the species on the side of the larger genera present varieties, than on the side of the smaller genera. Moreover, the species of the large genera which present any varieties, invariably present a larger average number of varieties than do the species of the small genera. Both these results follow when another division is made, and when all the smallest genera, with from only one to four species, are absolutely excluded from the tables. These facts are of plain signification on the view that species are only strongly marked and permanent varieties; for wherever many species of the same genus have been formed, or where, if we may use the expression, the manufactory of species has been active, we ought generally to find the manufactory still in action, more especially as we have every reason to believe the process of manufacturing new species to be a slow one. And this certainly is the case, if varieties be looked at as incipient species; for my tables clearly show as a general rule that, wherever many species of a genus have been formed, the species of that genus present a number of varieties, that is of incipient species, beyond the average. It is not that all large genera are now varying much, and are thus increasing in the number of their species, or that no small genera are now varying and increasing; for if this had been so, it would have been fatal to my theory; inasmuch as geology plainly tells us that small genera have in the lapse of time often increased greatly in size; and that large genera have often come to their maxima, declined, and disappeared. All that we want to show is, that where many species of a genus have been formed, on an average many are still forming; and this holds good.

There are other relations between the species of large genera and their recorded varieties which deserve notice. We have seen that there is no infallible criterion by which to distinguish species and well-marked varieties; and in those cases in which intermediate links have not been found between doubtful forms, naturalists are compelled to come to a determination by the amount of difference between them, judging by analogy whether or not the amount suffices to raise one or both to the rank of species. Hence the amount of difference is one very important criterion in settling whether two forms should be ranked as species or varieties. Now Fries has remarked in regard to plants, and Westwood in regard to insects, that in large genera the amount of difference between the species is often exceedingly small. I have endeavoured to test this numerically by averages, and, as far as my imperfect results go, they always confirm the view. I have also consulted some sagacious and most experienced observers, and, after deliberation, they concur in this view. In this respect, therefore, the species of the larger genera resemble varieties, more than do the species of the smaller genera. Or the case may be put in another way, and it may be said, that in the larger genera, in which a number of varieties or incipient species greater than the average are now manufacturing, many of the species already manufactured still to a certain extent resemble varieties, for they differ from each other by a less than usual amount of difference.

Moreover, the species of the large genera are related to each other, in the same manner as the varieties of any one species are related to each other. No naturalist pretends that all the species of a genus are equally distinct from each other; they may generally be divided into sub-genera, or sections, or lesser groups. As Fries has well remarked, little groups of species are generally clustered like satellites around certain other species. And what are varieties but groups of forms, unequally related to each other, and clustered round certain forms—that is, round their parent-species? Undoubtedly there is one most important point of difference between varieties and species; namely, that the amount of difference between varieties, when compared with each other or with their parent-species, is much less than that between the species of the same genus. But when we come to discuss the principle, as I call it, of Divergence of Character, we shall see how this may be explained, and how the lesser differences between varieties will tend to increase into the greater differences between species.

There is one other point which seems to me worth notice. Varieties generally have much restricted ranges: this statement is indeed scarcely more than a truism, for if a variety were found to have a wider range than that of its supposed parent-species, their denominations ought to be reversed. But there is also reason to believe, that those species which are very closely allied to other species, and in so far resemble varieties, often have much restricted ranges. For instance, Mr. H. C. Watson has marked for me in the well-sifted London Catalogue of plants (4th edition) 63 plants which are therein ranked as species, but which he considers as so closely allied to other species as to be of doubtful value: these 63 reputed species range on an average over 6.9 of the provinces into which Mr. Watson has divided Great Britain. Now, in this same catalogue, 53 acknowledged varieties are recorded, and these range over 7.7 provinces; whereas, the species to which these varieties belong range over 14.3 provinces. So that the acknowledged varieties have very nearly the same restricted average range, as have those very closely allied forms, marked for me by Mr. Watson as doubtful species, but which are almost universally ranked by British botanists as good and true species.

Finally, then, varieties have the same general characters as species, for they cannot be distinguished from species,—except, firstly, by the discovery of intermediate linking forms, and the occurrence of such links cannot affect the actual characters of the forms which they connect; and except, secondly, by a certain amount of difference, for two forms, if differing very little, are generally ranked as varieties, notwithstanding that intermediate linking forms have not been discovered; but the amount of difference considered necessary to give to two forms the rank of species is quite indefinite. In genera having more than the average number of species in any country, the species of these genera have more than the average number of varieties. In large genera the species are apt to be closely, but unequally, allied together, forming little clusters round certain species. Species very closely allied to other species apparently have restricted ranges. In all these several respects the species of large genera present a strong analogy with varieties. And we can clearly understand these analogies, if species have once existed as varieties, and have thus originated: whereas, these analogies are utterly inexplicable if each species has been independently created.

We have, also, seen that it is the most flourishing and dominant species of the larger genera which on an average vary most; and varieties, as we shall hereafter see, tend to become converted into new and distinct species. The larger genera thus tend to become larger; and throughout nature the forms of life which are now dominant tend to become still more dominant by leaving many modified and dominant descendants. But by steps hereafter to be explained, the larger genera also tend to break up into smaller genera. And thus, the forms of life throughout the universe become divided into groups subordinate to groups.

STRUGGLE FOR EXISTENCE.

Before entering on the subject of this chapter, I must make a few preliminary remarks, to show how the struggle for existence bears on Natural Selection. It has been seen in the last chapter that amongst organic beings in a state of nature there is some individual variability; indeed I am not aware that this has ever been disputed. It is immaterial for us whether a multitude of doubtful forms be called species or sub-species or varieties; what rank, for instance, the two or three hundred doubtful forms of British plants are entitled to hold, if the existence of any well-marked varieties be admitted. But the mere existence of individual variability and of some few well-marked varieties, though necessary as the foundation for the work, helps us but little in understanding how species arise in nature. How have all those exquisite adaptations of one part of the organisation to another part, and to the conditions of life, and of one distinct organic being to another being, been perfected? We see these beautiful co-adaptations most plainly in the woodpecker and missletoe; and only a little less plainly in the humblest parasite which clings to the hairs of a quadruped or feathers of a bird; in the structure of the beetle which dives through the water; in the plumed seed which is wafted by the gentlest breeze; in short, we see beautiful adaptations everywhere and in every part of the organic world.

Again, it may be asked, how is it that varieties, which I have called incipient species, become ultimately converted into good and distinct species, which in most cases obviously differ from each other far more than do the varieties of the same species? How do those groups of species, which constitute what are called distinct genera, and which differ from each other more than do the species of the same genus, arise? All these results, as we shall more fully see in the next chapter, follow inevitably from the struggle for life. Owing to this struggle for life, any variation, however slight and from whatever cause proceeding, if it be in any degree profitable to an individual of any species, in its infinitely complex relations to other organic beings and to external nature, will tend to the preservation of that individual, and will generally be inherited by its offspring. The offspring, also, will thus have a better chance of surviving, for, of the many individuals of any species which are periodically born, but a small number can survive. I have called this principle, by which each slight variation, if useful, is preserved, by the term of Natural Selection, in order to mark its relation to man's power of selection. We have seen that man by selection can certainly produce great results, and can adapt organic beings to his own uses, through the accumulation of slight but useful variations, given to him by the hand of Nature. But Natural Selection, as we shall hereafter see, is a power incessantly ready for action, and is as immeasurably superior to man's feeble efforts, as the works of Nature are to those of Art.

We will now discuss in a little more detail the struggle for existence. In my future work this subject shall be treated, as it well deserves, at much greater length. The elder De Candolle and Lyell have largely and philosophically shown that all organic beings are exposed to severe competition. In regard to plants, no one has treated this subject with more spirit and ability than W. Herbert, Dean of Manchester, evidently the result of his great horticultural knowledge. Nothing is easier than to admit in words the truth of the universal struggle for life, or more difficult—at least I have found it so—than constantly to bear this conclusion in mind. Yet unless it be thoroughly engrained in the mind, I am convinced that the whole economy of nature, with every fact on distribution, rarity, abundance, extinction, and variation, will be dimly seen or quite misunderstood. We behold the face of nature bright with gladness, we often see superabundance of food; we do not see, or we forget, that the birds which are idly singing round us mostly live on insects or seeds, and are thus constantly destroying life; or we forget how largely these songsters, or their eggs, or their nestlings, are destroyed by birds and beasts of prey; we do not always bear in mind, that though food may be now superabundant, it is not so at all seasons of each recurring year.

I should premise that I use the term Struggle for Existence in a large and metaphorical sense, including dependence of one being on another, and including (which is more important) not only the life of the individual, but success in leaving progeny. Two canine animals in a time of dearth, may be truly said to struggle with each other which shall get food and live. But a plant on the edge of a desert is said to struggle for life against the drought, though more properly it should be said to be dependent on the moisture. A plant which annually produces a thousand seeds, of which on an average only one comes to maturity, may be more truly said to struggle with the plants of the same and other kinds which already clothe the ground. The missletoe is dependent on the apple and a few other trees, but can only in a far-fetched sense be said to struggle with these trees, for if too many of these parasites grow on the same tree, it will languish and die. But several seedling missletoes, growing close together on the same branch, may more truly be said to struggle with each other. As the missletoe is disseminated by birds, its existence depends on birds; and it may metaphorically be said to struggle with other fruit-bearing plants, in order to tempt birds to devour and thus disseminate its seeds rather than those of other plants. In these several senses, which pass into each other, I use for convenience sake the general term of struggle for existence.

A struggle for existence inevitably follows from the high rate at which all organic beings tend to increase. Every being, which during its natural lifetime produces several eggs or seeds, must suffer destruction during some period of its life, and during some season or occasional year, otherwise, on the principle of geometrical increase, its numbers would quickly become so inordinately great that no country could support the product. Hence, as more individuals are produced than can possibly survive, there must in every case be a struggle for existence, either one individual with another of the same species, or with the individuals of distinct species, or with the physical conditions of life. It is the doctrine of Malthus applied with manifold force to the whole animal and vegetable kingdoms; for in this case there can be no artificial increase of food, and no prudential restraint from marriage. Although some species may be now increasing, more or less rapidly, in numbers, all cannot do so, for the world would not hold them.

There is no exception to the rule that every organic being naturally increases at so high a rate, that if not destroyed, the earth would soon be covered by the progeny of a single pair. Even slow-breeding man has doubled in twenty-five years, and at this rate, in a few thousand years, there would literally not be standing room for his progeny. Linnæus has calculated that if an annual plant produced only two seeds—and there is no plant so unproductive as this—and their seedlings next year produced two, and so on, then in twenty years there would be a million plants. The elephant is reckoned to be the slowest breeder of all known animals, and I have taken some pains to estimate its probable minimum rate of natural increase: it will be under the mark to assume that it breeds when thirty years old, and goes on breeding till ninety years old, bringing forth three pair of young in this interval; if this be so, at the end of the fifth century there would be alive fifteen million elephants, descended from the first pair.

But we have better evidence on this subject than mere theoretical calculations, namely, the numerous recorded cases of the astonishingly rapid increase of various animals in a state of nature, when circumstances have been favourable to them during two or three following seasons. Still more striking is the evidence from our domestic animals of many kinds which have run wild in several parts of the world: if the statements of the rate of increase of slow-breeding cattle and horses in South-America, and latterly in Australia, had not been well authenticated, they would have been quite incredible. So it is with plants: cases could be given of introduced plants which have become common throughout whole islands in a period of less than ten years. Several of the plants now most numerous over the wide plains of La Plata, clothing square leagues of surface almost to the exclusion of all other plants, have been introduced from Europe; and there are plants which now range in India, as I hear from Dr. Falconer, from Cape Comorin to the Himalaya, which have been imported from America since its discovery. In such cases, and endless instances could be given, no one supposes that the fertility of these animals or plants has been suddenly and temporarily increased in any sensible degree. The obvious explanation is that the conditions of life have been very favourable, and that there has consequently been less destruction of the old and young, and that nearly all the young have been enabled to breed. In such cases the geometrical ratio of increase, the result of which never fails to be surprising, simply explains the extraordinarily rapid increase and wide diffusion of naturalised productions in their new homes.

In a state of nature almost every plant produces seed, and amongst animals there are very few which do not annually pair. Hence we may confidently assert, that all plants and animals are tending to increase at a geometrical ratio, that all would most rapidly stock every station in which they could any how exist, and that the geometrical tendency to increase must be checked by destruction at some period of life. Our familiarity with the larger domestic animals tends, I think, to mislead us: we see no great destruction falling on them, and we forget that thousands are annually slaughtered for food, and that in a state of nature an equal number would have somehow to be disposed of.

The only difference between organisms which annually produce eggs or seeds by the thousand, and those which produce extremely few, is, that the slow-breeders would require a few more years to people, under favourable conditions, a whole district, let it be ever so large. The condor lays a couple of eggs and the ostrich a score, and yet in the same country the condor may be the more numerous of the two: the Fulmar petrel lays but one egg, yet it is believed to be the most numerous bird in the world. One fly deposits hundreds of eggs, and another, like the hippobosca, a single one; but this difference does not determine how many individuals of the two species can be supported in a district. A large number of eggs is of some importance to those species, which depend on a rapidly fluctuating amount of food, for it allows them rapidly to increase in number. But the real importance of a large number of eggs or seeds is to make up for much destruction at some period of life; and this period in the great majority of cases is an early one. If an animal can in any way protect its own eggs or young, a small number may be produced, and yet the average stock be fully kept up; but if many eggs or young are destroyed, many must be produced, or the species will become extinct. It would suffice to keep up the full number of a tree, which lived on an average for a thousand years, if a single seed were produced once in a thousand years, supposing that this seed were never destroyed, and could be ensured to germinate in a fitting place. So that in all cases, the average number of any animal or plant depends only indirectly on the number of its eggs or seeds.

In looking at Nature, it is most necessary to keep the foregoing considerations always in mind—never to forget that every single organic being around us may be said to be striving to the utmost to increase in numbers; that each lives by a struggle at some period of its life; that heavy destruction inevitably falls either on the young or old, during each generation or at recurrent intervals. Lighten any check, mitigate the destruction ever so little, and the number of the species will almost instantaneously increase to any amount. The face of Nature may be compared to a yielding surface, with ten thousand sharp wedges packed close together and driven inwards by incessant blows, sometimes one wedge being struck, and then another with greater force.

What checks the natural tendency of each species to increase in number is most obscure. Look at the most vigorous species; by as much as it swarms in numbers, by so much will its tendency to increase be still further increased. We know not exactly what the checks are in even one single instance. Nor will this surprise any one who reflects how ignorant we are on this head, even in regard to mankind, so incomparably better known than any other animal. This subject has been ably treated by several authors, and I shall, in my future work, discuss some of the checks at considerable length, more especially in regard to the feral animals of South America. Here I will make only a few remarks, just to recall to the reader's mind some of the chief points. Eggs or very young animals seem generally to suffer most, but this is not invariably the case. With plants there is a vast destruction of seeds, but, from some observations which I have made, I believe that it is the seedlings which suffer most from germinating in ground already thickly stocked with other plants. Seedlings, also, are destroyed in vast numbers by various enemies; for instance, on a piece of ground three feet long and two wide, dug and cleared, and where there could be no choking from other plants, I marked all the seedlings of our native weeds as they came up, and out of the 357 no less than 295 were destroyed, chiefly by slugs and insects. If turf which has long been mown, and the case would be the same with turf closely browsed by quadrupeds, be let to grow, the more vigorous plants gradually kill the less vigorous, though fully grown, plants: thus out of twenty species growing on a little plot of turf (three feet by four) nine species perished from the other species being allowed to grow up freely.

The amount of food for each species of course gives the extreme limit to which each can increase; but very frequently it is not the obtaining food, but the serving as prey to other animals, which determines the average numbers of a species. Thus, there seems to be little doubt that the stock of partridges, grouse, and hares on any large estate depends chiefly on the destruction of vermin. If not one head of game were shot during the next twenty years in England, and, at the same time, if no vermin were destroyed, there would, in all probability, be less game than at present, although hundreds of thousands of game animals are now annually killed. On the other hand, in some cases, as with the elephant and rhinoceros, none are destroyed by beasts of prey: even the tiger in India most rarely dares to attack a young elephant protected by its dam.

Climate plays an important part in determining the average numbers of a species, and periodical seasons of extreme cold or drought, I believe to be the most effective of all checks. I estimated that the winter of 1854–55 destroyed four-fifths of the birds in my own grounds; and this is a tremendous destruction, when we remember that ten per cent. is an extraordinarily severe mortality from epidemics with man. The action of climate seems at first sight to be quite independent of the struggle for existence; but in so far as climate chiefly acts in reducing food, it brings on the most severe struggle between the individuals, whether of the same or of distinct species, which subsist on the same kind of food. Even when climate, for instance extreme cold, acts directly, it will be the least vigorous, or those which have got least food through the advancing winter, which will suffer most. When we travel from south to north, or from a damp region to a dry, we invariably see some species gradually getting rarer and rarer, and finally disappearing; and the change of climate being conspicuous, we are tempted to attribute the whole effect to its direct action. But this is a very false view: we forget that each species, even where it most abounds, is constantly suffering enormous destruction at some period of its life, from enemies or from competitors for the same place and food; and if these enemies or competitors be in the least degree favoured by any slight change of climate, they will increase in numbers, and, as each area is already fully stocked with inhabitants, the other species will decrease. When we travel southward and see a species decreasing in numbers, we may feel sure that the cause lies quite as much in other species being favoured, as in this one being hurt. So it is when we travel northward, but in a somewhat lesser degree, for the number of species of all kinds, and therefore of competitors, decreases northwards; hence in going northward, or in ascending a mountain, we far oftener meet with stunted forms, due to the *directly* injurious action of climate, than we do in proceeding southwards or in descending a mountain. When we reach the Arctic regions, or snow-capped summits, or absolute deserts, the struggle for life is almost exclusively with the elements.

That climate acts in main part indirectly by favouring other species, we may clearly see in the prodigious number of plants in our gardens which can perfectly well endure our climate, but which never become naturalised, for they cannot compete with our native plants, nor resist destruction by our native animals.

When a species, owing to highly favourable circumstances, increases inordinately in numbers in a small tract, epidemics—at least, this seems generally to occur with our game animals—often ensue: and here we have a limiting check independent of the struggle for life. But even some of these so-called epidemics appear to be due to parasitic worms, which have from some cause, possibly in part through facility of diffusion amongst the crowded animals, been disproportionably favoured: and here comes in a sort of struggle between the parasite and its prey.

On the other hand, in many cases, a large stock of individuals of the same species, relatively to the numbers of its enemies, is absolutely necessary for its preservation. Thus we can easily raise plenty of corn and rape-seed, &c., in our fields, because the seeds are in great excess compared with the number of birds which feed on them; nor can the birds, though having a superabundance of food at this one season, increase in number proportionally to the supply of seed, as their numbers are checked during winter: but any one who has tried, knows how troublesome it is to get seed from a few wheat or other such plants in a garden; I have in this case lost every single seed. This view of the necessity of a large stock of the same species for its preservation, explains, I believe, some singular facts in nature, such as that of very rare plants being sometimes extremely abundant in the few spots where they do occur; and that of some social plants being social, that is, abounding in individuals, even on the extreme confines of their range. For in such cases, we may believe, that a plant could exist only where the conditions of its life were so favourable that many could exist together, and thus save each other from utter destruction. I should add that the good effects of frequent intercrossing, and the ill effects of close interbreeding, probably come into play in some of these cases; but on this intricate subject I will not here enlarge.

Many cases are on record showing how complex and unexpected are the checks and relations between organic beings, which have to struggle together in the same country. I will give only a single instance, which, though a simple one, has interested me. In Staffordshire, on the estate of a relation where I had ample means of investigation, there was a large and extremely barren heath, which had never been touched by the hand of man; but several hundred acres of exactly the same nature had been enclosed twenty-five years previously and planted with Scotch fir. The change in the native vegetation of the planted part of the heath was most remarkable, more than is generally seen in passing from one quite different soil to another: not only the proportional numbers of the heath-plants were wholly changed, but twelve species of plants (not counting grasses and carices) flourished in the plantations, which could not be found on the health. The effect on the insects must have been still greater, for six insectivorous birds were very common in the plantations, which were not to be seen on the heath; and the heath was frequented by two or three distinct insectivorous birds. Here we see how potent has been the effect of the introduction of a single tree, nothing whatever else having been done, with the exception that the land had been enclosed, so that cattle could not enter. But how important an element enclosure is, I plainly saw near Farnham, in Surrey. Here there are extensive heaths, with a few clumps of old Scotch firs on the distant hilltops: within the last ten years large spaces have been enclosed, and self-sown firs are now springing up in multitudes, so close together that all cannot live. When I ascertained that these young trees had not been sown or planted, I was so much surprised at their numbers that I went to several points of view, whence I could examine hundreds of acres of the unenclosed heath, and literally I could not see a single Scotch fir, except the old planted clumps. But on looking closely between the stems of the heath, I found a multitude of seedlings and little trees, which had been perpetually browsed down by the cattle. In one square yard, at a point some hundred yards distant from one of the old clumps, I counted thirty-two little trees; and one of them, judging from the rings of growth, had during twenty-six years tried to raise its head above the stems of the heath, and had failed. No wonder that, as soon as the land was enclosed, it became thickly clothed with vigorously growing young firs. Yet the heath was so extremely barren and so extensive that no one would ever have imagined that cattle would have so closely and effectually searched it for food.

Here we see that cattle absolutely determine the existence of the Scotch fir; but in several parts of the world insects determine the existence of cattle. Perhaps Paraguay offers the most curious instance of this; for here neither cattle nor horses nor dogs have ever run wild, though they swarm southward and northward in a feral state; and Azara and Rengger have shown that this is caused by the greater number in Paraguay of a certain fly, which lays its eggs in the navels of these animals when first born. The increase of these flies, numerous as they are, must be habitually checked by some means, probably by birds. Hence, if certain insectivorous birds (whose numbers are probably regulated by hawks or beasts of prey) were to increase in Paraguay, the flies would decrease—then cattle and horses would become feral, and this would certainly greatly alter (as indeed I have observed in parts of South America) the vegetation: this again would largely affect the insects; and this, as we just have seen in Staffordshire, the insectivorous birds, and so onwards in ever-increasing circles of complexity. We began this series by insectivorous birds, and we have ended with them. Not that in nature the relations can ever be as simple as this. Battle within battle must ever be recurring with varying success; and yet in the long-run the forces are so nicely balanced, that the face of nature remains uniform for long periods of time, though assuredly the merest trifle would often give the victory to one organic being over another. Nevertheless so profound is our ignorance, and so high our presumption, that we marvel when we hear of the extinction of an organic being; and as we do not see the cause, we invoke cataclysms to desolate the world, or invent laws on the duration of the forms of life!

I am tempted to give one more instance showing how plants and animals, most remote in the scale of nature, are bound together by a web of complex relations. I shall hereafter have occasion to show that the exotic Lobelia fulgens, in this part of England, is never visited by insects, and consequently, from its peculiar structure, never can set a seed. Many of our orchidaceous plants absolutely require the visits of moths to remove their pollen-masses and thus to fertilise them. I have, also, reason to believe that humble-bees are indispensable to the fertilisation of the heartsease (Viola tricolor), for other bees do not visit this flower. From experiments which I have tried, I have found that the visits of bees, if not indispensable, are at least highly beneficial to the fertilisation of our clovers; but humble-bees alone visit the common red clover (Trifolium pratense), as other bees cannot reach the nectar. Hence I have very little doubt, that if the whole genus of humble-bees became extinct or very rare in England, the heartsease and red clover would become very rare, or wholly disappear. The number of humble-bees in any district depends in a great degree on the number of field-mice, which destroys their combs and nests; and Mr. H. Newman, who has long attended to the habits of humble-bees, believes that "more than two-thirds of them are thus destroyed all over England." Now the number of mice is largely dependent, as every one knows, on the number of cats; and Mr. Newman says, "Near villages and small towns I have found the nests of humble-bees more numerous than elsewhere, which I attribute to the number of cats that destroy the mice." Hence it is quite credible that the presence of a feline animal in large numbers in a district might determine, through the intervention first of mice and then of bees, the frequency of certain flowers in that district!

In the case of every species, many different checks, acting at different periods of life, and during different seasons or years, probably come into play; some one check or some few being generally the most potent, but all concurring in determining the average number or even the existence of the species. In some cases it can be shown that widely-different checks act on the same species in different districts. When we look at the plants and bushes clothing an entangled bank, we are tempted to attribute their proportional numbers and kinds to what we call chance. But how false a view is this! Every one has heard that when an American forest is cut down, a very different vegetation springs up; but it has been observed that the trees now growing on the ancient Indian mounds, in the Southern United States, display the same beautiful diversity and proportion of kinds as in the surrounding virgin forests. What a struggle between the several kinds of trees must here have gone on during long centuries, each annually scattering its seeds by the thousand; what war between insect and insect—between insects, snails, and other animals with birds and beasts of prey—all striving to increase, and all feeding on each other or on the trees or their seeds and seedlings, or on the other plants which first clothed the ground and thus checked the growth of the trees! Throw up a handful of feathers, and all must fall to the ground according to definite laws; but how simple is this problem compared to the action and reaction of the innumerable plants and animals which have determined, in the course of centuries, the proportional numbers and kinds of trees now growing on the old Indian ruins!

The dependency of one organic being on another, as of a parasite on its prey, lies generally between beings remote in the scale of nature. This is often the case with those which may strictly be said to struggle with each other for existence, as in the case of locusts and grass-feeding quadrupeds. But the struggle almost invariably will be most severe between the individuals of the same species, for they frequent the same districts, require the same food, and are exposed to the same dangers. In the case of varieties of the same species, the struggle will generally be almost equally severe, and we sometimes see the contest soon decided: for instance, if several varieties of wheat be sown together, and the mixed seed be resown, some of the varieties which best suit the soil or climate, or are naturally the most fertile, will beat the others and so yield more seed, and will consequently in a few years quite supplant the other varieties. To keep up a mixed stock of even such extremely close varieties as the variously coloured sweet-peas, they must be each year harvested separately, and the seed then mixed in due proportion, otherwise the weaker kinds will steadily decrease in numbers and disappear. So again with the varieties of sheep: it has been asserted that certain mountain-varieties will starve out other mountain-varieties, so that they cannot be kept together. The same result has followed from keeping together different varieties of the medicinal leech. It may even be doubted whether the varieties of any one of our domestic plants or animals have so exactly the same strength, habits, and constitution, that the original proportions of a mixed stock could be kept up for half a dozen generations, if they were allowed to struggle together, like beings in a state of nature, and if the seed or young were not annually sorted.

As species of the same genus have usually, though by no means invariably, some similarity in habits and constitution, and always in structure, the struggle will generally be more severe between species of the same genus, when they come into competition with each other, than between species of distinct genera. We see this in the recent extension over parts of the United States of one species of swallow having caused the decrease of another species. The recent increase of the missel-thrush in parts of Scotland has caused the decrease of the song-thrush. How frequently we hear of one species of rat taking the place of another species under the most different climates! In Russia the small Asiatic cockroach has everywhere driven before it its great congener. One species of charlock will supplant another, and so in other cases. We can dimly see why the competition should be most severe between allied forms, which fill nearly the same place in the economy of nature; but probably in no one case could we precisely say why one species has been victorious over another in the great battle of life.

A corollary of the highest importance may be deduced from the foregoing remarks, namely, that the structure of every organic being is related, in the most essential yet often hidden manner, to that of all other organic beings, with which it comes into competition for food or residence, or from which it has to escape, or on which it preys. This is obvious in the structure of the teeth and talons of the tiger; and in that of the legs and claws of the parasite which clings to the hair on the tiger's body. But in the beautifully plumed seed of the dandelion, and in the flattened and fringed legs of the water-beetle, the relation seems at first confined to the elements of air and water. Yet the advantage of plumed seeds no doubt stands in the closest relation to the land being already thickly clothed by other plants; so that the seeds may be widely distributed and fall on unoccupied ground. In the water-beetle, the structure of its legs, so well adapted for diving, allows it to compete with other aquatic insects, to hunt for its own prey, and to escape serving as prey to other animals.

The store of nutriment laid up within the seeds of many plants seems at first sight to have no sort of relation to other plants. But from the strong growth of young plants produced from such seeds (as peas and beans), when sown in the midst of long grass, I suspect that the chief use of the nutriment in the seed is to favour the growth of the young seedling, whilst struggling with other plants growing vigorously all around.

Look at a plant in the midst of its range, why does it not double or quadruple its numbers? We know that it can perfectly well withstand a little more heat or cold, dampness or dryness, for elsewhere it ranges into slightly hotter or colder, damper or drier districts. In this case we can clearly see that if we wished in imagination to give the plant the power of increasing in number, we should have to give it some advantage over its competitors, or over the animals which preyed on it. On the confines of its geographical range, a change of constitution with respect to climate would clearly be an advantage to our plant; but we have reason to believe that only a few plants or animals range so far, that they are destroyed by the rigour of the climate alone. Not until we reach the extreme confines of life, in the arctic regions or on the borders of an utter desert, will competition cease. The land may be extremely cold or dry, yet there will be competition between some few species, or between the individuals of the same species, for the warmest or dampest spots.

Hence, also, we can see that when a plant or animal is placed in a new country amongst new competitors, though the climate may be exactly the same as in its former home, yet the conditions of its life will generally be changed in an essential manner. If we wished to increase its average numbers in its new home, we should have to modify it in a different way to what we should have done in its native country; for we should have to give it some advantage over a different set of competitors or enemies.

It is good thus to try in our imagination to give any form some advantage over another. Probably in no single instance should we know what to do, so as to succeed. It will convince us of our ignorance on the mutual relations of all organic beings; a conviction as necessary, as it seems to be difficult to acquire. All that we can do, is to keep steadily in mind that each organic being is striving to increase at a geometrical ratio; that each at some period of its life, during some season of the year, during each generation or at intervals, has to struggle for life, and to suffer great destruction. When we reflect on this struggle, we may console ourselves with the full belief, that the war of nature is not incessant, that no fear is felt, that death is generally prompt, and that the vigorous, the healthy, and the happy survive and multiply.

NATURAL SELECTION.

How will the struggle for existence, discussed too briefly in the last chapter, act in regard to variation? Can the principle of selection, which we have seen is so potent in the hands of man, apply in nature? I think we shall see that it can act most effectually. Let it be borne in mind in what an endless number of strange peculiarities our domestic productions, and, in a lesser degree, those under nature, vary; and how strong the hereditary tendency is. Under domestication, it may be truly said that the whole organisation becomes in some degree plastic. Let it be borne in mind how infinitely complex and close-fitting are the mutual relations of all organic beings to each other and to their physical conditions of life. Can it, then, be thought improbable, seeing that variations useful to man have undoubtedly occurred, that other variations useful in some way to each being in the great and complex battle of life, should sometimes occur in the course of thousands of generations? If such do occur, can we doubt (remembering that many more individuals are born than can possibly survive) that individuals having any advantage, however slight, over others, would have the best chance of surviving and of procreating their kind? On the other hand, we may feel sure that any variation in the least degree injurious would be rigidly destroyed. This preservation of favourable variations and the rejection of injurious variations, I call Natural Selection. Variations neither useful nor injurious would not be affected by natural selection, and would be left a fluctuating element, as perhaps we see in the species called polymorphic.

We shall best understand the probable course of natural selection by taking the case of a country undergoing some physical change, for instance, of climate. The proportional numbers of its inhabitants would almost immediately undergo a change, and some species might become extinct. We may conclude, from what we have seen of the intimate and complex manner in which the inhabitants of each country are bound together, that any change in the numerical proportions of some of the inhabitants, independently of the change of climate itself, would most seriously affect many of the others. If the country were open on its borders, new forms would certainly immigrate, and this also would seriously disturb the relations of some of the former inhabitants. Let it be remembered how powerful the influence of a single introduced tree or mammal has been shown to be. But in the case of an island, or of a country partly surrounded by barriers, into which new and better adapted forms could not freely enter, we should then have places in the economy of nature which would assuredly be better filled up, if some of the original inhabitants were in some manner modified; for, had the area been open to immigration, these same places would have been seized on by intruders. In such case, every slight modification, which in the course of ages chanced to arise, and which in any way favoured the individuals of any of the species, by better adapting them to their altered conditions, would tend to be preserved; and natural selection would thus have free scope for the work of improvement.

We have reason to believe, as stated in the first chapter, that a change in the conditions of life, by specially acting on the reproductive system, causes or increases variability; and in the foregoing case the conditions of life are supposed to have undergone a change, and this would manifestly be favourable to natural selection, by giving a better chance of profitable variations occurring; and unless profitable variations do occur, natural selection can do nothing. Not that, as I believe, any extreme amount of variability is necessary; as man can certainly produce great results by adding up in any given direction mere individual differences, so could Nature, but far more easily, from having incomparably longer time at her disposal. Nor do I believe that any great physical change, as of climate, or any unusual degree of isolation to check immigration, is actually necessary to produce new and unoccupied places for natural selection to fill up by modifying and improving some of the varying inhabitants. For as all the inhabitants of each country are struggling together with nicely balanced forces, extremely slight modifications in the structure or habits of one inhabitant would often give it an advantage over others; and still further modifications of the same kind would often still further increase the advantage. No country can be named in which all the native inhabitants are now so perfectly adapted to each other and to the physical conditions under which they live, that none of them could anyhow be improved; for in all countries, the natives have been so far conquered by naturalised productions, that they have allowed foreigners to take firm possession of the land. And as foreigners have thus everywhere beaten some of the natives, we may safely conclude that the natives might have been modified with advantage, so as to have better resisted such intruders.

As man can produce and certainly has produced a great result by his methodical and unconscious means of selection, what may not nature effect? Man can act only on external and visible characters: nature cares nothing for appearances, except in so far as they may be useful to any being. She can act on every internal organ, on every shade of constitutional difference, on the whole machinery of life. Man selects only for his own good; Nature only for that of the being which she tends. Every selected character is fully exercised by her; and the being is placed under well-suited conditions of life. Man keeps the natives of many climates in the same country; he seldom exercises each selected character in some peculiar and fitting manner; he feeds a long-

and a short-beaked pigeon on the same food; he does not exercise a long-backed or long-legged quadruped in any peculiar manner; he exposes sheep with long and short wool to the same climate. He does not allow the most vigorous males to struggle for the females. He does not rigidly destroy all inferior animals, but protects during each varying season, as far as lies in his power, all his productions. He often begins his selection by some half-monstrous form; or at least by some modification prominent enough to catch his eye, or to be plainly useful to him. Under nature, the slightest difference of structure or constitution may well turn the nicely-balanced scale in the struggle for life, and so be preserved. How fleeting are the wishes and efforts of man! how short his time! and consequently how poor will his products be, compared with those accumulated by nature during whole geological periods. Can we wonder, then, that nature's productions should be far "truer" in character than man's productions; that they should be infinitely better adapted to the most complex conditions of life, and should plainly bear the stamp of far higher workmanship?

It may be said that natural selection is daily and hourly scrutinising, throughout the world, every variation, even the slightest; rejecting that which is bad, preserving and adding up all that is good; silently and insensibly working, whenever and wherever opportunity offers, at the improvement of each organic being in relation to its organic and inorganic conditions of life. We see nothing of these slow changes in progress, until the hand of time has marked the long lapse of ages, and then so imperfect is our view into long past geological ages, that we only see that the forms of life are now different from what they formerly were.

Although natural selection can act only through and for the good of each being, yet characters and structures, which we are apt to consider as of very trifling importance, may thus be acted on. When we see leaf-eating insects green, and bark-feeders mottled-grey; the alpine ptarmigan white in winter, the red-grouse the colour of heather, and the black-grouse that of peaty earth, we must believe that these tints are of service to these birds and insects in preserving them from danger. Grouse, if not destroyed at some period of their lives, would increase in countless numbers; they are known to suffer largely from birds of prey; and hawks are guided by eyesight to their prey,—so much so, that on parts of the Continent persons are warned not to keep white pigeons, as being the most liable to destruction. Hence I can see no reason to doubt that natural selection might be most effective in giving the proper colour to each kind of grouse, and in keeping that colour, when once acquired, true and constant. Nor ought we to think that the occasional destruction of an animal of any particular colour would produce little effect: we should remember how essential it is in a flock of white sheep to destroy every lamb with the faintest trace of black. In plants the down on the fruit and the colour of the flesh are considered by botanists as characters of the most trifling importance: yet we hear from an excellent horticulturist, Downing, that in the United States smooth-skinned fruits suffer far more from a beetle, a curculio, than those with down; that purple plums suffer far more from a certain disease than yellow plums; whereas another disease attacks yellow-fleshed peaches far more than those with other coloured flesh. If, with all the aids of art, these slight differences make a great difference in cultivating the several varieties, assuredly, in a state of nature, where the trees would have to struggle with other trees and with a host of enemies, such differences would effectually settle which variety, whether a smooth or downy, a yellow- or purple-fleshed fruit, should succeed.

In looking at many small points of difference between species, which, as far as our ignorance permits us to judge, seem to be quite unimportant, we must not forget that climate, food, &c., probably produce some slight and direct effect. It is, however, far more necessary to bear in mind that there are many unknown laws of correlation of growth, which, when one part of the organisation is modified through variation, and the modifications are accumulated by natural selection for the good of the being, will cause other modifications, often of the most unexpected nature.

As we see that those variations which under domestication appear at any particular period of life, tend to reappear in the offspring at the same period;—for instance, in the seeds of the many varieties of our culinary and agricultural plants; in the caterpillar and cocoon stages of the varieties of the silkworm; in the eggs of poultry, and in the colour of the down of their chickens; in the horns of our sheep and cattle when nearly adult;—so in a state of nature, natural selection will be enabled to act on and modify organic beings at any age, by the accumulation of profitable variations at that age, and by their inheritance at a corresponding age. If it profit a plant to have its seeds more and more widely disseminated by the wind, I can see no greater difficulty in this being effected through natural selection, than in the cotton-planter increasing and improving by selection the down in the pods on his cotton-trees. Natural selection may modify and adapt the larva of an insect to a score of contingencies, wholly different from those which concern the mature insect. These modifications will no doubt affect, through the laws of correlation, the structure of the adult; and probably in the case of those insects which live only for a few hours, and which never feed, a large part of their structure is merely the correlated result of successive changes in the structure of their larvæ. So, conversely, modifications in the adult will probably often affect the structure of the larva; but in all cases natural selection will ensure that modifications consequent on other modifications at a different period of life, shall not be in the least degree injurious: for if they became so, they would cause the extinction of the species.

Natural selection will modify the structure of the young in relation to the parent, and of the parent in relation to the young. In social animals it will adapt the structure of each individual for the benefit of the community; if each in consequence profits by the selected change. What natural selection cannot do, is to modify the structure of one species, without giving it any advantage, for the good of another species; and though statements to this effect may be found in works of natural history, I cannot find one case which will bear investigation. A structure used only once in an animal's whole life, if of high importance to it, might be modified to any extent by natural selection; for instance, the great jaws possessed by certain insects, and used exclusively for opening the cocoon—or the hard tip of the beak of nestling birds, used for breaking the egg. It has been asserted, that of the best short-beaked tumbler-pigeons more perish in the egg than are able to get out of it; so that fanciers assist in the act of hatching. Now, if nature had to make the beak of a full-grown pigeon very short for the bird's own advantage, the process of modification would be very slow, and there would be simultaneously the most rigorous selection of the young birds within the egg, which had the most powerful and hardest beaks, for all with weak beaks would inevitably perish: or, more delicate and more easily broken shells might be selected, the thickness of the shell being known to vary like every other structure.

Sexual Selection. Inasmuch as peculiarities often appear under domestication in one sex and become hereditarily attached to that sex, the same fact probably occurs under nature, and if so, natural selection will be able to modify one sex in its functional relations to the other sex, or in relation to wholly different habits of life in the two sexes, as is sometimes the case with insects. And this leads me to say a few words on what I call Sexual Selection. This depends, not on a struggle for existence, but on a struggle between the males for possession of the females; the result is not death to the unsuccessful competitor, but few or no offspring. Sexual selection is, therefore, less rigorous than natural selection. Generally, the most vigorous males, those which are best fitted for their places in nature, will leave most progeny. But in many cases, victory will depend not on general vigour, but on having special weapons, confined to the male sex. A hornless stag or spurless cock would have a poor chance of leaving offspring. Sexual selection by always allowing the victor to breed might surely give indomitable courage, length to the spur, and strength to the wing to strike in the spurred leg, as well as the brutal cock-fighter, who knows well that he can improve his breed by careful selection of the best cocks. How low in the scale of nature this law of battle descends, I know not; male alligators have been described as fighting, bellowing, and whirling round, like Indians in a war-dance, for the possession of the females; male salmons have been seen fighting all day long; male stag-beetles often bear wounds from the huge mandibles of other males. The war is, perhaps, severest between the males of polygamous animals, and these seem oftenest provided with special weapons. The males of carnivorous animals are already well armed; though to them and to others, special means of defence may be given through means of sexual selection, as the mane to the lion, the shoulder-pad to the boar, and the hooked jaw to the male salmon; for the shield may be as important for victory, as the sword or spear.

Amongst birds, the contest is often of a more peaceful character. All those who have attended to the subject, believe that there is the severest rivalry between the males of many species to attract by singing the females. The rock-thrush of Guiana, birds of Paradise, and some others, congregate; and successive males display their gorgeous plumage and perform strange antics before the females, which standing by as spectators, at last choose the most attractive partner. Those who have closely attended to birds in confinement well know that they often take individual preferences and dislikes: thus Sir R. Heron has described how one pied peacock was eminently attractive to all his hen birds. It may appear childish to attribute any effect to such apparently weak means: I cannot here enter on the details necessary to support this view; but if man can in a short time give elegant carriage and beauty to his bantams, according to his standard of beauty, I can see no good reason to doubt that female birds, by selecting, during thousands of generations, the most melodious or beautiful males, according to their standard of beauty, might produce a marked effect. I strongly suspect that some well-known laws with respect to the plumage of male and female birds, in comparison with the plumage of the young, can be explained on the view of plumage having been chiefly modified by sexual selection, acting when the birds have come to the breeding age or during the breeding season; the modifications thus produced being inherited at corresponding ages or seasons, either by the males alone, or by the males and females; but I have not space here to enter on this subject.

Thus it is, as I believe, that when the males and females of any animal have the same general habits of life, but differ in structure, colour, or ornament, such differences have been mainly caused by sexual selection; that is, individual males have had, in successive generations, some slight advantage over other males, in their weapons, means of defence, or charms; and have transmitted these advantages to their male offspring. Yet, I would not wish to attribute all such sexual differences to this agency: for we see peculiarities arising and becoming attached to the male sex in our domestic animals (as the wattle in male carriers, horn-like protuberances in the cocks of certain fowls, &c.), which we cannot believe to be either useful to the males in battle, or attractive to the females. We see analogous cases under nature, for instance, the tuft of hair on the breast of the turkey-cock, which can hardly be either useful or ornamental to this bird;—indeed, had the tuft appeared under domestication, it would have been called a monstrosity.

Illustrations of the action of Natural Selection.—In order to make it clear how, as I believe, natural selection acts, I must beg permission to give one or two imaginary illustrations. Let us take the case of a wolf, which preys on various animals, securing some by craft, some by strength, and some by fleetness; and let us suppose that the fleetest prey, a deer for instance, had from any change in the country increased in numbers, or that other prey had decreased in numbers, during that season of the year when the wolf is hardest pressed for food. I can under such circumstances see no reason to doubt that the swiftest and slimmest wolves would have the best chance of surviving, and so be preserved or selected,—provided always that they retained strength to master their prey at this or at some other period of the year, when they might be compelled to prey on other animals. I can see no more reason to doubt this, than that man can improve the fleetness of his greyhounds by careful and methodical selection, or by that unconscious selection which results from each man trying to keep the best dogs without any thought of modifying the breed.

Even without any change in the proportional numbers of the animals on which our wolf preyed, a cub might be born with an innate tendency to pursue certain kinds of prey. Nor can this be thought very improbable; for we often observe great differences in the natural tendencies of our domestic animals; one cat, for instance, taking to catch rats, another mice; one cat, according to Mr. St. John, bringing home winged game, another hares or rabbits, and another hunting on marshy ground and almost nightly catching woodcocks or snipes. The tendency to catch rats rather than mice is known to be inherited. Now, if any slight innate change of habit or of structure benefited an individual wolf, it would have the best chance of surviving and of leaving offspring. Some of its young would probably inherit the same habits or structure, and by the repetition of this process, a new variety might be formed which would either supplant or coexist with the parent-form of wolf. Or, again, the wolves inhabiting a mountainous district, and those frequenting the lowlands, would naturally be forced to hunt different prey; and from the continued preservation of the individuals best fitted for the two sites, two varieties might slowly be formed. These varieties would cross and blend where they met; but to this subject of intercrossing we shall soon have to return. I may add, that, according to Mr. Pierce, there are two varieties of the wolf inhabiting the Catskill Mountains in the United States, one with a light greyhound-like form, which pursues deer, and the other more bulky, with shorter legs, which more frequently attacks the shepherd's flocks.

Let us now take a more complex case. Certain plants excrete a sweet juice, apparently for the sake of eliminating something injurious from their sap: this is effected by glands at the base of the stipules in some Leguminosæ, and at the back of the leaf of the common laurel. This juice, though small in quantity, is greedily sought by insects. Let us now suppose a little sweet juice or nectar to be excreted by the inner bases of the petals of a flower. In this case insects in seeking the nectar would get dusted with pollen, and would certainly often transport the pollen from one flower to the stigma of another flower. The flowers of two distinct individuals of the same species would thus get crossed; and the act of crossing, we have good reason to believe (as will hereafter be more fully alluded to), would produce very vigorous seedlings, which consequently would have the best chance of flourishing and surviving. Some of these seedlings would probably inherit the nectar-excreting power. Those individual flowers which had the largest glands or nectaries, and which excreted most nectar, would be oftenest visited by insects, and would be oftenest crossed; and so in the long-run would gain the upper hand. Those flowers, also, which had their stamens and pistils placed, in relation to the size and habits of the particular insects which visited them, so as to favour in any degree the transportal of their pollen from flower to flower, would likewise be favoured or selected. We might have taken the case of insects visiting flowers for the sake of collecting pollen instead of nectar; and as pollen is formed for the sole object of fertilisation, its destruction appears a simple loss to the plant; yet if a little pollen were carried, at first occasionally and then habitually, by the pollen-devouring insects from flower to flower, and a cross thus effected, although nine-tenths of the pollen were destroyed, it might still be a great gain to the plant; and those individuals which produced more and more pollen, and had larger and larger anthers, would be selected.

When our plant, by this process of the continued preservation or natural selection of more and more attractive flowers, had been rendered highly attractive to insects, they would, unintentionally on their part, regularly carry pollen from flower to flower; and that they can most effectually do this, I could easily show by many striking instances. I will give only one—not as a very striking case, but as likewise illustrating one step in the separation of the sexes of plants, presently to be alluded to. Some holly-trees bear only male flowers, which have four stamens producing rather a small quantity of pollen, and a rudimentary pistil; other holly-trees bear only female flowers; these have a full-sized pistil, and four stamens with shrivelled anthers, in which not a grain of pollen can be detected. Having found a female tree exactly sixty yards from a male tree, I put the stigmas of twenty flowers, taken from different branches, under the microscope, and on all, without exception, there were pollen-grains, and on some a profusion of pollen. As the wind had set for several days from the female to the male tree, the pollen could not thus have been carried. The weather had been cold and boisterous, and therefore not favourable to bees, nevertheless every female flower which I examined had been effectually fertilised by the bees, accidentally dusted with pollen, having flown from tree to tree in search of nectar. But to return to our imaginary case: as soon as the plant had been rendered so highly attractive to insects that pollen was regularly carried from flower to flower, another process might commence. No naturalist doubts the advantage of what has been called the "physiological division of labour;" hence we may believe that it would be advantageous to a plant to produce stamens alone in one flower or on one whole plant, and pistils alone in another flower or on another plant. In plants under culture and placed under new conditions of life, sometimes the male organs and sometimes the female organs become more or less impotent; now if we suppose this to occur in ever so slight a degree under nature, then as pollen is already carried regularly from flower to flower, and as a more complete separation of the sexes of our plant would be advantageous on the principle of the division of labour, individuals with this tendency more and more increased, would be continually favoured or selected, until at last a complete separation of the sexes would be effected.

Let us now turn to the nectar-feeding insects in our imaginary case: we may suppose the plant of which we have been slowly increasing the nectar by continued selection, to be a common plant; and that certain insects depended in main part on its nectar for food. I could give many facts, showing how anxious bees are to save time; for instance, their habit of cutting holes and sucking the nectar at the bases of certain flowers, which they can, with a very little more trouble, enter by the mouth. Bearing such facts in mind, I can see no reason to doubt that an accidental deviation in the size and form of the body, or in the curvature and length of the proboscis, &c., far too slight to be appreciated by us, might profit a bee or other insect, so that an individual so characterised would be able to obtain its food more quickly, and so have a better chance of living and leaving descendants. Its descendants would probably inherit a tendency to a similar slight deviation of structure. The tubes of the corollas of the common red and incarnate clovers (Trifolium pratense and incarnatum) do not on a hasty glance appear to differ in length; yet the hive-bee can easily suck the nectar out of the incarnate clover, but not out of the common red clover, which is visited by humble-bees alone; so that whole fields of the red clover offer in vain an abundant supply of precious nectar to the hive-bee. Thus it might be a great advantage to the hive-bee to have a slightly longer or differently constructed proboscis. On the other hand, I have found by experiment that the fertility of clover greatly depends on bees visiting and moving parts of the corolla, so as to push the pollen on to the stigmatic surface. Hence, again, if humble-bees were to become rare in any country, it might be a great advantage to the red clover to have a shorter or more deeply divided tube to its corolla, so that the hive-bee could visit its flowers. Thus I can understand how a flower and a bee might slowly become, either simultaneously or one after the other, modified and adapted in the most perfect manner to each other, by the continued preservation of individuals presenting mutual and slightly favourable deviations of structure.

I am well aware that this doctrine of natural selection, exemplified in the above imaginary instances, is open to the same objections which were at first urged against Sir Charles Lyell's noble views on "the modern changes of the earth, as illustrative of geology;" but we now very seldom hear the action, for instance, of the coast-waves, called a trifling and insignificant cause, when applied to the excavation of gigantic valleys or to the formation of the longest lines of inland cliffs. Natural selection can act only by the preservation and accumulation of infinitesimally small inherited modifications, each profitable to the preserved being; and as modern geology has almost banished such views as the excavation of a great valley by a single diluvial wave, so will natural selection, if it be a true principle, banish the belief of the continued creation of new organic beings, or of any great and sudden modification in their structure.

On the Intercrossing of Individuals.— I must here introduce a short digression. In the case of animals and plants with separated sexes, it is of course obvious that two individuals must always unite for each birth; but in the case of hermaphrodites this is far from obvious. Nevertheless I am strongly inclined to believe that with all hermaphrodites two individuals, either occasionally or habitually, concur for the reproduction of their kind. This view, I may add, was first suggested by Andrew Knight. We shall presently see its importance; but I must here treat the subject with extreme brevity, though I have the materials prepared for an ample discussion. All vertebrate animals, all insects, and some other large groups of animals, pair for each birth. Modern research has much diminished the number of supposed hermaphrodites, and of real hermaphrodites a large number pair; that is, two individuals regularly unite for reproduction, which is all that concerns us. But still there are many hermaphrodite animals which certainly do not habitually pair, and a vast majority of plants are hermaphrodites. What reason, it may be asked, is there for supposing in these cases that two individuals ever concur in reproduction? As it is impossible here to enter on details, I must trust to some general considerations alone.

In the first place, I have collected so large a body of facts, showing, in accordance with the almost universal belief of breeders, that with animals and plants a cross between different varieties, or between individuals of the same variety but of another strain, gives vigour and fertility to the offspring; and on the other hand, that *close* interbreeding diminishes vigour and fertility; that these facts alone incline me to believe that it is a general law of nature (utterly ignorant though we be of the meaning of the law) that no organic being self-fertilises itself for an eternity of generations; but that a cross with another individual is occasionally—perhaps at very long intervals—indispensable.

On the belief that this is a law of nature, we can, I think, understand several large classes of facts, such as the following, which on any other view are inexplicable. Every hybridizer knows how unfavourable exposure to wet is to the fertilisation of a flower, yet what a multitude of flowers have their anthers and stigmas fully exposed to the weather! but if an occasional cross be indispensable, the fullest freedom for the entrance of pollen from another individual will explain this state of exposure, more especially as the plant's own anthers and pistil generally stand so close together that self-fertilisation seems almost inevitable. Many flowers, on the other hand, have their organs of fructification closely enclosed, as in the great papilionaceous or pea-family; but in several, perhaps in all, such flowers, there is a very curious adaptation between the structure of the flower and the manner in which bees suck the nectar; for, in doing this, they either push the flower's own pollen on the stigma, or bring pollen from another flower. So necessary are the visits of bees to papilionaceous flowers, that I have found, by experiments published elsewhere, that their fertility is greatly diminished if these visits be prevented. Now, it is scarcely possible that bees should fly from flower to flower, and not carry pollen from one to the other, to the great good, as I believe, of the plant. Bees will act like a camel-hair pencil, and it is quite sufficient just to touch the anthers of one flower and then the stigma of another with the same brush to ensure fertilisation; but it must not be supposed that bees would thus produce a multitude of hybrids between distinct species; for if you bring on the same brush a plant's own pollen and pollen from another species, the former will have such a prepotent effect, that it will invariably and completely destroy, as has been shown by Gärtner, any influence from the foreign pollen.

When the stamens of a flower suddenly spring towards the pistil, or slowly move one after the other towards it, the contrivance seems adapted solely to ensure self-fertilisation; and no doubt it is useful for this end: but, the agency of insects is often required to cause the stamens to spring forward, as Kölreuter has shown to be the case with the barberry; and curiously in this very genus, which seems to have a special contrivance for self-fertilisation, it is well known that if very closely-allied forms or varieties are planted near each other, it is hardly possible to raise pure seedlings, so largely do they naturally cross. In many other cases, far from there being any aids for self-fertilisation, there are special contrivances, as I could show from the writings of C. C. Sprengel and from my own observations, which effectually prevent the stigma receiving pollen from its own flower: for instance, in Lobelia fulgens, there is a really beautiful and elaborate contrivance by which every one of the infinitely numerous pollen-granules are swept out of the conjoined anthers of each flower, before the stigma of that individual flower is ready to receive them; and as this flower is never visited, at least in my garden, by insects, it never sets a seed, though by placing pollen from one flower on the stigma of another, I raised plenty of seedlings; and whilst another species of Lobelia growing close by, which is visited by bees, seeds freely. In very many other cases, though there be no special mechanical contrivance to prevent the stigma of a flower receiving its own pollen, yet, as C. C. Sprengel has shown, and as I can confirm, either the anthers burst before the stigma is ready for fertilisation, or the stigma is ready before the pollen of that flower is ready, so that these plants have in fact separated sexes, and must habitually be crossed. How strange are these facts! How strange that the pollen and stigmatic surface of the same flower, though placed so close together, as if for the very purpose of self-fertilisation, should in so many cases be mutually useless to each other! How simply are these facts explained on the view of an occasional cross with a distinct individual being advantageous or indispensable!

If several varieties of the cabbage, radish, onion, and of some other plants, be allowed to seed near each other, a large majority, as I have found, of the seedlings thus raised will turn out mongrels: for instance, I raised 233 seedling cabbages from some plants of different varieties growing near each other, and of these only 78 were true to their kind, and some even of these were not perfectly true. Yet the pistil of each cabbage-flower is surrounded not only by its own six stamens, but by those of the many other flowers on the same plant. How, then, comes it that such a vast number of the seedlings are mongrelized? I suspect that it must arise from the pollen of a distinct *variety* having a prepotent effect over a flower's own pollen; and that this is part of the general law of good being derived from the intercrossing of distinct individuals of the same species. When distinct *species* are crossed the case is directly the reverse, for a plant's own pollen is always prepotent over foreign pollen; but to this subject we shall return in a future chapter.

In the case of a gigantic tree covered with innumerable flowers, it may be objected that pollen could seldom be carried from tree to tree, and at most only from flower to flower on the same tree, and that flowers on the same tree can be considered as distinct individuals only in a limited sense. I believe this objection to be valid, but that nature has largely provided against it by giving to trees a strong tendency to bear flowers with separated sexes. When the sexes are separated, although the male and female flowers may be produced on the same tree, we can see that pollen must be regularly carried from flower to flower; and this will give a better chance of pollen being occasionally carried from tree to tree. That trees belonging to all Orders have their sexes more often separated than other plants, I find to be the case in this country; and at my request Dr. Hooker tabulated the trees of New Zealand, and Dr. Asa Gray those of the United States, and the result was as I anticipated. On the other hand, Dr. Hooker has recently informed me that he finds that the rule does not hold in Australia; and I have made these few remarks on the sexes of trees simply to call attention to the subject.

Turning for a very brief space to animals: on the land there are some hermaphrodites, as land-mollusca and earth-worms; but these all pair. As yet I have not found a single case of a terrestrial animal which fertilises itself. We can understand this remarkable fact, which offers so strong a contrast with terrestrial plants, on the view of an occasional cross being indispensable, by considering the medium in which terrestrial animals live, and the nature of the fertilising element; for we know of no means, analogous to the action of insects and of the wind in the case of plants, by which an occasional cross could be effected with terrestrial animals without the concurrence of two individuals. Of aquatic animals, there are many self-fertilising hermaphrodites; but here currents in the water offer an obvious means for an occasional cross. And, as in the case of flowers, I have as yet failed, after consultation with one of the highest authorities, namely, Professor Huxley, to discover a single case of an hermaphrodite animal with the organs of reproduction so perfectly enclosed within the body, that access from without and the occasional influence of a distinct individual can be shown to be physically impossible. Cirripedes long appeared to me to present a case of very great difficulty under this point of view; but I have been enabled, by a fortunate chance, elsewhere to prove that two individuals, though both are self-fertilising hermaphrodites, do sometimes cross.

It must have struck most naturalists as a strange anomaly that, in the case of both animals and plants, species of the same family and even of the same genus, though agreeing closely with each other in almost their whole organisation, yet are not rarely, some of them hermaphrodites, and some of them unisexual. But if, in fact, all hermaphrodites do occasionally intercross with other individuals, the difference between hermaphrodites and unisexual species, as far as function is concerned, becomes very small.

From these several considerations and from the many special facts which I have collected, but which I am not here able to give, I am strongly inclined to suspect that, both in the vegetable and animal kingdoms, an occasional intercross with a distinct individual is a law of nature. I am well aware that there are, on this view, many cases of difficulty, some of which I am trying to investigate. Finally then, we may conclude that in many organic beings, a cross between two individuals is an obvious necessity for each birth; in many others it occurs perhaps only at long intervals; but in none, as I suspect, can self-fertilisation go on for perpetuity.

Circumstances Favorable to Natural Selection.—This is an extremely intricate subject. A large amount of inheritable and diversified variability is favourable, but I believe mere individual differences suffice for the work. A large number of individuals, by giving a better chance for the appearance within any given period of profitable variations, will compensate for a lesser amount of variability in each individual, and is, I believe, an extremely important element of success. Though nature grants vast periods of time for the work of natural selection, she does not grant an indefinite period; for as all organic beings are striving, it may be said, to seize on each place in the economy of nature, if any one species does not become modified and improved in a corresponding degree with its competitors, it will soon be exterminated.

In man's methodical selection, a breeder selects for some definite object, and free intercrossing will wholly stop his work. But when many men, without intending to alter the breed, have a nearly common standard of perfection, and all try to get and breed from the best animals, much improvement and modification surely but slowly follow from this unconscious process of selection, notwithstanding a large amount of crossing with inferior animals. Thus it will be in nature; for within a confined area, with some place in its polity not so perfectly occupied as might be, natural selection will always tend to preserve all the individuals varying in the right direction, though in different degrees, so as better to fill up the unoccupied place. But if the area be large, its several districts will almost certainly present different conditions of life; and then if natural selection be modifying and improving a species in the several districts, there will be intercrossing with the other individuals of the same species on the confines of each. And in this case the effects of intercrossing can hardly be counterbalanced by natural selection always tending to modify all the individuals in each district in exactly the same manner to the conditions of each; for in a continuous area, the conditions will generally graduate away insensibly from one district to another. The intercrossing will most affect those animals which unite for each birth, which wander much, and which do not breed at a very quick rate. Hence in animals of this nature, for instance in birds, varieties will generally be confined to separated countries; and this I believe to be the case. In hermaphrodite organisms which cross only occasionally, and likewise in animals which unite for each birth, but which wander little and which can increase at a very rapid rate, a new and improved variety might be quickly formed on any one spot, and might there maintain itself in a body, so that whatever intercrossing took place would be chiefly between the individuals of the same new variety. A local variety when once thus formed might subsequently slowly spread to other districts. On the above principle, nurserymen always prefer getting seed from a large body of plants of the same variety, as the chance of intercrossing with other varieties is thus lessened.

Even in the case of slow-breeding animals, which unite for each birth, we must not overrate the effects of intercrosses in retarding natural selection; for I can bring a considerable catalogue of facts, showing that within the same area, varieties of the same animal can long remain distinct, from haunting different stations, from breeding at slightly different seasons, or from varieties of the same kind preferring to pair together.

Intercrossing plays a very important part in nature in keeping the individuals of the same species, or of the same variety, true and uniform in character. It will obviously thus act far more efficiently with those animals which unite for each birth; but I have already attempted to show that we have reason to believe that occasional intercrosses take place with all animals and with all plants. Even if these take place only at long intervals, I am convinced that the young thus produced will gain so much in vigour and fertility over the offspring from long-continued self-fertilisation, that they will have a better chance of surviving and propagating their kind; and thus, in the long run, the influence of intercrosses, even at rare intervals, will be great. If there exist organic beings which never intercross, uniformity of character can be retained amongst them, as long as their conditions of life remain the same, only through the principle of inheritance, and through natural selection destroying any which depart from the proper type; but if their conditions of life change and they undergo modification, uniformity of character can be given to their modified offspring, solely by natural selection preserving the same favourable variations.

Isolation, also, is an important element in the process of natural selection. In a confined or isolated area, if not very large, the organic and inorganic conditions of life will generally be in a great degree uniform; so that natural selection will tend to modify all the individuals of a varying species throughout the area in the same manner in relation to the same conditions. Intercrosses, also, with the individuals of the same species, which otherwise would have inhabited the surrounding and differently circumstanced districts, will be prevented. But isolation probably acts more efficiently in checking the immigration of better adapted organisms, after any physical change, such as of climate or elevation of the land, &c.; and thus new places in the natural economy of the country are left open for the old inhabitants to struggle for, and become adapted to, through modifications in their structure and constitution. Lastly, isolation, by checking immigration and consequently competition, will give time for any new variety to be slowly improved; and this may sometimes be of importance in the production of new species. If, however, an isolated area be very small, either from being surrounded by barriers, or from having very peculiar physical conditions, the total number of the individuals supported on it will necessarily be very small; and fewness of individuals will greatly retard the production of new species through natural selection, by decreasing the chance of the appearance of favourable variations.

If we turn to nature to test the truth of these remarks, and look at any small isolated area, such as an oceanic island, although the total number of the species inhabiting it, will be found to be small, as we shall see in our chapter on geographical distribution; yet of these species a very large proportion are endemic,—that is, have been produced there, and nowhere else. Hence an oceanic island at first sight seems to have been highly favourable for the production of new species. But we may thus greatly deceive ourselves, for to ascertain whether a small isolated area, or a large open area like a continent, has been most favourable for the production of new organic forms, we ought to make the comparison within equal times; and this we are incapable of doing.

Although I do not doubt that isolation is of considerable importance in the production of new species, on the whole I am inclined to believe that largeness of area is of more importance, more especially in the production of species, which will prove capable of enduring for a long period, and of spreading widely. Throughout a great and open area, not only will there be a better chance of favourable variations arising from the large number of individuals of the same species there supported, but the conditions of life are infinitely complex from the large number of already existing species; and if some of these many species become modified and improved, others will have to be improved in a corresponding degree or they will be exterminated. Each new form, also, as soon as it has been much improved, will be able to spread over the open and continuous area, and will thus come into competition with many others. Hence more new places will be formed, and the competition to fill them will be more severe, on a large than on a small and isolated area. Moreover, great areas, though now continuous, owing to oscillations of level, will often have recently existed in a broken condition, so that the good effects of isolation will generally, to a certain extent, have concurred. Finally, I conclude that, although small isolated areas probably have been in some respects highly favourable for the production of new species, yet that the course of modification will generally have been more rapid on large areas; and what is more important, that the new forms produced on large areas, which already have been victorious over many competitors, will be those that will spread most widely, will give rise to most new varieties and species, and will thus play an important part in the changing history of the organic world.

We can, perhaps, on these views, understand some facts which will be again alluded to in our chapter on geographical distribution; for instance, that the productions of the smaller continent of Australia have formerly yielded, and apparently are now yielding, before those of the larger Europæo-Asiatic area. Thus, also, it is that continental productions have everywhere become so largely naturalised on islands. On a small island, the race for life will have been less severe, and there will have been less modification and less extermination. Hence, perhaps, it comes that the flora of Madeira, according to Oswald Heer, resembles the extinct tertiary flora of Europe. All fresh-water basins, taken together, make a small area compared with that of the sea or of the land; and, consequently, the competition between fresh-water productions will have been less severe than elsewhere; new forms will have been more slowly formed, and old forms more slowly exterminated. And it is in fresh water that we find seven genera of Ganoid fishes, remnants of a once preponderant order: and in fresh water we find some of the most anomalous forms now known in the world, as the Ornithorhynchus and Lepidosiren, which, like fossils, connect to a certain extent orders now widely separated in the natural scale. These anomalous forms may almost be called living fossils; they have endured to the present day, from having inhabited a confined area, and from having thus been exposed to less severe competition.

To sum up the circumstances favourable and unfavourable to natural selection, as far as the extreme intricacy of the subject permits. I conclude, looking to the future, that for terrestrial productions a large continental area, which will probably undergo many oscillations of level, and which consequently will exist for long periods in a broken condition, will be the most favourable for the production of many new forms of life, likely to endure long and to spread widely. For the area will first have existed as a continent, and the inhabitants, at this period numerous in individuals and kinds, will have been subjected to very severe competition. When converted by subsidence into large separate islands, there will still exist many individuals of the same species on each island: intercrossing on the confines of the range of each species will thus be checked: after physical changes of any kind, immigration will be prevented, so that new places in the polity of each island will have to be filled up by modifications of the old inhabitants; and time will be allowed for the varieties in each to become well modified and perfected. When, by renewed elevation, the islands shall be reconverted into a continental area, there will again be severe competition: the most favoured or improved varieties will be enabled to spread: there will be much extinction of the less improved forms, and the relative proportional numbers of the various inhabitants of the renewed continent will again be changed; and again there will be a fair field for natural selection to improve still further the inhabitants, and thus produce new species.

That natural selection will always act with extreme slowness, I fully admit. Its action depends on there being places in the polity of nature, which can be better occupied by some of the inhabitants of the country undergoing modification of some kind. The existence of such places will often depend on physical changes, which are generally very slow, and on the immigration of better adapted forms having been checked. But the action of natural selection will probably still oftener depend on some of the inhabitants becoming slowly modified; the mutual relations of many of the other inhabitants being thus disturbed. Nothing can be effected, unless favourable variations occur, and variation itself is apparently always a very slow process. The process will often be greatly retarded by free intercrossing. Many will exclaim that these several causes are amply sufficient wholly to stop the action of natural selection. I do not believe so. On the other hand, I do believe that natural selection will always act very slowly, often only at long intervals of time, and generally on only a very few of the inhabitants of the same region at the same time. I further believe, that this very slow, intermittent action of natural selection accords perfectly well with what geology tells us of the rate and manner at which the inhabitants of this world have changed.

Slow though the process of selection may be, if feeble man can do much by his powers of artificial selection, I can see no limit to the amount of change, to the beauty and infinite complexity of the co-adaptations between all organic beings, one with another and with their physical conditions of life, which may be effected in the long course of time by nature's power of selection.

Extinction.—This subject will be more fully discussed in our chapter on Geology; but it must be here alluded to from being intimately connected with natural selection. Natural selection acts solely through the preservation of variations in some way advantageous, which consequently endure. But as from the high geometrical powers of increase of all organic beings, each area is already fully stocked with inhabitants, it follows that as each selected and favoured form increases in number, so will the less favoured forms decrease and become rare. Rarity, as geology tells us, is the precursor to extinction. We can, also, see that any form represented by few individuals will, during fluctuations in the seasons or in the number of its enemies, run a good chance of utter extinction. But we may go further than this; for as new forms are continually and slowly being produced, unless we believe that the number of specific forms goes on perpetually and almost indefinitely increasing, numbers inevitably must become extinct. That the number of specific forms has not indefinitely increased, geology shows us plainly; and indeed we can see reason why they should not have thus increased, for the number of places in the polity of nature is not indefinitely great,—not that we have any means of knowing that any one region has as yet got its maximum of species. Probably no region is as yet fully stocked, for at the Cape of Good Hope, where more species of plants are crowded together than in any other quarter of the world, some foreign plants have become naturalised, without causing, as far as we know, the extinction of any natives.

Furthermore, the species which are most numerous in individuals will have the best chance of producing within any given period favourable variations. We have evidence of this, in the facts given in the second chapter, showing that it is the common species which afford the greatest number of recorded varieties, or incipient species. Hence, rare species will be less quickly modified or improved within any given period, and they will consequently be beaten in the race for life by the modified descendants of the commoner species.

From these several considerations I think it inevitably follows, that as new species in the course of time are formed through natural selection, others will become rarer and rarer, and finally extinct. The forms which stand in closest competition with those undergoing modification and improvement, will naturally suffer most. And we have seen in the chapter on the Struggle for Existence that it is the most closely-allied forms,—varieties of the same species, and species of the same genus or of related genera,—which, from having nearly the same structure, constitution, and habits, generally come into the severest competition with each other. Consequently, each new variety or species, during the progress of its formation, will generally press hardest on its nearest kindred, and tend to exterminate them. We see the same process of extermination amongst our domesticated productions, through the selection of improved forms by man. Many curious instances could be given showing how quickly new breeds of cattle, sheep, and other animals, and varieties of flowers, take the place of older and inferior kinds. In Yorkshire, it is historically known that the ancient black cattle were displaced by the long-horns, and that these "were swept away by the short-horns" (I quote the words of an agricultural writer) "as if by some murderous pestilence."

Divergence of Character.—The principle, which I have designated by this term, is of high importance on my theory, and explains, as I believe, several important facts. In the first place, varieties, even strongly-marked ones, though having somewhat of the character of species—as is shown by the hopeless doubts in many cases how to rank them—yet certainly differ from each other far less than do good and distinct species. Nevertheless, according to my view, varieties are species in the process of formation, or are, as I have called them, incipient species. How, then, does the lesser difference between varieties become augmented into the greater difference between species? That this does habitually happen, we must infer from most of the innumerable species throughout nature presenting well-marked differences; whereas varieties, the supposed prototypes and parents of future well-marked species, present slight and ill-defined differences. Mere chance, as we may call it, might cause one variety to differ in some character from its parents, and the offspring of this variety again to differ from its parent in the very same character and in a greater degree; but this alone would never account for so habitual and large an amount of difference as that between varieties of the same species and species of the same genus.

As has always been my practice, let us seek light on this head from our domestic productions. We shall here find something analogous. A fancier is struck by a pigeon having a slightly shorter beak; another fancier is struck by a pigeon having a rather longer beak; and on the acknowledged principle that "fanciers do not and will not admire a medium standard, but like extremes," they both go on (as has actually occurred with tumbler-pigeons) choosing and breeding from birds with longer and longer beaks, or with shorter and shorter beaks. Again, we may suppose that at an early period one man preferred swifter horses; another stronger and more bulky horses. The early differences would be very slight; in the course of time, from the continued selection of swifter horses by some breeders, and of stronger ones by others, the differences would become greater, and would be noted as forming two sub-breeds; finally, after the lapse of centuries, the sub-breeds would become converted into two well-established and distinct breeds. As the differences slowly become greater, the inferior animals with intermediate characters, being neither very swift nor very strong, will have been neglected, and will have tended to disappear. Here, then, we see in man's productions the action of what may be called the principle of divergence, causing differences, at first barely appreciable, steadily to increase, and the breeds to diverge in character both from each other and from their common parent.

But how, it may be asked, can any analogous principle apply in nature? I believe it can and does apply most efficiently, from the simple circumstance that the more diversified the descendants from any one species become in structure, constitution, and habits, by so much will they be better enabled to seize on many and widely diversified places in the polity of nature, and so be enabled to increase in numbers.

We can clearly see this in the case of animals with simple habits. Take the case of a carnivorous quadruped, of which the number that can be supported in any country has long ago arrived at its full average. If its natural powers of increase be allowed to act, it can succeed in increasing (the country not undergoing any change in its conditions) only by its varying descendants seizing on places at present occupied by other animals: some of them, for instance, being enabled to feed on new kinds of prey, either dead or alive; some inhabiting new stations, climbing trees, frequenting water, and some perhaps becoming less carnivorous. The more diversified in habits and structure the descendants of our carnivorous animal became, the more places they would be enabled to occupy. What applies to one animal will apply throughout all time to all animals—that is, if they vary—for otherwise natural selection can do nothing. So it will be with plants. It has been experimentally proved, that if a plot of ground be sown with one species of grass, and a similar plot be sown with several distinct genera of grasses, a greater number of plants and a greater weight of dry herbage can thus be raised. The same has been found to hold good when first one variety and then several mixed varieties of wheat have been sown on equal spaces of ground. Hence, if any one species of grass were to go on varying, and those varieties were continually selected which differed from each other in at all the same manner as distinct species and genera of grasses differ from each other, a greater number of individual plants of this species of grass, including its modified descendants, would succeed in living on the same piece of ground. And we well know that each species and each variety of grass is annually sowing almost countless seeds; and thus, as it may be said, is striving its utmost to increase its numbers. Consequently, I cannot doubt that in the course of many thousands of generations, the most distinct varieties of any one species of grass would always have the best chance of succeeding and of increasing in numbers, and thus of supplanting the less distinct varieties; and varieties, when rendered very distinct from each other, take the rank of species.

The truth of the principle, that the greatest amount of life can be supported by great diversification of structure, is seen under many natural circumstances. In an extremely small area, especially if freely open to immigration, and where the contest between individual and individual must be severe, we always find great diversity in its inhabitants. For instance, I found that a piece of turf, three feet by four in size, which had been exposed for many years to exactly the same conditions, supported twenty species of plants, and these belonged to eighteen genera and to eight orders, which shows how much these plants differed from each other. So it is with the plants and insects on small and uniform islets; and so in small ponds of fresh water. Farmers find that they can raise most food by a rotation of plants belonging to the most different orders: nature follows what may be called a simultaneous rotation. Most of the animals and plants which live close round any small piece of ground, could live on it (supposing it not to be in any way peculiar in its nature), and may be said to be striving to the utmost to live there; but, it is seen, that where they come into the closest competition with each other, the advantages of diversification of structure, with the accompanying differences of habit and constitution, determine that the inhabitants, which thus jostle each other most closely, shall, as a general rule, belong to what we call different genera and orders.

The same principle is seen in the naturalisation of plants through man's agency in foreign lands. It might have been expected that the plants which have succeeded in becoming naturalised in any land would generally have been closely allied to the indigenes; for these are commonly looked at as specially created and adapted for their own country. It might, also, perhaps have been expected that naturalised plants would have belonged to a few groups more especially adapted to certain stations in their new homes. But the case is very different; and Alph. De Candolle has well remarked in his great and admirable work, that floras gain by naturalisation, proportionally with the number of the native genera and species, far more in new genera than in new species. To give a single instance: in the last edition of Dr. Asa Gray's 'Manual of the Flora of the Northern United States,' 260 naturalised plants are enumerated, and these belong to 162 genera. We thus see that these naturalised plants are of a highly diversified nature. They differ, moreover, to a large extent from the indigenes, for out of the 162 genera, no less than 100 genera are not there indigenous, and thus a large proportional addition is made to the genera of these States.

By considering the nature of the plants or animals which have struggled successfully with the indigenes of any country, and have there become naturalised, we can gain some crude idea in what manner some of the natives would have had to be modified, in order to have gained an advantage over the other natives; and we may, I think, at least safely infer that diversification of structure, amounting to new generic differences, would have been profitable to them.

The advantage of diversification in the inhabitants of the same region is, in fact, the same as that of the physiological division of labour in the organs of the same individual body—a subject so well elucidated by Milne Edwards. No physiologist doubts that a stomach by being adapted to digest vegetable matter alone, or flesh alone, draws most nutriment from these substances. So in the general economy of any land, the more widely and perfectly the animals and plants are diversified for different habits of life, so will a greater number of individuals be capable of there supporting themselves. A set of animals, with their organisation but little diversified, could hardly compete with a set more perfectly diversified in structure. It may be doubted, for instance, whether the Australian marsupials, which are divided into groups differing but little from each other, and feebly representing, as Mr. Waterhouse and others have remarked, our carnivorous, ruminant, and rodent mammals, could successfully compete with these well-pronounced orders. In the Australian mammals, we see the process of diversification in an early and incomplete stage of development.

After the foregoing discussion, which ought to have been much amplified, we may, I think, assume that the modified descendants of any one species will succeed by so much the better as they become more diversified in structure, and are thus enabled to encroach on places occupied by other beings. Now let us see how this principle of great benefit being derived from divergence of character, combined with the principles of natural selection and of extinction, will tend to act.

The accompanying diagram will aid us in understanding this rather perplexing subject. Let A to L represent the species of a genus large in its own country; these species are supposed to resemble each other in unequal degrees, as is so generally the case in nature, and as is represented in the diagram by the letters standing at unequal distances. I have said a large genus, because we have seen in the second chapter, that on an average more of the species of large genera vary than of small genera; and the varying species of the large genera present a greater number of varieties. We have, also, seen that the species, which are the commonest and the most widely-diffused, vary more than rare species with restricted ranges. Let (A) be a common, widely-diffused, and varying species, belonging to a genus large in its own country. The little fan of diverging dotted lines of unequal lengths proceeding from (A), may represent its varying offspring. The variations are supposed to be extremely slight, but of the most diversified nature; they are not supposed all to appear simultaneously, but often after long intervals of time; nor are they all supposed to endure for equal periods. Only those variations which are in some way profitable will be preserved or naturally selected. And here the importance of the principle of benefit being derived from divergence of character comes in; for this will generally lead to the most different or divergent variations (represented by the outer dotted lines) being preserved and accumulated by natural selection. When a dotted line reaches one of the horizontal lines, and is there marked by a small numbered letter, a sufficient amount of variation is supposed to have been accumulated

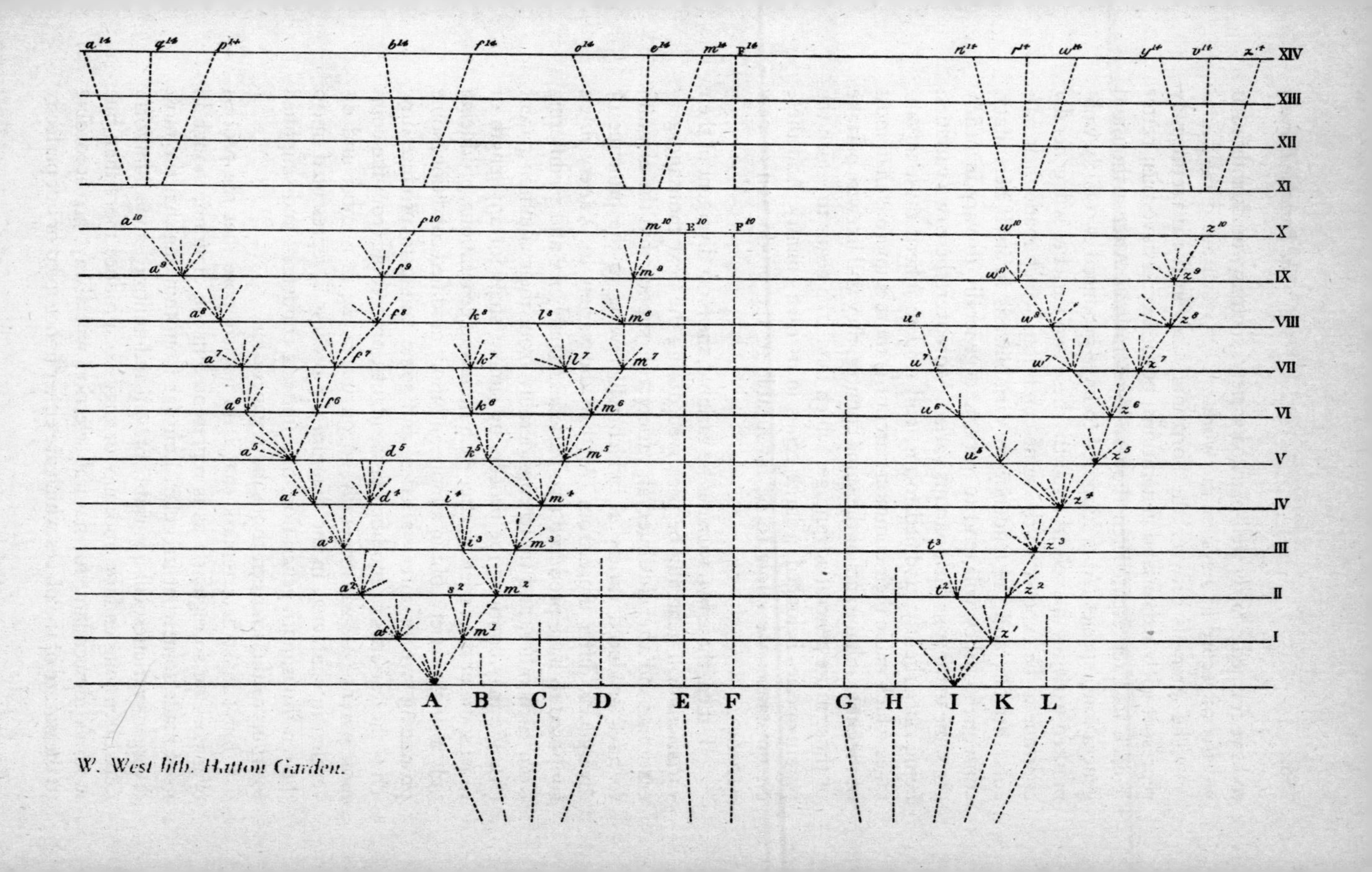

a14 q14 p14 b14 f14 o14 e14 m14 F14 n14 r14 w14 y14 v14 z14
XIV
XIII
XII
XI
a10 f10 m10 E10 F10 w10 z10
X
IX
VIII
VII
VI
V
IV
III
II
I
A B C D E F G H I K L
W. West lith. Hatton Garden.

to have formed a fairly well-marked variety, such as would be thought worthy of record in a systematic work.

The intervals between the horizontal lines in the diagram, may represent each a thousand generations; but it would have been better if each had represented ten thousand generations. After a thousand generations, species (A) is supposed to have produced two fairly well-marked varieties, namely a^1 and m^1. These two varieties will generally continue to be exposed to the same conditions which made their parents variable, and the tendency to variability is in itself hereditary, consequently they will tend to vary, and generally to vary in nearly the same manner as their parents varied. Moreover, these two varieties, being only slightly modified forms, will tend to inherit those advantages which made their common parent (A) more numerous than most of the other inhabitants of the same country; they will likewise partake of those more general advantages which made the genus to which the parent-species belonged, a large genus in its own country. And these circumstances we know to be favourable to the production of new varieties.

If, then, these two varieties be variable, the most divergent of their variations will generally be preserved during the next thousand generations. And after this interval, variety a^1 is supposed in the diagram to have produced variety a^2, which will, owing to the principle of divergence, differ more from (A) than did variety a^1. Variety m^1 is supposed to have produced two varieties, namely m^2 and s^2, differing from each other, and more considerably from their common parent (A). We may continue the process by similar steps for any length of time; some of the varieties, after each thousand generations, producing only a single variety, but in a more and more modified condition, some producing two or three varieties, and some failing to produce any. Thus the varieties or modified descendants, proceeding from the common parent (A), will generally go on increasing in number and diverging in character. In the diagram the process is represented up to the ten-thousandth generation, and under a condensed and simplified form up to the fourteen-thousandth generation.

But I must here remark that I do not suppose that the process ever goes on so regularly as is represented in the diagram, though in itself made somewhat irregular. I am far from thinking that the most divergent varieties will invariably prevail and multiply: a medium form may often long endure, and may or may not produce more than one modified descendant; for natural selection will always act according to the nature of the places which are either unoccupied or not perfectly occupied by other beings; and this will depend on infinitely complex relations. But as a general rule, the more diversified in structure the descendants from any one species can be rendered, the more places they will be enabled to seize on, and the more their modified progeny will be increased. In our diagram the line of succession is broken at regular intervals by small numbered letters marking the successive forms which have become sufficiently distinct to be recorded as varieties. But these breaks are imaginary, and might have been inserted anywhere, after intervals long enough to have allowed the accumulation of a considerable amount of divergent variation.

As all the modified descendants from a common and widely-diffused species, belonging to a large genus, will tend to partake of the same advantages which made their parent successful in life, they will generally go on multiplying in number as well as diverging in character: this is represented in the diagram by the several divergent branches proceeding from (A). The modified offspring from the later and more highly improved branches in the lines of descent, will, it is probable, often take the place of, and so destroy, the earlier and less improved branches: this is represented in the diagram by some of the lower branches not reaching to the upper horizontal lines. In some cases I do not doubt that the process of modification will be confined to a single line of descent, and the number of the descendants will not be increased; although the amount of divergent modification may have been increased in the successive generations. This case would be represented in the diagram, if all the lines proceeding from (A) were removed, excepting that from a^1 to a^{10}. In the same way, for instance, the English race-horse and English pointer have apparently both gone on slowly diverging in character from their original stocks, without either having given off any fresh branches or races.

After ten thousand generations, species (A) is supposed to have produced three forms, a^{10}, f^{10}, and m^{10}, which, from having diverged in character during the successive generations, will have come to differ largely, but perhaps unequally, from each other and from their common parent. If we suppose the amount of change between each horizontal line in our diagram to be excessively small, these three forms may still be only well-marked varieties; or they may have arrived at the doubtful category of sub-species; but we have only to suppose the steps in the process of modification to be more numerous or greater in amount, to convert these three forms into well-defined species: thus the diagram illustrates the steps by which the small differences distinguishing varieties are increased into the larger differences distinguishing species. By continuing the same process for a greater number of generations (as shown in the diagram in a condensed and simplified manner), we get eight species, marked by the letters between a^{14} and m^{14}, all descended from (A). Thus, as I believe, species are multiplied and genera are formed.

In a large genus it is probable that more than one species would vary. In the diagram I have assumed that a second species (I) has produced, by analogous steps, after ten thousand generations, either two well-marked varieties (w^{10} and z^{10}) or two species, according to the amount of change supposed to be represented between the horizontal lines. After fourteen thousand generations, six new species, marked by the letters n^{14} to z^{14}, are supposed to have been produced. In each genus, the species, which are already extremely different in character, will generally tend to produce the greatest number of modified descendants; for these will have the best chance of filling new and widely different places in the polity of nature: hence in the diagram I have chosen the extreme species (A), and the nearly extreme species (I), as those which have largely varied, and have given rise to new varieties and species. The other nine species (marked by capital letters) of our original genus, may for a long period continue transmitting unaltered descendants; and this is shown in the diagram by the dotted lines not prolonged far upwards from want of space.

But during the process of modification, represented in the diagram, another of our principles, namely that of extinction, will have played an important part. As in each fully stocked country natural selection necessarily acts by the selected form having some advantage in the struggle for life over other forms, there will be a constant tendency in the improved descendants of any one species to supplant and exterminate in each stage of descent their predecessors and their original parent. For it should be remembered that the competition will generally be most severe between those forms which are most nearly related to each other in habits, constitution, and structure. Hence all the intermediate forms between the earlier and later states, that is between the less and more improved state of a species, as well as the original parent-species itself, will generally tend to become extinct. So it probably will be with many whole collateral lines of descent, which will be conquered by later and improved lines of descent. If, however, the modified offspring of a species get into some distinct country, or become quickly adapted to some quite new station, in which child and parent do not come into competition, both may continue to exist.

If then our diagram be assumed to represent a considerable

amount of modification, species (A) and all the earlier varieties will have become extinct, having been replaced by eight new species (a^{14} to m^{14}); and (I) will have been replaced by six (n^{14} to z^{14}) new species.

But we may go further than this. The original species of our genus were supposed to resemble each other in unequal degrees, as is so generally the case in nature; species (A) being more nearly related to B, C, and D, than to the other species; and species (I) more to G, H, K, L, than to the others. These two species (A) and (I), were also supposed to be very common and widely diffused species, so that they must originally have had some advantage over most of the other species of the genus. Their modified descendants, fourteen in number at the fourteen-thousandth generation, will probably have inherited some of the same advantages: they have also been modified and improved in a diversified manner at each stage of descent, so as to have become adapted to many related places in the natural economy of their country. It seems, therefore, to me extremely probable that they will have taken the places of, and thus exterminated, not only their parents (A) and (I), but likewise some of the original species which were most nearly related to their parents. Hence very few of the original species will have transmitted offspring to the fourteen-thousandth generation. We may suppose that only one (F), of the two species which were least closely related to the other nine original species, has transmitted descendants to this late stage of descent.

The new species in our diagram descended from the original eleven species, will now be fifteen in number. Owing to the divergent tendency of natural selection, the extreme amount of difference in character between species a^{14} and z^{14} will be much greater than that between the most different of the original eleven species. The new species, moreover, will be allied to each other in a widely different manner. Of the eight descendants from (A) the three marked a^{14}, q^{14}, p^{14}, will be nearly related from having recently branched off from a^{10}; b^{14} and f^{14}, from having diverged at an earlier period from a^5, will be in some degree distinct from the three first-named species; and lastly, o^{14}, e^{14}, and m^{14} will be nearly related one to the other, but from having diverged at the first commencement of the process of modification, will be widely different from the other five species, and may constitute a sub-genus or even a distinct genus.

The six descendants from (I) will form two sub-genera or even genera. But as the original species (I) differed largely from (A), standing nearly at the extreme points of the original genus, the six descendants from (I) will, owing to inheritance, differ considerably from the eight descendants from (A); the two groups, moreover, are supposed to have gone on diverging in different directions. The intermediate species, also (and this is a very important consideration), which connected the original species (A) and (I), have all become, excepting (F), extinct, and have left no descendants. Hence the six new species descended from (I), and the eight descended from (A), will have to be ranked as very distinct genera, or even as distinct sub-families.

Thus it is, as I believe, that two or more genera are produced by descent, with modification, from two or more species of the same genus. And the two or more parent-species are supposed to have descended from some one species of an earlier genus. In our diagram, this is indicated by the broken lines, beneath the capital letters, converging in sub-branches downwards towards a single point; this point representing a single species, the supposed single parent of our several new sub-genera and genera.

It is worth while to reflect for a moment on the character of the new species F^{14}, which is supposed not to have diverged much in character, but to have retained the form of (F), either unaltered or altered only in a slight degree. In this case, its affinities to the other fourteen new species will be of a curious and circuitous nature. Having descended from a form which stood between the two parent-species (A) and (I), now supposed to be extinct and unknown, it will be in some degree intermediate in character between the two groups descended from these species. But as these two groups have gone on diverging in character from the type of their parents, the new species (F^{14}) will not be directly intermediate between them, but rather between types of the two groups; and every naturalist will be able to bring some such case before his mind.

In the diagram, each horizontal line has hitherto been supposed to represent a thousand generations, but each may represent a million or hundred million generations, and likewise a section of the successive strata of the earth's crust including extinct remains. We shall, when we come to our chapter on Geology, have to refer again to this subject, and I think we shall then see that the diagram throws light on the affinities of extinct beings, which, though generally belonging to the same orders, or families, or genera, with those now living, yet are often, in some degree, intermediate in character between existing groups; and we can understand this fact, for the extinct species lived at very ancient epochs when the branching lines of descent had diverged less.

I see no reason to limit the process of modification, as now explained, to the formation of genera alone. If, in our diagram, we suppose the amount of change represented by each successive group of diverging dotted lines to be very great, the forms marked a^{14} to p^{14}, those marked b^{14} and f^{14}, and those marked o^{14} to m^{14}, will form three very distinct genera. We shall also have two very distinct genera descended from (I); and as these latter two genera, both from continued divergence of character and from inheritance from a different parent, will differ widely from the three genera descended from (A), the two little groups of genera will form two distinct families, or even orders, according to the amount of divergent modification supposed to be represented in the diagram. And the two new families, or orders, will have descended from two species of the original genus; and these two species are supposed to have descended from one species of a still more ancient and unknown genus.

We have seen that in each country it is the species of the larger genera which oftenest present varieties or incipient species. This, indeed, might have been expected; for as natural selection acts through one form having some advantage over other forms in the struggle for existence, it will chiefly act on those which already have some advantage; and the largeness of any group shows that its species have inherited from a common ancestor some advantage in common. Hence, the struggle for the production of new and modified descendants, will mainly lie between the larger groups, which are all trying to increase in number. One large group will slowly conquer another large group, reduce its numbers, and thus lessen its chance of further variation and improvement. Within the same large group, the later and more highly perfected sub-groups, from branching out and seizing on many new places in the polity of Nature, will constantly tend to supplant and destroy the earlier and less improved sub-groups. Small and broken groups and sub-groups will finally tend to disappear. Looking to the future, we can predict that the groups of organic beings which are now large and triumphant, and which are least broken up, that is, which as yet have suffered least extinction, will for a long period continue to increase. But which groups will ultimately prevail, no man can predict; for we well know that many groups, formerly most extensively developed, have now become extinct. Looking still more remotely to the future, we may predict that, owing to the continued and steady increase of the larger groups, a multitude of smaller groups will become utterly extinct, and leave no modified descendants; and consequently that of the species living at any one period, extremely few will transmit descendants to a remote futurity. I shall have to return to this subject in the chapter on Classification, but I may add that on this view of extremely few of the more ancient species having transmitted descendants, and on the view of all the descendants of the same species making a class, we can understand how it is that there exist but very few classes in each main division of the animal and vegetable kingdoms. Although extremely few of the most ancient species may now have living and modified descendants, yet at the most remote geological period, the earth may have been as well peopled with many species of many genera, families, orders, and classes, as at the present day.

Summary of Chapter.— If during the long course of ages and under varying conditions of life, organic beings vary at all in the several parts of their organisation, and I think this cannot be disputed; if there be, owing to the high geometrical powers of increase of each species, at some age, season, or year, a severe struggle for life, and this certainly cannot be disputed; then, considering the infinite complexity of the relations of all organic beings to each other and to their conditions of existence, causing an infinite diversity in structure, constitution, and habits, to be advantageous to them, I think it would be a most extraordinary fact if no variation ever had occurred useful to each being's own welfare, in the same way as so many variations have occurred useful to man. But if variations useful to any organic being do occur, assuredly individuals thus characterised will have the best chance of being preserved in the struggle for life; and from the strong principle of inheritance they will tend to produce offspring similarly characterised. This principle of preservation, I have called, for the sake of brevity, Natural Selection. Natural selection, on the principle of qualities being inherited at corresponding ages, can modify the egg, seed, or young, as easily as the adult. Amongst many animals, sexual selection will give its aid to ordinary selection, by assuring to the most vigorous and best adapted males the greatest number of offspring. Sexual selection will also give characters useful to the males alone, in their struggles with other males.

Whether natural selection has really thus acted in nature, in modifying and adapting the various forms of life to their several conditions and stations, must be judged of by the general tenour and balance of evidence given in the following chapters. But we already see how it entails extinction; and how largely extinction has acted in the world's history, geology plainly declares. Natural selection, also, leads to divergence of character; for more living beings can be supported on the same area the more they diverge in structure, habits, and constitution, of which we see proof by looking at the inhabitants of any small spot or at naturalised productions. Therefore during the modification of the descendants of any one species, and during the incessant struggle of all species to increase in numbers, the more diversified these descendants become, the better will be their chance of succeeding in the battle of life. Thus the small differences distinguishing varieties of the same species, will steadily tend to increase till they come to equal the greater differences between species of the same genus, or even of distinct genera.

We have seen that it is the common, the widely-diffused, and widely-ranging species, belonging to the larger genera, which vary most; and these will tend to transmit to their modified offspring that superiority which now makes them dominant in their own countries. Natural selection, as has just been remarked, leads to divergence of character and to much extinction of the less improved and intermediate forms of life. On these principles, I believe, the nature of the affinities of all organic beings may be explained. It is a truly wonderful fact—the wonder of which we are apt to overlook from familiarity—that all animals and all plants throughout all time and space should be related to each other in group subordinate to group, in the manner which we everywhere behold—namely, varieties of the same species most closely related together, species of the same genus less closely and unequally related together, forming sections and sub-genera, species of distinct genera much less closely related, and genera related in different degrees, forming sub-families, families, orders, sub-classes, and classes. The several subordinate groups in any class cannot be ranked in a single file, but seem rather to be clustered round points, and these round other points, and so on in almost endless cycles. On the view that each species has been independently created, I can see no explanation of this great fact in the classification of all organic beings; but, to the best of my judgment, it is explained through inheritance and the complex action of natural selection, entailing extinction and divergence of character, as we have seen illustrated in the diagram.

The affinities of all the beings of the same class have sometimes been represented by a great tree. I believe this simile largely speaks the truth. The green and budding twigs may represent existing species; and those produced during each former year may represent the long succession of extinct species. At each period of growth all the growing twigs have tried to branch out on all sides, and to overtop and kill the surrounding twigs and branches, in the same manner as species and groups of species have tried to overmaster other species in the great battle for life. The limbs divided into great branches, and these into lesser and lesser branches, were themselves once, when the tree was small, budding twigs; and this connexion of the former and present buds by ramifying branches may well represent the classification of all extinct and living species in groups subordinate to groups. Of the many twigs which flourished when the tree was a mere bush, only two or three, now grown into great branches, yet survive and bear all the other branches; so with the species which lived during long-past geological periods, very few now have living and modified descendants. From the first growth of the tree, many a limb and branch has decayed and dropped off; and these lost branches of various sizes may represent those whole orders, families, and genera which have now no living representatives, and which are known to us only from having been found in a fossil state. As we here and there see a thin straggling branch springing from a fork low down in a tree, and which by some chance has been favoured and is still alive on its summit, so we occasionally see an animal like the Ornithorhynchus or Lepidosiren, which in some small degree connects by its affinities two large branches of life, and which has apparently been saved from fatal competition by having inhabited a protected station. As buds give rise by growth to fresh buds, and these, if vigorous, branch out and overtop on all sides many a feebler branch, so by generation I believe it has been with the great Tree of Life, which fills with its dead and broken branches the crust of the earth, and covers the surface with its ever branching and beautiful ramifications.

RECAPITULATION AND CONCLUSION.

As this whole volume is one long argument, it may be convenient to the reader to have the leading facts and inferences briefly recapitulated.

That many and grave objections may be advanced against the theory of descent with modification through natural selection, I do not deny. I have endeavoured to give to them their full force. Nothing at first can appear more difficult to believe than that the more complex organs and instincts should have been perfected, not by means superior to, though analogous with, human reason, but by the accumulation of innumerable slight variations, each good for the individual possessor. Nevertheless, this difficulty, though appearing to our imagination insuperably great, cannot be considered real if we admit the following propositions, namely,—that gradations in the perfection of any organ or instinct, which we may consider, either do now exist or could have existed, each good of its kind,—that all organs and instincts are, in ever so slight a degree, variable,—and, lastly, that there is a struggle for existence leading to the preservation of each profitable deviation of structure or instinct. The truth of these propositions cannot, I think, be disputed.

It is, no doubt, extremely difficult even to conjecture by what gradations many structures have been perfected, more especially amongst broken and failing groups of organic beings; but we see so many strange gradations in nature, as is proclaimed by the canon, "Natura non facit saltum," that we ought to be extremely cautious in saying that any organ or instinct, or any whole being, could not have arrived at its present state by many graduated steps. There are, it must be admitted, cases of special difficulty on the theory of natural selection; and one of the most curious of these is the existence of two or three defined castes of workers or sterile females in the same community of ants; but I have attempted to show how this difficulty can be mastered.

With respect to the almost universal sterility of species when first crossed, which forms so remarkable a contrast with the almost universal fertility of varieties when crossed, I must refer the reader to the recapitulation of the facts given at the end of the eighth chapter, which seem to me conclusively to show that this sterility is no more a special endowment than is the incapacity of two trees to be grafted together; but that it is incidental on constitutional differences in the reproductive systems of the intercrossed species. We see the truth of this conclusion in the vast difference in the result, when the same two species are crossed reciprocally; that is, when one species is first used as the father and then as the mother.

The fertility of varieties when intercrossed and of their mongrel offspring cannot be considered as universal; nor is their very general fertility surprising when we remember that it is not likely that either their constitutions or their reproductive systems should have been profoundly modified. Moreover, most of the varieties which have been experimentised on have been produced under domestication; and as domestication apparently tends to eliminate sterility, we ought not to expect it also to produce sterility.

The sterility of hybrids is a very different case from that of first crosses, for their reproductive organs are more or less functionally impotent; whereas in first crosses the organs on both sides are in a perfect condition. As we continually see that organisms of all kinds are rendered in some degree sterile from their constitutions having been disturbed by slightly different and new conditions of life, we need not feel surprise at hybrids being in some degree sterile, for their constitutions can hardly fail to have been disturbed from being compounded of two distinct organisations. This parallelism is supported by another parallel, but directly opposite, class of facts; namely, that the vigour and fertility of all organic beings are increased by slight changes in their conditions of life, and that the offspring of slightly modified forms or varieties acquire from being crossed increased vigour and fertility. So that, on the one hand, considerable changes in the conditions of life and crosses between greatly modified forms, lessen fertility; and on the other hand, lesser changes in the conditions of life and crosses between less modified forms, increase fertility.

Turning to geographical distribution, the difficulties encountered on the theory of descent with modification are grave enough. All the individuals of the same species, and all the species of the same genus, or even higher group, must have descended from common parents; and therefore, in however distant and isolated parts of the world they are now found, they must in the course of successive generations have passed from some one part to the others. We are often wholly unable even to conjecture how this could have been effected. Yet, as we have reason to believe that some species have retained the same specific form for very long periods, enormously long as measured by years, too much stress ought not to be laid on the occasional wide diffusion of the same species; for during very long periods of time there will always be a good chance for wide migration by many means. A broken or interrupted range may often be accounted for by the extinction of the species in the intermediate regions. It cannot be denied that we are as yet very ignorant of the full extent of the various climatal and geographical changes which have affected the earth during modern periods; and such changes will obviously have greatly facilitated migration. As an example, I have attempted to show how potent has been the influence of the Glacial period on the distribution both of the same and of representative species throughout the world. We are as yet profoundly ignorant of the many occasional means of transport. With respect to distinct species of the same genus inhabiting very distant and isolated regions, as the process of modification has necessarily been slow, all the means of migration will have been possible during a very long period; and consequently the difficulty of the wide diffusion of species of the same genus is in some degree lessened.

As on the theory of natural selection an interminable number of intermediate forms must have existed, linking together all the species in each group by gradations as fine as our present varieties, it may be asked, Why do we not see these linking forms all around us? Why are not all organic beings blended together in an inextricable chaos? With respect to existing forms, we should remember that we have no right to expect (excepting in rare cases) to discover *directly* connecting links between them, but only between each and some extinct and supplanted form. Even on a wide area, which has during a long period remained continuous, and of which the climate and other conditions of life change insensibly in going from a district occupied by one species into another district occupied by a closely allied species, we have no just right to expect often to find intermediate varieties in the intermediate zone. For we have reason to believe that only a few species are undergoing change at any one period; and all changes are slowly effected. I have also shown that the intermediate varieties which will at first probably exist in the intermediate zones, will be liable to be supplanted by the allied forms on either hand; and the latter, from existing in greater numbers, will generally be modified and improved at a quicker rate than the intermediate varieties, which exist in lesser numbers; so that the intermediate varieties will, in the long run, be supplanted and exterminated.

On this doctrine of the extermination of an infinitude of connecting links, between the living and extinct inhabitants of the world, and at each successive period between the extinct and still older species, why is not every geological formation charged with such links? Why does not every collection of fossil remains afford plain evidence of the gradation and mutation of the forms of life? We meet with no such evidence, and this is the most obvious and forcible of the many objections which may be urged against my theory. Why, again, do whole groups of allied species appear, though certainly they often falsely appear, to have come in suddenly on the several geological stages? Why do we not find great piles of strata beneath the Silurian system, stored with the remains of the progenitors of the Silurian groups of fossils? For certainly on my theory such strata must somewhere have been deposited at these ancient and utterly unknown epochs in the world's history.

I can answer these questions and grave objections only on the supposition that the geological record is far more imperfect than most geologists believe. It cannot be objected that there has not been time sufficient for any amount of organic change; for the lapse of time has been so great as to be utterly inappreciable by the human intellect. The number of specimens in all our museums is absolutely as nothing compared with the countless generations of countless species which certainly have existed. We should not be able to recognize a species as the parent of any one or more species if we were to examine them ever so closely, unless we likewise possessed many of the intermediate links between their past or parent and present states; and these many links we could hardly ever expect to discover, owing to the imperfection of the geological record. Numerous existing doubtful forms could be named which are probably varieties; but who will pretend that in future ages so many fossil links will be discovered, that naturalists will be able to decide, on the common view, whether or not these doubtful forms are varieties? As long as most of the links between any two species are unknown, if any one link or intermediate variety be discovered, it will simply be classed as another and distinct species. Only a small portion of the world has been geologically explored. Only organic beings of certain classes can be preserved in a fossil condition, at least in any great number. Widely ranging species vary most, and varieties are often at first local,—both causes rendering the discovery of intermediate links less likely. Local varieties will not spread into other and distant regions until they are considerably modified and improved; and when they do spread, if discovered in a geological formation, they will appear as if suddenly created there, and will be simply classed as new species. Most formations have been intermittent in their accumulation; and their duration, I am inclined to believe, has been shorter than the average duration of specific forms. Successive formations are separated from each other by enormous blank intervals of time; for fossiliferous formations, thick enough to resist future degradation, can be accumulated only where much sediment is deposited on the subsiding bed of the sea. During the alternate periods of elevation and of stationary level the record will be blank. During these latter periods there will probably be more variability in the forms of life; during periods of subsidence, more extinction.

With respect to the absence of fossiliferous formations beneath the lowest Silurian strata, I can only recur to the hypothesis given in the ninth chapter. That the geological record is imperfect all will admit; but that it is imperfect to the degree which I require, few will be inclined to admit. If we look to long enough intervals of time, geology plainly declares that all species have changed; and they have changed in the manner which my theory requires, for they have changed slowly and in a graduated manner. We clearly see this in the fossil remains from consecutive formations invariably being much more closely related to each other, than are the fossils from formations distant from each other in time.

Such is the sum of the several chief objections and difficulties which may justly be urged against my theory; and I have now briefly recapitulated the answers and explanations which can be given to them. I have felt these difficulties far too heavily during many years to doubt their weight. But it deserves especial notice that the more important objections relate to questions on which we are confessedly ignorant; nor do we know how ignorant we are. We do not know all the possible transitional gradations between the simplest and the most perfect organs; it cannot be pretended that we know all the varied means of Distribution during the long lapse of years, or that we know how imperfect the Geological Record is. Grave as these several difficulties are, in my judgment they do not overthrow the theory of descent with modification.

Now let us turn to the other side of the argument. Under domestication we see much variability. This seems to be mainly due to the reproductive system being eminently susceptible to changes in the conditions of life; so that this system, when not rendered impotent, fails to reproduce offspring exactly like the parent-form. Variability is governed by many complex laws,—by correlation of growth, by use and disuse, and by the direct action of the physical conditions of life. There is much difficulty in ascertaining how much modification our domestic productions have undergone; but we may safely infer that the amount has been large, and that modifications can be inherited for long periods. As long as the conditions of life remain the same, we have reason to believe that a modification, which has already been inherited for many generations, may continue to be inherited for an almost infinite number of generations. On the other hand we have evidence that variability, when it has once come into play, does not wholly cease; for new varieties are still occasionally produced by our most anciently domesticated productions.

Man does not actually produce variability; he only unintentionally exposes organic beings to new conditions of life, and then nature acts on the organisation, and causes variability. But man can and does select the variations given to him by nature, and thus accumulate them in any desired manner. He thus adapts animals and plants for his own benefit or pleasure. He may do this methodically, or he may do it unconsciously by preserving the individuals most useful to him at the time, without any thought of altering the breed. It is certain that he can largely influence the character of a breed by selecting, in each successive generation, individual differences so slight as to be quite inappreciable by an uneducated eye. This process of selection has been the great agency in the production of the most distinct and useful domestic breeds. That many of the breeds produced by man have to a large extent the character of natural species, is shown by the inextricable doubts whether very many of them are varieties or aboriginal species.

There is no obvious reason why the principles which have acted so efficiently under domestication should not have acted under nature. In the preservation of favoured individuals and races, during the constantly-recurrent Struggle for Existence, we see the most powerful and ever-acting means of selection. The struggle for existence inevitably follows from the high geometrical ratio of increase which is common to all organic beings. This high rate of increase is proved by calculation, by the effects of a succession of peculiar seasons, and by the results of naturalisation, as explained in the third chapter. More individuals are born than can possibly survive. A grain in the balance will determine which individual shall live and which shall die,—which variety or species shall increase in number, and which shall decrease, or finally become extinct. As the individuals of the same species come in all respects into the closest competition with each other, the struggle will generally be most severe between them; it will be almost equally severe between the varieties of the same species, and next in severity between the species of the same genus. But the struggle will often be very severe between beings most remote in the scale of nature. The slightest advantage in one being, at any age or during any season, over those with which it comes into competition, or better adaptation in however slight a degree to the surrounding physical conditions, will turn the balance.

With animals having separated sexes there will in most cases be a struggle between the males for possession of the females. The most vigorous individuals, or those which have most successfully struggled with their conditions of life, will generally leave most progeny. But success will often depend on having special weapons or means of defence, or on the charms of the males; and the slightest advantage will lead to victory.

As geology plainly proclaims that each land has undergone great physical changes, we might have expected that organic beings would have varied under nature, in the same way as they generally have varied under the changed conditions of domestication. And if there be any variability under nature, it would be an unaccountable fact if natural selection had not come into play. It has often been asserted, but the assertion is quite incapable of proof, that the amount of variation under nature is a strictly limited quantity. Man, though acting on external characters alone and often capriciously, can produce within a short period a great result by adding up mere individual differences in his domestic productions; and every one admits that there are at least individual differences in species under nature. But, besides such differences, all naturalists have admitted the existence of varieties, which they think sufficiently distinct to be worthy of record in systematic works. No one can draw any clear distinction between individual differences and slight varieties; or between more plainly marked varieties and sub-species, and species. Let it be observed how naturalists differ in the rank which they assign to the many representative forms in Europe and North America.

If then we have under nature variability and a powerful agent always ready to act and select, why should we doubt that variations in any way useful to beings, under their excessively complex relations of life, would be preserved, accumulated, and inherited? Why, if man can by patience select variations most useful to himself, should nature fail in selecting variations useful, under changing conditions of life, to her living products? What limit can be put to this power, acting during long ages and rigidly scrutinising the whole constitution, structure, and habits of each creature,—favouring the good and rejecting the bad? I can see no limit to this power, in slowly and beautifully adapting each form to the most complex relations of life. The theory of natural selection, even if we looked no further than this, seems to me to be in itself probable. I have already recapitulated, as fairly as I could, the opposed difficulties and objections: now let us turn to the special facts and arguments in favour of the theory.

On the view that species are only strongly marked and permanent varieties, and that each species first existed as a variety, we can see why it is that no line of demarcation can be drawn between species, commonly supposed to have been produced by special acts of creation, and varieties which are acknowledged to have been produced by secondary laws. On this same view we can understand how it is that in each region where many species of a genus have been produced, and where they now flourish, these same species should present many varieties; for where the manufactory of species has been active, we might expect, as a general rule, to find it still in action; and this is the case if varieties be incipient species. Moreover, the species of the larger genera, which afford the greater number of varieties or incipient species, retain to a certain degree the character of varieties; for they differ from each other by a less amount of difference than do the species of smaller genera. The closely allied species also of the larger genera apparently have restricted ranges, and they are clustered in little groups round other species—in which respects they resemble varieties. These are strange relations on the view of each species having been independently created, but are intelligible if all species first existed as varieties.

As each species tends by its geometrical ratio of reproduction to increase inordinately in number; and as the modified descendants of each species will be enabled to increase by so much the more as they become more diversified in habits and structure, so as to be enabled to seize on many and widely different places in the economy of nature, there will be a constant tendency in natural selection to preserve the most divergent offspring of any one species. Hence during a long-continued course of modification, the slight differences, characteristic of varieties of the same species, tend to be augmented into the greater differences characteristic of species of the same genus. New and improved varieties will inevitably supplant and exterminate the older, less improved and intermediate varieties; and thus species are rendered to a large extent defined and distinct objects. Dominant species belonging to the larger groups tend to give birth to new and dominant forms; so that each large group tends to become still larger, and at the same time more divergent in character. But as all groups cannot thus succeed in increasing in size, for the world would not hold them, the more dominant groups beat the less dominant. This tendency in the large groups to go on increasing in size and diverging in character, together with the almost inevitable contingency of much extinction, explains the arrangement of all the forms of life, in groups subordinate to groups, all within a few great classes, which we now see everywhere around us, and which has prevailed throughout all time. This grand fact of the grouping of all organic beings seems to me utterly inexplicable on the theory of creation.

As natural selection acts solely by accumulating slight, successive, favourable variations, it can produce no great or sudden modification; it can act only by very short and slow steps. Hence the canon of "Natura non facit saltum," which every fresh addition to our knowledge tends to make more strictly correct, is on this theory simply intelligible. We can plainly see why nature is prodigal in variety, though niggard in innovation. But why this should be a law of nature if each species has been independently created, no man can explain.

Many other facts are, as it seems to me, explicable on this theory. How strange it is that a bird, under the form of woodpecker, should have been created to prey on insects on the ground; that upland geese, which never or rarely swim, should have been created with webbed feet; that a thrush should have been created to dive and feed on sub-aquatic insects; and that a petrel should have been created with habits and structure fitting it for the life of an auk or grebe! and so on in endless other cases. But on the view of each species constantly trying to increase in number, with natural selection always ready to adapt the slowly varying descendants of each to any unoccupied or ill-occupied place in nature, these facts cease to be strange, or perhaps might even have been anticipated.

As natural selection acts by competition, it adapts the inhabitants of each country only in relation to the degree of perfection of their associates; so that we need feel no surprise at the inhabitants of any one country, although on the ordinary view supposed to have been specially created and adapted for that country, being beaten and supplanted by the naturalised productions from another land. Nor ought we to marvel if all the contrivances in nature be not, as far as we can judge, absolutely perfect; and if some of them be abhorrent to our ideas of fitness. We need not marvel at the sting of the bee causing the bee's own death; at drones being produced in such vast numbers for one single act, and being then slaughtered by their sterile sisters; at the astonishing waste of pollen by our fir-trees; at the instinctive hatred of the queen bee for her own fertile daughters; at ichneumonidæ) feeding within the live bodies of caterpillars; and at other such cases. The wonder indeed is, on the theory of natural selection, that more cases of the want of absolute perfection have not been observed.

The complex and little known laws governing variation are the same, as far as we can see, with the laws which have governed the production of so-called specific forms. In both cases physical conditions seem to have produced but little direct effect; yet when varieties enter any zone, they occasionally assume some of the characters of the species proper to that zone. In both varieties and species, use and disuse seem to have produced some effect; for it is difficult to resist this conclusion when we look, for instance, at the logger-headed duck, which has wings incapable of flight, in nearly the same condition as in the domestic duck; or when we look at the burrowing tucutucu, which is occasionally blind, and then at certain moles, which are habitually blind and have their eyes covered with skin; or when we look at the blind animals inhabiting the dark caves of America and Europe. In both varieties and species correlation of growth seems to have played a most important part, so that when one part has been modified other parts are necessarily modified. In both varieties and species reversions to long-lost characters occur. How inexplicable on the theory of creation is the occasional appearance of stripes on the shoulder and legs of the several species of the horse-genus and in their hybrids! How simply is this fact explained if we believe that these species have descended from a striped progenitor, in the same manner as the several domestic breeds of pigeon have descended from the blue and barred rock-pigeon!

On the ordinary view of each species having been independently created, why should the specific characters, or those by which the species of the same genus differ from each other, be more variable than the generic characters in which they all agree? Why, for instance, should the colour of a flower be more likely to vary in any one species of a genus, if the other species, supposed to have been created independently, have differently coloured flowers, than if all the species of the genus have the same coloured flowers? If species are only well-marked varieties, of which the characters have become in a high degree permanent, we can understand this fact; for they have already varied since they branched off from a common progenitor in certain characters, by which they have come to be specifically distinct from each other; and therefore these same characters would be more likely still to be variable than the generic characters which have been inherited without change for an enormous period. It is inexplicable on the theory of creation why a part developed in a very unusual manner in any one species of a genus, and therefore, as we may naturally infer, of great importance to the species, should be eminently liable to variation; but, on my view, this part has undergone, since the several species branched off from a common progenitor, an unusual amount of variability and modification, and therefore we might expect this part generally to be still variable. But a part may be developed in the most unusual manner, like the wing of a bat, and yet not be more variable than any other structure, if the part be common to many subordinate forms, that is, if it has been inherited for a very long period; for in this case it will have been rendered constant by long-continued natural selection.

Glancing at instincts, marvellous as some are, they offer no greater difficulty than does corporeal structure on the theory of the natural selection of successive, slight, but profitable modifications. We can thus understand why nature moves by graduated steps in endowing different animals of the same class with their several instincts. I have attempted to show how much light the principle of gradation throws on the admirable architectural powers of the hive-bee. Habit no doubt sometimes comes into play in modifying instincts; but it certainly is not indispensable, as we see, in the case of neuter insects, which leave no progeny to inherit the effects of long-continued habit. On the view of all the species of the same genus having descended from a common parent, and having inherited much in common, we can understand how it is that allied species, when placed under considerably different conditions of life, yet should follow nearly the same instincts; why the thrush of South America, for instance, lines her nest with mud like our British species. On the view of instincts having been slowly acquired through natural selection we need not marvel at some instincts being apparently not perfect and liable to mistakes, and at many instincts causing other animals to suffer.

If species be only well-marked and permanent varieties, we can at once see why their crossed offspring should follow the same complex laws in their degrees and kinds of resemblance to their parents,—in being absorbed into each other by successive crosses, and in other such points,—as do the crossed offspring of acknowledged varieties. On the other hand, these would be strange facts if species have been independently created, and varieties have been produced by secondary laws.

If we admit that the geological record is imperfect in an extreme degree, then such facts as the record gives, support the theory of descent with modification. New species have come on the stage slowly and at successive intervals; and the amount of change, after equal intervals of time, is widely different in different groups. The extinction of species and of whole groups of species, which has played so conspicuous a part in the history of the organic world, almost inevitably follows on the principle of natural selection; for old forms will be supplanted by new and improved forms. Neither single species nor groups of species reappear when the chain of ordinary generation has once been broken. The gradual diffusion of dominant forms, with the slow modification of their descendants, causes the forms of life, after long intervals of time, to appear as if they had changed simultaneously throughout the world. The fact of the fossil remains of each formation being in some degree intermediate in character between the fossils in the formations above and below, is simply explained by their intermediate position in the chain of descent. The grand fact that all extinct organic beings belong to the same system with recent beings, falling either into the same or into intermediate groups, follows from the living and the extinct being the offspring of common parents. As the groups which have descended from an ancient progenitor have generally diverged in character, the progenitor with its early descendants will often be intermediate in character in comparison with its later descendants; and thus we can see why the more ancient a fossil is, the oftener it stands in some degree intermediate between existing and allied groups. Recent forms are generally looked at as being, in some vague sense, higher than ancient and extinct forms; and they are in so far higher as the later and more improved forms have conquered the older and less improved organic beings in the struggle for life. Lastly, the law of the long endurance of allied forms on the same continent, —of marsupials in Australia, of edentata in America, and other such cases,—is intelligible, for within a confined country, the recent and the extinct will naturally be allied by descent.

Looking to geographical distribution, if we admit that there has been during the long course of ages much migration from one part of the world to another, owing to former climatal and geographical changes and to the many occasional and unknown means of dispersal, then we can understand, on the theory of descent with modification, most of the great leading facts in Distribution. We can see why there should be so striking a parallelism in the distribution of organic beings throughout space, and in their geological succession throughout time; for in both cases the beings have been connected by the bond of ordinary generation, and the means of modification have been the same. We see the full meaning of the wonderful fact, which must have struck every traveller, namely, that on the same continent, under the most diverse conditions, under heat and cold, on mountain and lowland, on deserts and marshes, most of the inhabitants within each great class are plainly related; for they will generally be descendants of the same progenitors and early colonists. On this same principle of former migration, combined in most cases with modification, we can understand, by the aid of the Glacial period, the identity of some few plants, and the close alliance of many others, on the most distant mountains, under the most different climates; and likewise the close alliance of some of the inhabitants of the sea in the northern and southern temperate zones, though separated by the whole intertropical ocean. Although two areas may present the same physical conditions of life, we need feel no surprise at their inhabitants being widely different, if they have been for a long period completely separated from each other; for as the relation of organism to organism is the most important of all relations, and as the two areas will have received colonists from some third source or from each other, at various periods and in different proportions, the course of modification in the two areas will inevitably be different.

On this view of migration, with subsequent modification, we can see why oceanic islands should be inhabited by few species, but of these, that many should be peculiar. We can clearly see why those animals which cannot cross wide spaces of ocean, as frogs and terrestrial mammals, should not inhabit oceanic islands; and why, on the other hand, new and peculiar species of bats, which can traverse the ocean, should so often be found on islands far distant from any continent. Such facts as the presence of peculiar species of bats, and the absence of all other mammals, on oceanic islands, are utterly inexplicable on the theory of independent acts of creation.

The existence of closely allied or representative species in any two areas, implies, on the theory of descent with modification, that the same parents formerly inhabited both areas; and we almost invariably find that wherever many closely allied species inhabit two areas, some identical species common to both still exist. Wherever many closely allied yet distinct species occur, many doubtful forms and varieties of the same species likewise occur. It is a rule of high generality that the inhabitants of each area are related to the inhabitants of the nearest source whence immigrants might have been derived. We see this in nearly all the plants and animals of the Galapagos archipelago, of Juan Fernandez, and of the other American islands being related in the most striking manner to the plants and animals of the neighbouring American mainland; and those of the Cape de Verde archipelago and other African islands to the African mainland. It must be admitted that these facts receive no explanation on the theory of creation.

The fact, as we have seen, that all past and present organic beings constitute one grand natural system, with group subordinate to group, and with extinct groups often falling in between recent groups, is intelligible on the theory of natural selection with its contingencies of extinction and divergence of character. On these same principles we see how it is, that the mutual affinities of the species and genera within each class are so complex and circuitous. We see why certain characters are far more serviceable than others for classification;—why adaptive characters, though of paramount importance to the being, are of hardly any importance in classification; why characters derived from rudimentary parts, though of no service to the being, are often of high classificatory value; and why embryological characters are the most valuable of all. The real affinities of all organic beings are due to inheritance or community of descent. The natural system is a genealogical arrangement, in which we have to discover the lines of descent by the most permanent characters, however slight their vital importance may be.

The framework of bones being the same in the hand of a man, wing of a bat, fin of the porpoise, and leg of the horse,—the same number of vertebræ forming the neck of the giraffe and of the elephant,—and innumerable other such facts, at once explain themselves on the theory of descent with slow and slight successive modifications. The similarity of pattern in the wing and leg of a bat, though used for such different purpose,—in the jaws and legs of a crab,—in the petals, stamens, and pistils of a flower, is likewise intelligible on the view of the gradual modification of parts or organs, which were alike in the early progenitor of each class. On the principle of successive variations not always supervening at an early age, and being inherited at a corresponding not early period of life, we can clearly see why the embryos of mammals, birds, reptiles, and fishes should be so closely alike, and should be so unlike the adult forms. We may cease marvelling at the embryo of an air-breathing mammal or bird having branchial slits and arteries running in loops, like those in a fish which has to breathe the air dissolved in water, by the aid of well-developed branchiæ.

Disuse, aided sometimes by natural selection, will often tend to reduce an organ, when it has become useless by changed habits or under changed conditions of life; and we can clearly understand on this view the meaning of rudimentary organs. But disuse and selection will generally act on each creature, when it has come to maturity and has to play its full part in the struggle for existence, and will thus have little power of acting on an organ during early life; hence the organ will not be much reduced or rendered rudimentary at this early age. The calf, for instance, has inherited teeth, which never cut through the gums of the upper jaw, from an early progenitor having well-developed teeth; and we may believe, that the teeth in the mature animal were reduced, during successive generations, by disuse or by the tongue and palate having been fitted by natural selection to browse without their aid; whereas in the calf, the teeth have been left untouched by selection or disuse, and on the principle of inheritance at corresponding ages have been inherited from a remote period to the present day. On the view of each organic being and each separate organ having been specially created, how utterly inexplicable it is that parts, like the teeth in the embryonic calf or like the shrivelled wings under the soldered wing-covers of some beetles, should thus so frequently bear the plain stamp of inutility! Nature may be said to have taken pains to reveal, by rudimentary organs and by homologous structures, her scheme of modification, which it seems that we wilfully will not understand.

I have now recapitulated the chief facts and considerations which have thoroughly convinced me that species have changed, and are still slowly changing by the preservation and accumulation of successive slight favourable variations. Why, it may be asked, have all the most eminent living naturalists and geologists rejected this view of the mutability of species? It cannot be asserted that organic beings in a state of nature are subject to no variation; it cannot be proved that the amount of variation in the course of long ages is a limited quantity; no clear distinction has been, or can be, drawn between species and well-marked varieties. It cannot be maintained that species when intercrossed are invariably sterile, and varieties invariably fertile; or that sterility is a special endowment and sign of creation. The belief that species were immutable productions was almost unavoidable as long as the history of the world was thought to be of short duration; and now that we have acquired some idea of the lapse of time, we are too apt to assume, without proof, that the geological record is so perfect that it would have afforded us plain evidence of the mutation of species, if they had undergone mutation.

But the chief cause of our natural unwillingness to admit that one species has given birth to other and distinct species, is that we are always slow in admitting any great change of which we do not see the intermediate steps. The difficulty is the same as that felt by so many geologists, when Lyell first insisted that long lines of inland cliffs had been formed, and great valleys excavated, by the slow action of the coast-waves. The mind cannot possibly grasp the full meaning of the term of a hundred million years; it cannot add up and perceive the full effects of many slight variations, accumulated during an almost infinite number of generations.

Although I am fully convinced of the truth of the views given in this volume under the form of an abstract, I by no means expect to convince experienced naturalists whose minds are stocked with a multitude of facts all viewed, during a long course of years, from a point of view directly opposite to mine. It is so easy to hide our ignorance under such expressions as the "plan of creation," "unity of design," &c., and to think that we give an explanation when we only restate a fact. Any one whose disposition leads him to attach more weight to unexplained difficulties than to the explanation of a certain number of facts will certainly reject my theory. A few naturalists, endowed with much flexibility of mind, and who have already begun to doubt on the immutability of species, may be influenced by this volume; but I look with confidence to the future, to young and rising naturalists, who will be able to view both sides of the question with impartiality. Whoever is led to believe that species are mutable will do good service by conscientiously expressing his conviction; for only thus can the load of prejudice by which this subject is overwhelmed be removed.

Several eminent naturalists have of late published their belief that a multitude of reputed species in each genus are not real species; but that other species are real, that is, have been independently created. This seems to me a strange conclusion to arrive at. They admit that a multitude of forms, which till lately they themselves thought were special creations, and which are still thus looked at by the majority of naturalists, and which consequently have every external characteristic feature of true species,—they admit that these have been produced by variation, but they refuse to extend the same view to other and very slightly different forms. Nevertheless they do not pretend that they can define, or even conjecture, which are the created forms of life, and which are those produced by secondary laws. They admit variation as a *vera causa* in one case, they arbitrarily reject it in another, without assigning any distinction in the two cases. The day will come when this will be given as a curious illustration of the blindness of preconceived opinion. These authors seem no more startled at a miraculous act of creation than at an ordinary birth. But do they really believe that at innumerable periods in the earth's history certain elemental atoms have been commanded suddenly to flash into living tissues? Do they believe that at each supposed act of creation one individual or many were produced? Were all the infinitely numerous kinds of animals and plants created as eggs or seed, or as full grown? and in the case of mammals, were they created bearing the false marks of nourishment from the mother's womb? Although naturalists very properly demand a full explanation of every difficulty from those who believe in the mutability of species, on their own side they ignore the whole subject of the first appearance of species in what they consider reverent silence.

It may be asked how far I extend the doctrine of the modification of species. The question is difficult to answer, because the more distinct the forms are which we may consider, by so much the arguments fall away in force. But some arguments of the greatest weight extend very far. All the members of whole classes can be connected together by chains of affinities, and all can be classified on the same principle, in groups subordinate to groups. Fossil remains sometimes tend to fill up very wide intervals between existing orders. Organs in a rudimentary condition plainly show that an early progenitor had the organ in a fully developed state; and this in some instances necessarily implies an enormous amount of modification in the descendants. Throughout whole classes various structures are formed on the same pattern, and at an embryonic age the species closely resemble each other. Therefore I cannot doubt that the theory of descent with modification embraces all the members of the same class. I believe that animals have descended from at most only four or five progenitors, and plants from an equal or lesser number.

Analogy would lead me one step further, namely, to the belief that all animals and plants have descended from some one prototype. But analogy may be a deceitful guide. Nevertheless all living things have much in common, in their chemical composition, their germinal vesicles, their cellular structure, and their laws of growth and reproduction. We see this even in so trifling a circumstance as that the same poison often similarly affects plants and animals; or that the poison secreted by the gall-fly produces monstrous growths on the wild rose or oak-tree. Therefore I should infer from analogy that probably all the organic beings which have ever lived on this earth have descended from some one primordial form, into which life was first breathed.

When the views entertained in this volume on the origin of species, or when analogous views are generally admitted, we can dimly foresee that there will be a considerable revolution in natural history. Systematists will be able to pursue their labours as at present; but they will not be incessantly haunted by the shadowy doubt whether this or that form be in essence a species. This I feel sure, and I speak after experience, will be no slight relief. The endless disputes whether or not some fifty species of British brambles are true species will cease. Systematists will have only to decide (not that this will be easy) whether any form be sufficiently constant and distinct from other forms, to be capable of definition; and if definable, whether the differences be sufficiently important to deserve a specific name. This latter point will become a far more essential consideration than it is at present; for differences, however slight, between any two forms, if not blended by intermediate gradations, are looked at by most naturalists as sufficient to raise both forms to the rank of species. Hereafter we shall be compelled to acknowledge that the only distinction between species and well-marked varieties is, that the latter are known, or believed, to be connected at the present day by intermediate gradations, whereas species were formerly thus connected. Hence, without quite rejecting the consideration of the present existence of intermediate gradations between any two forms, we shall be led to weigh more carefully and to value higher the actual amount of difference between them. It is quite possible that forms now generally acknowledged to be merely varieties may hereafter be thought worthy of specific names, as with the primrose and cowslip; and in this case scientific and common language will come into accordance. In short, we shall have to treat species in the same manner as those naturalists treat genera, who admit that genera are merely artificial combinations made for convenience. This may not be a cheering prospect; but we shall at least be freed from the vain search for the undiscovered and undiscoverable essence of the term species.

The other and more general departments of natural history will rise greatly in interest. The terms used by naturalists of affinity, relationship, community of type, paternity, morphology, adaptive characters, rudimentary and aborted organs, &c., will cease to be metaphorical, and will have a plain signification. When we no longer look at an organic being as a savage looks at a ship, as at something wholly beyond his comprehension; when we regard every production of nature as one which has had a history; when we contemplate every complex structure and instinct as the summing up of many contrivances, each useful to the possessor, nearly in the same way as when we look at any great mechanical invention as the summing up of the labour, the experience, the reason, and even the blunders of numerous workmen; when we thus view each organic being, how far more interesting, I speak from experience, will the study of natural history become!

A grand and almost untrodden field of inquiry will be opened, on the causes and laws of variation, on correlation of growth, on the effects of use and disuse, on the direct action of external conditions, and so forth. The study of domestic productions will rise immensely in value. A new variety raised by man will be a far more important and interesting subject for study than one more species added to the infinitude of already recorded species. Our classifications will come to be, as far as they can be so made, genealogies; and will then truly give what may be called the plan of creation. The rules for classifying will no doubt become simpler when we have a definite object in view. We possess no pedigrees or armorial bearings; and we have to discover and trace the many diverging lines of descent in our natural genealogies, by characters of any kind which have long been inherited. Rudimentary organs will speak infallibly with respect to the nature of long-lost structures. Species and groups of species, which are called aberrant, and which may fancifully be called living fossils, will aid us in forming a picture of the ancient forms of life. Embryology will reveal to us the structure, in some degree obscured, of the prototypes of each great class.

When we can feel assured that all the individuals of the same species, and all the closely allied species of most genera, have within a not very remote period descended from one parent, and have migrated from some one birthplace; and when we better know the many means of migration, then, by the light which geology now throws, and will continue to throw, on former changes of climate and of the level of the land, we shall surely be enabled to trace in an admirable manner the former migrations of the inhabitants of the whole world. Even at present, by comparing the differences of the inhabitants of the sea on the opposite sides of a continent, and the nature of the various inhabitants of that continent in relation to their apparent means of immigration, some light can be thrown on ancient geography.

The noble science of Geology loses glory from the extreme imperfection of the record. The crust of the earth with its embedded remains must not be looked at as a well-filled museum, but as a poor collection made at hazard and at rare intervals. The accumulation of each great fossiliferous formation will be recognised as having depended on an unusual concurrence of circumstances, and the blank intervals between the successive stages as having been of vast duration. But we shall be able to gauge with some security the duration of these intervals by a comparison of the preceding and succeeding organic forms. We must be cautious in attempting to correlate as strictly contemporaneous two formations, which include few identical species, by the general succession of their forms of life. As species are produced and exterminated by slowly acting and still existing causes, and not by miraculous acts of creation and by catastrophes; and as the most important of all causes of organic change is one which is almost independent of altered and perhaps suddenly altered physical conditions, namely, the mutual relation of organism to organism,—the improvement of one being entailing the improvement or the extermination of others; it follows, that the amount of organic change in the fossils of consecutive formations probably serves as a fair measure of the lapse of actual time. A number of species, however, keeping in a body might remain for a long period unchanged, whilst within this same period, several of these species, by migrating into new countries and coming into competition with foreign associates, might become modified; so that we must not overrate the accuracy of organic change as a measure of time. During early periods of the earth's history, when the forms of life were probably fewer and simpler, the rate of change was probably slower; and at the first dawn of life, when very few forms of the simplest structure existed, the rate of change may have been slow in an extreme degree. The whole history of the world, as at present known, although of a length quite incomprehensible by us, will hereafter be recognised as a mere fragment of time, compared with the ages which have elapsed since the first creature, the progenitor of innumerable extinct and living descendants, was created.

In the distant future I see open fields for far more important researches. Psychology will be based on a new foundation, that of the necessary acquirement of each mental power and capacity by gradation. Light will be thrown on the origin of man and his history.

Authors of the highest eminence seem to be fully satisfied with the view that each species has been independently created. To my mind it accords better with what we know of the laws impressed on matter by the Creator, that the production and extinction of the past and present inhabitants of the world should have been due to secondary causes, like those determining the birth and death of the individual. When I view all beings not as special creations, but as the lineal descendants of some few beings which lived long before the first bed of the Silurian system was deposited, they seem to me to become ennobled. Judging from the past, we may safely infer that not one living species will transmit its unaltered likeness to a distant futurity. And of the species now living very few will transmit progeny of any kind to a far distant futurity; for the manner in which all organic beings are grouped, shows that the greater number of species of each genus, and all the species of many genera, have left no descendants, but have become utterly extinct. We can so far take a prophetic glance into futurity as to foretell that it will be the common and widely spread species, belonging to the larger and dominant groups, which will ultimately prevail and procreate new and dominant species. As all the living forms of life are the lineal descendants of those which lived long before the Silurian epoch, we may feel certain that the ordinary succession by generation has never once been broken, and that no cataclysm has desolated the whole world. Hence we may look with some confidence to a secure future of equally inappreciable length. And as natural selection works solely by and for the good of each being, all corporeal and mental endowments will tend to progress towards perfection.

It is interesting to contemplate an entangled bank, clothed with many plants of many kinds, with birds singing on the bushes, with various insects flitting about, and with worms crawling through the damp earth, and to reflect that these elaborately constructed forms, so different from each other, and dependent on each other in so complex a manner, have all been produced by laws acting around us. These laws, taken in the largest sense, being Growth with Reproduction; Inheritance which is almost implied by reproduction; Variability from the indirect and direct action of the external conditions of life, and from use and disuse; a Ratio of Increase so high as to lead to a Struggle for Life, and as a consequence to Natural Selection, entailing Divergence of Character and the Extinction of less-improved forms. Thus, from the war of nature, from famine and death, the most exalted object which we are capable of conceiving, namely, the production of the higher animals, directly follows. There is grandeur in this view of life, with its several powers, having been originally breathed into a few forms or into one; and that, whilst this planet has gone cycling on according to the fixed law of gravity, from so simple a beginning endless forms most beautiful and most wonderful have been, and are being, evolved.

Natural Selection. [1860]

THE ORIGIN OF SPECIES *met with many reviews, both favorable and unfavorable. What Darwin had not expected was to face the charge that someone else had earlier formulated the principle of natural selection. Nonetheless, a Scottish author of political and agricultural books, Patrick Matthew, made just that assertion in the April 7, 1860, issue of* Gardeners' Chronicle, *where he claimed to have discussed the idea in the appendix of his obscure book* Naval Timber and Arboriculture, *published in 1831. Darwin reacted by publishing the following note two weeks later and by adding a historical sketch to the third and all subsequent editions of the* Origin. *The sketch was meant to counter further criticism that he had not sufficiently acknowledged his predecessors in thinking and writing about evolution. In 1865, it was pointed out to Darwin that William Charles Wells, an American-born physician living in London, had also anticipated him in a paper read to the Royal Society in 1813 and published in 1818. Both Wells and Matthew, however, only briefly mention ideas that can be interpreted as describing evolution by means of natural selection. Only Darwin and Wallace show how it explains the data. Darwin always stressed that a theory that explained so many facts in so many fields must have something to it.*

I have been much interested by Mr. Patrick Matthew's communication in the Number of your Paper, dated April 7th. I freely acknowledge that Mr. Matthew has anticipated by many years the explanation which I have offered of the origin of species, under the name of natural selection. I think that no one will feel surprised that neither I, nor apparently any other naturalist, had heard of Mr. Matthew's views, considering how briefly they are given, and that they appeared in the appendix to a work on Naval Timber and Arboriculture. I can do no more than offer my apologies to Mr. Matthew for my entire ignorance of his publication. If another edition of my work is called for, I will insert a notice to the foregoing effect.

Fertilisation of Orchids by Insect Agency. [1860]

HERE AGAIN, in a letter published June 9, 1860, we find Darwin seeking information from the readers of the Gardeners' Chronicle, *this time for observations on pollination in orchids. Darwin uses the term "fertilisation" here and elsewhere, but "pollination" is the proper term for what he is describing. Pollination involves the placement of pollen grains onto the stigma of a flower; fertilization is the fusion of the sperm from the pollen grain with the egg in the ovule to form the seed, a phenomenon taking place after pollination and the growth of the pollen tube.*

According to his Autobiography, *Darwin began to gather information on the role of insects in pollination during the summer of 1838 or 1839. He "attended to the subject more or less during every subsequent summer." The incremental observations and experiments led to a series of papers and several books. Darwin's first comments on orchid pollination in particular appear in a June 15, 1855, letter to J. D. Hooker, which contains several of the observations described in the following selection.*

The Bee Orchis and Fly Orchis alluded to in Darwin's letter to the Gardeners' Chronicle *are called by these names because their flowers resemble these insects. We now know that the Fly Orchis is pollinated by a species of small burrowing wasp. The male of this species attempts to mate with the flower, which resembles the female. This method of pollination is accordingly termed "pseudocopulation." It is likely that the Bee Orchis, with flowers resembling female bumblebees, was originally pollinated in a similar way by the males of a bumblebee species. This phenomenon has never been observed in the British Isles, though; and the Bee Orchis is now self-pollinated in the manner Darwin describes.*

Robert Brown, mentioned in the letter, was keeper of botany at the British Museum. An expert on the taxonomy and anatomy of orchids, he had advised Darwin on plants both before and after the Beagle *voyage. Brown's death in June 1858 may have played a role in the absence of discussion of the Darwin-Wallace "joint paper" upon its first presentation on the first of July: much of that evening's meeting of the Linnean Society was spent eulogizing Brown, a longtime member of the society's council.*

One of the readers who heeded Darwin's plea for information was Alexander Goodman More, a botanist on the Isle of Wight off the south coast of England. More, who became one of Darwin's regular correspondents, is cited several times in both editions of Darwin's orchid book.

I should be extremely much obliged to any person living where the Bee or Fly Orchis is tolerably common, if he will have the kindness to make a few simple observations on their manner of fertilisation. To render the subject clear to those who know nothing of botany, I must briefly describe what takes place in our common British Orchids. The pollen-grains form two pear-shaped masses; each borne on a foot-stalk, with a sticky gland at the end. The pollen-masses are hidden in little pouches open in front. When an insect visits a flower, it almost necessarily, owing to the position of the parts, uncovers and touches the sticky glands. These firmly adhere to the head or body of the insect, and thus the pollen-masses are drawn out of their pouches, are dragged over the humid stigmatic surface, and the plant is fertilised. So beautifully are the relative degrees of adhesiveness of the gland, and of the grains of pollen to each other and to the stigmatic surface mutually adapted, that an insect with an adherent pollen-mass will drag it over the stigmas of several flowers, and leave granules of pollen on each. The contrivance by which the sticky glands are prevented from drying, and so kept always viscid and ready for action, is even still more curious; they lie suspended (at least in the two species which I have examined) in a little hemispherical cup, full of liquid, and formed of such delicate membrance, that the side projecting over the gangway into the nectary is ruptured transversely and depressed by the slightest touch; and then the glands, sticky and fresh out of their bath, immediately and almost inevitably come into contact with and adhere to the body which has just ruptured the cup. It is certain that with most of our common Orchids insects are absolutely necessary for their fertilisation; for without their agency, the pollen-masses are never removed and wither within their pouches. I have proved this in the case of *Orchis morio* and *mascula* by covering up plants under a bell-glass, leaving other adjoining plants uncovered; in the latter I found every morning, as the flowers became fully expanded, some of the pollen-masses removed, whereas in the plants under the glass all the pollen-masses remained enclosed in their pouches.

Robert Brown, however, has remarked that the fact of all the capsules in a dense spike of certain Orchids producing seed seems hardly reconcileable with their fertilisation having been accidentally effected by insects. But I could give many facts showing how effectually insects do their work; two cases will here suffice; in a plant of *Orchis maculata* with 44 flowers open, the 12 upper ones, which were not quite mature, had not one pollen-mass removed, whereas every one of the 32 lower flowers had one or both pollen-masses removed; in a plant of *Gymnadenia conopsea* with 54 open flowers, 52 had their pollen-masses removed. I have repeatedly observed in various Orchids grains of pollen, and in one case *three* whole pollen-masses on the stigmatic surface of a flower, which still retained its own two pollen-masses; and as often, or even oftener, I have found flowers with the pollen-masses removed, but with no pollen on their stigmas. These facts clearly show that each flower is often, or even generally, fertilised by the pollen brought by insects from another flower or plant. I may add that after observing our Orchids during many years, I have never seen a bee or any other diurnal insect (excepting once a butterfly) visit them; therefore I have no doubt that moths are the priests who perform the marriage ceremony. The structure, indeed, of some Orchids leads to this same conclusion; for no insect without a very long and extremely fine proboscis could possibly reach the nectar at the bottom of the extremely long and narrow nectary of the Butterfly-Orchis; and entomologists have occasionally captured moths with pollen-masses adhering to them. If any entomologist reads this, and can remember positively having caught a moth thus furnished, I hope he will give its name, and describe exactly to which part of the moth's body the sticky gland adhered.

We may now turn to the genus Ophrys; in the Fly Orchis (*Ophrys muscifera*), the pollen-masses, furnished with sticky glands, do not naturally fall out of their pouches, nor can they be shaken out; so that insect-agency is necessary, as with the species of the other genera, for their fertilisation. But insects here do their work far less effectually than with common Orchids; during several years, previously to 1858, I kept a record of the state of the pollen-masses in well-opened flowers of those plants which I examined, and out of 102 flowers I found either one or both pollen-masses removed in only 13 flowers. But in 1858 I found 17 plants growing near each other and bearing 57 flowers and of these 30 flowers had one or both pollen-masses removed; and as all the remaining 27 flowers were the upper and younger flowers, they probably would subsequently have had most of their pollen-masses removed, and thus have been fertilised. I should much like to hear how the case stands with the Fly Orchis in other districts; for it seems a strange fact that a plant should grow pretty well, as it does in this part of Kent, and yet during several years seldom be fertilised.

We now come to the Bee Orchis (*Ophrys apifera*), which presents a very different case; the pollen-masses are furnished with sticky glands, but differently from in all the foregoing Orchids, they naturally fall out of their pouches; and from being of the proper length, though still retained at the gland-end, they fall on the stigmatic surface, and the plant is thus self-fertilised. During several years I have examined many flowers, and never in a single instance found even one of the pollen-masses carried away by insects, or ever saw the flower's own pollen-masses fail to fall on the stigma. Robert Brown consequently believed that the visits of insects would be injurious to the fertilisation of this Orchis; and rather fancifully imagined that the flower resembled a bee in order to deter their visits. We must admit that the natural falling out of the pollen-masses of this Orchis is a special contrivance for this self-fertilisation; and as far as my experience goes, a perfectly successful contrivance, for I have always found this plant self-fertilised; nevertheless a long course of observation has made me greatly doubt whether the flowers of any kind of plant are for a perpetuity of generations fertilised by their own pollen. And what are we to say with respect to the sticky glands of the Bee Orchis, the use and efficiency of which glands in all other British Orchids are so manifest? Are we to conclude that this one species is provided with these organs for no use? I cannot think so; but would rather infer that, during some years or in some other districts, insects do visit the Bee Orchis and occasionally transport pollen from one flower to another, and thus give it the advantage of an occasional cross. We have seen that the Fly Orchis is not in this part of the country by any means sufficiently often visited by insects, though the visits of insects are indispensable to its fertilisation. So with the Bee Orchis, though its self-fertilisation is specially provided for, it may not exist here under the most favourable conditions of life; and in other districts or during particular seasons it may be visited by insects, and in this case, as its pollen-masses are furnished with sticky glands, it would almost certainly receive the benefit of an occasional cross impregnation. It is this curious apparent contradiction in the structure of the Bee Orchis—one part, namely the sticky glands, being adapted for fertilisation by insect agency—another part, namely the natural falling out of the pollen-masses, being adapted for self-fertilisation without insect agency—which make me anxious to hear what happens to the pollen-masses of the Bee Orchis in other districts or parts of England. I should be extremely much obliged to any one who will take the trouble to observe this point and to communicate the result to the *Gardeners' Chronicle* or to me.

On the Various Contrivances by Which British and Foreign Orchids are Fertilised by Insects, and on the Good Effects of Intercrossing. [1862]

ALL OF DARWIN'S BOOKS following the publication of the Origin *pursue the study of evolutionary adaptations in various organisms, mostly in plants, which lend themselves to experimentation more readily than do animals. The first post-*Origin *book describes the elaborate mechanisms that have evolved to ensure that orchids will be successfully pollinated and form seeds.*

As early as the late 1830s, Darwin had begun observing native orchids in the Kentish countryside around Down. Later, he grew orchids in his home to observe them more closely, and he borrowed many living specimens from his friend J. D. Hooker at the Royal Botanic Gardens, Kew. Much of the correspondence between the two men in the early 1860s was given over to discussions of orchids. As was to prove true in other cases, Darwin seemed to accumulate data on orchids until it was obvious to him that he should write a book on the subject.

He wrote The Fertilisation of Orchids *over a ten-month period in 1861 and 1862, and his publisher John Murray brought out the book on May 15, 1862. The previous September Darwin had written to Murray, "I think this little volume will do good to the 'Origin', as it will show that I have worked hard at details." Following its appearance, he wrote, "The Botanists praise my Orchid-book to the skies." In spite of such favorable reception and a much-expanded second edition in 1877, the book had sold only about six thousand copies by the end of the century.*

The Orchid-book's introduction is followed by a section explain-

ing the terms used for the highly modified structures of these extraordinary flowers. Darwin describes the adaptions for pollination in a number of flowers. We include four such examples: Orchis pyramidalis *(Pyramidal Orchid), pollinated by both day- and night-flying moths;* Ophrys apifera *(Bee Ophrys), usually a self-pollinator, though crossing can take place;* Spiranthes autumnalis *(Autumn Ladies' Tresses), protected against self-pollination; and* Listera ovata *(Common Twayblade), with its exploding rostellum. The importance of cross-fertilization in evolution becomes apparent here and in later publications.*

Fertilisation of Orchids *marks Darwin as a pioneer in the study of pollination ecology, which is still an exciting and expanding field. Another pioneering aspect of* Orchids *is the chapter on the homologies of orchid flowers. Here Darwin follows the vascular traces ("spiral vessels") from the base to the apex of the orchid flower to determine the relationships of the various parts. To our knowledge, no one has ever called Darwin a plant anatomist; but his anatomical technique is still used to study the relationships of flower parts and of species throughout the angiosperms. Francis Darwin states in his edition of his father's* Life and Letters *that Asa Gray "pointed out that if the Orchid-book (with a few trifling omissions) had appeared before the 'Origin', the author would have been canonised rather than anathematised by the natural theologians." Gray considered the adaptive mechanism to be evidence of design, which explains why natural theologians would "canonise" Darwin, but the claim also has more general relevance: the Orchid-book contains some of Darwin's sharpest and subtlest observations and some of his most engaging prose.*

INTRODUCTION.

The object of the following work is to show that the contrivances by which Orchids are fertilised, are as varied and almost as perfect as any of the most beautiful adaptations in the animal kingdom; and, secondly, to show that these contrivances have for their main object the fertilisation of each flower by the pollen of another flower. In my volume 'On the Origin of Species' I have given only general reasons for my belief that it is apparently a universal law of nature that organic beings require an occasional cross with another individual; or, which is almost the same thing, that no hermaphrodite fertilises itself for a perpetuity of generations. Having been blamed for propounding this doctrine without giving ample facts, for which I had not, in that work, sufficient space, I wish to show that I have not spoken without having gone into details.

I have been led to publish this little treatise separately, as it has become inconveniently large to be incorporated with the rest of the discussion on the same subject. And I have thought, that, as Orchids are universally acknowledged to rank amongst the most singular and most modified forms in the vegetable kingdom, the facts to be presently given might lead some observers to look more curiously into the habits of our several native species. An examination of their many beautiful contrivances will exalt the whole vegetable kingdom in most persons' estimation. I fear, however, that the necessary details will be too minute and complex for any one who has not a strong taste for Natural History. This treatise affords me also an opportunity of attempting to show that the study of organic beings may be as interesting to an observer who is fully convinced that the structure of each is due to secondary laws, as to one who views every trifling detail of structure as the result of the direct interposition of the Creator.

I must premise that Christian Konrad Sprengel, in his curious and valuable work, 'Das entdeckte Geheimniss der Natur,' published in 1793, gave an excellent outline of the action of the several parts in Orchids; for he well knew the position of the stigma; and he discovered that insects were necessary to remove the pollen-masses, by pushing open the pouch and coming into contact with the enclosed sticky glands. But he overlooked many curious contrivances,—a consequence, apparently, of his belief that the stigma generally receives the pollen of the same flower. Sprengel, likewise, has partially described the structure of Epipactis; but in the case of Listera he entirely misunderstood the remarkable phenomena characteristic of that genus, which has been so well described by Dr. Hooker in the 'Philosophical Transactions' for 1854. Dr. Hooker has given a full and accurate account, with drawings, of the structure of the parts, and of what takes place; but from not having attended to the agency of insects, he has not fully understood the object gained. Robert Brown, in his celebrated paper in the 'Linnæan Transactions,' expresses his belief that insects are necessary for the fructification of most Orchids; but adds, that the fact of all the capsules on a dense spike not infrequently producing seed, seems hardly reconcileable with this belief: we shall hereafter see that this doubt is groundless. Many other authors have given facts and expressed their belief, more or less fully, on the necessity of insect-agency in the fertilisation of Orchids.

In the course of the following work I shall have the pleasure of expressing my deep obligation to several gentlemen for their unremitting kindness in sending me fresh specimens, without which aid this work would have been impossible. The trouble which several of my kind assistants have taken has been extraordinary: I have never once expressed a wish for aid or for information which has not been granted, as far as possible, in the most liberal spirit.

EXPLANATION OF TERMS.

In case any one should look at this treatise who has never attended to Botany, it may be convenient to explain the meaning of the common terms used. In most flowers the stamens, or male organs, surround in a ring the one or more female organs, called the pistils. In all common Orchids there is only one stamen, and this is confluent with the pistil forming the *Column*. The stamens consist of a filament, or supporting thread (rarely seen in British Orchids), which carries the anther; and within the anther the pollen, the male vivifying element, is included. The anther is divided into two cells, which are very distinct in most Orchids, so much so as to appear in some species like two separate anthers. The pollen in all common plants consists of fine granular powder: but in most Orchids the grains cohere in masses, which are often supported by a very curious appendage, called the *Caudicle*; as will hereafter be more fully explained. The pollen-masses, with their caudicles and other appendages, are called the *Pollinia*.

There are properly in Orchids three united pistils, or female organs. The upper part of the pistil has its anterior surface soft and viscid, which forms the stigma. The two lower stigmas are often completely confluent, so as to appear as one. The stigma in the act of fertilisation is penetrated by long tubes emitted by the pollen-grains, which carry the contents of the grains down to the ovules, or young seeds in the ovarium.

Of the three pistils, which ought to be present, the stigma of the upper one has been modified into an extraordinary organ, called the *Rostellum*, which in many Orchids presents no resemblance to a true stigma. The rostellum either includes or is formed of viscid matter; and in very many Orchids the pollen-masses are firmly attached to a portion of its exterior membrane, which is removed, together with the pollen-masses, by insects. This removeable portion consists in most British Orchids of a small piece of membrane, with a layer or ball of viscid matter underneath, and I shall call it the "*viscid disc*"; but in many exotic Orchids the portion removed is so large and so important, that one part must be called, as before, the viscid disc, and the other part the *pedicel* of the rostellum, to the end of which pedicel the pollen-masses are attached. Authors have called that portion of the rostellum which is removed the "gland," or the "retinaculum," from its apparent function of retaining the pollen-masses in place. The pedicel, or prolongation of the rostellum, to which in many exotic Orchids the pollen-masses are attached, seems generally to have been confounded, under the name of caudicle, with the true caudicle of the pollen-masses, though their nature and origin are totally different. The part of the rostellum which is not removed, and which includes the viscid matter, is sometimes called the "bursicula," or "fovea," or "pouch." But it will be found most convenient to avoid all these terms, and to call the whole modified stigma the rostellum—sometimes adding an adjective to define its shape; and to call that portion of the rostellum which is attached to and removed with the pollen masses the *viscid disc*, together in some cases with its *pedicel*.

Lastly, the three outer divisions of the flower are called *Sepals*, and form the calyx; but, instead of being green, as in most common flowers, they are generally coloured, like the three inner divisions or *Petals* of the flower. The one petal which commonly stands lowest is larger than the others, and often assumes most singular shapes; it is called the lower lip, or *Labellum*. It secretes nectar, in order to attract insects, and is often produced into a long spur-like nectary.

• • •

ORCHIS PYRAMIDALIS.

As in no other plant, or indeed in hardly any animal, can adaptations of one part to another, and of the whole to other organised beings widely remote in the scale of nature, be named more perfect than those presented by this Orchis, it may be worth while briefly to sum them up. As the flowers are visited both by day and night-flying Lepidoptera, I do not think that it is fanciful to believe that the bright-purple tint (whether or not specially developed for this purpose) attracts the day-fliers, and the strong foxy odour the night-fliers. The upper sepal and two upper petals form a hood protecting the anther and stigmatic surfaces from the weather. The labellum is developed into a long nectary in order to attract Lepidoptera, and we shall presently give reasons for suspecting that the nectar is purposely so lodged that it can be sucked only slowly (very differently from in most flowers of other families), in order to give time for the curious chemical quality of the viscid matter on the under side of the saddle setting hard and dry. He who will insert a fine and flexible bristle into the expanded mouth of the sloping ridges on the labellum will not doubt that they serve as guides; and that they effectually prevent the bristle or proboscis from being inserted obliquely into the nectary. This circumstance is of manifest importance, for, if the proboscis were inserted obliquely, the saddle-formed disc would become attached obliquely, and after the compounded movement of the pollinia they could not strike the two lateral stigmatic surfaces.

Then we have the rostellum partially closing the mouth of the nectary, like a trap placed in a run for game; and the trap so complex and perfect, with its symmetrical lines of rupture forming the saddle-shaped disc above, and the lip of the pouch below; and, lastly, this lip so easily depressed that the proboscis of a moth could hardly fail to uncover the viscid disc and adhere to it. But if this did fail to occur, the elastic lip would rise again and re-cover and keep damp the viscid surface. We see the viscid matter within the rostellum attached to the saddle-shaped disc alone, and surrounded by fluid, so that the viscid matter does not set hard till the disc is withdrawn. Then we have the upper surface of the saddle, with its attached caudicles, also kept damp within the bases of the anther-cells, until withdrawn, when the curious clasping movement instantly commences, causing the pollinia to diverge, followed by the movement of depression, which compounded movements together are exactly fitted to cause the ends of the two pollinia to strike the two stigmatic surfaces. These stigmatic surfaces are sticky enough not to tear off the whole pollinium from the proboscis of the moth, but by rupturing the elastic threads to secure a few packets of pollen, leaving plenty for other flowers.

But let it be observed that, although the moth probably takes a considerable time to suck the nectar of any one flower, yet the movement of depression in the pollinia does not commence (as I know by trial) until the pollinia are fairly withdrawn out of their cells; nor will the movement be completed, and the pollinia be fitted to strike the stigmatic surfaces, until about half a minute has elapsed, which will give ample time for the moth to fly to another plant, and thus effect a union between two distinct individuals. Lastly, we have the wonderful growth of the pollen-tubes and their penetration of the stigma, as well as the mysteries of germination, though these are common to all phanerogamic plants.

• • •

OPHRYS APIFERA.

In the Bee Ophrys we meet with widely different means of fertilisation as compared with the other species of the genus, and, indeed, as far as I know, with all other Orchids. The two pouch-formed rostellums, the viscid discs, and the position of the stigma, are nearly the same as in other species of Ophrys; but to my surprise, I have observed that the distance of the two pouches from each other, and the shape of the mass of pollen-grains, are variable. The caudicles of the pollinia are remarkably long, thin, and flexible, instead of being, as in all the other Ophreæ, rigid enough to stand upright. They are necessarily curved forward at their upper ends, owing to the shape of the anther-cells; and the pear-shaped masses of pollen lie embedded high above and directly over the stigma. The anther-cells naturally open soon after the flower is fully expanded, and the thick ends of the pollinia fall out, the viscid discs still remaining in their pouches. Slight as is the weight of the pollen, yet the caudicle is so thin, and soon becomes so flexible, that, in the course of a few hours, they sink down, until they hang freely in the air exactly opposite to and in front of the stigmatic surface. When in this position a breath of air, acting on the expanded petals, sets the flexible and elastic caudicles vibrating, and they almost immediately strike the viscid stigma, and, being there secured, impregnation is effected. To make sure that no other aid was requisite, though the experiment was superfluous, I covered up a plant under a net, so that some wind, but no insects, could pass in, and in a few days the pollinia had become attached to the stigmas; but the pollinia of a spike kept in water in a still room, remained free, suspended in front of the stigma.

Robert Brown first observed that the structure of the Bee Ophrys is adapted for self-fertilisation. When we consider the unusual and perfectly-adapted length, as well as the remarkable thinness, of the caudicles of the pollinia; when we see that the anther-cells naturally open, and that the masses of pollen, from their weight, slowly fall down to the exact level of the stigmatic surface, and are there made to vibrate to and fro by the slightest breath of wind till the stigma is struck; it is impossible to doubt that these points of structure and function, which occur in no other British Orchid, are specially adapted for self-fertilisation.

The result is what might have been anticipated. I have often noticed that the spikes of the Bee Ophrys apparently produced as many seed-capsules as flowers; and near Torquay I carefully examined many dozen plants, some time after the flowering season; and on all I found from one to four, and occasionally five, fine capsules, that is, as many capsules as there had been flowers; in extremely few cases (excepting a few deformed flowers, generally on the summit of the spike) could a flower be found which had not produced a capsule. Let it be observed what a contrast this case presents with that of the Fly Ophrys, which requires insect agency, and which from forty-nine flowers produced only seven capsules!

From what I had seen of other British Orchids, I was so much surprised at the self-fertilisation of this species, that during many years I have looked at the state of the pollen-masses in hundreds of flowers, and I have never seen, in a single instance, reason to believe that pollen had been brought from one flower to another. Excepting in a few monstrous flowers, I have never seen an instance of the pollinia failing to reach their own stigmas. In a very few cases I have found one pollinium removed, but in some of these cases the marks of slime led me to suppose that slugs had devoured them. For instance, in 1860, I examined in North Kent twelve spikes bearing thirty-nine flowers, and three of these had one pollinium removed, all the other pollinia being glued to their own stigmas. In another lot, from another locality, however, I found the unparalleled case of two flowers with both pollinia removed, and two others with one removed. I have examined some flowers from South Kent with the same result. Near Torquay I examined twelve spikes bearing thirty-eight flowers, and in these one single pollinium alone had been removed. We must not forget that blows from animals or storms of wind might occasionally cause the loss of a pollinium.

In the Isle of Wight Mr. A. G. More was so kind as to examine carefully a large number of flowers. He observed that in plants growing singly both pollinia were invariably present. But on taking home several plants, from a large number growing in two places, and *selecting plants* which seemed to have had some pollinia removed, he examined 136 flowers: of these ten had lost both pollinia, and fourteen had lost one; so here there seems at first evidence of the pollinia having been removed by their adhesion to insects; but then Mr. More found no less than eleven pollinia (not included in the above cases of removal) with their caudicles cut or gnawed through, but with their viscid discs still in their pouches, and this proves that some other animals, not insects, probably slugs, had been at work. Three of the flowers were much gnawed. Two pollinia, which had apparently been thrown out by strong wind, were sticking to the sepals, and three pollinia were found loose in his collecting box, so that it is very doubtful whether many, or indeed any, of the pollinia had been removed by adhesion to insects. I will only add that I have never seen an insect visit these flowers. Robert Brown imagined that the flowers resembled bees in order to deter insects from visiting them; I cannot think this probable. The equal or greater resemblance of the Fly Ophrys to an insect does not deter the visits of some unknown insect, which, in that species, are indispensable for the act of fertilisation.

Whether we look to the structure of the several parts of the flower as far as hitherto described, or to the actual state of the pollinia in numerous plants taken during different seasons from different localities, or to the number of seed-capsules produced, the evidence seems conclusive that we here have a plant which is self-fertilised for perpetuity. But now let us look to the other side of the case. When an object is pushed (as in the case of the Fly Ophrys) right against one of the pouches of the rostellum, the lip is depressed, and the large extremely viscid disc adheres firmly to the object, and the pollinium is removed. Even after the pollinia have naturally fallen out of their cells and are glued to the stigma, their removal can sometimes be thus effected. As soon as the disc is drawn out of the pouch the movement of depression commences by which the pollinium would be brought in front of an insect's head ready to strike the stigma. When a pollen-mass is placed on the stigma and then withdrawn, the elastic threads by which the packets are tied together, break, and leave several packets on the viscid surface. In all other Orchids the meaning of these several contrivances—namely, the downward movement of the lip of the rostellum when gently pushed—the viscidity of the disc—the act of depression of the caudicle after the disc has been removed—the rupturing of the elastic threads by the viscidity of the stigma, so that pollen may be left on several stigmas—is unmistakably clear. Are we to believe that these contrivances in the Bee Ophrys are absolutely purposeless, as would certainly be the case if this species is perpetually self-fertilised? If the discs had been small or only viscid in a slight degree, if the other related contrivances had been imperfect in any degree, we might have concluded that they had begun to abort; that Nature, if I may use the expression, seeing that the Fly and Spider Ophrys were imperfectly fertilised and produced few seed capsules, had changed her plan and effected complete and perpetual self-fertilisation, in order that more seeds might be produced. The case is perplexing in an unparalleled degree, for in the same flower we apparently have elaborate contrivances for directly opposed objects.

We have already seen many curious structures and movements, as in Orchis pyramidalis, which evidently lead to the fertilisation of one flower by the pollen of another flower, and we shall meet with numerous other and very different contrivances for the same object throughout the whole great Orchidean Family. Hence it is impossible to doubt that some great good is derived from the union of two distinct flowers, often borne on distinct plants; but the good in the case of the Fly and Spider Ophrys is gained at the expense of much lessened fertility. In the Bee Ophrys great fertility is gained at the expense of apparently perpetual self-fertilisation; but the contrivances are still present which are assuredly adapted to give an occasional cross with another individual; and the safest conclusion, as it seems to me, is, that under certain unknown circumstances, and perhaps at very long intervals of time, one individual of the Bee Ophrys is crossed by another. Thus the generative functions of this plant would be brought into harmony with those of other Orchidaceæ, and, indeed, with those of all other plants, as far as I have been able to make out their structure.

• • •

We have now finished with the Ophreæ; but before passing on to the next Division, I will recapitulate the chief facts on the movements of the pollinia, all due to the nicely regulated contraction of that small portion of membrane (together with the pedicel in the case of Habenaria) lying between the layer or ball of adhesive matter and the extremity of the caudicle. In most of the species of Orchis the stigma lies directly beneath the anther-cells, and the pollinia simply move vertically downwards. In Orchis pyramidalis and in Gymnadenia there are two lateral and inferior stigmas, and the pollinia move downwards and outwards, diverging at the proper angle (by a different mechanism in the two cases), so as to strike the two lateral stigmas; in Habenaria the stigmatic surface lies beneath and between the two widely separated anther-cells, and the pollinia here again move downwards, but, at the same time, converge. A poet might imagine, that whilst the pollinia are borne from flower to flower through the air, adhering to a moth's body, they voluntarily and eagerly place themselves, in each case, in that exact position in which alone they can hope to gain their wish and perpetuate their race.

• • •

SPIRANTHES AUTUMNALIS.

With most Orchids the flowers remain open for some time before they are visited by insects; but with Spiranthes I have generally found the boat-formed disc removed very soon after the expansion of the flower. For example, in the two last spikes which I happened to examine there were in one numerous buds on the summit, with the seven lowest flowers alone expanded, of which six had their discs and pollinia removed; the other spike had eight expanded flowers, with the pollinia of all removed. We have seen that when the flower first opens it would be attractive to insects, for the receptacle already contains nectar; and the rostellum at this period lies so close to the channelled labellum that a bee or moth could not pass down its proboscis without touching the medial furrow of the rostellum. This I know to be the case by repeated trials with a bristle.

Let it be observed how beautifully everything is contrived that the pollinia should be withdrawn by an insect visiting the flower. The pollinia are already attached to the disc by their threads, and, from the early withering of the anther-cells, they hang loosely suspended but protected within the clinandrum. The touch of the proboscis causes the rostellum to split in front and behind, and frees the long, narrow, boat-formed disc, which is loaded with extremely viscid matter, sure to adhere longitudinally to the proboscis. When the bee flies away, so surely will it carry away the pollinia. As the pollinia are attached parallel to the disc, they will become attached parallel to the proboscis. Here, however, appears a difficulty; when the flower first opens and is best adapted for the removal of the pollinia, the labellum lies so close to the rostellum, that the pollinia, when attached to the proboscis of an insect, could not possibly be forced into the flower so as to reach the stigma; they would be either upturned or broken off; but we have seen that after two or three days the labellum becomes more reflexed and moves from the rostellum,—a wider passage being thus left. When in this condition, I have tried with the pollinia attached to a fine bristle; and by inserting the bristle into the nectar-receptacle it is pretty to see how sheets of pollen are left adhering to the viscid stigma. It may be observed that the orifice into the nectar-receptacle lies, owing to the projection of the stigma, close to the lower side of the flower; hence insects would insert their probosces on this side, and an open space is thus afforded for the attached pollinia to be carried down to the stigma, without being brushed off. The stigma evidently projects so that the ends of the pollinia may strike against it.

Hence, in Spiranthes, not only the pollen must be carried from one flower to another, as in most Orchids, but a lately expanded flower, which has its pollinia in the best state for removal, cannot then be fertilised. Generally old flowers will be fertilised by the pollen of younger flowers, borne, as we shall see, on a separate plant. In conformity with this I observed that the stigmatic surfaces of the older flowers were far more viscid than those of the younger flowers. Nevertheless, a flower which in its early state had not been visited by insects would not necessarily, in its later and more expanded condition, have its pollen wasted; for insects, in inserting and withdrawing their probosces, bow them forwards, and would thus often strike the furrow in the rostellum. I imitated this action with a bristle, and often succeeded in withdrawing the pollinia. I was led to try this from at first choosing older flowers for examination; and passing a bristle, or fine culm of grass, straight down into the nectary, the pollinia were never withdrawn; but when I bowed it forward, I succeeded. Those flowers which have not their own pollinia removed can of course be fertilised, and I have seen not a few cases of flowers with their pollinia still in place, with sheets of pollen on their stigmas.

At Torquay I watched a number of these flowers growing together for about half an hour, and saw three humble-bees of two kinds visit them. I caught one and examined its proboscis: on the superior lamina, some little way from the tip, two perfect pollinia were attached, and three other boat-formed discs without pollen; so that this bee had removed the pollinia from five flowers, and had probably left the pollen of three of them on the stigmas of other flowers. The next day I watched the same flowers for a quarter of an hour, and caught another humble-bee at work; one perfect pollinium and four boat-formed discs adhered to its proboscis, one on the top of the other, showing how exactly the same part had each time touched the rostellum.

The bees always alighted at the bottom of the spike, and, crawling spirally up it, sucked one flower after the other. I believe humble-bees generally act thus when visiting a dense spike of flowers, as it is most convenient for them; in the same manner as a woodpecker always climbs up a tree in search of insects. This seems a most insignificant observation; but see the result. In the early morning, when the bee starts on her rounds, let us suppose that she alighted on the summit of the spike; she would surely extract the pollinia from the uppermost and last opened flowers; but when visiting the next succeeding flower, of which the labellum in all probability would not as yet have moved from the column (for this is slowly and very gradually effected), the pollen-masses would often be brushed off her proboscis and be wasted. But nature suffers no such waste. The bee goes first to the lowest flower, and, crawling spirally up the spike, effects nothing on the first spike which she visits till she reaches the upper flowers, then she withdraws the pollinia: she soon flies to another plant, and, alighting on the lowest and oldest flower, into which there will be a wide passage from the greater reflexion of the labellum, the pollinia will strike the protuberant stigma: if the stigma of the lowest flower has already been fully fertilised, little or no pollen will be left on its dried surface; but on the next succeeding flower, of which the stigma is viscid, large sheets of pollen will be left. Then as soon as the bee arrives near the summit of the spike she will again withdraw fresh pollinia, will fly to the lower flowers on another plant, and fertilise them; and thus, as she goes her rounds and adds to her store of honey, she will continually fertilise fresh flowers and perpetuate the race of our autumnal Spiranthes, which will yield honey to future generations of bees.

• • •

LISTERA OVATA, OR TWAY-BLADE.

The structure and action of the rostellum of this Orchid has been the subject of a highly remarkable paper in the Philosophical Transactions, by Dr. Hooker, who has described minutely, and of course correctly, its curious structure; he did not, however, attend to the part which insects play in the fertilisation of this flower. C. K. Sprengel well knew the importance of insect-agency, but he misunderstood both the structure and the action of the rostellum.

• • •

As soon as the flower opens, if the crest of the rostellum be touched ever so lightly, a large drop of viscid fluid is instantaneously expelled; this, as Dr. Hooker has shown, is formed by the coalescence of two drops proceeding from the two depressed spaces on each side of the crest. A good proof of this fact was afforded by some specimens kept in weak spirits of wine, which had apparently expelled the viscid matter slowly, and in which two separate little spherical balls of viscid matter had hardened and had become attached to the two pollinia. The fluid is at first slightly opaque and milky; but as soon as exposed to the air, in less time than a second, a film is formed over it, and in two or three seconds the whole drop sets hard, and soon assumes a purplish-brown tint. So exquisitely sensitive is the rostellum, that a touch from the thinnest human hair suffices to cause the explosion. It will take place under water. Exposure to the vapour of chloroform for about one minute also causes the explosion. The viscid fluid when pressed between two plates of glass before it has set hard is seen to be structureless; but it assumes a reticulated appearance, perhaps caused by the presence and union of globules of a denser fluid immersed in a thinner fluid. As the pointed tips of the loose pollinia lie on the crest of the rostellum, they are always caught by the exploded drop: I have never once seen this fail. So rapid is the explosion and so viscid the fluid, that it is difficult to touch the rostellum with a needle quickly enough not to catch the pollinia already attached to the partially-hardened drop. Hence, if a bunch of flowers be carried home in the hand, some of the sepals or petals, being shaken, will be almost sure to touch the rostellum and withdraw the pollinia, which will cause the false appearance of their having been violently ejected to a distance.

After the anther-cells have opened and have left the naked pollinia resting on the concave back of the rostellum, the rostellum curves a little forwards, and perhaps the anther also moves a little backwards. This movement is of much importance, as the tips of the anther would otherwise be caught by the exploded viscid matter, and the pollinia would for ever be locked up and rendered useless. I once found an injured flower which had been pressed and had exploded before fully expanding, and the anther, with the enclosed pollen-masses, was permanently glued down to the crest of the rostellum. At the moment of the explosion the rostellum, which already stands arched over the stigma, quickly bends forwards and downwards so as to stand at right angles to its surface. The pollinia, if not removed by the touching object that causes the explosion, become fixed to the rostellum, and by its movement are drawn a little forward. If their lower ends are now freed by a needle from the anther-cells, they spring up; but they are not by this movement placed on the stigma. In the course of some hours, or of a day, the rostellum not only slowly recovers its original slightly-arched position, but becomes quite straight and parallel to the stigmatic surface. This backward movement of the rostellum is of importance; for if after the explosion it permanently remained projecting at right angles close over the stigma, pollen could with difficulty have been left on its viscid surface. When the rostellum is touched so quickly that the pollinia are not removed, they are, as I have said, drawn at the moment of explosion a little forward by the movement of the rostellum; but by its subsequent backward movement the pollinia are pushed back into their original position.

From the account here given we may safely infer how the fertilisation of this Orchid is effected. Small insects alight on the broad lower end of the labellum for the sake of the nectar copiously secreted by it; as they lick up the nectar they slowly crawl up its narrowed surface until their heads stand directly under the over-arched crest of the rostellum; as they raise their heads they touch the crest, which explodes, and the pollinia become firmly cemented to them. The insect in flying away withdraws the pollinia, carries them to another flower, and leaves masses of the friable pollen on its viscid stigma.

In order to witness what I felt sure would take place, I watched a group of plants on two or three occasions for an hour; each day I saw numerous specimens of two small Hymenopterous insects, namely, a Hæmiteles and a Cryptus, flying about the plants and licking up the nectar; most of the flowers, which were visited over and over again, had already had their pollinia removed, but at last I saw both these insect-species crawl into younger flowers, and suddenly retreat with a pair of bright yellow pollinia sticking to their foreheads; I caught them, and found the point of attachment was to the inner edge of the eye; on the other eye of one specimen there was a ball of the hardened viscid matter, showing that it had previously removed another pair of pollinia, and had subsequently in all probability left them on the stigma of one of the flowers. As I caught these insects, I did not witness the act of fertilisation; but C. K. Sprengel actually saw a Hymenopterous insect leave its pollen-mass on the stigma. My son watched another bed of this Orchid at some miles' distance, and brought me home the same Hymenopterous insects with attached pollinia, and he saw Diptera also visiting the flowers. He was struck with the number of spider-webs spread over these plants, as if the spiders were aware how attractive the Listera was to insects, and how necessary they were to its fertilisation.

To show how delicate a touch suffices to cause the rostellum to explode, I may mention that I found an extremely minute Hymenopterous insect vainly struggling with its whole head buried in the hardened viscid matter, thus cemented to the crest of the rostellum and to the tips of pollinia; the insect was not so large as one of the pollinia, and after causing the explosion it had not force to remove them, and was thus punished for attempting a work beyond its strength, and perished miserably.

In Spiranthes the young flowers, which have their pollinia in the best state for removal, cannot possibly be fertilised; they must remain in a virgin condition until they are a little older and the labellum has moved from the column. Here the same thing apparently occurs, for the stigmas of the older flowers were more viscid than those of the younger flowers. These latter have their pollinia quite ready for removal; but immediately after the explosion the rostellum, as we have seen, curls forwards and downwards, thus protecting the stigma for a time, until the rostellum slowly becomes quite straight, when the stigma in its more mature condition is left freely exposed, and is ready for fertilisation.

I was curious to ascertain whether the rostellum would ultimately explode, if never touched; but I have found it difficult to ascertain this fact, as the flowers are so attractive to insects, and the touch of such minute insects suffices to cause the explosion, that it was scarcely possible to exclude them.

I have covered up plants several times, and left them till long after all uncovered plants had set their pods; and without going into unnecessary details, I may positively state that the rostella in several flowers had not exploded, though the stigma was withered, the pollen quite mouldy and incapable of removal. Some few of the very old flowers, however, when roughly touched, were still capable of feeble explosion. Other flowers under the nets had exploded, and the tips of their pollinia were fixed to the crest of the rostellum; but whether these had been touched by some excessively minute insect, or had exploded spontaneously, it is impossible to say. In the case of these latter flowers, it is of some importance to note that not a grain of pollen had got on their stigmas (and I looked carefully), and their ovaria had not swollen. These several facts clearly show that the removal of the pollinia by insect-agency is absolutely necessary to the fertilisation of this species.

That insects do their work effectually, the following cases show: —A rather young spike, with many buds in the upper part, had the pollinia unremoved from the seven upper flowers, but they had been completely removed from the ten lower flowers: there was pollen on the stigmas of six of these ten lower flowers. In two spikes taken together, the twenty-seven lower flowers had their pollinia removed, with pollen on the stigmas of all; these were succeeded by five open flowers with pollinia not removed, and with no pollen on their stigmas; and these were succeeded by eighteen buds. Lastly, in an older spike with forty-four flowers, all fully expanded, the pollinia had been removed from every one; and on all the stigmas that I examined there was pollen, generally in large quantity.

It will perhaps be worth while to recapitulate the several special adaptations for the fertilisation of this Orchid. The anther-cells open early, leaving the pollen-masses quite loose, with their tips resting on the concave crest of the rostellum. The rostellum then slowly curves over the stigmatic surface, so that its explosive crest stands at a little distance from the anther; and this is very necessary, otherwise the anther would be caught by the viscid matter, and the pollen for ever locked up. This curvature of the rostellum over the stigma and base of the labellum is excellently well adapted to favour an insect striking the crest when it raises its head, after having crawled up the labellum, and licked up the last drop of nectar at its base. The labellum, as C. K. Sprengel has remarked, becomes narrower where it joins the column beneath the rostellum, so that there is no risk of the insect going too much to either side. The crest of the rostellum is so exquisitely sensitive, that a touch from a most minute insect causes it to rupture at two points, and instantaneously two drops of viscid fluid are expelled, which coalesce. This viscid fluid sets hard in so wonderfully rapid a manner that it rarely fails to cement the tips of the pollinia, nicely laid on the crest of the rostellum, to the insect's forehead. As soon as the rostellum has exploded it suddenly curves downwards till it projects at right angles over the stigma, protecting it in its early state from impregnation, in the same manner as the stigma of Spiranthes is protected by the labellum clasping the column. But as in Spiranthes the labellum after a time moves from the column, leaving a free passage for the introduction of the pollinia, so here the rostellum moves back, and not only recovers its former arched position, but stands upright, leaving the stigmatic surface, now become more viscid, perfectly free for pollen to be left on it. The pollen-masses, when once cemented to an insect's forehead, will generally remain firmly attached to it until the viscid stigma of a mature flower removes these encumbrances from the insect, by rupturing the weak elastic threads by which the grains are tied together—receiving at the same time the benefit of fertilisation.

• • •

HOMOLOGIES OF ORCHID FLOWERS.

The theoretical structure of few flowers has been so largely discussed as that of Orchids; nor is this surprising, seeing how unlike they are to common flowers. No group of organic beings can be well understood until their homologies are made out; that is, until the general pattern, or, as it is often called, the ideal type, of the several members of the group is intelligible. No one member may now exist exhibiting the full pattern; but this does not make the subject less important to the naturalist,—probably makes it more important for the full understanding of the group.

The homologies of any being, or group of beings, can be most surely made out by tracing their embryological development when that is possible; or by the discovery of organs in a rudimentary condition; or by tracing, through a long series of beings, a close gradation from one part to another, until the two parts, or organs, employed for widely different functions, and most unlike each other, can be joined by a succession of short links. No instance is known of a close gradation between two organs, unless they be homologically one and the same organ.

The importance of the science of Homology rests on its giving us the key-note of the possible amount of difference in plan within any group; it allows us to class under proper heads the most diversified organs; it shows us gradations which would otherwise have been overlooked, and thus aids us in our classification; it explains many monstrosities; it leads to the detection of obscure and hidden parts, or mere vestiges of parts, and shows us the meaning of rudiments. Besides these practical uses, to the naturalist who believes in the gradual modification of organic beings, the science of Homology clears away the mist from such terms as the scheme of nature, ideal types, archetypal patterns or ideas, &c.; for these terms come to express real facts. The naturalist, thus guided, sees that all homologous parts or organs, however much diversified, are modifications of one and the same ancestral organ; in tracing existing gradations he gains a clue in tracing, as far as that is possible, the probable course of modification during a long line of generations. He may feel assured that, whether he follows embryological development, or searches for the merest rudiments, or traces gradations between the most different beings, he is pursuing the same object by different routes, and is tending towards the knowledge of the actual progenitor of the group, as it once grew and lived. Thus the subject of Homology gains largely in interest.

Although this subject, under whatever aspect it be viewed, will always be most interesting to the student of nature, it is very doubtful whether the following details on the homological nature of the flowers of Orchids will be endured by the general reader. If, indeed, he should care to see how much light, though far from perfect, homology throws on a subject, this will, perhaps, be nearly as good an instance as could be given. He will see how curiously a flower may be moulded out of many separate organs,—how perfect the cohesion of primordially distinct parts may become,—how organs may be used for purposes widely different from their proper function,—how other organs may be entirely suppressed, or leave mere useless emblems of their former existence. Finally, he will see how enormous has been the total amount of change from the simple parental or typical structure which these flowers have undergone.

Robert Brown first clearly discussed the homologies of Orchids, and left, as might be expected, little to be done. Guided by the general structure of monocotyledonous plants, and by various considerations, he propounded the doctrine that the flower properly consists of three sepals, three petals, six anthers in two whorls or circles (of which only one anther belonging to the outer whorl is perfect in all common forms), and of three pistils, with one of them modified into the rostellum. These fifteen organs are arranged as usual, alternately, three within three, in five whorls. Of the existence of three of the anthers in two whorls, R. Brown offers no sufficient evidence, but believes that they are combined with the labellum, whenever that organ presents crests or ridges. In these views Brown is followed by the greatest living authority on Orchids, namely, Lindley.

Brown traced the spiral vessels in the flower by making transverse sections, and only occasionally, as far as it appears, by longitudinal sections. As spiral vessels are developed at a very early period of growth, which always gives much value to an organ in making out homologies; and as they are apparently of high functional importance, though their function is not well known, it appeared to me, guided also by the advice of Dr. Hooker, to be worth while to trace upwards all the spiral vessels from the six groups surrounding the ovarium. Of the six ovarian groups of vessels, I will call (though not correctly) that under the labellum the anterior group; that under the upper sepal the posterior group; and the two groups on both sides of the ovarium the antero-lateral and postero-lateral groups.

The result of my dissections is given in the following diagram. The fifteen little circles represent so many groups of spiral vessels, in every case traced down to one of the six large ovarian groups. They alternate in five whorls, as represented; but I have not attempted to give the actual distances at which they stand apart. In order to guide the eye, the three central groups running to the three pistils are connected by a triangle.

Five groups of vessels run into the three sepals and two upper petals; three enter the labellum; and seven run up the great central column. These vessels are arranged, as may be seen, in rays proceeding from the axis of the flower; and all on the same ray invariably run into the same ovarian group: thus the vessels supplying the upper sepal, the fertile anther (A_1), and the upper pistil or stigma (*i.e.* rostellum S_r), all unite and form the posterior ovarian group. Again, the vessels supplying one of the lower sepals, the corner of the labellum, and one of the two stigmas (S), unite and form the antero-lateral group; and so with all the other vessels.

Hence, if the existence of groups of spiral vessels can be trusted, and Dr. Hooker informs me that he has never known them to speak falsely, the flower of an Orchid certainly consists of fifteen organs, in a much modified and confluent condition. We see three stigmas, with the two lower ones generally confluent, and with the upper one modified into the rostellum. We see six stamens, arranged in two whorls, with one alone (A_1) generally fertile. In Cypripedium, however, two stamens of the inner whorl (a_1 and a_2) are fertile, and in other Orchids these two are represented in various ways more plainly than the remaining stamens. The third stamen of the inner whorl (a_3), when its vessels can be traced, forms the front of the column: Brown thought that it often formed a medial excrescence, or ridge, cohering to the

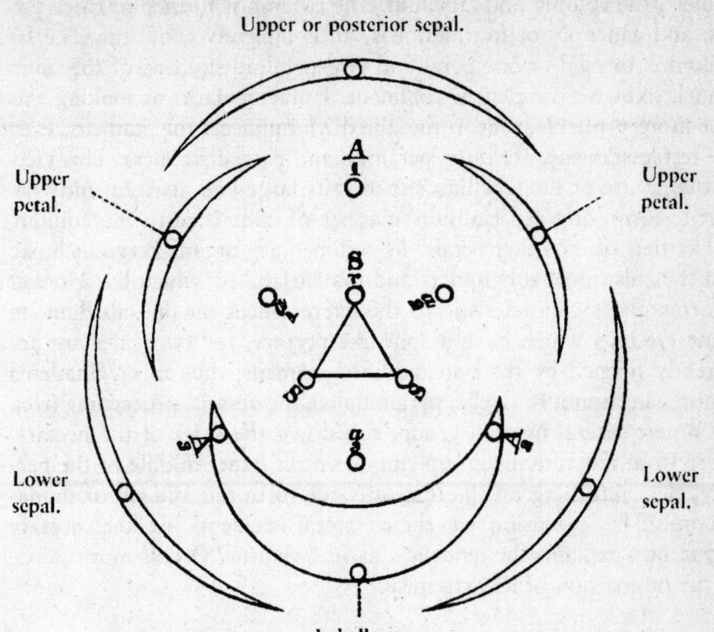

SECTION OF THE FLOWER OF AN ORCHID.

The little circles show the position of the spiral vessels.

S S. Stigmas; S_r, stigma modified into the rostellum. A_1. Fertile anther of the outer whorl; A_2 A_3, anthers of the same whorl combined with the lower petal, forming the labellum. a_1 a_2. Rudimentary anthers of the inner whorl (fertile in Cypripedium), generally forming the clinandrum; a_3, third anther of the same whorl, when present, forming the front of the column.

labellum; or, in the case of Glossodia, a filamentous organ, freely projecting in front of the labellum. The former conclusion does not agree with my dissections; about Glossodia I know nothing. The two infertile stamens of the outer whorl (A_2, A_3) were believed by Brown to be sometimes represented by lateral excrescences on the labellum; I find these vessels invariably present in the labellum of every Orchid examined,—even when the labellum is very narrow, or quite simple, as in Malaxis, Herminium, and Habenaria.

We thus see that an Orchid-flower consists of five simple parts, namely, three sepals and two petals; and of two compounded parts, namely, the column and labellum. The column is formed of three pistils, and generally of four stamens, all completely confluent. The labellum is formed of one petal and two petaloid stamens of the outer whorl, likewise completely confluent. I may remark, as making this fact more probable, that in the allied Marantaceæ the stamens, even the fertile stamens, are often petaloid, and partially cohere. This view of the nature of the labellum explains its large size, its frequently tripartite form, and especially its manner of coherence to the column, unlike that of the other petals. As rudimentary organs vary much, we can thus also probably understand the variability, which Dr. Hooker informs me is characteristic of the excrescences on the labellum. In some Orchids which have a spur-like nectary, the two sides are apparently formed by the two modified stamens; thus in Gymnadenia conopsea (but not in Orchis pyramidalis), the vessels, proceeding from the antero-lateral ovarian group, run down the sides of the nectary; those from the anterior group run down the exact middle of the nectary, then returning up the opposite side form the mid-rib of the labellum. The extension of these lateral elements of the nectary apparently explains the tendency, as in Calanthe, Orchis morio, &c., to the bifurcation of its extremity.

• • •

We have now finished with the general homologies of the flowers of Orchids. It is interesting to look at one of the magnificent exotic species, or, indeed, at one of our humblest forms, and observe how profoundly it has been modified, as compared with all ordinary flowers, —with its usually great labellum, formed of one petal and two petaloid stamens,—with its singular pollen-masses, presently to be referred to,—with its column formed of seven cohering organs, of which three alone perform their proper function, namely, one anther and two generally confluent stigmas,—with the third stigma incapable of fertilisation and modified into the wonderful rostellum—with three of the anthers no longer capable of producing pollen, but serving either to protect the pollen of the fertile anther, or to strengthen the column, or existing as mere rudiments, or entirely suppressed. What an amount of modification, change of function, cohesion, and abortion do we here see! Yet hidden in that column, with its surrounding petals and sepals, we know that there are fifteen groups of vessels, arranged three within three, in alternating order, which probably have been preserved to the present time from having been developed in each flower at a very early period of growth, before the shape or existence of this or that part signified to the well-being of the plant.

Can we, in truth, feel satisfied by saying that each Orchid was created, exactly as we now see it, on a certain "ideal type;" that the Omnipotent Creator, having fixed on one plan for the whole Order, did not please to depart from this plan; that He, therefore, made the same organ to perform diverse functions—often of trifling importance compared with their proper function—converted other organs into mere purposeless rudiments, and arranged all as if they had to stand separate, and then made them cohere? Is it not a more simple and intelligible view that all Orchids owe what they have in common to descent from some monocotyledonous plant, which, like so many other plants of the same division, possessed fifteen organs, arranged alternately three within three in five whorls; and that the now wonderfully changed structure of the flower is due to a long course of slow modification,—each modification having been preserved which was useful to each plant, during the incessant changes to which the organic and the inorganic world has been exposed?

• • •

The final end of the whole flower, with all its parts, is the production of seed; and these are produced by Orchids in vast profusion. Not that this is anything to boast of in the order; for the production of an almost infinite number of eggs, or seeds, is undoubtedly a sign of lowness of organisation. That a plant, not an annual, should escape destruction at some period of its life simply by the production of a vast number of seeds or seedlings, shows a poverty of contrivance, or a want of some fitting protection against some danger. I was curious to estimate the number of seeds produced by Orchids; so I took a ripe capsule of Cephalanthera grandiflora, and arranged the seeds as equably as I could in a narrow hillock, on a long ruled line; and then counted the seeds in a length, accurately measured, of one-tenth of an inch. They were 83 in number, and this would give for the whole capsule 6020 seeds; and for the four capsules borne by the plant 24,000 seeds. Estimating in the same manner the smaller seeds in Orchis maculata, I found the number nearly the same, viz. 6200; and, as I have often seen above 30 capsules on the same plant, the total amount will be 186,300,—a prodigious number for one small plant to bear. As this Orchid is perennial, and cannot in most places be increasing in number, one seed alone of this large number, once in every few years, produces a mature plant. I examined many seeds of the Cephalanthcra, and very few seemed bad.

To give an idea what the above figures really mean, I will briefly show the possible rate of increase of O. maculata: an acre of land would hold 174,240 plants, each having a space of six inches square, which is rather closer than they could flourish together; so that, allowing twelve thousand bad seeds, an acre would be thickly clothed by the progeny of a single plant. At the same rate of increase, the grandchildren would cover a space slightly exceeding the island of Anglesea; and the great grandchildren of a single plant would nearly (in the proportion of 47 to 50) clothe with one uniform green carpet the entire surface of the land throughout the globe.

What checks this unlimited multiplication cannot be told. The minute seeds within their light coats are well fitted for wide dissemination; and I have several times observed seedlings in my orchard, and in a newly-planted wood, which must have come from some little distance. Yet it is notorious that Orchids are sparingly distributed; for instance, this district is highly favourable to the order, for within a mile of my house nine genera, including thirteen species, grow; but of these one alone, Orchis morio, is sufficiently abundant to make a conspicuous feature in the vegetation; as is O. maculata in a lesser degree in open woodlands. Most of the other species, though not deserving to be called rare, are sparingly distributed; yet, if their seeds or seedlings were not largely and habitually destroyed, any one of them would, as we have just seen, immediately cover the whole land.

I have now nearly finished this too lengthy volume. It has, I think, been shown that Orchids exhibit an almost endless diversity of beautiful adaptations. When this or that part has been spoken of as contrived for some special purpose, it must not be supposed that it was originally always formed for this sole purpose. The regular course of events seems to be, that a part which originally served for one purpose, by slow changes becomes adapted for widely different purposes. To give an instance: in all the Ophreæ, the long and nearly rigid caudicle manifestly serves for the application of the pollen-grains to the stigma, when the pollinium attached to an insect is transported from flower to flower; and the anther opens widely that the pollinium may be easily withdrawn; but in the Bee Ophrys, the caudicle, by a slight increase in length, and decrease in thickness, and by the anther opening a little more widely, becomes specially adapted for the very different purpose of self-fertilisation, through the combined aid of the gravity of the pollen-mass and the vibration of the flower. Every gradation between these two states would be possible,—of which we have seen partial proof in O. arachnites.

Again the elasticity of the pedicel of the pollinium in some Vandeæ is adapted to free the pollen-masses out of their anther-cases; but by further slight modifications, the elasticity of the pedicel becomes specially adapted to shoot out the pollinia to a distance. The great cavity in the labellum of many Vandeae serves to attract insects, but in Mormodes ignea it is greatly reduced in size, and only serves to keep the labellum in its proper position on the summit of the column. From the analogy of many plants we may infer that a long spur-like nectary is primarily adapted to secrete and hold a store of nectar; but in many Orchids it has so far lost this function, as only to contain fluid between its two coats. In those Orchids, in which the nectary contains both free nectar and fluid in the intercellular spaces, we can see how a passage from one state to the other could have been effected, namely, by less and less nectar being secreted from the inner membrane, and more and more being retained within the intercellular spaces. Other analogous cases could be given.

Although an organ may not have been originally formed for some special purpose, if it now serves for this end we are justified in saying that it is specially contrived for it. On the same principle, if a man were to make a machine for some special purpose, but were to use old wheels, springs, and pulleys, only slightly altered, the whole machine, with all its parts, might be said to be specially contrived for that purpose. Thus throughout nature almost every part of each living being has probably served, in a slightly modified condition, for diverse purposes, and has acted in the living machinery of many ancient and distinct specific forms.

In my examination of Orchids, hardly any fact has so much struck me as the endless diversity of structure,—the prodigality of resources,—for gaining the very same end, namely, the fertilisation of one flower by the pollen of another. The fact to a certain extent is intelligible on the principle of natural selection. As all the parts of a flower are co-ordinated, if slight variations in any one part are preserved from being beneficial to the plant, then the other parts will generally have to be modified in some corresponding manner. But certain parts may not vary at all, or may not vary in the simplest corresponding manner, and those variations, whatever their nature may be, which will bring all the parts into more perfect harmony with each other, will be seized on and preserved by natural selection.

To give a simple illustration: in many Orchids the ovarium (but sometimes the footstalk) becomes for a period twisted, causing the labellum to hang downwards, so that insects can easily visit the flower; but from slow changes in the form and position of the petals, or from new sorts of insects visiting the flower, it might become advantageous to the plant that the labellum should resume its normal upward position, as is actually the case with Malaxis paludosa; this change, it is obvious, might be simply effected by the continued selection of varieties which had their ovarium a little less twisted; but if the plant only afforded varieties with the ovarium more twisted, the same end could be attained by their selection until the flower had turned completely round on its axis: this seems to have occurred with the Malaxis, for the labellum has acquired its present upward position, and the ovarium is twisted to excess.

Again, we have seen that in most Vandeae there is a plain relation between the depth of the stigmatic chamber and the length of the pedicel, by which the pollen-masses are inserted; now if the chamber became slightly less deep from any change in the form of the column or any other unknown cause, the shortening of the pedicel would be the simplest corresponding change; but if the pedicel did not happen to vary in length, any the slightest tendency to an upward curvature from elasticity as in Phalænopsis, or to a backward hygrometric movement as in one of the Maxillarias, would be preserved, and the tendency would be continually augmented by selection; thus the pedicel, as far as its action is concerned, would be modified in the same manner as if it had been shortened. Such processes carried on during many thousand generations in various ways, with the several parts of the flower, would create an endless diversity of coadapted structures for the same general purpose. This view affords, I believe, the key which partly solves the problem of the vast diversity of structure adapted for closely analogous ends in many large groups of organic beings.

The more I study nature, the more I become impressed with ever-increasing force with the conclusion, that the contrivances and beautiful adaptations slowly acquired through each part occasionally varying in a slight degree but in many ways, with the preservation or natural selection of those variations which are beneficial to the organism under the complex and ever-varying conditions of life, transcend in an incomparable degree the contrivances and adaptations which the most fertile imagination of the most imaginative man could suggest with unlimited time at his disposal.

The use of each trifling detail of structure is far from a barren search to those who believe in natural selection. When a naturalist casually takes up an organic being, and does not study its whole life (imperfect though that study will ever be), he naturally doubts whether each trifling point can be of any use, or indeed whether it be due to any general law. Some naturalists believe that numberless structures have been created for the sake of mere variety and beauty,—much as a workman would make a set of different patterns. I, for one, have often and often doubted whether this or that detail of structure could be of any service; yet, if of no good, these structures could not have been modelled by the natural preservation of useful variations; such details could only be vaguely accounted for by the direct action of the conditions of life, or the mysterious laws of correlation of growth.

To give nearly all the instances of trifling details of structure in the flowers of Orchids, which are certainly of high importance, would be to recapitulate a great portion of this volume. But I will recall to the reader's memory a few cases. I do not here refer to the fundamental framework of the plant, such as the remnants of the fifteen primary organs arranged alternately in the five whorls; for nearly all those who believe in the modification of organic beings will admit that their presence is due to inheritance from a remote parent-form. A series of facts with respect to the use of the variously shaped and placed petals and sepals has just been enumerated. So, again, the importance of the slight differences in the shape of the caudicle of the pollinium of the Bee Ophrys, compared with that of the other species of the genus, has just been referred to: to this might be added the doubly-bent caudicle of the Fly Ophrys: indeed, the important relation of the length and shape of the caudicle, with reference to the position of the stigma, might be cited throughout whole tribes. The solid projecting knob of the anther in Epipactis palustris, which does not include pollen, when moved by insects, liberates the pollen-masses. In Cephalanthera grandiflora, the upright position of the flower, and its almost closed condition, protect from disturbance the slightly coherent pillars of pollen. The length and elasticity of the filament of the anther in certain species of Dendrobium apparently serves for self-fertilisation, if insects fail to transport the pollen-masses. The slight forward inclination of the crest of the rostellum in Listera prevents the anther-case being caught when the viscid matter explodes. The elasticity of the lip of the rostellum in Orchis causes it to spring up again when one pollen-mass is removed, thus keeping ready for action the second viscid disc, which otherwise would be wasted. The two nectar-secreting spots in the Frog Orchis, beneath the viscid discs at the base of the labellum, and the medial nectary in front of the stigma, are apparently necessary for the fertilisation of the flower. No one who had not studied Orchids would have suspected that these and many other small details of structure were of the highest importance to each species; and that consequently, if the species were exposed to new conditions of life, and the structure of the several parts varied ever so little, such small details of structure might be modified by natural selection. These cases afford a good lesson of caution with respect to the importance, in other organic beings, of apparently trifling particulars of structure.

It may naturally be inquired, why do Orchids exhibit so many perfect contrivances? From the observations of C. K. Sprengel, and from my own, I am sure that many other plants offer, in the means of fertilisation, analogous adaptations of high perfection; but it seems that they are really more numerous with Orchids than with most other plants. To a certain extent the inquiry can be answered. As each ovule requires at least one, probably more than one, pollen-grain, and as the seeds produced by Orchids are so inordinately numerous, we can see that large masses of pollen would have to be left on the stigma of each flower. Even in the Neotteæ, which have granular pollen, with the grains tied together only by weak threads, I have observed that considerable masses of pollen are generally left on the stigmas. Hence we can perhaps understand the use of the grains cohering in large waxy masses, as they do in so many tribes, so as to prevent waste in the act of transportal. Most plants produce pollen enough to fertilise several flowers, even when each flower has several stigmas. But as the two confluent stigmas of Orchids require so much pollen, its elaboration, if proportional in amount to that in most other plants, would have been extravagant in the highest degree, and exhausting to the individual. To save this waste and exhaustion, special and admirable contrivances were necessary for safely placing the pollen-masses on the stigma; and thus we can partially understand why Orchids have been more highly endowed in this respect than most other plants.

The simple fact that many Vandeae have only two pollen-masses, and that, from their coherence, some Malaxeæ have only a single pollen-mass, necessitates that extraordinary pains should have been taken in their fertilisation, otherwise these plants would have been barren. The existence of a single pollen-mass, so that all the pollen-grains from one flower cannot possibly fertilise more than a single stigma, occurs, I believe, in no other plants. The case is partially analogous with that of seeds: many flowers produce a multitude of seeds, several produce a single seed; very many flowers produce a countless number of pollen-grains; some Orchid-flowers produce, as far as the power of fertilising other flowers is concerned, a single pollen-mass, though really consisting of a multitude of pollen-grains.

Notwithstanding that such care has been taken that the pollen of Orchids should not be wasted, we see that throughout the vast Orchidean order,—including, according to Lindley, 433 genera, and probably about 6000 species,—the act of fertilisation is almost invariably left to insects. This assertion can hardly be considered rash, after the examination of so many British and exotic genera scattered through the main Tribes, which generally have a nearly uniform structure. In all plants in which insects play an important part in the act of fertilisation, there will be a good chance of pollen being carried from one flower to another. But in Orchids we have seen numerous adaptations,—such as the movements of the pollinia after their removal in order to acquire a proper position,—the slow movement of the labellum or rostellum to allow the entrance of the pollen-masses,—the separation of the sexes in some instances—which render it certain that in these cases the pollen of one flower or of one plant is habitually transported to another flower or plant. As this transportal increases the risk of loss, it necessitates and still further explains the extraordinary care bestowed on the contrivances for fertilisation.

Self-fertilisation is a rare event with Orchids. In Cephalanthera grandiflora it occurs, but in a very imperfect degree; and the early penetration of the stigma by the flower's own pollen-tubes seems to be fully as much determined by the support thus given to the pillars of pollen, as by the production of a small proportion of seed: certainly the fertilisation of this Orchid is aided by insects. In some species of Dendrobium self-fertilisation apparently occurs, but only if insects accidentally fail in removing the flower's own single pollen-mass. In Cypripedium, the Frog Orchis, and perhaps in a few other cases, it will depend on the manner (at present unknown) in which insects first insert their probosces by the one or the other entrance,—whether the flower's own pollen, or that of another flower, is habitually placed on the stigma; but in these cases there will assuredly always be a good chance of the stigma being fertilised by pollen brought from another flower. In the Bee Ophrys alone, as far as I have seen, there are special and perfectly efficient contrivances for self-fertilisation, combined, however, in the most paradoxical manner, with manifest adaptations for the occasional transport by insects of the pollinia from one flower to another, as in the other species of the same genus.

Considering how precious the pollen of Orchids evidently is, and what care has been bestowed on its organisation and on the accessory parts;—considering that the anther always stands close behind or above the stigma, self-fertilisation would have been an incomparably safer process than the transportal of the pollen from flower to flower. It is an astonishing fact that self-fertilisation should not have been an habitual occurrence. It apparently demonstrates to us that there must be something injurious in the process. Nature thus tells us, in the most emphatic manner, that she abhors perpetual self-fertilisation. This conclusion seems to be of high importance, and perhaps justifies the lengthy details given in this volume. For may we not further infer as probable, in accordance with the belief of the vast majority of the breeders of our domestic productions, that marriage between near relations is likewise in some way injurious,—that some unknown great good is derived from the union of individuals which have been kept distinct for many generations?

Variations Effected by Cultivation. [1862]

Darwin began accumulating information for his book on variation under domestication early in 1860. However, he was ill during much of the 1860s and found that he could work on his "hobbies" such as orchids and climbing plants when he was unable to work on Variation. *His "hobbies" took only a few hours per day.* The Variation of Animals and Plants under Domestication *was not published until 1868. Its two volumes are full of footnotes and textual commentary acknowledging correspondents who had replied to questions such as the following, published on December 2, 1862, in the* Journal of Horticulture and Cottage Gardener, *one of the periodicals in which a number of such letters from and to Darwin appeared during the 1860s. The correspondence yielded much information for* Variation. *Gardeners, nurserymen, stockmen, bird fanciers, and gentlemen farmers contributed, for in spite of its title the* Journal of Horticulture and Cottage Gardener *had sections covering farm stock and fowl as well. Thus Darwin's request for data reached a larger audience than we might imagine. Many of his informants became regular correspondents.*

The abbreviation "dwt" found in Darwin's letter refers to a pennyweight, one-twentieth of an ounce.

As you have been so obliging as to insert my query on the crossing of Strawberries, perhaps you will grant me a favour to insert two or three other questions, for the chance of some one having the kindness to answer them. I am writing a book on "Variation under Domestication," in which I treat chiefly on animals; but I wish to give some few facts on the changes of cultivated plants.

1st. The fruit of the wild Gooseberry is said to weigh about 5 dwts. (I am surprised that it is so heavy), and from various records I find that towards the close of the last century the fruit had doubled in weight; in 1817, a weight of 26 dwts. 17 grs. was obtained; in 1825, 31 dwts. 13 grs.; in 1841, "Wonderful" weighed 32 dwts. 16 grs.; in 1845, "London" reached the astonishing weight of 36 dwts. 16 grs., or 880 grains. I find in the "Gooseberry Register" for 1862, that this famous kind attained only the weight of 29 dwts. 8 grs., and was beaten by "Antagonist." Will any one have the kindness to inform me whether it is authentically known that the weight of 36 dwts. 16 grs., has, since the year 1845, been ever excelled?

2nd. Is any record kept of the diameter attained by the largest Pansies? I have read of one above 2 inches in diameter, which is a surprising size compared with the flowers of the wild Viola tricolor, and the allied species or varieties.

3rd. How early does any variety of the Dahlia flower? Mr. Salisbury, writing in 1808, shortly after the first introduction of this plant into England, speaks of their flowering from September, or the end of September, to November. Whereas, Mr. J. Wells, in Loudon's "Gardener's Magazine" for 1828, states that some of his dwarf kinds began flowering in June. I presume the end of June. Do any of the varieties now regularly flower as early as June? Have any varieties been observed to withstand frost better than other varieties?

If any one will give me information on these small points, I shall feel greatly obliged.

Recollections of Professor Henslow. [1862]

The Reverend John Stevens Henslow, Darwin's mentor in Cambridge days, died in May 1861 at the age of sixty-five. During his thirty-eight-year Cambridge career, first as professor of mineralogy and then as professor of botany, Henslow was loved and respected by colleagues, friends, and students. The Reverend Leonard Jenyns, a fellow Cambridge student with Henslow and later his brother-in-law, edited a volume dedicated to his memory in 1862. The Memoir of the Rev. John Stevens Henslow M.A., F.L.S., F.G.S., F.C.P.S., late Rector of Hitcham and Professor of Botany in the University of Cambridge, *containing contributions by Darwin and others, shows what a powerful influence Henslow had on the lives he touched. His field excursions and weekly open houses for student naturalists were renowned at Cambridge. In 1837, when he became rector of Hitcham, Suffolk, and was unable to continue them, the open houses were formalized as the Cambridge Ray Club. Characteristically, the club's name commemorates John Ray, a famous Cambridge botanist of an earlier generation, not the modest Henslow.*

We have earlier seen Henslow's part in getting Darwin the opportunity to sail with HMS Beagle. *Of all Darwin's correspondents during the voyage, Henslow received the fullest scientific accounts; and to him Darwin sent the* Beagle *collections. When those specimens and their collector arrived in England, Henslow took on the task of identifying the plants. In the* Autobiography *Darwin recalls that during the latter half of his time at Cambridge he and Henslow habitually took long daily walks together "so that I was called by some of the dons 'the man who walks with Henslow.'" But Darwin's granddaugh-*

ter Nora Barlow, introducing her edition of the correspondence between the two men, hit on a phrase that more precisely characterizes their relation as the wider world has come to see it—Henslow is now best remembered as "the man to whom Darwin wrote."

I went to Cambridge early in the year 1828, and soon became acquainted, through some of my brother entomologists, with Professor Henslow, for all who cared for any branch of natural history were equally encouraged by him. Nothing could be more simple, cordial, and unpretending than the encouragement which he afforded to all young naturalists. I soon became intimate with him, for he had a remarkable power of making the young feel completely at ease with him; though we were all awe-struck with the amount of his knowledge. Before I saw him, I heard one young man sum up his attainments by simply saying that he knew everything. When I reflect how immediately we felt at perfect ease with a man older and in every way so immensely our superior, I think it was as much owing to the transparent sincerity of his character, as to his kindness of heart; and, perhaps, even still more to a highly remarkable absence in him of all self-consciousness. One perceived at once that he never thought of his own varied knowledge or clear intellect, but solely on the subject in hand. Another charm, which must have struck every one, was that his manner to old and distinguished persons and to the youngest student was exactly the same: to all he showed the same winning courtesy. He would receive with interest the most trifling observation in any branch of natural history; and however absurd a blunder one might make, he pointed it out so clearly and kindly, that one left him no way disheartened, but only determined to be more accurate the next time. In short, no man could be better formed to win the entire confidence of the young, and to encourage them in their pursuits.

His Lectures on Botany were universally popular, and as clear as daylight. So popular were they, that several of the older members of the University attended successive courses. Once every week he kept open house in the evening, and all who cared for natural history attended these parties, which, by thus favouring intercommunication, did the same good in Cambridge, in a very pleasant manner, as the Scientific Societies do in London. At these parties many of the most distinguished members of the University occasionally attended; and when only a few were present, I have listened to the great men of those days, conversing on all sorts of subjects, with the most varied and brilliant powers. This was no small advantage to some of the younger men, as it stimulated their mental activity and ambition. Two or three times in each session he took excursions with his botanical class; either a long walk to the habitat of some rare plant, or in a barge down the river to the fens, or in coaches to some more distant place, as to Gamlingay, to see the wild lily of the valley, and to catch on the heath the rare natter-jack. These excursions have left a delightful impression on my mind. He was, on such occasions, in as good spirits as a boy, and laughed as heartily as a boy at the misadventures of those who chased the splendid swallow-tail butterflies across the broken and treacherous fens. He used to pause every now and then and lecture on some plant or other object; and something he could tell us on every insect, shell, or fossil collected, for he had attended to every branch of natural history. After our day's work we used to dine at some inn or house, and most jovial we then were. I believe all who joined these excursions will agree with me that they left an enduring impression of delight on our minds.

As time passed on at Cambridge I became very intimate with Professor Henslow, and his kindness was unbounded; he continually asked me to his house, and allowed me to accompany him in his walks. He talked on all subjects, including his deep sense of religion, and was entirely open. I owe more than I can express to this excellent man. His kindness was steady: when Captain FitzRoy offered to give up part of his own cabin to any naturalist who would join the expedition in H.M.S. *Beagle*, Professor Henslow recommended me, as one who knew very little, but who, he thought, would work. I was strongly attached to natural history, and this attachment I owed, in large part, to him. During the five years' voyage, he regularly corresponded with me and guided my efforts; he received, opened, and took care of all the specimens sent home in many large boxes; but I firmly believe that, during these five years, it never once crossed his mind that he was acting towards me with unusual and generous kindness.

During the years when I associated so much with Professor Henslow, I never once saw his temper even ruffled. He never took an ill-natured view of any one's character, though very far from blind to the foibles of others. It always struck me that his mind could not be even touched by any paltry feeling of vanity, envy, or jealousy. With all this equability of temper and remarkable benevolence, there was no insipidity of character. A man must have been blind not to have perceived that beneath this placid exterior there was a vigorous and determined will. When principle came into play, no power on earth could have turned him one hair's breadth.

After the year 1842, when I left London, I saw Professor Henslow only at long intervals; but to the last, he continued in all respects the same man. I think he cared somewhat less about science, and more for his parishioners. When speaking of his allotments, his parish children, and plans of amusing and instructing them, he would always kindle up with interest and enjoyment. I remember one trifling fact which seemed to me highly characteristic of the man: in one of the bad years for the potato, I asked him how his crop had fared; but after a little talk I perceived that in fact, he knew nothing about his own potatoes, but seemed to know exactly what sort of crop there was in the garden of almost every poor man in his parish.

In intellect, as far as I could judge, accurate powers of observation, sound sense, and cautious judgement seemed predominant. Nothing seemed to give him so much enjoyment, as drawing conclusions from minute observations. But his admirable memoir on the geology of Anglesea, shows his capacity for extended observations and broad views. Reflecting over his character with gratitude and reverence, his moral attributes rise, as they should do in the highest character, in preeminence over his intellect.

The Variation of Animals and Plants under Domestication. [1868]

We have already seen that Natural Selection, *the big evolutionary work that had engrossed Darwin between 1856 and 1858, was interrupted by the arrival of Wallace's letter and manuscript and further delayed by the writing of* The Origin of Species. Natural Selection *was never completed; but following the publication of the second edition of the* Origin *in January 1860, Darwin planned to mine the unfinished manuscript for three books documenting his assertions in the* Origin. *Two of these books, one on "Variability of organic beings in a state of nature" and the other on "The principle of natural selection," never were written, though their titles can be deduced from discussion in the third book,* The Variation of Animals and Plants under Domestication. *Based on the first two chapters of* Natural Selection *(much augmented and updated),* Variation *proved to be Darwin's longest book. The two-volume first edition of 1868 ran to 899 pages, and the extensively altered second edition of 1875 was even longer. Darwin considered* Variation *one of his most important works, but it is one of the few Darwin books little cited by scientists today.*

Like many of Darwin's books, Variation *arose from investigations he had been pursuing for decades. In 1839 Darwin had distributed a sheet titled "Questions about the breeding of Animals." The answers he received from farmers and landowners were incorporated into* Variation *almost thirty years later. Primarily a study of how domesticated plants and animals have arisen through a process of artificial (that is to say, human) selection,* Variation *builds mainly on the published research of other investigators, information Darwin assim-*

ilates and discusses in his own distinctive way. Observations on the plants and animals he has raised appear here and there: several chapters, for instance, are devoted to Darwin's research on domesticated pigeons.

Much of the second volume of Variation *has to do with heredity. Since 1837, when he was first beginning his quest for an explanation of evolution, Darwin had realized that the key lay in understanding inheritance—and in* Variation *his objective is to trace the workings of heredity in domestic plants and animals, a process that he (unlike Wallace) saw as corresponding to what takes place in nature. Darwin realized that inheritance must be particulate, that particles of some sort must pass on traits from generation to generation. Thanks to research beginning with the experimental studies of the Moravian monk Gregor Mendel, we now know that these particles are genes on chromosomes. Although Mendel published his first paper on inheritance in garden peas in 1865, Darwin did not know of his work; indeed the scientific community did not recognize its significance until 1900.*

Darwin's own hypothesis of inheritance, which he called "pangenesis," was that each cell, tissue, and organ in the body of a plant or animal contained or formed particles that he called "gemmules." The gemmules migrated in some way through the tissues to the reproductive organs, where they all entered the egg or sperm. Union of the gemmules of two parents produced the traits of their offspring. Darwin, like many others of his day, believed that sexual reproduction resulted in the blending of inherited traits—that the characteristics of offspring would be intermediate between those of the two equally contributing parents. Therefore the hereditary contribution of each ancestor would be halved every generation. Because of blending, children would have half the gemmules of each parent, grandchildren one-fourth, great-grandchildren one-eighth, and so on. Darwin accounted for deviations from this process by asserting that blending inheritance was counteracted by the effects of internal and environmental factors, which stimulated fresh variation to appear in each generation.

Pangenesis and blending inheritance were ingenious explanations in an era when chromosomes and genes were unknown. But both hypotheses were dead wrong. Although pangenesis was advocated by some geneticists later in the nineteenth century, among all Darwin's ideas it proved second only to the theory of natural selection in provoking detractors. Despite some discredited hypotheses, however, Variation *remains a goldmine of particular information on domestic plants and animals. Below we reprint Darwin's introduction, several of his discussions of variation in specific plants and animals, remarks on natural and artificial selection, and the hypothesis of pangenesis.*

INTRODUCTION.

The object of this work is not to describe all the many races of animals which have been domesticated by man, and of the plants which have been cultivated by him; even if I possessed the requisite knowledge, so gigantic an undertaking would be here superfluous. It is my intention to give under the head of each species only such facts as I have been able to collect or observe, showing the amount and nature of the changes which animals and plants have undergone whilst under man's dominion, or which bear on the general principles of variation. In one case alone, namely in that of the domestic pigeon, I will describe fully all the chief races, their history, the amount and nature of their differences, and the probable steps by which they have been formed. I have selected this case, because, as we shall hereafter see, the materials are better than in any other; and one case fully described will in fact illustrate all others. But I shall also describe domesticated rabbits, fowls, and ducks, with considerable fullness.

The subjects discussed in this volume are so connected that it is not a little difficult to decide how they can be best arranged. I have determined in the first part to give, under the heads of the various animals and plants, a large body of facts, some of which may at first appear but little related to our subject, and to devote the latter part to general discussions. Whenever I have found it necessary to give numerous details, in support of any proposition or conclusion, small type has been used. The reader will, I think, find this plan a convenience, for, if he does not doubt the conclusion or care about the details, he can easily pass them over; yet I may be permitted to say that some of the discussions thus printed deserve attention, at least from the professed naturalist.

It may be useful to those who have read nothing about Natural Selection, if I here give a brief sketch of the whole subject and of its bearing on the origin of species.* This is the more desirable, as it is

* To anyone who has attentively read my *Origin of Species* this Introduction will be superfluous. As I stated in that work that I should soon publish the facts on which the conclusions given in it were founded, I here beg permission to remark

impossible in the present work to avoid many allusions to questions which will be fully discussed in future volumes.

From a remote period, in all parts of the world, man has subjected many animals and plants to domestication or culture. Man has no power of altering the absolute conditions of life; he cannot change the climate of any country; he adds no new element to the soil; but he can remove an animal or plant from one climate or soil to another, and give it food on which it did not subsist in its natural state. It is an error to speak of man "tampering with nature" and causing variability. If organic beings had not possessed an inherent tendency to vary, man could have done nothing. He unintentionally exposes his animals and plants to various conditions of life, and variability supervenes, which he cannot even prevent or check. Consider the simple case of a plant which has been cultivated during a long time in its native country, and which consequently has not been subjected to any change of climate. It has been protected to a certain extent from the competing roots of plants of other kinds; it has generally been grown in manured soil, but probably not richer than that of many an alluvial flat; and lastly, it has been exposed to changes in its conditions, being grown sometimes in one district and sometimes in another, in different soils. Under such circumstances, scarcely a plant can be named, though cultivated in the rudest manner, which has not given birth to several varieties. It can hardly be maintained that during the many changes which this earth has undergone, and during the natural migrations of plants from one land or island to another, tenanted by different species, that such plants will not often have been subjected to changes in their conditions analogous to those which almost inevitably cause cultivated plants to vary. No doubt man selects varying individuals, sows their seeds, and again selects their varying offspring. But the initial variation on which man works, and without which he can do nothing, is caused by slight changes in the conditions of life, which must often have occurred under nature. Man, therefore, may be said to have been trying an experiment on a gigantic scale; and it is an experiment which nature during the long lapse of time has incessantly tried. Hence it follows that the principles of domestication are important for us. The main result is that organic beings thus treated have varied largely, and the variations have been inherited. This has apparently been one chief

---

that the great delay in publishing this first work has been caused by continued ill-health.

cause of the belief long held by some few naturalists that species in a state of nature undergo change.

I shall in this volume treat, as fully as my materials permit, the whole subject of variation under domestication. We may thus hope to obtain some light, little though it be, on the causes of variability,—on the laws which govern it, such as the direct action of climate and food, the effects of use and disuse, and of correlation of growth,—and on the amount of change to which domesticated organisms are liable. We shall learn something on the laws of inheritance, on the effects of crossing different breeds, and on that sterility which often supervenes when organic beings are removed from their natural conditions of life, and likewise when they are too closely interbred. During this investigation we shall see that the principle of Selection is all important. Although man does not cause variability and cannot even prevent it, he can select, preserve, and accumulate the variations given to him by the hand of nature in any way which he chooses; and thus he can certainly produce a great result. Selection may be followed either methodically and intentionally, or unconsciously and unintentionally. Man may select and preserve each successive variation, with the distinct intention of improving and altering a breed, in accordance with a preconceived idea; and by thus adding up variations, often so slight as to be imperceptible by an uneducated eye, he has effected wonderful changes and improvements. It can, also, be clearly shown that man, without any intention or thought of improving the breed, by preserving in each successive generation the individuals which he prizes most, and by destroying the worthless individuals, slowly, though surely, induces great changes. As the will of man thus comes into play, we can understand how it is that domesticated breeds show adaptation to his wants and pleasures. We can further understand how it is that domestic races of animals and cultivated races of plants often exhibit an abnormal character, as compared with natural species; for they have been modified not for their own benefit, but for that of man.

In a second work I shall discuss the variability of organic beings in a state of nature; namely, the individual differences presented by animals and plants, and those slightly greater and generally inherited differences which are ranked by naturalists as varieties or geographical races. We shall see how difficult, or rather how impossible it often is, to distinguish between races and sub-species, as the less well-marked forms have sometimes been denominated; and again between sub-species and true species. I shall further attempt to show that it is the common and widely ranging, or, as they may be called, the dominant species, which most frequently vary; and that it is the large and flourishing genera which include the greatest number of varying species. Varieties, as we shall see, may justly be called incipient species.

But it may be urged, granting that organic beings in a state of nature present some varieties,—that their organization is in some slight degree plastic; granting that many animals and plants have varied greatly under domestication, and that man by his power of selection has gone on accumulating such variations until he has made strongly marked and firmly inherited races; granting all this, how, it may be asked, have species arisen in a state of nature? The differences between natural varieties are slight; whereas the differences are considerable between the species of the same genus, and great between the species of distinct genera. How do these lesser differences become augmented into the greater difference? How do varieties, or as I have called them incipient species, become converted into true and well-defined species? How has each new species been adapted to the surrounding physical conditions, and to the other forms of life on which it in any way depends? We see on every side of us innumerable adaptations and contrivances, which have justly excited in the mind of every observer the highest admiration. There is, for instance, a fly (Cecidomyia) which deposits its eggs within the stamens of a Scrophularia, and secretes a poison which produces a gall, on which the larva feeds; but there is another insect (Misocampus) which deposits its eggs within the body of the larva within the gall, and is thus nourished by its living prey; so that here a hymenopterous insect depends on a dipterous insect, and this depends on its power of producing a monstrous growth in a particular organ of a particular plant. So it is, in a more or less plainly marked manner, in thousands and tens of thousands of cases, with the lowest as well as with the highest productions of nature.

This problem of the conversion of varieties into species,—that is, the augmentation of the slight differences characteristic of varieties into the greater differences characteristic of species and genera, including the admirable adaptations of each being to its complex organic and inorganic conditions of life,—will form the main subject of my second work. We shall therein see that all organic beings, without exception, tend to increase at so high a ratio, that no district, no station, not even the whole surface of the land or the whole ocean, would hold the progeny of a single pair after a certain number of generations. The inevitable result is an ever-recurrent Struggle for Existence. It has truly been said that all nature is at war; the strongest ultimately prevail, the weakest fail; and we well know that myriads of forms have disappeared from the face of the earth. If then organic beings in a state of nature vary even in a slight degree, owing to changes in the surrounding conditions, of which we have abundant geological evidence, or from any other cause; if, in the long course of ages, inheritable variations ever arise in any way advantageous to any being under its excessively complex and changing relations of life; and it would be a strange fact if beneficial variations did never arise, seeing how many have arisen which man has taken advantage of for his own profit or pleasure; if then these contingencies ever occur, and I do not see how the probability of their occurrence can be doubted, then the severe and often-recurrent struggle for existence will determine that those variations, however slight, which are favourable shall be preserved or selected, and those which are unfavourable shall be destroyed.

This preservation, during the battle for life, of varieties which possess any advantage in structure, constitution, or instinct, I have called Natural Selection; and Mr. Herbert Spencer has well expressed the same idea by the Survival of the Fittest. The term "natural selection" is in some respects a bad one, as it seems to imply conscious choice; but this will be disregarded after a little familiarity. No one objects to chemists speaking of "elective affinity;" and certainly an acid has no more choice in combining with a base, than the conditions of life have in determining whether or not a new form be selected or preserved. The term is so far a good one as it brings into connection the production of domestic races by man's power of selection, and the natural preservation of varieties and species in a state of nature. For brevity sake I sometimes speak of natural selection as an intelligent power;—in the same way as astronomers speak of the attraction of gravity as ruling the movements of the planets, or as agriculturists speak of man making domestic races by his power of selection. In the one case, as in the other, selection does nothing without variability, and this depends in some manner on the action of the surrounding circumstances on the organism. I have, also, often personified the word Nature; for I have found it difficult to avoid this ambiguity; but I mean by nature only the aggregate action and product of many natural laws,—and by laws only the ascertained sequence of events.

In the chapter devoted to natural selection I shall show from experiment and from a multitude of facts, that the greatest amount of life can be supported on each spot by great diversification or divergence in the structure and constitution of its inhabitants. We shall, also, see that the continued production of new forms through natural selection, which implies that each new variety has some advantage over others, almost inevitably leads to the extermination of the older and less improved forms. These latter are almost necessarily intermediate in structure as well as in descent between the last-produced forms and their original parent-species. Now, if we suppose a species to produce two or more varieties, and these in the course of time to produce other varieties, the principle of good being derived from diversification of structure will generally lead to the preservation of the most divergent varieties; thus the lesser differences characteristic of varieties come to be augmented into the greater differences characteristic of species, and, by the extermination of the older intermediate forms, new species come to be distinctly defined objects. Thus, also, we shall see how it is that organic beings can be classed by what is called a natural method in distinct groups—species under genera, and genera under families.

As all the inhabitants of each country may be said, owing to their high rate of reproduction, to be striving to increase in numbers; as each form is related to many other forms in the struggle for life,—for destroy any one and its place will be seized by others; as every part of the organization occasionally varies in some slight degree, and as natural selection acts exclusively by the preservation of variations which are advantageous under the excessively complex conditions to which each being is exposed, no limit exists to the number, singularity, and perfection of the contrivances and co-adaptations which may thus be produced. An animal or a plant may thus slowly become related in its structure and habits in the most intricate manner to many other animals and plants, and to the physical conditions of its home. Variations in the organization will in some cases be aided by habit, or by the use and disuse of parts, and they will be governed by the direct action of the surrounding physical conditions and by correlation of growth.

On the principles here briefly sketched out, there is no innate or necessary tendency in each being to its own advancement in the scale of organization. We are almost compelled to look at the specialization or differentiation of parts or organs for different functions as the best or even sole standard of advancement; for by such division of labour each function of body and mind is better performed. And, as natural selection acts exclusively through the preservation of profitable modifications of structure, and as the conditions of life in each area generally become more and more complex, from the increasing number of different forms which inhabit it and from most of these forms acquiring a more and more perfect structure, we may confidently believe, that, on the whole, organization advances. Nevertheless a very simple form fitted for very simple conditions of life might remain for indefinite ages unaltered or unimproved; for what would it profit an infusorial animalcule, for instance, or an intestinal worm, to become highly organized? Members of a high group might even become, and this apparently has occurred, fitted for simpler conditions of life; and in this case natural selection would tend to simplify or degrade the organization, for complicated mechanism for simple actions would be useless or even disadvantageous.

In a second work, after treating of the Variation of organisms in a state of nature, of the Struggle for Existence and the principle of Natural Selection, I shall discuss the difficulties which are opposed to the theory. These difficulties may be classed under the following heads: —the apparent impossibility in some cases of a very simple organ graduating by small steps into a highly perfect organ; the marvellous facts of Instinct; the whole question of Hybridity; and, lastly, the absence, at the present time and in our geological formations, of innumerable links connecting all allied species. Although some of these difficulties are of great weight, we shall see that many of them are explicable on the theory of natural selection, and are otherwise inexplicable.

In scientific investigations it is permitted to invent any hypothesis, and if it explains various large and independent classes of facts it rises to the rank of a well-grounded theory. The undulations of the ether and even its existence are hypothetical, yet every one now admits the undulatory theory of light. The principle of natural selection may be looked at as a mere hypothesis, but rendered in some degree probable by what we positively know of the variability of organic beings in a state of nature,—by what we positively know of the struggle for existence, and the consequent almost inevitable preservation of favourable variations,—and from the analogical formation of domestic races. Now this hypothesis may be tested,—and this seems to me the only fair and legitimate manner of considering the whole question,—by trying whether it explains several large and independent classes of facts; such as the geological succession of organic beings, their distribution in past and present times, and their mutual affinities and homologies. If the principle of natural selection does explain these and other large bodies of facts, it ought to be received. On the ordinary view of each species having been independently created, we gain no scientific explanation of any one of these facts. We can only say that it has so pleased the Creator to command that the past and present inhabitants of the world should appear in a certain order and in certain areas; that He has impressed on them the most extraordinary resemblances, and has classed them in groups subordinate to groups. But by such statements we gain no new knowledge; we do not connect together facts and laws; we explain nothing.

In a third work I shall try the principle of natural selection by seeing how far it will give a fair explanation of the several classes of facts just alluded to. It was the consideration of these facts which first led me to take up the present subject. When I visited, during the voyage of H.M.S. *Beagle*, the Galapagos Archipelago, situated in the Pacific Ocean about 500 miles from the shore of South America, I found myself surrounded by peculiar species of birds, reptiles, and plants, existing nowhere else in the world. Yet they nearly all bore an American stamp. In the song of the mocking-thrush, in the harsh cry of the carrion-hawk, in the great candlestick-like opuntias, I clearly perceived the neighbourhood of America, though the islands were separated by so many miles of ocean from the mainland, and differed much from it in their geological constitution and climate. Still more surprising was the fact that most of the inhabitants of each separate island in this small archipelago were specifically different, though most closely related to each other. The archipelago, with its innumerable craters and bare streams of lava, appeared to be of recent origin; and thus I fancied myself brought near to the very act of creation. I often asked myself how these many peculiar animals and plants had been produced: the simplest answer seemed to be that the inhabitants of the several islands had descended from each other, undergoing modification in the course of their descent; and that all the inhabitants of the archipelago had descended from those of the nearest land, namely America, whence colonists would naturally have been derived. But it long remained to me an inexplicable problem how the necessary degree of modification could have been effected, and it would have thus remained for ever, had I not studied domestic productions, and thus acquired a just idea of the power of Selection. As soon as I had fully realized this idea, I saw, on reading Malthus on Population, that Natural Selection was the inevitable result of the rapid increase of all organic beings; for I was prepared to appreciate the struggle for existence by having long studied the habits of animals.

Before visiting the Galapagos I had collected many animals whilst travelling from north to south on both sides of America, and everywhere, under conditions of life as different as it is possible to conceive, American forms were met with—species replacing species of the same peculiar genera. Thus it was when the Cordilleras were ascended, or the thick tropical forests penetrated, or the fresh waters of America searched. Subsequently I visited other countries, which in all the conditions of life were incomparably more like to parts of South America, than the different parts of that continent were to each other; yet in these countries, as in Australia or Southern Africa, the traveller cannot fail to be struck with the entire difference of their productions. Again the reflection was forced on me that community of descent from the early inhabitants or colonists of South America would alone explain the wide prevalence of American types of structure throughout that immense area.

To exhume with one's own hands the bones of extinct and gigantic quadrupeds brings the whole question of the succession of species vividly before one's mind; and I had found in South America great pieces of tesselated armour exactly like, but on a magnificent scale, that covering the pigmy armadillo; I had found great teeth like those of the living sloth, and bones like those of the cavy. An analogous succession of allied forms had been previously observed in Australia. Here then we see the prevalence, as if by descent, in time as in space, of the same types in the same areas; and in neither case does the similarity of the conditions by any means seem sufficient to account for the similarity of the forms of life. It is notorious that the fossil remains of closely consecutive formations are closely allied in structure, and we can at once understand the fact if they are likewise closely allied by descent. The succession of the many distinct species of the same genus throughout the long series of geological formations seems to have been unbroken or continuous. New species come in gradually one by one. Ancient and extinct forms of life often show combined or intermediate characters, like the words of a dead language with respect to its several offshoots or living tongues. All these and other such facts seemed to me to point to descent with modification as the method of production of new groups of species.

The innumerable past and present inhabitants of the world are connected together by the most singular and complex affinities, and can be classed in groups under groups, in the same manner as varieties can be classed under species and sub-varieties under varieties, but with much higher grades of difference. It will be seen in my third work that these complex affinities and the rules for classification receive a rational explanation on the principle of descent, together with modifications acquired through natural selection, entailing divergence of character and the extinction of intermediate forms. How inexplicable is the similar pattern of the hand of a man, the foot of a dog, the wing of a bat, the flipper of a seal, on the doctrine of independent acts of creation! how simply explained on the principle of the natural selection of successive slight variations in the diverging descendants from a single progenitor! So it is, if we look to the structure of an individual animal or plant, when we see the fore and hind limbs, the skull and vertebræ, the jaws and legs of a crab, the petals, stamens, and pistils of a flower, built on the same type or pattern. During the many changes to which in the course of time all organic beings have been subjected, certain organs or parts have occasionally become at first of little use and ultimately superfluous; and the retention of such parts in a rudimentary and utterly useless condition can, on the descent-theory, be simply understood. On the principle of modifications being inherited at the same age in the child, at which each successive variation first appeared in the parent, we shall see why rudimentary parts and organs are generally well developed in the individual at a very early age. On the same principle of inheritance at corresponding ages, and on the principle of variations not generally supervening at a very early period of embryonic growth (and both these principles can be shown to be probable from direct evidence), that most wonderful fact in the whole round of natural history, namely, the similarity of members of the same great class in their embryonic condition,—the embryo, for instance, of a mammal, bird, reptile, and fish being barely distinguishable,—becomes simply intelligible.

It is the consideration and explanation of such facts as these which has convinced me that the theory of descent with modification by means of natural selection is in the main true. These facts have as yet received no explanation on the theory of independent Creations; they cannot be grouped together under one point of view, but each has to be considered as an ultimate fact. As the first origin of life on this earth, as well as the continued life of each individual, is at present quite beyond the scope of science, I do not wish to lay much stress on the greater simplicity of the view of a few forms, or of only one form, having been originally created, instead of innumerable miraculous creations having been necessary at innumerable periods; though this more simple view accords well with Maupertuis's philosophical axiom "of least action."

In considering how far the theory of natural selection may be extended,—that is, in determining from how many progenitors the inhabitants of the world have descended,—we may conclude that at least all the members of the same class have descended from a single ancestor. A number of organic beings are included in the same class, because they present, independently of their habits of life, the same fundamental type of structure, and because they graduate into each other. Moreover, members of the same class can in most cases be shown to be closely alike at an early embryonic age. These facts can be explained on the belief of their descent from a common form; therefore it may be safely admitted that all the members of the same class have descended from one progenitor. But as the members of quite distinct classes have something in common in structure and much in common in constitution, analogy and the simplicity of the view would lead us one step further, and to infer as probable that all living creatures have descended from a single prototype.

I hope that the reader will pause before coming to any final and hostile conclusion on the theory of natural selection. It is the facts and views to be hereafter given which have convinced me of the truth of the theory. The reader may consult my 'Origin of Species,' for a general sketch of the whole subject; but in that work he has to take many statements on trust. In considering the theory of natural selection, he will assuredly meet with weighty difficulties, but these difficulties relate chiefly to subjects—such as the degree of perfection of the geological record, the means of distribution, the possibility of transitions in organs, &c.—on which we are confessedly ignorant; nor do we know how ignorant we are. If we are much more ignorant than is generally supposed, most of these difficulties wholly disappear. Let the reader reflect on the difficulty of looking at whole classes of facts from a new point of view. Let him observe how slowly, but surely, the noble views of Lyell on the gradual changes now in progress on the earth's surface have been accepted as sufficient to account for all that we see in its past history. The present action of natural selection may seem more or less probable; but I believe in the truth of the theory, because it collects under one point of view, and gives a rational explanation of, many apparently independent classes of facts.*

* In treating the several subjects included in the present and succeeding works I have continually been led to ask for information from many zoologists, botanists, geologists, breeders of animals, and horticulturists, and I have invariably received from them the most generous assistance. Without such aid I could have effected little. I have repeatedly applied for information and specimens to foreigners, and to British merchants and officers of the Government residing in distant lands, and, with the rarest exceptions, I have received prompt, open-handed, and valuable assistance. I cannot express too strongly my obligations to the many persons who have assisted me, and who, I am convinced, would be equally willing to assist others in any scientific investigation.

• • •

DOMESTIC PIGEONS.

A difficulty with respect to the power of selection will perhaps already have occurred to the reader, namely, what could have led fanciers first to attempt to make such singular breeds as pouters, fantails, carriers, &c.? But it is this very difficulty which the principle of unconscious selection removes. Undoubtedly no fancier ever did intentionally make such an attempt. All that we need suppose is that a variation occurred sufficiently marked to catch the discriminating eye of some ancient fancier, and then unconscious selection carried on for many generations, that is, the wish of succeeding fanciers to excel their rivals, would do the rest. In the case of the fantail we may suppose that the first progenitor of the breed had a tail only slightly erected, as may now be seen in certain runts, with some increase in the number of the tail-feathers, as now occasionally occurs with nuns. In the case of the pouter we may suppose that some bird inflated its crop a little more than other pigeons, as is now the case in a slight degree with the œsophagus of the turbit. We do not in the least know the origin of the common tumbler, but we may suppose that a bird was born with some affection of the brain, leading it to make somersaults in the air; and the difficulty in this case is lessened, as we know that, before the year 1600, in India, pigeons remarkable for their diversified manner of flight were much valued, and by the order of the Emperor Akber Khan were sedulously trained and carefully matched.

In the foregoing cases we have supposed that a sudden variation, conspicuous enough to catch a fancier's eye, first appeared; but even this degree of abruptness in the process of variation is not necessary for the formation of a new breed. When the same kind of pigeon has been kept pure, and has been bred during a long period by two or more fanciers, slight differences in the strain can often be recognised. Thus I have seen first-rate jacobins in one man's possession which certainly differed slightly in several characters from those kept by another. I possessed some excellent barbs descended from a pair which had won a prize, and another lot descended from a stock formerly kept by that famous fancier Sir John Sebright, and these plainly differed in the form of the beak; but the differences were so slight, that they could hardly be described by words. Again, the common English and Dutch tumbler differ in a somewhat greater degree, both in length of beak and shape of head. What first caused these slight differences cannot be explained any more than why one man has a long nose and another a short one. In the strains long kept distinct by different fanciers, such differences are so common that they cannot be accounted for by the accident of the birds first chosen for breeding having been originally as different as they now are. The explanation no doubt lies in selection of a slightly different nature having been applied in each case; for no two fanciers have exactly the same taste, and consequently no two, in choosing and carefully matching their birds, prefer or select exactly the same. As each man naturally admires his own birds, he goes on continually exaggerating by selection whatever slight peculiarities they may possess. This will more especially happen with fanciers living in different countries, who do not compare their stocks and aim at a common standard of perfection. Thus, when a mere strain has once been formed, unconscious selection steadily tends to augment the amount of difference, and thus converts the strain into a sub-breed, and this ultimately into a well-marked breed or race.

The principle of correlation of growth should never be lost sight of. Most pigeons have small feet, apparently caused by their lessened use, and from correlation, as it would appear, their beaks have likewise become reduced in length. The beak is a conspicuous organ, and, as soon as it had thus become perceptibly shortened, fanciers would almost certainly strive to reduce it still more by the continued selection of birds with the shortest beaks; whilst at the same time other fanciers, as we know has actually been the case, would, in other sub-breeds, strive to increase its length. With the increased length of the beak, the tongue would become greatly lengthened, as would the eyelids with the increased development of the eye-wattles; with the reduced or increased size of the feet the number of the scutellæ would vary; with the length of the wing the number of the primary wing-feathers would differ; and with the increased length of the body in the pouter the number of the sacral vertebræ would be augmented. These important and correlated differences of structure do not invariably characterise any breed; but if they had been attended to and selected with as much care as the more conspicuous external differences, there can hardly be a doubt that they would have been rendered constant. Fanciers could assuredly have made a race of tumblers with nine instead of ten primary wing-feathers, seeing how often the number nine appears without any wish on their part, and indeed in the case of the white-winged varieties in opposition to their wish. In a similar manner, if the vertebræ had been visible and had been attended to by fanciers, assuredly an additional number might easily have been fixed in the pouter. If these latter characters had once been rendered constant we should never have suspected that they had at first been highly variable, or that they had arisen from correlation, in the one case with the shortness of the wings, and in the other case with the length of the body.

In order to understand how the chief domestic races have become distinctly separated from each other, it is important to bear in mind, that fanciers constantly try to breed from the best birds, and consequently that those which are inferior in the requisite qualities are in each generation neglected; so that after a time the less improved parent-stocks and many subsequently formed intermediate grades become extinct. This has occurred in the case of the pouter, turbit, and trumpeter, for these highly improved breeds are now left without any links closely connecting them either with each other or with the aboriginal rock-pigeon. In other countries, indeed, where the same care has not been applied, or where the same fashion has not prevailed, the earlier forms may long remain unaltered or altered only in a slight degree, and we are thus sometimes enabled to recover the connecting links. This is the case in Persia and India with the tumbler and carrier, which there differ but slightly from the rock-pigeon in the proportions of their beaks. So again in Java, the fantail sometimes has only fourteen caudal feathers, and the tail is much less elevated and expanded than in our improved birds; so that the Java bird forms a link between a first-rate fantail and the rock-pigeon.

Occasionally a breed may be retained for some particular quality in a nearly unaltered condition in the same country, together with highly modified offshoots or sub-breeds, which are valued for some distinct property. We see this exemplified in England, where the common tumbler, which is valued only for its flight, does not differ much from its parent-form, the Eastern tumbler; whereas the short-faced tumbler has been prodigiously modified, from being valued, not for its flight, but for other qualities. But the common-flying tumbler of Europe has already begun to branch out into slightly different sub-breeds, such as the common English tumbler, the Dutch roller, the Glasgow house-tumbler, and the long-faced beard tumbler, &c.; and in the course of centuries, unless fashions greatly change, these sub-breeds will diverge through the slow and insensible process of unconscious selection, and become modified, in a greater and greater degree. After a time the perfectly graduated links, which now connect all these sub-breeds together, will be lost, for there would be no object and much difficulty in retaining such a host of intermediate sub-varieties.

The principle of divergence, together with the extinction of the many previously existing intermediate forms, is so important for understanding the origin of domestic races, as well as of species in a state of nature, that I will enlarge a little more on this subject. Our third main group includes carriers, barbs, and runts, which are plainly related to each other, yet wonderfully distinct in several important characters. According to the view given in the last chapter, these three races have probably descended from an unknown race having an intermediate character, and this from the rock-pigeon. Their characteristic differences are believed to be due to different breeders having at an early period admired different points of structure; and then, on the acknowledged principle of admiring extremes, having gone on breeding, without any thought of the future, as good birds as they could, —carrier-fanciers preferring long beaks with much wattle,—barb-fanciers preferring short thick beaks with much eye-wattle,—and runt-fanciers not caring about the beak or wattle, but only for the size and weight of the body. This process will have led to the neglect and final extinction of the earlier, inferior, and intermediate birds; and thus it has come to pass, that in Europe these three races are now so extraordinarily distinct from each other. But in the East, whence they were originally brought, the fashion has been different, and we there see breeds which connect the highly modified English carrier with the rock-pigeon, and others which to a certain extent connect carriers and runts. Looking back to the time of Aldrovandi, we find that there existed in Europe, before the year 1600, four breeds which were closely allied to carriers and barbs, but which competent authorities cannot now identify with our present barbs and carriers; nor can Aldrovandi's runts be identified with our present runts. These four breeds certainly did not differ from each other nearly so much as do our existing English carriers, barbs, and runts. All this is exactly what might have been anticipated. If we could collect all the pigeons which have ever lived, from before the time of the Romans to the present day, we should be able to group them in several lines, diverging from the parent rock-pigeon. Each line would consist of almost insensible steps, occasionally broken by some slightly greater variation or sport, and each would culminate in one of our present highly modified forms. Of the many former connecting links, some would be found to have become absolutely extinct without having left any issue, whilst others though extinct would be seen to be the progenitors of the existing races.

I have heard it remarked as a strange circumstance that we occasionally hear of the local or complete extinction of domestic races, whilst we hear nothing of their origin. How, it has been asked, can these losses be compensated, and more than compensated, for we know that with almost all domesticated animals the races have largely increased in number since the time of the Romans? But on the view here given, we can understand this apparent contradiction. The extinction of a race within historical times is an event likely to be noticed; but its gradual and scarcely sensible modification through unconscious selection, and its subsequent divergence, either in the same or more commonly in distant countries, into two or more strains, and their gradual conversion into sub-breeds, and these into well-marked breeds, are events which would rarely be noticed. The death of a tree, that has attained gigantic dimensions, is recorded; the slow growth of smaller trees and their increase in number excite no attention.

In accordance with the belief of the great power of selection, and of the little direct power of changed conditions of life, except in causing general variability or plasticity of organisation, it is not surprising that dovecot-pigeons have remained unaltered from time immemorial; and that some toy-pigeons, which differ in little else besides colour from the dovecot-pigeon, have retained the same character for several centuries. For when one of these toy-pigeons had once become beautifully and symmetrically coloured,—when, for instance, a Spot had been produced with the crown of its head, its tail, and tail-coverts of a uniform colour, the rest of the body being snow-white,—no alteration or improvement would be desired. On the other hand, it is not surprising that during this same interval of time our highly-bred pigeons have undergone an astonishing amount of change; for in regard to them there is no defined limit to the wish of the fancier, and there is no known limit to the variability of their characters. What is there to stop the fancier desiring to give to his carrier a longer and longer beak, or to his tumbler a shorter and shorter beak? nor has the extreme limit of variability in the beak, if there be any such limit, as yet been reached. Notwithstanding the great improvement effected within recent times in the short-faced almond tumbler, Mr. Eaton remarks, "the field is still as open for fresh competitors as it was one hundred years ago;" but this is perhaps an exaggerated assertion, for the young of all highly improved fancy birds are extremely liable to disease and death.

I have heard it objected that the formation of the several domestic races of the pigeon throws no light on the origin of the wild species of the Columbidæ, because their differences are not of the same nature. The domestic races for instance do not differ, or differ hardly at all, in the relative lengths and shapes of the primary wing-feathers, in the relative length of the hind toe, or in habits of life, as in roosting and building in trees. But the above objection shows how completely the principle of selection has been misunderstood. It is not likely that characters selected by the caprice of man should resemble differences preserved under natural conditions, either from being of direct service to each species, or from standing in correlation with other modified and serviceable structures. Until man selects birds differing in the relative length of the wing-feathers or toes, &c., no sensible change in these parts should be expected. Nor could man do anything unless these parts happened to vary under domestication: I do not positively assert that this is the case, although I have seen traces of such variability in the wing-feathers, and certainly in the tail-feathers. It would be a strange fact if the relative length of the hind toe should never vary, seeing how variable the foot is both in size and in the number of the scutellæ. With respect to the domestic races not roosting or building in trees, it is obvious that fanciers would never attend to or select such changes in habits; but we have seen that the pigeons in Egypt, which do not for some reason like settling on the low mud hovels of the natives, are led, apparently by compulsion, to perch in crowds on the trees. We may even affirm that, if our domestic races had become greatly modified in any of the above specified respects, and it could be shown that fanciers had never attended to such points, or that they did not stand in correlation with other selected characters, the fact, on the principles advocated in this chapter, would have offered a serious difficulty.

Let us briefly sum up the last two chapters on the pigeon. We may conclude with confidence that all the domestic races, notwithstanding their great amount of difference, are descended from the *Columba livia*, including under this name certain wild races. But the differences between these latter forms throw no light whatever on the characters which distinguish the domestic races. In each breed or sub-breed the individual birds are more variable than birds in a state of nature; and occasionally they vary in a sudden and strongly-marked manner. This plasticity of organisation apparently results from changed conditions of life. Disuse has reduced certain parts of the body. Correlation of growth so ties the organisation together, that when one part varies other parts vary at the same time. When several breeds have once been formed, their intercrossing aids the progress of modification, and has even produced new sub-breeds. But as, in the construction of a building, mere stones or bricks are of little avail without the builder's art, so, in the production of new races, selection has been the presiding power. Fanciers can act by selection on excessively slight individual differences, as well as on those greater differences which are called sports. Selection is followed methodically when the fancier tries to improve and modify a breed according to a prefixed standard of excellence; or he acts unmethodically and unconsciously, by merely trying to rear as good birds as he can, without any wish or intention to alter the breed. The progress of selection almost inevitably leads to the neglect and ultimate extinction of the earlier and less improved forms, as well as of many intermediate links in each long line of descent. Thus it has come to pass that most of our present races are so marvellously distinct from each other, and from the aboriginal rock-pigeon.

FOWLS.

As some naturalists may not be familiar with the chief breeds of the fowl, it will be advisable to give a condensed description of them. From what I have read and seen of specimens brought from several quarters of the world, I believe that most of the chief kinds have been imported into England, but many sub-breeds are probably still here unknown. The following discussion on the origin of the various breeds and on their characteristic differences does not pretend to completeness, but may be of some interest to the naturalist. The classification of the breeds cannot, as far as I can see, be made natural. They differ from each other in different degrees, and do not afford characters in subordination to each other, by which they can be ranked in group under group. They seem all to have diverged by independent and different roads from a single type. Each chief breed includes differently coloured sub-varieties, most of which can be truly propagated, but it would be superfluous to describe them. I have classed the various crested fowls as sub-breeds under the Polish fowl; but I have great doubts whether this is a natural arrangement, showing true affinity or blood relationship. It is scarcely possible to avoid laying stress on the commonness of a breed; and if certain foreign sub-breeds had been largely kept in this country they would perhaps have been raised to the rank of main-breeds. Several breeds are abnormal in character; that is, they differ in certain points from all wild Gallinaceous birds. At first I made a division of the breeds into normal and abnormal, but the result was wholly unsatisfactory.

• • •

From this synopsis we see that the several breeds differ considerably, and they would have been nearly as interesting for us as pigeons, if there had been equally good evidence that all had descended from one parent-species. Most fanciers believe that they are descended from several primitive stocks. The Rev. E. S. Dixon argues strongly on this side of the question; and one fancier even denounces the opposite conclusion by asking, "Do we not perceive pervading this spirit, the spirit of the *Deist*?" Most naturalists, with the exception of a few, such as Temminck, believe that all the breeds have proceeded from a single species; but authority on such a point goes for little. Fanciers look to all parts of the world as the possible sources of their unknown stocks; thus ignoring the laws of geographical distribution. They know well that the several kinds breed truly even in colour. They assert, but, as we shall see, on very weak grounds, that most of the breeds are extremely ancient. They are strongly impressed with the great difference between the chief kinds, and they ask with force, can differences in climate, food, or treatment have produced birds so different as the black stately Spanish, the diminutive elegant Bantam, the heavy Cochin with its many peculiarities, and the Polish fowl with its great top-knot and protuberant skull? But fanciers, whilst admitting and even overrating the effects of crossing the various breeds, do not sufficiently regard the probability of the occasional birth, during the course of centuries, of birds with abnormal and hereditary peculiarities; they overlook the effects of correlation of growth—of the long-continued use and disuse of parts, and of some direct result from changed food and climate, though on this latter head I have found no sufficient evidence; and lastly, they all, as far as I know, entirely overlook the all-important subject of unconscious or unmethodical selection, though they are well aware that their birds differ individually, and that by selecting the best birds for a few generations they can improve their stocks.

• • •

THE TURKEY.

It seems fairly well established by Mr. Gould, that the turkey, in accordance with the history of its first introduction, is descended from a wild Mexican species (*Meleagris Mexicana*) which had been already domesticated by the natives before the discovery of America, and which differs specifically, as it is generally thought, from the common wild species of the United States. Some naturalists, however, think that these two forms should be ranked only as well-marked geographical races. However this may be, the case deserves notice because in the United States wild male turkeys sometimes court the domestic hens, which are descended from the Mexican form, "and are generally received by them with great pleasure." Several accounts have likewise been published of young birds, reared in the United States from the eggs of the wild species, crossing and commingling with the common breed. In England, also, this same species has been kept in several parks; from two of which the Rev. W. D. Fox procured birds, and they crossed freely with the common domestic kind, and during many years afterwards, as he informs me, the turkeys in his neighbourhood clearly showed traces of their crossed parentage. We here have an instance of a domestic race being modified by a cross with a distinct species or wild race. F. Michaux suspected in 1802 that the common domestic turkey was not descended from the United States species alone, but likewise from a southern form, and he went so far as to believe that English and French turkeys differed from having different proportions of the blood of the two parent-forms.

English turkeys are smaller than either wild form. They have not varied in any great degree; but there are some breeds which can be distinguished—as Norfolks, Suffolks, Whites, and Copper-coloured (or Cambridge), all of which, if precluded from crossing with other breeds, propagate their kind truly. Of these kinds, the most distinct is the small, hardy, dull-black Norfolk turkey, of which the chickens are black, with occasionally white patches about the head. The other breeds scarcely differ except in colour, and their chickens are generally mottled all over with brownish-grey. The tuft of hair on the breast, which is proper to the male alone, occasionally appears on the breast of the domesticated female. The inferior tail-coverts vary in number, and according to a German superstition the hen lays as many eggs as the cock has feathers of this kind. In Holland there was formerly, according to Temminck, a beautiful buff-yellow breed, furnished with an ample white topknot. Mr. Wilmot has described a white turkey-cock with a crest formed of "feathers about four inches long, with bare quills, and a tuft of soft white down growing at the end." Many of the young birds whilst young inherited this kind of crest, but afterwards it either fell off or was pecked out by the other birds. This is an interesting case, as with care a new breed might probably have been formed; and a topknot of this nature would have been to a certain extent analogous to that borne by the males in several allied genera, such as Euplocomus, Lophophorus, and Pavo.

Wild turkeys, believed in every instance to have been imported from the United States, have been kept in the parks of Lords Powis, Leicester, Hill, and Derby. The Rev. W. D. Fox procured birds from the two first-named parks, and he informs me that they certainly differed a little from each other in the shape of their bodies and in the barred plumage on their wings. These birds likewise differed from Lord Hill's stock. Some of the latter kept at Oulton by Sir P. Egerton, though precluded from crossing with common turkeys, occasionally produced much paler-coloured birds, and one that was almost white, but not an albino. These half-wild turkeys in thus slightly differing from each other present an analogous case with the wild cattle kept in the several British parks. We must suppose that the differences have resulted from the prevention of free intercrossing between birds ranging over a wide area, and from the changed conditions to which they have been exposed in England. In India the climate has apparently wrought a still greater change in the turkey, for it is described by Mr. Blyth as being much degenerated in size, "utterly incapable of rising on the wing," of a black colour, and "with the long pendulous appendages over the beak enormously developed."

• • •

GOLD-FISH.

Besides mammals and birds, few animals belonging to the other great classes have been domesticated; but to show that it is an almost universal law that animals, when removed from their natural conditions of life, vary, and that races can be formed when selection is applied, it is necessary to say a few words on gold-fish, bees, and silk-moths.

Gold-fish (*Cyprinus auratus*) were introduced into Europe only two or three centuries ago; but it is believed that they have been kept in confinement from an ancient period in China. Mr. Blyth suspects from the analogous variation of other fishes that golden-coloured fish do not occur in a state of nature. These fishes frequently live under the most unnatural conditions, and their variability in colour, size, and in some important points of structure is very great. M. Sauvigny has described and given coloured drawings of no less than eighty-nine varieties. Many of the varieties, however, such as triple tail-fins, &c., ought to be called monstrosities; but it is difficult to draw any distinct line between a variation and a monstrosity. As gold-fish are kept for ornament or curiosity, and as "the Chinese are just the people to have secluded a chance variety of any kind, and to have matched and paired from it," we may feel nearly confident that selection has been largely practised in the formation of new breeds. It is however a singular fact that some of the monstrosities or variations are not inherited; for Sir R. Heron kept many of these fishes, and placed all the deformed fishes, namely those destitute of dorsal fins, and those furnished with a double anal fin, or triple tail, in a pond by themselves; but they did "not produce a greater proportion of deformed offspring than the perfect fishes."

Passing over an almost infinite diversity of colour, we meet with the most extraordinary modifications of structure. Thus, out of about two dozen specimens bought in London, Mr. Yarrell observed some with the dorsal fin extending along more than half the length of the back; others with this fin reduced to only five or six rays; and one with no dorsal fin. The anal fins are sometimes double, and the tail is often triple. This latter deviation of structure seems generally to occur "at the expense of the whole or part of some other fin;" but Bory de Saint Vincent saw at Madrid gold-fish furnished with a dorsal fin and a triple tail. One variety is characterized by a hump on its back near the head; and the Rev. L. Jenyns has described a most singular variety, imported from China, almost globular in form like a Diodon, with "the fleshy part of the tail as if entirely cut away; the caudal fin being set on a little behind the dorsal and immediately above the anal." In this fish the anal and caudal fins were double; the anal fin being attached to the body in a vertical line: the eyes also were enormously large and protuberant.

HIVE-BEES.

Bees have been domesticated from an ancient period; if indeed their state can be considered one of domestication, for they search for their own food, with the exception of a little generally given to them during the winter. Their habitation is a hive instead of a hole in a tree. Bees, however, have been transported into almost every quarter of the world, so that climate ought to have produced whatever direct effect it is capable of producing. It is frequently asserted that the bees in different parts of Great Britain differ in size, colour, and temper; and Godron says that they are generally larger in the south than in other parts of France; it has also been asserted that the little brown bees of High Burgundy, when transported to La Bresse, become large and yellow in the second generation. But these statements require confirmation. As far as size is concerned, it is known that bees produced in very old combs are smaller, owing to the cells having become smaller from the successive old cocoons. The best authorities concur that, with the exception of the Ligurian race or species, presently to be mentioned, distinct breeds do not exist in Britain or on the Continent. There is, however, even in the same stock, some variability in colour. Thus Mr. Woodbury states that he has several times seen queen bees of the common kind annulated with yellow like Ligurian queens, and the latter dark-coloured like common bees. He has also observed variations in the colour of the drones, without any corresponding difference in the queens or workers of the same hive. The great apiarian Dzierzon, in answer to my queries on this subject, says that in Germany bees of some stocks are decidedly dark, whilst others are remarkable for their yellow colour. Bees also seem to differ in habits in different districts, for Dzierzon adds, "If many stocks with their offspring are more inclined to swarm, whilst others are richer in honey, so that some beekeepers even distinguish between swarming and honey-gathering bees, this is a habit which has become second nature, caused by the customary mode of keeping the bees and the pasturage of the district. For example; what a difference in this respect one may perceive to exist between the bees of the Lüneburg heath and those of this country! . . . Removing an old queen and substituting a young one of the current year is here an infallible mode of keeping the strongest stock from swarming and preventing drone-breeding; whilst the same means if adopted in Hanover would certainly be of no avail." I procured a hive full of dead bees from Jamaica, where they have long been naturalised, and, on carefully comparing them under the microscope with my own bees, I could detect not a trace of difference.

This remarkable uniformity in the hive-bee, wherever kept, may probably be accounted for by the great difficulty, or rather impossibility, of bringing selection into play by pairing particular queens and drones, for these insects unite only during flight. Nor is there any record, with a single partial exception, of any person having separated and bred from a hive in which the workers presented some appreciable difference. In order to form a new breed, seclusion from other bees would, as we now know, be indispensable; for since the introduction of the Ligurian bee into Germany and England, it has been found that the drones wander at least two miles from their own hives, and often cross with the queens of the common bee. The Ligurian bee, although perfectly fertile when crossed with the common kind, is ranked by most naturalists as a distinct species, whilst by others it is ranked as a natural variety: but this form need not here be noticed, as there is no reason to believe that it is the product of domestication. The Egyptian and some other bees are likewise ranked by Dr. Gerstäcker, but not by other highly competent judges, as geographical races; and he grounds his conclusion in chief part on the fact that in certain districts, as in the Crimea and Rhodes, the hive-bee varies so much in colour, that the several geographical races can be closely connected by intermediate forms.

I have alluded to a single instance of the separation and preservation of a particular stock of bees. Mr. Lowe procured some bees from a cottager a few miles from Edinburgh, and perceived that they differed from the common bee in the hairs on the head and thorax being lighter coloured and more profuse in quantity. From the date of the introduction of the Ligurian bee into Great Britain we may feel sure that these bees had not been crossed with this form. Mr. Lowe propagated this variety, but unfortunately did not separate the stock from his other bees, and after three generations the new character was almost completely lost. Nevertheless, as he adds, "a great number of the bees still retain traces, though faint, of the original colony." This case shows us what could probably be effected by careful and long-continued selection applied exclusively to the workers, for, as we have seen, queens and drones cannot be selected and paired.

CULTIVATED PLANTS: CEREAL AND CULINARY PLANTS.

I shall not enter into so much detail on the variability of cultivated plants, as in the case of domesticated animals. The subject is involved in much difficulty. Botanists have generally neglected cultivated varieties, as beneath their notice. In several cases the wild prototype is unknown or doubtfully known; and in other cases it is hardly possible to distinguish between escaped seedlings and truly wild plants, so that there is no safe standard of comparison by which to judge of any supposed amount of change. Not a few botanists believe that several of our anciently cultivated plants have become so profoundly modified that it is not possible now to recognise their aboriginal parent-forms. Equally perplexing are the doubts whether some of them are descended from one species, or from several inextricably commingled by crossing and variation. Variations often pass into, and cannot be distinguished from, monstrosities; and monstrosities are of little significance for our purpose. Many varieties are propagated solely by grafts, buds, layers, bulbs, &c., and frequently it is not known how far their peculiarities can be transmitted by seminal generation. Nevertheless some facts of value can be gleaned; and other facts will hereafter be incidentally given. One chief object in the two following chapters is to show how generally almost every character in our cultivated plants has become variable.

Before entering on details a few general remarks on the origin of cultivated plants may be introduced. M. Alph. De Candolle in an admirable discussion on this subject, in which he displays a wonderful amount of knowledge, gives a list of 157 of the most useful cultivated plants. Of these he believes that 85 are almost certainly known in their wild state; but on this head other competent judges entertain great doubts. Of 40 of them, the origin is admitted by M. De Candolle to be doubtful, either from a certain amount of dissimilarity which they present when compared with their nearest allies in a wild state, or from the probability of the latter not being truly wild plants, but seedlings escaped from culture. Of the entire 157, 32 alone are ranked by M. De Candolle as quite unknown in their aboriginal condition. But it should be observed that he does not include in his list several plants which present ill-defined characters, namely, the various forms of pumpkins, millet, sorghum, kidney-bean, dolichos, capsicum, and indigo. Nor does he include flowers; and several of the more anciently cultivated flowers, such as certain roses, the common Imperial lily, the tuberose, and even the lilac, are said not to be known in the wild state.

From the relative numbers above given, and from other arguments of much weight, M. De Candolle concludes that plants have rarely been so much modified by culture that they cannot be identified with their wild prototypes. But on this view, considering that savages probably would not have chosen rare plants for cultivation, that useful plants are generally conspicuous, and that they could not have been the inhabitants of deserts or of remote and recently discovered islands, it appears strange to me that so many of our cultivated plants should be still unknown or only doubtfully known in the wild state. If, on the other hand, many of these plants have been profoundly modified by culture, the difficulty disappears. Their extermination during the progress of civilisation would likewise remove the difficulty; but M. De Candolle has shown that this probably has seldom occurred. As soon as a plant became cultivated in any country, the half-civilised inhabitants would no longer have need to search the whole surface of the land for it, and thus lead to its extirpation; and even if this did occur during a famine, dormant seeds would be left in the ground. In tropical countries the wild luxuriance of nature, as was long ago remarked by Humboldt, overpowers the feeble efforts of man. In anciently civilised temperate countries, where the whole face of the land has been greatly changed, it can hardly be doubted that some plants have been exterminated; nevertheless De Candolle has shown that all the plants historically known to have been first cultivated in Europe still exist here in the wild state.

MM. Loiseleur-Deslongchamps and De Candolle have remarked that our cultivated plants, more especially the cereals, must originally have existed in nearly their present state; for otherwise they would not have been noticed and valued as objects of food. But these authors apparently have not considered the many accounts given by travellers of the wretched food collected by savages. I have read an account of the savages of Australia cooking, during a dearth, many vegetables in various ways, in the hopes of rendering them innocuous and more nutritious. Dr. Hooker found the half-starved inhabitants of a village in Sikhim suffering greatly from having eaten arum-roots, which they had pounded and left for several days to ferment, so as partially to destroy their poisonous nature; and he adds that they cooked and ate many other deleterious plants. Sir Andrew Smith informs me that in South Africa a large number of fruits and succulent leaves, and especially roots, are used in times of scarcity. The natives, indeed, know the properties of a long catalogue of plants, some having been found during famines to be eatable, others injurious to health, or even destructive to life. He met a party of Baquanas who, having been expelled by the conquering Zulus, had lived for years on any roots or leaves which afforded some little nutriment, and distended their stomachs, so as to relieve the pangs of hunger. They looked like walking skeletons, and suffered fearfully from constipation. Sir Andrew Smith also informs me that on such occasions the natives observe as a guide for themselves, what the wild animals, especially baboons and monkeys, eat.

From innumerable experiments made through dire necessity by the savages of every land, with the results handed down by tradition, the nutritious, stimulating, and medicinal properties of the most unpromising plants were probably first discovered. It appears, for instance, at first an inexplicable fact that untutored man, in three distant quarters of the world, should have discovered amongst a host of native plants that the leaves of the tea-plant and mattee, and the berries of the coffee, all included a stimulating and nutritious essence, now known to be chemically the same. We can also see that savages suffering from severe constipation would naturally observe whether any of the roots which they devoured acted as aperients. We probably owe our knowledge of the uses of almost all plants to man having originally existed in a barbarous state, and having been often compelled by severe want to try as food almost everything which he could chew and swallow.

From what we know of the habits of savages in many quarters of the world, there is no reason to suppose that our cereal plants originally existed in their present state so valuable to man. Let us look to one continent alone, namely, Africa: Barth states that the slaves over a large part of the central region regularly collect the seeds of a wild grass, the *Pennisetum distichum*; in another district he saw women collecting the seeds of a Poa by swinging a sort of basket through the rich meadow-land. Near Tete Livingstone observed the natives collecting the seeds of a wild grass; and farther south, as Andersson informs me, the natives largely use the seeds of a grass of about the size of canary-seed, which they boil in water. They eat also the roots of certain reeds, and every one has read of the Bushmen prowling about and digging up with a fire-hardened stake various roots. Similar facts with respect to the collection of seeds of wild grasses in other parts of the world could be given.

Accustomed as we are to our excellent vegetables and luscious fruits, we can hardly persuade ourselves that the stringy roots of the wild carrot and parsnip, or the little shoots of the wild asparagus, or crabs, sloes, &c., should ever have been valued; yet, from what we know of the habits of Australian and South African savages, we need feel no doubt on this head. The inhabitants of Switzerland during the Stone-period largely collected wild crabs, sloes, bullaces, hips of roses, elderberries, beech-mast, and other wild berries and fruit. Jemmy Button, a Fuegian on board the *Beagle*, remarked to me that the poor and acid black-currants of Tierra del Fuego were too sweet for his taste.

The savage inhabitants of each land, having found out by many and hard trials what plants were useful, or could be rendered useful by various cooking processes, would after a time take the first step in cultivation by planting them near their usual abodes. Livingstone states that the savage Batokas sometimes left wild fruit-trees standing in their gardens, and occasionally even planted them, "a practice seen nowhere else amongst the natives." But Du Chaillu saw a palm and some other wild fruit-trees which had been planted; and these trees were considered private property. The next step in cultivation, and this would require but little forethought, would be to sow the seeds of useful plants; and as the soil near the hovels of the natives would often be in some degree manured, improved varieties would sooner or later arise. Or a wild and unusually good variety of a native plant might attract the attention of some wise old savage; and he would transplant it, or sow its seed. That superior varieties of wild fruit-trees occasionally are found is certain, as in the case of the American species of hawthorns, plums, cherries, grapes, and hickories, specified by Professor Asa Gray. Downing also refers to certain wild varieties of the hickory, as being "of much larger size and finer flavor than the common species." I have referred to American fruit-trees, because we are not in this case troubled with doubts whether or not the varieties are seedlings which have escaped from cultivation. Transplanting any superior variety, or sowing its seeds, hardly implies more forethought than might be expected at an early and rude period of civilisation. Even the Australian barbarians "have a law that no plant bearing seeds is to be dug up after it has flowered;" and Sir G. Grey never saw this law, evidently framed for the preservation of the plant, violated. We see the same spirit in the superstitious belief of the Fuegians, that killing water-fowl whilst very young will be followed by "much rain, snow, blow much." I may add, as showing forethought in the lowest barbarians, that the Fuegians when they find a stranded whale bury large portions in the sand, and during the often-recurrent famines travel from great distances for the remnants of the half-putrid mass.

It has often been remarked that we do not owe a single useful plant to Australia or the Cape of Good Hope,—countries abounding to an unparalleled degree with endemic species,—or to New Zealand, or to America south of the Plata; and, according to some authors, not to America northward of Mexico. I do not believe that any edible or valuable plant, except the canary-grass, has been derived from an oceanic or uninhabited island. If nearly all our useful plants, natives of Europe, Asia, and South America, had originally existed in their present condition, the complete absence of similarly useful plants in the great countries just named would indeed be a surprising fact. But if these plants have been so greatly modified and improved by culture as no longer closely to resemble any natural species, we can understand why the above-named countries have given us no useful plants, for they were either inhabited by men who did not cultivate the ground at all, as in Australia and the Cape of Good Hope, or who cultivated it very imperfectly, as in some parts of America. These countries do yield plants which are useful to savage man; and Dr. Hooker enumerates no less than 107 such species in Australia alone; but these plants have not been improved, and consequently cannot compete with those which have been cultivated and improved during thousands of years in the civilised world.

The case of New Zealand, to which fine island we as yet owe no widely cultivated plant, may seem opposed to this view; for, when first discovered, the natives cultivated several plants; but all inquirers believe, in accordance with the traditions of the natives, that the early Polynesian colonists brought with them seeds and roots, as well as the dog, which had all been wisely preserved during their long voyage. The Polynesians are so frequently lost on the ocean, that this degree of prudence would occur to any wandering party: hence the early colonists of New Zealand, like the later European colonists, would not have had any strong inducement to cultivate the aboriginal plants. According to De Candolle we owe thirty-three useful plants to Mexico, Peru, and Chile; nor is this surprising when we remember the civilised state of the inhabitants, as shown by the fact of their having practised artificial irrigation and made tunnels through hard rocks without the use of iron or gunpowder, and who, as we shall see in a future chapter, fully recognised, as far as animals were concerned, and therefore probably in the case of plants, the important principle of selection. We owe some plants to Brazil; and the early voyagers, namely Vespucius and Cabral, describe the country as thickly peopled and cultivated. In North America the natives cultivated maize, pumpkins, gourds, beans, and peas, "all different from ours," and tobacco; and we are hardly justified in assuming that none of our present plants are descended from these North American forms. Had North America been civilised for as long a period, and as thickly peopled, as Asia or Europe, it is probable that the native vines, walnuts, mulberries, crabs, and plums, would have given rise, after a long course of cultivation, to a multitude of varieties, some extremely different from their parent-stocks; and escaped seedlings would have caused in the New, as in the Old World, much perplexity with respect to their specific distinctness and parentage.

• • •

SELECTION BY MAN.

The power of Selection, whether exercised by man, or brought into play under nature through the struggle for existence and the consequent survival of the fittest, absolutely depends on the variability of organic beings. Without variability nothing can be effected; slight individual differences, however, suffice for the work, and are probably the sole differences which are effective in the production of new species. Hence our discussion on the causes and laws of variability ought in strict order to have preceded our present subject, as well as the previous subjects of inheritance, crossing, &c.; but practically the present arrangement has been found the most convenient. Man does not attempt to cause variability; though he unintentionally effects this by exposing organisms to new conditions of life, and by crossing breeds already formed. But variability being granted, he works wonders. Unless some degree of selection be exercised, the free commingling of the individuals of the same variety soon obliterates, as we have previously seen, the slight differences which may arise, and gives to the whole body of individuals uniformity of character. In separated districts, long-continued exposure to different conditions of life may perhaps produce new races without the aid of selection; but to this difficult subject of the direct action of the conditions of life we shall in a future chapter recur.

When animals or plants are born with some conspicuous and firmly inherited new character, selection is reduced to the preservation of such individuals, and to the subsequent prevention of crosses; so that nothing more need be said on the subject. But in the great majority of cases a new character, or some superiority in an old character, is at first faintly pronounced, and is not strongly inherited; and then the full difficulty of selection is experienced. Indomitable patience, the finest powers of discrimination, and sound judgment must be exercised during many years. A clearly predetermined object must be kept steadily in view. Few men are endowed with all these qualities, especially with that of discriminating very slight differences; judgment can be acquired only by long experience; but if any of these qualities be wanting, the labour of a life may be thrown away. I have been astonished when celebrated breeders, whose skill and judgment have been proved by their success at exhibitions, have shown me their animals, which appeared all alike, and have assigned their reasons for matching this and that individual. The importance of the great principle of Selection mainly lies in this power of selecting scarcely appreciable differences, which nevertheless are found to be transmissible, and which can be accumulated until the result is made manifest to the eyes of every beholder.

The principle of selection may be conveniently divided into three kinds. *Methodical selection* is that which guides a man who systematically endeavours to modify a breed according to some predetermined standard. *Unconscious selection* is that which follows from men naturally preserving the most valued and destroying the less valued individuals, without any thought of altering the breed; and undoubtedly this process slowly works great changes. Unconscious selection graduates into methodical, and only extreme cases can be distinctly separated; for he who preserves a useful or perfect animal will generally breed from it with the hope of getting offspring of the same character; but as long as he has not a predetermined purpose to improve the breed, he may be said to be selecting unconsciously. Lastly, we have *Natural selection*, which implies that the individuals which are best fitted for the complex, and in the course of ages changing conditions to which they are exposed, generally survive and procreate their kind. With domestic productions, with which alone we are here strictly concerned, natural selection comes to a certain extent into action, independently of, and even in opposition to, the will of man.

• • •

I will conclude this chapter by some remarks on an important subject. With animals such as the giraffe, of which the whole structure is admirably co-ordinated for certain purposes, it has been supposed that all the parts must have been simultaneously modified; and it has been argued that, on the principle of natural selection, this is scarcely possible. But in thus arguing, it has been tacitly assumed that the variations must have been abrupt and great. No doubt, if the neck of a ruminant were suddenly to become greatly elongated, the fore-limbs and back would have to be simultaneously strengthened and modified; but it cannot be denied that an animal might have its neck, or head, or tongue, or fore-limbs elongated a very little without any corresponding modification in other parts of the body; and animals thus slightly modified would, during a dearth, have a slight advantage, and be enabled to browse on higher twigs, and thus survive. A few mouthfuls more or less every day would make all the difference between life and death. By the repetition of the same process, and by the occasional intercrossing of the survivors, there would be some progress, slow and fluctuating though it would be, towards the admirably co-ordinated structure of the giraffe. If the short-faced tumbler-pigeon, with its small conical beak, globular head, rounded body, short wings, and small feet—characters which appear all in harmony—had been a natural species, its whole structure would have been viewed as well fitted for its life; but in this case we know that inexperienced breeders are urged to attend to point after point, and not to attempt improving the whole structure at the same time. Look at the greyhound, that perfect image of grace, symmetry, and vigour; no natural species can boast of a more admirably co-ordinated structure, with its tapering head, slim body, deep chest, tucked-up abdomen, rat-like tail, and long muscular limbs, all adapted for extreme fleetness, and for running down weak prey. Now, from what we see of the variability of animals, and from what we know of the method which different men follow in improving their stock—some chiefly attending to one point, others to another point, others again correcting defects by crosses, and so forth—we may feel assured that if we could see the long line of ancestors of a first-rate greyhound, up to its wild wolf-like progenitor, we should behold an infinite number of the finest gradations, sometimes in one character and sometimes in another, but all leading towards our present perfect type. By small and doubtful steps such as these, nature, as we may confidently believe, has progressed on her grand march of improvement and development.

A similar line of reasoning is as applicable to separate organs as to the whole organisation. A writer has recently maintained that "it is probably no exaggeration to suppose that, in order to improve such an organ as the eye at all, it must be improved in ten different ways at once. And the improbability of any complex organ being produced and brought to perfection in any such way is an improbability of the same kind and degree as that of producing a poem or a mathematical demonstration by throwing letters at random on a table." If the eye were abruptly and greatly modified, no doubt many parts would have to be simultaneously altered, in order that the organ should remain serviceable.

But is this the case with smaller changes? There are persons who can see distinctly only in a dull light, and this condition depends, I believe, on the abnormal sensitiveness of the retina, and is known to be inherited. Now, if a bird, for instance, received some great advantage from seeing well in the twilight, all the individuals with the most sensitive retina would succeed best and be the most likely to survive; and why should not all those which happened to have the eye itself a little larger, or the pupil capable of greater dilatation, be likewise preserved, whether or not these modifications were strictly simultaneous? These individuals would subsequently intercross and blend their respective advantages. By such slight successive changes, the eye of a diurnal bird would be brought into the condition of that of an owl, which has often been advanced as an excellent instance of adaptation. Short-sight, which is often inherited, permits a person to see distinctly a minute object at so near a distance that it would be indistinct to ordinary eyes; and here we have a capacity which might be serviceable under certain conditions, abruptly gained. The Fuegians on board the Beagle could certainly see distant objects more distinctly than our sailors with all their long practice; I do not know whether this depends on nervous sensitiveness or on the power of adjustment in the focus; but this capacity for distant vision might, it is probable, be slightly augmented by successive modifications of either kind. Amphibious animals, which are enabled to see both in the water and in the air, require and possess, as M. Plateau has shown, eyes constructed on the following plan: "the cornea is always flat, or at least much flattened in front of the crystalline and over a space equal to the diameter of that lens, whilst the lateral portions may be much curved." The crystalline is very nearly a sphere, and the humours have nearly the same density as water. Now, as a terrestrial animal slowly became more and more aquatic in its habits, very slight changes, first in the curvature of the cornea or crystalline, and then in the density of the humours, or conversely, might successively occur, and would be advantageous to the animal whilst under water, without serious detriment to its power of vision in the air. It is of course impossible to conjecture by what steps the fundamental structure of the eye in the Vertebrata was originally acquired, for we know absolutely nothing about this organ in the first progenitors of the class. With respect to the lowest animals in the scale, the transitional states through which the eye at first probably passed, can by the aid of analogy be indicated, as I have attempted to show in my 'Origin of Species.'

SELECTION, *CONTINUED*.

Natural Selection, or the Survival of the Fittest, as affecting domestic productions.—We know little on this head. But as animals kept by savages have to provide their own food, either entirely or to a large extent, throughout the year, it can hardly be doubted that, in different countries, varieties differing in constitution and in various characters would succeed best, and so be naturally selected. Hence perhaps it is that the few domesticated animals kept by savages partake, as has been remarked by more than one writer, of the wild appearance of their masters, and likewise resemble natural species. Even in long-civilised countries, at least in the wilder parts, natural selection must act on our domestic races. It is obvious that varieties, having very different habits, constitution, and structure, would succeed best on mountains and on rich lowland pastures. For example, the improved Leicester sheep were formerly taken to the Lammermuir Hills; but an intelligent sheep-master reported that "our coarse lean pastures were unequal to the task of supporting such heavy-bodied sheep; and they gradually dwindled away into less and less bulk: each generation was inferior to the preceding one; and when the spring was severe, seldom more than two-thirds of the lambs survived the ravages of the storms." So with the mountain cattle of North Wales and the Hebrides, it has been found that they could not withstand being crossed with the larger and more delicate lowland breeds. Two French naturalists, in describing the horses of Circassia, remark that, subjected as they are to extreme vicissitudes of climate, having to search for scanty pasture, and exposed to constant danger from wolves, the strongest and most vigourous alone survive.

Every one must have been struck with the surpassing grace, strength, and vigour of the Game-cock, with its bold and confident air, its long, yet firm neck, compact body, powerful and closely pressed wings, muscular thighs, strong beak massive at the base, dense and sharp spurs set low on the legs for delivering the fatal blow, and its compact, glossy, and mail-like plumage serving as a defence. Now the English game-cock has not only been improved during many years by man's careful selection, but in addition, as Mr. Tegetmeier has remarked, by a kind of natural selection, for the strongest, most active and courageous birds have stricken down their antagonists in the cock-pit, generation after generation, and have subsequently served as the progenitors of their kind.

In Great Britain, in former times, almost every district had its own breed of cattle and sheep; "they were indigenous to the soil, climate, and pasturage of the locality on which they grazed: they seemed to have been formed for it and by it." But in this case we are quite unable to disentangle the effects of the direct action of the conditions of life,—of use or habit—of natural selection—and of that kind of selection which we have seen is occasionally and unconsciously followed by man even during the rudest periods of history.

Let us now look to the action of natural selection on special characters. Although nature is difficult to resist, yet man often strives against her power, and sometimes, as we shall see, with success. From the facts to be given, it will also be seen that natural selection would powerfully affect many of our domestic productions if left unprotected. This is a point of much interest, for we thus learn that differences apparently of very slight importance would certainly determine the survival of a form when forced to struggle for its own existence. It may have occurred to some naturalists, as it formerly did to me, that, though selection acting under natural conditions would determine the structure of all important organs, yet that it could not affect characters which are esteemed by us of little importance; but this is an error to which we are eminently liable, from our ignorance of what characters are of real value to each living creature.

When man attempts to breed an animal with some serious defect in structure, or in the mutual relation of parts, he will either partially or completely fail, or encounter much difficulty; and this is in fact a form of natural selection. We have seen that the attempt was once made in Yorkshire to breed cattle with enormous buttocks, but the cows perished so often in bringing forth their calves, that the attempt had to be given up. In rearing short-faced tumblers, Mr. Eaton says, "I am convinced that better head and beak birds have perished in the shell than ever were hatched; the reason being that the amazingly short-faced bird cannot reach and break the shell with its beak, and so perishes." Here is a more curious case, in which natural selection comes into play only at long intervals of time: during ordinary seasons the Niata cattle can graze as well as others, but occasionally, as from 1827 to 1830, the plains of La Plata suffer from long-continued droughts and the pasture is burnt up; at such times common cattle and horses perish by the thousand, but many survive by browsing on twigs, reeds, &c.; this the Niata cattle cannot so well effect from their upturned jaws and the shape of their lips; consequently, if not attended to, they perish before the other cattle. In Colombia, according to Roulin, there is a breed of nearly hairless cattle, called Pelones; these succeed in their native hot district, but are found too tender for the Cordillera; in this case, natural selection determines only the range of the variety. It is obvious that a host of artificial races could never survive in a state of nature;—such as Italian greyhounds,—hairless and almost toothless Turkish dogs,—fantail pigeons, which cannot fly well against a strong wind,—barbs with their vision impeded by their eye-wattle,—Polish fowls with their vision impeded by their great top-knots,—hornless bulls and rams which consequently cannot cope with other males, and thus have a poor chance of leaving offspring,—seedless plants, and many other such cases.

Colour is generally esteemed by the systematic naturalist as unimportant: let us, therefore, see how far it indirectly affects our domestic productions, and how far it would affect them if they were left exposed to the full force of natural selection. In a future chapter I shall have to show that constitutional peculiarities of the strangest kind, entailing liability to the action of certain poisons, are correlated with the colour of the skin. I will here give a single case, on the high authority of Professor Wyman; he informs me that, being surprised at all the pigs in a part of Virginia being black, he made inquiries, and ascertained that these animals feed on the roots of the *Lachnanthes tinctoria*, which colours their bones pink, and, excepting in the case of the black varieties, causes the hoofs to drop off. Hence, as one of the squatters remarked, "we select the black members of the litter for raising, as they alone have a good chance of living." So that here we have artificial and natural selection working hand in hand. I may add that in the Tarentino the inhabitants keep black sheep alone, because the *Hypericum crispum* abounds there; and this plant does not injure black sheep, but kills the white ones in about a fortnight's time.

Complexion, and liability to certain diseases, are believed to run together in man and the lower animals. Thus white terriers suffer more than terriers of any other colour from the fatal Distemper. In North America plum-trees are liable to a disease which Downing believes is not caused by insects; the kinds bearing purple fruit are most affected, "and we have never known the green or yellow fruited varieties infected until the other sorts had first become filled with the knots." On the other hand, peaches in North America suffer much from a disease called the *yellows*, which seems to be peculiar to that continent, and "more than nine-tenths of the victims, when the disease first appeared, were the yellow-fleshed peaches. The white-fleshed kinds are much more rarely attacked; in some parts of the country never." In Mauritius, the white sugar-canes have of late years been so severely attacked by a disease, that many planters have been compelled to give up growing this variety (although fresh plants were imported from China for trial), and cultivate only red canes. Now, if these plants had been forced to struggle with other competing plants and enemies, there cannot be a doubt that the colour of the flesh or skin of the fruit, unimportant as these characters are considered, would have rigorously determined their existence.

Liability to the attacks of parasites is also connected with colour. It appears that white chickens are certainly more subject than dark-coloured chickens to the *gapes*, which is caused by a parasitic worm in the trachea. On the other hand, experience has shown that in France the caterpillars which produce white cocoons resist the deadly fungus better than those producing yellow cocoons. Analogous facts have been observed with plants: a new and beautiful white onion, imported from France, though planted close to other kinds, was alone attacked by a parasitic fungus. White verbenas are especially liable to mildew. Near Malaga, during an early period of the vine-disease, the green sorts suffered most; "and red and black grapes, even when interwoven with the sick plants, suffered not at all." In France whole groups of varieties were comparatively free, and others, such as the Chasselas, did not afford a single fortunate exception; but I do not know whether any correlation between colour and liability to disease was here observed. In a former chapter it was shown how curiously liable one variety of the strawberry is to mildew.

It is certain that insects regulate in many cases the range and even the existence of the higher animals, whilst living under their natural conditions. Under domestication light-coloured animals suffer most: in Thuringia the inhabitants do not like grey, white, or pale cattle, because they are much more troubled by various kinds of flies than the brown, red, or black cattle. An Albino negro, it has been remarked, was peculiarly sensitive to the bites of insects. In the West Indies it is said that "the only horned cattle fit for work are those which have a good deal of black in them. The white are terribly tormented by the insects; and they are weak and sluggish in proportion to the white."

In Devonshire there is a prejudice against white pigs, because it is believed that the sun blisters them when turned out; and I knew a man who would not keep white pigs in Kent, for the same reason. The scorching of flowers by the sun seems likewise to depend much on colour; thus, dark pelargoniums suffer most; and from various accounts it is clear that the cloth-of-gold variety will not withstand a degree of exposure to sunshine which other varieties enjoy. Another amateur asserts that not only all dark-coloured verbenas, but likewise scarlets, suffer from the sun; "the paler kinds stand better, and pale blue is perhaps the best of all." So again with the heartsease (*Viola tricolour*); hot weather suits the blotched sorts, whilst it destroys the beautiful markings of some other kinds. During one extremely cold season in Holland all red-flowered hyacinths were observed to be very inferior in quality. It is believed by many agriculturists that red wheat is hardier in northern climates than white wheat.

With animals, white varieties from being conspicuous are the most liable to be attacked by beasts and birds of prey. In parts of France and Germany where hawks abound, persons are advised not to keep white pigeons; for, as Parmentier says, "it is certain that in a flock the white always first fall victims to the kite." In Belgium, where so many societies have been established for the flight of carrier-pigeons, white is the one colour which for the same reason is disliked. On the other hand, it is said that the sea-eagle (*Falco ossifragus*, Linn.) on the west coast of Ireland picks out the black fowls, so that "the villagers avoid as much as possible rearing birds of that colour." M. Daudin, speaking of white rabbits kept in warrens in Russia, remarks that their colour is a great disadvantage, as they are thus more exposed to attack, and can be seen during bright nights from a distance. A gentleman in Kent, who failed to stock his woods with a nearly white and hardy kind of rabbit, accounted in the same manner for their early disappearance. Any one who will watch a white cat prowling after her prey will soon perceive under what a disadvantage she lies.

The white Tartarian cherry, "owing either to its colour being so much like that of the leaves, or to the fruit always appearing from a distance unripe," is not so readily attacked by birds as other sorts. The yellow-fruited raspberry, which generally comes nearly true by seed, "is very little molested by birds, who evidently are not fond of it; so that nets may be dispensed with in places where nothing else will protect the red fruit." This immunity, though a benefit to the gardener, would be a disadvantage in a state of nature both to the cherry and raspberry, as their dissemination depends on birds. I noticed during several winters that some trees of the yellow-berried holly, which were raised from seed from a wild tree found by my father, remained covered with fruit, whilst not a scarlet berry could be seen on the adjoining trees of the common kind. A friend informs me that a mountain-ash (*Pyrus aucuparia*) growing in his garden bears berries which, though not differently coloured, are always devoured by birds before those on the other trees. This variety of the mountain-ash would thus be more freely disseminated, and the yellow-berried variety of the holly less freely, than the common varieties of these two trees.

Independently of colour, other trifling differences are sometimes found to be of importance to plants under cultivation, and would be of paramount importance if they had to fight their own battle and to struggle with many competitors. The thin-shelled peas, called *pois sans parchemin*, are attacked by birds much more than common peas. On the other hand, the purple-podded pea, which has a hard shell, escaped the attacks of tomtits (*Parus major*) in my garden far better than any other kind. The thin-shelled walnut likewise suffers greatly from the tomtit. These same birds have been observed to pass over and thus favour the filbert, destroying only the other kinds of nuts which grow in the same orchard.

Certain varieties of the pear have soft bark, and these suffer severely from boring wood-beetles; whilst other varieties are known to resist their attacks much better. In North America the smoothness, or absence of down on the fruit, makes a great difference in the attacks of the weevil, "which is the uncompromising foe of all smooth stone-fruits;" and the cultivator "has the frequent mortification of seeing nearly all, or indeed often the whole crop, fall from the trees when half or two-thirds grown." Hence the nectarine suffers more than the peach. A particular variety of the Morello cherry, raised in North America, is without any assignable cause more liable to be injured by this same insect than other cherry-trees. From some unknown cause, the Winter Majetin apple enjoys the great advantage of not being infested by the coccus. On the other hand, a particular case has been recorded in which aphides confined themselves to the Winter Nelis pear, and touched no other kind in an extensive orchard. The existence of minute glands on the leaves of peaches, nectarines, and apricots, would not be esteemed by botanists as a character of the least importance, for they are present or absent in closely related sub-varieties, descended from the same parent-tree; yet there is good evidence that the absence of glands leads to mildew, which is highly injurious to these trees.

A difference either in flavour or in the amount of nutriment in certain varieties causes them to be more eagerly attacked by various enemies than other varieties of the same species. Bull-finches (*Pyrrhula vulgaris*) injure our fruit-trees by devouring the flower-buds, and a pair of these birds have been seen "to denude a large plum-tree in a couple of days of almost every bud;" but certain varieties of the apple and thorn (*Cratægus oxycantha*) are more especially liable to be attacked. A striking instance of this was observed in Mr. Rivers's garden, in which two rows of a particular variety of plum had to be carefully protected, as they were usually stripped of all their buds during the winter, whilst other sorts growing near them escaped. The root (or enlarged stem) of Laing's Swedish turnip is preferred by hares, and therefore suffers more than other varieties. Hares and rabbits eat down common rye before St. John's day-rye, when both grow together. In the South of France, when an orchard of almond-trees is formed, the nuts of the bitter variety are sown, "in order that they may not be devoured by field-mice;" so we see the use of the bitter principle in almonds.

Other slight differences, which would be thought quite unimportant, are no doubt sometimes of great service both to plants and animals. The Whitesmith's gooseberry, as formerly stated, produces its leaves later than other varieties, and, as the flowers are thus left unprotected, the fruit often fails. In one variety of the cherry, according to Mr. Rivers, the petals are much curled backwards, and in consequence of this the stigmas were observed to be killed by a severe frost; whilst at the same time, in another variety with incurved petals, the stigmas were not in the least injured. The straw of the Fenton wheat is remarkably unequal in height; and a competent observer believes that this variety is highly productive, partly because the ears, from being distributed at various heights above the ground, are less crowded together. The same observer maintains that in the upright varieties the divergent awns are serviceable by breaking the shocks when the ears are dashed together by the wind. If several varieties of a plant are grown together, and the seed is indiscriminately harvested, it is clear that the hardier and more productive kinds will, by a sort of natural selection, gradually prevail over the others; this takes place, as Colonel Le Couteur believes, in our wheat-fields, for, as formerly shown, no variety is quite uniform in character. The same thing, as I am assured by nurserymen, would take place in our flower-gardens, if the seed of the different varieties were not separately sayed. When the eggs of the wild and tame duck are hatched together, the young wild ducks almost invariably perish, from being of smaller size and not getting their fair share of food.

Facts in sufficient number have now been given showing that natural selection often checks, but occasionally favours, man's power of selection. These facts teach us, in addition, a valuable lesson, namely, that we ought to be extremely cautious in judging what characters are of importance in a state of nature to animals and plants, which have to struggle from the hour of their birth to that of their death for existence,—their existence depending on conditions, about which we are profoundly ignorant.

• • •

It is not surprising that domestic races should generally present a different aspect from natural species. Man selects and propagates modifications solely for his own use or fancy, and not for the creature's own good. His attention is struck by strongly marked modifications, which have appeared suddenly, due to some great disturbing cause in the organisation. He attends almost exclusively to external characters; and when he succeeds in modifying internal organs,—when for instance he reduces the bones and offal, or loads the viscera with fat, or gives early maturity, &c.,—the chances are strong that he will at the same time weaken the constitution. On the other hand, when an animal has to struggle throughout its life with many competitors and enemies, under circumstances inconceivably complex and liable to change, modifications of the most varied nature—in the internal organs as well as in external characters, in the functions and mutual relations of parts—will be rigorously tested, preserved, or rejected. Natural selection often checks man's comparatively feeble and capricious attempts at improvement; and if this were not so, the result of his work, and of nature's work, would be even still more different. Nevertheless, we must not overrate the amount of difference between natural species and domestic races; the most experienced naturalists have often disputed whether the latter are descended from one or from several aboriginal stocks, and this clearly shows that there is no palpable difference between species and races.

Domestic races propagate their kind far more truly, and endure for much longer periods, than most naturalists are willing to admit. Breeders feel no doubt on this head; ask a man who has long reared Shorthorn or Hereford cattle, Leicester or Southdown sheep, Spanish or Game poultry, tumbler or carrier-pigeons, whether these races may not have been derived from common progenitors, and he will probably laugh you to scorn. The breeder admits that he may hope to produce sheep with finer or longer wool and with better carcases, or handsomer fowls, or carrier-pigeons with beaks just perceptibly longer to the practised eye, and thus be successful at an exhibition. Thus far he will go, but no farther. He does not reflect on what follows from adding up during a long course of time many, slight, successive modifications; nor does he reflect on the former existence of numerous varieties, connecting the links in each divergent line of descent. He concludes, as was shown in the earlier chapters, that all the chief breeds to which he has long attended are aboriginal productions. The systematic naturalist, on the other hand, who generally knows nothing of the art of breeding, who does not pretend to know how and when the several domestic races were formed, who cannot have seen the intermediate gradations, for they do not now exist, nevertheless feels no doubt that these races are sprung from a single source. But ask him whether the closely allied natural species which he has studied may not have descended from a common progenitor, and he in his turn will perhaps reject the notion with scorn. Thus the naturalist and breeder may mutually learn a useful lesson from each other.

Summary on Selection by Man.—There can be no doubt that methodical selection has effected and will effect wonderful results. It was occasionally practised in ancient times, and is still practised by semi-civilised people. Characters of the highest importance, and others of trifling value, have been attended to, and modified. I need not here repeat what has been so often said on the part which unconscious selection has played: we see its power in the difference between flocks which have been separately bred, and in the slow changes, as circumstances have slowly changed, which many animals have undergone in the same country, or when transported into a foreign land. We see the combined effects of methodical and unconscious selection in the great amount of difference between varieties in those parts or qualities which are valued by man, in comparison with those which are not valued, and consequently have not been attended to. Natural selection often determines man's power of selection. We sometimes err in imagining that characters, which are considered as unimportant by the systematic naturalist, could not be affected by the struggle for existence, and therefore be acted on by natural selection; but striking cases have been given, showing how great an error this is.

The possibility of selection coming into action rests on variability; and this is mainly caused, as we shall hereafter see, by changes in the conditions of life. Selection is sometimes rendered difficult, or even impossible, by the conditions being opposed to the desired character or quality. It is sometimes checked by the lessened fertility and weakened constitution which follow from long-continued close interbreeding. That methodical selection may be successful, the closest attention and discernment, combined with unwearied patience, are absolutely necessary; and these same qualities, though not indispensable, are highly serviceable in the case of unconscious selection. It is almost necessary that a large number of individuals should be reared; for thus there will be a fair chance of variations of the desired nature arising, and every individual with the slightest blemish or in any degree inferior may be freely rejected. Hence length of time is an important element of success. Thus, also, propagation at an early age and at short intervals favours the work. Facility in pairing animals, or their inhabiting a confined area, is advantageous as a check to free crossing. Whenever and wherever selection is not practised, distinct races are not formed. When any one part of the body or quality is not attended to, it remains either unchanged or varies in a fluctuating manner, whilst at the same time other parts and other qualities may become permanently and greatly modified. But from the tendency to reversion and to continued variability, those parts or organs which are now undergoing rapid improvement through selection, are likewise found to vary much. Consequently highly-bred animals, when neglected, soon degenerate; but we have no reason to believe that the effects of long-continued selection would, if the conditions of life remained the same, be soon and completely lost.

Man always tends to go to an extreme point in the selection, whether methodical or unconscious, of all useful and pleasing qualities. This is an important principle, as it leads to continued divergence, and in some rare cases to convergence of character. The possibility of continued divergence rests on the tendency in each part or organ to go on varying in the same manner in which it has already varied; and that this occurs, is proved by the steady and gradual improvement of many animals and plants during lengthened periods. The principle of divergence of character, combined with the neglect and final extinction of all previous, less-valued, and intermediate varieties, explains the amount of difference and the distinctness of our several races. Although we may have reached the utmost limit to which certain characters can be modified, yet we are far from having reached, as we have good reason to believe, the limit in the majority of cases. Finally, from the difference between selection as carried on by man and by nature, we can understand how it is that domestic races often, though by no means always, differ in general aspect from closely allied natural species.

Throughout this chapter and elsewhere I have spoken of selection as the paramount power, yet its action absolutely depends on what we in our ignorance call spontaneous or accidental variability. Let an architect be compelled to build an edifice with uncut stones, fallen from a precipice. The shape of each fragment may be called accidental; yet the shape of each has been determined by the force of gravity, the nature of the rock, and the slope of the precipice,—events and circumstances, all of which depend on natural laws; but there is no relation between these laws and the purpose for which each fragment is used by the builder. In the same manner the variations of each creature are determined by fixed and immutable laws; but these bear no relation to the living structure which is slowly built up through the power of selection, whether this be natural or artificial selection.

If our architect succeeded in rearing a noble edifice, using the rough wedge-shaped fragments for the arches, the longer stones for the lintels, and so forth, we should admire his skill even in a higher degree than if he had used stones shaped for the purpose. So it is with selection, whether applied by man or by nature; for though variability is indispensably necessary, yet, when we look at some highly complex and excellently adapted organism, variability sinks to a quite subordinate position in importance in comparison with selection, in the same manner as the shape of each fragment used by our supposed architect is unimportant in comparison with his skill.

PROVISIONAL HYPOTHESIS OF PANGENESIS.

In the previous chapters large classes of facts, such as those bearing on bud-variation, the various forms of inheritance, the causes and laws of variation, have been discussed; and it is obvious that these subjects, as well as the several modes of reproduction, stand in some sort of relation to each other. I have been led, or rather forced, to form a view which to a certain extent connects these facts by a tangible method. Every one would wish to explain to himself, even in an imperfect manner, how it is possible for a character possessed by some remote ancestor suddenly to reappear in the offspring; how the effects of increased or decreased use of a limb can be transmitted to the child; how the male sexual element can act not solely on the ovule, but occasionally on the mother-form; how a limb can be reproduced on the exact line of amputation, with neither too much nor too little added; how the various modes of reproduction are connected, and so forth. I am aware that my view is merely a provisional hypothesis or speculation; but until a better one be advanced, it may be serviceable by bringing together a multitude of facts which are at present left disconnected by any efficient cause. As Whewell, the historian of the inductive sciences, remarks:—"Hypotheses may often be of service to science, when they involve a certain portion of incompleteness, and even of error." Under this point of view I venture to advance the hypothesis of Pangenesis, which implies that the whole organisation, in the sense of every separate atom or unit, reproduces itself. Hence ovules and pollen-grains,—the fertilised seed or egg, as well as buds, —include and consist of a multitude of germs thrown off from each separate atom of the organism.

In the First Part I will enumerate as briefly as I can the groups of facts which seem to demand connection; but certain subjects, not hitherto discussed, must be treated at disproportionate length. In the Second Part the hypothesis will be given; and we shall see, after considering how far the necessary assumptions are in themselves improbable, whether it serves to bring under a single point of view the various facts.

PART I.

Reproduction may be divided into two main classes, namely, sexual and asexual. The latter is effected in many ways—by gemmation, that is by the formation of buds of various kinds, and by fissiparous generation, that is by spontaneous or artificial division. It is notorious that some of the lower animals, when cut into many pieces, reproduce so many perfect individuals: Lyonnet cut a Nais or freshwater worm into nearly forty pieces, and these all reproduced perfect animals. It is probable that segmentation could be carried much further in some of the protozoa, and with some of the lowest plants each cell will reproduce the parent-form. Johannes Müller thought that there was an important distinction between gemmation and fission; for in the latter case the divided portion, however small, is more perfectly organised; but most physiologists are now convinced that the two processes are essentially alike. Prof. Huxley remarks, "fission is little more than a peculiar mode of budding," and Prof. H. J. Clark, who has especially attended to this subject, shows in detail that there is sometimes "a compromise between self-division and budding." When a limb is amputated, or when the whole body is bisected, the cut extremities are said to bud forth; and as the papilla, which is first formed, consists of undeveloped cellular tissue like that forming an ordinary bud, the expression is apparently correct. We see the connection of the two processes in another way; for Trembley observed that with the hydra the reproduction of the head after amputation was checked as soon as the animal began to bud.

Between the production, by fissiparous generation, of two or more complete individuals, and the repair of even a very slight injury, we have, as remarked in a former chapter, so perfect and insensible a gradation, that it is impossible to doubt that they are connected processes. Between the power which repairs a trifling injury in any part, and the power which previously "was occupied in its maintenance by the continued mutation of its particles," there cannot be any great difference; and we may follow Mr. Paget in believing them to be the selfsame power. As at each stage of growth an amputated part is replaced by one in the same state of development, we must likewise follow Mr. Paget in admitting "that the powers of development from the embryo are identical with those exercised for the restoration from injuries: in other words, that the powers are the same by which perfection is first achieved, and by which, when lost, it is recovered." Finally, we may conclude that the several forms of gemmation, and of fissiparous generation, the repair of injuries, the maintenance of each part in its proper state, and the growth or progressive development of the whole structure of the embryo, are all essentially the results of one and the same great power.

• • •

PART II.

I have now enumerated the chief facts which every one would desire to connect by some intelligible bond. This can be done, as it seems to me, if we make the following assumptions; if the first and chief one be not rejected, the others, from being supported by various physiological considerations, will not appear very improbable. It is almost universally admitted that cells, or the units of the body, propagate themselves by self-division or proliferation, retaining the same nature, and ultimately becoming converted into the various tissues and substances of the body. But besides this means of increase I assume that cells, before their conversion into completely passive or "formed material," throw off minute granules or atoms, which circulate freely throughout the system, and when supplied with proper nutriment multiply by self-division, subsequently becoming developed into cells like those from which they were derived. These granules for the sake of distinctness may be called cell-gemmules, or, as the cellular theory is not fully established, simply gemmules. They are supposed to be transmitted from the parents to the offspring, and are generally developed in the generation which immediately succeeds, but are often transmitted in a dormant state during many generations and are then developed. Their development is supposed to depend on their union with other partially developed cells or gemmules which precede them in the regular course of growth. Why I use the term union, will be seen when we discuss the direct action of pollen on the tissues of the mother-plant. Gemmules are supposed to be thrown off by every cell or unit, not only during the adult state, but during all the stages of development. Lastly, I assume that the gemmules in their dormant state have a mutual affinity for each other, leading to their aggregation either into buds or into the sexual elements. Hence, speaking strictly, it is not the reproductive elements, nor the buds, which generate new organisms, but the cells themselves throughout the body. These assumptions constitute the provisional hypothesis which I have called Pangenesis. Nearly similar views have been propounded, as I find, by other authors, more especially by Mr. Herbert Spencer; but they are here modified and amplified.

Before proceeding to show, firstly, how far these assumptions are in themselves probable, and secondly, how far they connect and explain the various groups of facts with which we are concerned, it may be useful to give an illustration of the hypothesis. If one of the simplest Protozoa be formed, as appears under the microscope, of a small mass of homogeneous gelatinous matter, a minute atom thrown off from any part and nourished under favourable circumstances would naturally reproduce the whole; but if the upper and lower surfaces were to differ in texture from the central portion, then all three parts would have to throw off atoms or gemmules, which when aggregated by mutual affinity would form either buds or the sexual elements. Precisely the same view may be extended to one of the higher animals;

although in this case many thousand gemmules must be thrown off from the various parts of the body. Now, when the leg, for instance, of a salamander is cut off, a slight crust forms over the wound, and beneath this crust the uninjured cells or units of bone, muscle, nerves, &c., are supposed to unite with the diffused gemmules of those cells which in the perfect leg come next in order; and these as they become slightly developed unite with others, and so on until a papilla of soft cellular tissue, the "budding leg," is formed, and in time a perfect leg. Thus, that portion of the leg which had been cut off, neither more nor less, would be reproduced. If the tail or leg of a young animal had been cut off, a young tail or leg would have been reproduced, as actually occurs with the amputated tail of the tadpole; for gemmules of all the units which compose the tail are diffused throughout the body at all ages. But during the adult state the gemmules of the larval tail would remain dormant, for they would not meet with preexisting cells in a proper state of development with which to unite. If from changed conditions or any other cause any part of the body should become permanently modified, the gemmules, which are merely minute portions of the contents of the cells forming the part, would naturally reproduce the same modification. But gemmules previously derived from the same part before it had undergone any change, would still be diffused throughout the organisation, and would be transmitted from generation to generation, so that under favourable circumstances they might be redeveloped, and then the new modification would be for a time or for ever lost. The aggregation of gemmules derived from every part of the body, through their mutual affinity, would form buds, and their aggregation in some special manner, apparently in small quantity, together probably with the presence of gemmules of certain primordial cells, would constitute the sexual elements. By means of these illustrations the hypothesis of pangenesis has, I hope, been rendered intelligible.

Physiologists maintain, as we have seen, that each cell, though to a large extent dependent on others, is likewise, to a certain extent, independent or autonomous. I go one small step further, and assume that each cell casts off a free gemmule, which is capable of reproducing a similar cell. There is some analogy between this view and what we see in compound animals and in the flower-buds on the same tree; for these are distinct individuals capable of true or seminal reproduction, yet have parts in common and are dependent on each other; thus the tree has its bark and trunk, and certain corals, as the Virgularia, have not only parts, but movements in common.

The existence of free gemmules is a gratuitous assumption, yet can hardly be considered as very improbable, seeing that cells have the power of multiplication through the self-division of their contents. Gemmules differ from true ovules or buds inasmuch as they are supposed to be capable of multiplication in their undeveloped state. No one probably will object to this capacity as improbable. The blastema within the egg has been known to divide and give birth to two embryos; and Thuret has seen the zoospore of an alga divide itself, and both halves germinate. An atom of small-pox matter, so minute as to be borne by the wind, must multiply itself many thousandfold in a person thus inoculated. It has recently been ascertained that a minute portion of the mucous discharge from an animal affected with rinderpest, if placed in the blood of a healthy ox, increases so fast that in a short space of time "the whole mass of blood, weighing many pounds, is infected, and every small particle of that blood contains enough poison to give, within less than forty-eight hours, the disease to another animal."

The retention of free and undeveloped gemmules in the same body from early youth to old age may appear improbable, but we should remember how long seeds lie dormant in the earth and buds in the bark of a tree. Their transmission from generation to generation may appear still more improbable; but here again we should remember that many rudimentary and useless organs are transmitted and have been transmitted during an indefinite number of generations. We shall presently see how well the long-continued transmission of undeveloped gemmules explains many facts.

As each unit, or group of similar units throughout the body, casts off its gemmules, and as all are contained within the smallest egg or seed, and within each spermatozoon or pollen-grain, their number and minuteness must be something inconceivable. I shall hereafter recur to this objection, which at first appears so formidable; but it may here be remarked that a cod-fish has been found to produce 4,872,000 eggs, a single Ascaris about 64,000,000 eggs, and a single Orchidaceous plant probably as many million seeds. In these several cases, the spermatozoa and pollen-grains must exist in considerably larger numbers. Now, when we have to deal with numbers such as these, which the human intellect cannot grasp, there is no good reason for rejecting our present hypothesis on account of the assumed existence of cell-gemmules a few thousand times more numerous.

The gemmules in each organism must be thoroughly diffused; nor does this seem improbable considering their minuteness, and the steady circulation of fluids throughout the body. So it must be with the gemmules of plants, for with certain kinds even a minute fragment of a leaf will reproduce the whole. But a difficulty here occurs; it would appear that with plants, and probably with compound animals, such as corals, the gemmules do not spread from bud to bud, but only through the tissues developed from each separate bud. We are led to this conclusion from the stock being rarely affected by the insertion of a bud or graft from a distinct variety. This non-diffusion of the gemmules is still more plainly shown in the case of ferns; for Mr. Bridgman has proved that, when spores (which it should be remembered are of the nature of buds) are taken from a monstrous part of a frond, and others from an ordinary part, each reproduces the form of the part whence derived. But this non-diffusion of the gemmules from bud to bud may be only apparent, depending, as we shall hereafter see, on the nature of the first-formed cells in the buds.

The assumed elective affinity of each gemmule for that particular cell which precedes it in the order of development is supported by many analogies. In all ordinary cases of sexual reproduction the male and female elements have a mutual affinity for each other: thus, it is believed that about ten thousand species of Compositæ exist, and there can be no doubt that if the pollen of all these species could be, simultaneously or successively, placed on the stigma of any one species, this one would elect with unerring certainty its own pollen. This elective capacity is all the more wonderful, as it must have been acquired since the many species of this great group of plants branched off from a common progenitor. On any view of the nature of sexual reproduction, the protoplasm contained within the ovules and within the sperm-cells (or the "spermatic force" of the latter, if so vague a term be preferred) must act on each other by some law of special affinity, either during or subsequently to impregnation, so that corresponding parts alone affect each other; thus, a calf produced from a short-horned cow by a long-horned bull has its horns and not its horny hoofs affected by the union of the two forms, and the offspring from two birds with differently coloured tails have their tails and not their whole plumage affected.

The various tissues of the body plainly show, as many physiologists have insisted, an affinity for special organic substances, whether natural or foreign to the body. We see this in the cells of the kidneys attracting urea from the blood; in the worrara poison affecting the nerves; upas and digitalis the muscles; the Lytta vesicatoria the kidneys; and in the poisonous matter of many diseases, as small-pox, scarlet-fever, hooping cough, glanders, cancer, and hydrophobia, affecting certain definite parts of the body or certain tissues or glands.

The affinity of various parts of the body for each other during their early development was shown in the last chapter, when discussing the tendency to fusion in homologous parts. This affinity displays itself in the normal fusion of organs which are separate at an early embryonic age, and still more plainly in those marvellous cases of double monsters in which each bone, muscle, vessel, and nerve in the one embryo, blends with the corresponding part in the other. The affinity between homologous organs may come into action with single parts, or with the entire individual, as in the case of flowers or fruits which are symmetrically blended together with all their parts doubled, but without any other trace of fusion.

It has also been assumed that the development of each gemmule depends on its union with another cell or unit which has just commenced its development, and which, from preceding it in order of growth, is of a somewhat different nature. Nor is it a very improbable assumption that the development of a gemmule is determined by its union with a cell slightly different in nature, for abundant evidence was given in the seventeenth chapter, showing that a slight degree of differentiation in the male and female sexual elements favours in a marked manner their union and subsequent development. But what determines the development of the gemmules of the first-formed or primordial cell in the unimpregnated ovule, is beyond conjecture.

It must also be admitted that analogy fails to guide us towards any determination on several other points: for instance, whether cells, derived from the same parent-cell, may, in the regular course of growth, become developed into different structures, from absorbing peculiar kinds of nutriment, independently of their union with distinct gemmules. We shall appreciate this difficulty if we call to mind, what complex yet symmetrical growths the cells of plants yield when they are inoculated by the poison of a gall-insect. With animals various polypoid excrescences and tumours are now generally admitted to be the direct product, through proliferation, of normal cells which have become abnormal. In the regular growth and repair of bones, the tissues undergo, as Virchow remarks, a whole series of permutations and substitutions. "The cartilage-cells may be converted by a direct transformation into marrow-cells, and continue as such; or they may first be converted into osseous and then into medullary tissue; or lastly, they may first be converted into marrow and then into bone. So variable are the permutations of these tissues, in themselves so nearly allied, and yet in their external appearance so completely distinct." But as these tissues thus change their nature at any age, without any obvious change in their nutrition, we must suppose in accordance with our hypothesis that gemmules derived from one kind of tissue combine with the cells of another kind, and cause the successive modifications.

It is useless to speculate at what period of development each organic unit casts off its gemmules; for the whole subject of the development of the various elemental tissues is as yet involved in much doubt. Some physiologists, for instance, maintain that muscle or nerve-fibres are developed from cells, which are afterwards nourished by their own proper powers of absorption; whilst other physiologists deny their cellular origin; and Beale maintains that such fibres are renovated exclusively by the conversion of fresh germinal matter (that is the so-called nuclei) into "formed material." However this may be, it appears probable that all external agencies, such as changed nutrition, increased use or disuse, &c., which induced any permanent modification in a structure, would at the same time or previously act on the cells, nuclei, germinal or formative matter, from which the structures in question were developed, and consequently would act on the gemmules or cast-off atoms.

There is another point on which it is useless to speculate, namely, whether all gemmules are free and separate, or whether some are from the first united into small aggregates. A feather, for instance, is a complex structure, and, as each separate part is liable to inherited variations, I conclude that each feather certainly generates a large number of gemmules; but it is possible that these may be aggregated into a compound gemmule. The same remark applies to the petals of a flower, which in some cases are highly complex, with each ridge and hollow contrived for special purposes, so that each part must have been separately modified, and the modifications transmitted; consequently, separate gemmules, according to our hypothesis, must have been thrown off from each cell or part. But, as we sometimes see half an anther or a small portion of a filament becoming petaliform, or parts or mere stripes of the calyx assuming the colour and texture of the corolla, it is probable that with petals the gemmules of each cell are not aggregated together into a compound gemmule, but are freely and separately diffused.

• • •

CONCLUSION.

The hypothesis of Pangenesis, as applied to the several great classes of facts just discussed, no doubt is extremely complex; but so assuredly are the facts. The assumptions, however, on which the hypothesis rests cannot be considered as complex in any extreme degree—namely, that all organic units, besides having the power, as is generally admitted, of growing by self-division, throw off free and minute atoms of their contents, that is gemmules. These multiply and aggregate themselves into buds and the sexual elements; their development depends on their union with other nascent cells or units; and they are capable of transmission in a dormant state to successive generations.

In a highly organised and complex animal, the gemmules thrown off from each different cell or unit throughout the body must be inconceivably numerous and minute. Each unit of each part, as it changes during development, and we know that some insects undergo at least twenty metamorphoses, must throw off its gemmules. All organic beings, moreover, include many dormant gemmules derived from their grandparents and more remote progenitors, but not from all their progenitors. These almost infinitely numerous and minute gemmules must be included in each bud, ovule, spermatozoon, and pollen-grain. Such an admission will be declared impossible; but, as previously remarked, number and size are only relative difficulties, and the eggs or seeds produced by certain animals or plants are so numerous that they cannot be grasped by the intellect.

The organic particles with which the wind is tainted over miles of space by certain offensive animals must be infinitely minute and numerous; yet they strongly affect the olfactory nerves. An analogy more appropriate is afforded by the contagious particles of certain diseases, which are so minute that they float in the atmosphere and adhere to smooth paper; yet we know how largely they increase within the human body, and how powerfully they act. Independent organisms exist which are barely visible under the highest powers of our recently-improved microscopes, and which probably are fully as large as the cells or units in one of the higher animals; yet these organisms no doubt reproduce themselves by germs of extreme minuteness, relatively to their own minute size. Hence the difficulty, which at first appears insurmountable, of believing in the existence of gemmules so numerous and so small as they must be according to our hypothesis, has really little weight.

The cells or units of the body are generally admitted by physiologists to be autonomous, like the buds on a tree, but in a less degree. I go one step further and assume that they throw off reproductive gemmules. Thus an animal does not, as a whole, generate its kind through the sole agency of the reproductive system, but each separate cell generates its kind. It has often been said by naturalists that each cell of a plant has the actual or potential capacity of reproducing the whole plant; but it has this power only in virtue of containing gemmules derived from every part. If our hypothesis be provisionally accepted, we must look at all the forms of asexual reproduction, whether occurring at maturity or as in the case of alternate generation during youth, as fundamentally the same, and dependent on the mutual aggregation and multiplication of the gemmules. The regrowth of an amputated limb or the healing of a wound is the same process partially carried out. Sexual generation differs in some important respects, chiefly, as it would appear, in an insufficient number of gemmules being aggregated within the separate sexual elements, and probably in the presence of certain primordial cells. The development of each being, including all the forms of metamorphosis and metagenesis, as well as the so-called growth of the higher animals, in which structure changes though not in a striking manner, depends on the presence of gemmules thrown off at each period of life, and on their development, at a corresponding period, in union with preceding cells. Such cells may be said to be fertilised by the gemmules which come next in the order of development. Thus the ordinary act of impregnation and the development of each being are closely analogous processes. The child, strictly speaking, does not grow into the man, but includes germs which slowly and successively become developed and form the man. In the child, as well as in the adult, each part generates the same part for the next generation. Inheritance must be looked at as merely a form of growth, like the self-division of a lowly-organised unicellular plant. Reversion depends on the transmission from the forefather to his descendants of dormant gemmules, which occasionally become developed under certain known or unknown conditions. Each animal and plant may be compared to a bed of mould full of seeds, most of which soon germinate, some lie for a period dormant, whilst others perish. When we hear it said that a man carries in his constitution the seeds of an inherited disease, there is much literal truth in the expression. Finally, the power of propagation possessed by each separate cell, using the term in its largest sense, determines the reproduction, the variability, the development and renovation of each living organism.

No other attempt, as far as I am aware, has been made, imperfect as this confessedly is, to connect under one point of view these several grand classes of facts. We cannot fathom the marvellous complexity of an organic being; but on the hypothesis here advanced this complexity is much increased. Each living creature must be looked at as a microcosm—a little universe, formed of a host of self-propagating organisms, inconceivably minute and as numerous as the stars in heaven.

• • •

CONCLUDING REMARKS.

In accordance with the views maintained by me in this work and elsewhere, not only the various domestic races, but the most distinct genera and orders within the same great class,—for instance, whales, mice, birds, and fishes—are all the descendants of one common progenitor, and we must admit that the whole vast amount of difference between these forms of life has primarily arisen from simple variability. To consider the subject under this point of view is enough to strike one dumb with amazement. But our amazement ought to be lessened when we reflect that beings, almost infinite in number, during an almost infinite lapse of time, have often had their whole organisation rendered in some degree plastic, and that each slight modification of structure which was in any way beneficial under excessively complex conditions of life, will have been preserved, whilst each which was in any way injurious will have been rigorously destroyed. And the long-continued accumulation of beneficial variations will infallibly lead to structures as diversified, as beautifully adapted for various purposes, and as excellently co-ordinated, as we see in the animals and plants all around us. Hence I have spoken of selection as the paramount power, whether applied by man to the formation of domestic breeds, or by nature to the production of species. I may recur to the metaphor given in a former chapter: if an architect were to rear a noble and commodious edifice, without the use of cut stone, by selecting from the fragments at the base of a precipice wedge-formed stones for his arches, elongated stones for his lintels, and flat stones for his roof, we should admire his skill and regard him as the paramount power. Now, the fragments of stone, though indispensable to the architect, bear to the edifice built by him the same relation which the fluctuating variations of each organic being bear to the varied and admirable structures ultimately acquired by its modified descendants.

Some authors have declared that natural selection explains nothing, unless the precise cause of each slight individual difference be made clear. Now, if it were explained to a savage utterly ignorant of the art of building, how the edifice had been raised stone upon stone, and why wedge-formed fragments were used for the arches, flat stones for the roof, &c.; and if the use of each part and of the whole building were pointed out, it would be unreasonable if he declared that nothing had been made clear to him, because the precise cause of the shape of each fragment could not be given. But this is a nearly parallel case with the objection that selection explains nothing, because we know not the cause of each individual difference in the structure of each being.

The shape of the fragments of stone at the base of our precipice may be called accidental, but this is not strictly correct; for the shape of each depends on a long sequence of events, all obeying natural laws; on the nature of the rock, on the lines of deposition or cleavage, on the form of the mountain which depends on its upheaval and subsequent denudation, and lastly on the storm or earthquake which threw down the fragments. But in regard to the use to which the fragments may be put, their shape may be strictly said to be accidental. And here we are led to face a great difficulty, in alluding to which I am aware that I am travelling beyond my proper province. An omniscient Creator must have foreseen every consequence which results from the laws imposed by Him. But can it be reasonably maintained that the Creator intentionally ordered, if we use the words in any ordinary sense, that certain fragments of rock should assume certain shapes so that the builder might erect his edifice? If the various laws which have determined the shape of each fragment were not predetermined for the builder's sake, can it with any greater probability be maintained that He specially ordained for the sake of the breeder each of the innumerable variations in our domestic animals and plants,—many of these variations being of no service to man, and not beneficial, far more often injurious, to the creatures themselves? Did He ordain that the crop and tail-feathers of the pigeon should vary in order that the fancier might make his grotesque pouter and fantail breeds? Did He cause the frame and mental qualities of the dog to vary in order that a breed might be formed of indomitable ferocity, with jaws fitted to pin down the bull for man's brutal sport? But if we give up the principle in one case,—if we do not admit that the variations of the primeval dog were intentionally guided in order that the greyhound, for instance, that perfect image of symmetry and vigour, might be formed,—no shadow of reason can be assigned for the belief that variations, alike in nature and the result of the same general laws, which have been the groundwork through natural selection of the formation of the most perfectly adapted animals in the world, man included, were intentionally and specially guided. However much we may wish it, we can hardly follow Professor Asa Gray in his belief "that variation has been led along certain beneficial lines," like a stream "along definite and useful lines of irrigation." If we assume that each particular variation was from the beginning of all time preordained, the plasticity of organisation, which leads to many injurious deviations of structure, as well as that redundant power of reproduction which inevitably leads to a struggle for existence, and, as a consequence, to the natural selection or survival of the fittest, must appear to us superfluous laws of nature. On the other hand, an omnipotent and omniscient Creator ordains everything and foresees everything. Thus we are brought face to face with a difficulty as insoluble as is that of free will and predestination.

Origin of Species. [1869]

Darwin was not usually willing to respond directly to criticisms of the Origin. *When he found an objection convincing, he characteristically made corrections or amplifications in the book's next edition. When, however, he considered a specific criticism dubious, he would question it in print, as occurs in the following two letters concerned with the increase of elephants. These letters of June 26 and July 17, 1869, appeared in the* Athenaeum: Journal of Literature, Science, and the Fine Arts, *an elite intellectual weekly with a circulation of over 15,000.*

Several references in the two Athenaeum *letters warrant glossing. The correspondent called "Ponderer" remains unidentified, but the "late Dr. Falconer" is Hugh Falconer, paleontologist and expert on elephants. The "good arithmetician" may have been Darwin's oldest son, William. The "Cambridge mathematician" may have been Joseph Wolstenholme, William's tutor at Christ's College. The "learned Professor" is probably Ernst Haeckel, professor of zoology at Jena. Edward L. Garbett had, in the July 3 issue of the* Athenaeum, *indicated that there would be 2.5 million elephants after five centuries and 50 million after six.*

The correct number of descendants after twenty-five generations is 5,111,514, the figure Darwin gave in Natural Selection, *as opposed to the figure 15,000,000 given in the* Origin. *Although two of Darwin's sons graduated from Cambridge in mathematics and one of them, George, even attained the honor of being Second Wrangler (second in his class), they did not inherit their mathematical talents from their father. Darwin himself always had trouble solving problems of this sort.*

FIRST LETTER.

I am much obliged to your Correspondent of June 5 for having pointed out a great error in my "Origin of Species," on the possible rate of increase of the elephant. I inquired from the late Dr. Falconer with respect to the age of breeding, &c., and understated the data obtained from him, with the intention, vain as it has proved, of not exaggerating the result. Finding that the calculation was difficult, I applied to a good arithmetician; but he did not know any formula by which a result could easily be obtained; and he now informs me that I then applied to some Cambridge mathematician. Who this was I cannot remember, and therefore cannot find out how the error arose. From the many familiar instances of rapid geometrical increase, I confess that, if the answer had been thirty or sixty million elephants, I should not have felt much surprise; but I ought not to have relied so implicitly on my mathematical friend. I have misled your Correspondent by using language which implies that the elephant produces a pair of young at each birth; but the calculation by this assumption is rendered easier and the result but little different. A friend has extended your Correspondent's calculation to a further period of years. Commencing with a pair of elephants, at the age of thirty, and assuming that they would in each generation survive ten years after the last period of breeding—namely, when ninety years old—there would be, after a period of 750 to 760 years (instead of after 500 years, as I stated in "The Origin of Species"), considerably more than fifteen million elephants alive, namely, 18,803,080. At the next succeeding period of 780 to 790 years there would be alive no less than 34,584,256 elephants.

SECOND LETTER.

I have received a letter from Germany on the increase of the elephant, in which a learned Professor arrives at a totally different result from that of Mr. Garbett, both of which differ from that of your Correspondent "Ponderer." Hence you may perhaps think it worth while to publish a rule by which my son, Mr. George Darwin, finds that the product for any number of generations may easily be calculated:—

"The supposition is that each pair of elephants begins to breed when aged 30, breeds at 60, and again, for the last time, at 90, and dies when aged 100, bringing forth a pair at each birth. We start, then, in the year 0 with a pair of elephants, aged 30. They produce a pair in the year 0, a pair in the year 30, a pair in the year 60, and die in the year 70. In the year 60, then, there will be the following pairs alive, viz.: one aged 90, one aged 60, two aged 30, four aged 0. The last three sets are the only ones which will breed in the year 90. At each breeding a pair produces a pair, so that the number of pairs produced in the year 90 will be the sum of the three numbers 1, 2, 4, i.e. 7. Henceforward, at each period, there will be sets of pairs, aged 30, 60, 90 respectively, which breed. These sets will consist of the pairs born at the three preceding periods respectively. Thus the number of pairs born at any period will be the sum of the three preceding numbers in the series, which gives the number of births at each period; and because the first three terms of this series are 1, 2, 4, therefore the series is 1, 2, 4, 7, 13, 24, 44, &c. These are the numbers given by 'Ponderer.' At any period, the whole number of pairs of elephants consists of the young elephants together with the three sets of parents; but since the sum of the three sets of parents is equal in number to the number of young ones, therefore the whole number of pairs is twice the number of young ones, and therefore the whole number of elephants at this period (and for ten years onwards) is four times the corresponding number in the series. In order to obtain the general term of the series, it is necessary to solve an easy equation by the Calculus of Finite Differences."

The Descent of Man, and Selection in Relation to Sex. [1871]

Darwin only briefly alludes to the possibility of human evolution in The Origin of Species. *Nevertheless, his earliest notebooks from 1837 to 1839 show that he was considering human evolution right from the start. Following the publication of the* Origin *in 1859, the big question for many of Darwin's readers became not so much whether evolution had taken place but what its implications might be for the place of humans in the universe, a matter bearing, of course, on relations between humanity and a deity. In* The Descent of Man, and Selection in Relation to Sex, *Darwin points out the many physical characteristics humans share with other animals, especially the great apes, and argues that the most distinctive human attributes have arisen through natural selection. He also sets out to show that there is no great gap between the mental powers of humans and those of other animals, that the difference is a matter of degree, not of kind. In contrast Wallace, in 1864, had written that whereas the human body could be accounted for by natural selection, the human mind was created by some "higher intelligence."*

The two volumes of The Descent of Man, *published in 1871, consist of 250 pages on the descent of man and 578 pages on sexual selection. Some readers have claimed that the two parts have no obvious or necessary relation to each other. Although it is true that Darwin could have published two separate books on the topics, he does skilfully interweave them—especially in the final three chapters, which discuss the role of sexual selection in humans. Full of interesting and often innovative discussions relevant to biology, anthropology, psychology, sociology, philosophy, ethology, and sociobiology, the* Descent *did not provoke the sensation caused by the publication of the*

Origin, *even though its subject was potentially even more controversial. Not only does the work tie humans to the other animals, but it also argues that there are few differences among the races of humans, a position cherished in the liberal and abolitionist Darwin-Wedgwood families since the eighteenth century and here given scientific support. But public, and published, discussion of the relevance of evolution to human development had already begun in books that had appeared in 1863, Huxley's* Evidence as to Man's Place in Nature *and Lyell's* The Geological Evidence of the Antiquity of Man. *In fact, Lyell's reluctance to acknowledge the role of natural selection in the evolution of humanity—instead, he had to go "the whole orang" with Lamarck, as he memorably phrased it in a letter to Darwin—was one reason for Darwin's writing* The Descent of Man. *A second, one-volume edition of* Descent *appeared in 1874, extensively revised, as usual. Today the book has been reprinted over two hundred times and translated into 18 languages.*

The selections below consist of the book's most important points on humans. Regrettably, space limitations have not allowed the inclusion of any of the many interesting chapters discussing sexual selection in the nonhuman part of the animal kingdom. To pursue this topic, readers should turn to the first edition of the Descent.

INTRODUCTION.

The nature of the following work will be best understood by a brief account of how it came to be written. During many years I collected notes on the origin or descent of man, without any intention of publishing on the subject, but rather with the determination not to publish, as I thought that I should thus only add to the prejudices against my views. It seemed to me sufficient to indicate, in the first edition of my 'Origin of Species,' that by this work "light would be thrown on the origin of man and his history;" and this implies that man must be included with other organic beings in any general conclusion respecting his manner of appearance on this earth. Now the case wears a wholly different aspect. When a naturalist like Carl Vogt ventures to say in his address as President of the National Institution of Geneva (1869), "personne, en Europe au moins, n'ose plus soutenir la création indépendante et de toutes pièces, des espèces," ["no one, in Europe at least, any longer dares entirely to support the independent creation of species"—editors], it is manifest that at least a large number of naturalists must admit that species are the modified descendants of other species; and this especially holds good with the younger and rising naturalists. The greater number accept the agency of natural selection; though some urge, whether with justice the future must decide, that I have greatly overrated its importance. Of the older and honoured chiefs in natural science, many unfortunately are still opposed to evolution in every form.

In consequence of the views now adopted by most naturalists, and which will ultimately, as in every other case, be followed by other men, I have been led to put together my notes, so as to see how far the general conclusions arrived at in my former works were applicable to man. This seemed all the more desirable as I had never deliberately applied these views to a species taken singly. When we confine our attention to any one form, we are deprived of the weighty arguments derived from the nature of the affinities which connect together whole groups of organisms—their geographical distribution in past and present times, and their geological succession. The homological structure, embryological development, and rudimentary organs of a species, whether it be man or any other animal, to which our attention may be directed, remain to be considered; but these great classes of facts afford, as it appears to me, ample and conclusive evidence in favour of the principle of gradual evolution. The strong support derived from the other arguments should, however, always be kept before the mind.

The sole object of this work is to consider, firstly, whether man, like every other species, is descended from some pre-existing form; secondly, the manner of his development; and thirdly, the value of the differences between the so-called races of man. As I shall confine myself to these points, it will not be necessary to describe in detail the differences between the several races—an enormous subject which has been fully discussed in many valuable works. The high antiquity of man has recently been demonstrated by the labours of a host of eminent men, beginning with M. Boucher de Perthes; and this is the indispensable basis for understanding his origin. I shall, therefore, take this conclusion for granted, and may refer my readers to the admirable treatises of Sir Charles Lyell, Sir John Lubbock, and others. Nor shall I have occasion to do more than to allude to the amount of difference between man and the anthropomorphous apes; for Prof. Huxley, in the opinion of most competent judges, has conclusively shewn that in every single visible character man differs less from the higher apes than these do from the lower members of the same order of Primates.

This work contains hardly any original facts in regard to man; but as the conclusions at which I arrived, after drawing up a rough draft, appeared to me interesting, I thought that they might interest others. It has often and confidently been asserted, that man's origin can never be known: but ignorance more frequently begets confidence than does knowledge: it is those who know little, and not those who know much, who so positively assert that this, or that problem will never be solved by science. The conclusion that man is the co-descendant with other species of some ancient, lower, and extinct form, is not in any degree new. Lamarck long ago came to this conclusion, which has lately been maintained by several eminent naturalists and philosophers; for instance by Wallace, Huxley, Lyell, Vogt, Lubbock, Büchner, Rolle, &c., and especially by Häckel. This last naturalist, besides his great work, 'Generelle Morphologie' (1866), has recently (1868, with a second edit. in 1870), published his 'Natürliche Schöpfungsgeschichte,' in which he fully discusses the genealogy of man. If this work had appeared before my essay had been written, I should probably never have completed it. Almost all the conclusions at which I have arrived I find confirmed by this naturalist, whose knowledge on many points is much fuller than mine. Wherever I have added any fact or view from Prof. Häckel's writings, I give his authority in the text, other statements I leave as they originally stood in my manuscript, occasionally giving in the foot-notes references to his works, as a confirmation of the more doubtful or interesting points.

During many years it has seemed to me highly probable that sexual selection has played an important part in differentiating the races of man; but in my 'Origin of Species' (first edition, p. 199) I contented myself by merely alluding to this belief. When I came to apply this view to man, I found it indispensable to treat the whole subject in full detail.* Consequently the second part of the present work, treating of sexual selection, has extended to an inordinate length, compared with the first part; but this could not be avoided.

I had intended adding to the present volumes an essay on the expression of the various emotions by man and the lower animals. My attention was called to this subject many years ago by Sir Charles Bell's admirable work. This illustrious anatomist maintains that man is endowed with certain muscles solely for the sake of expressing his emo-

* Prof. Häckel is the sole author who, since the publication of the 'Origin,' has discussed, in his various works, in a very able manner, the subject of sexual selection, and has seen its full importance.

tions. As this view is obviously opposed to the belief that man is descended from some other and lower form, it was necessary for me to consider it. I likewise wished to ascertain how far the emotions are expressed in the same manner by the different races of man. But owing to the length of the present work, I have thought it better to reserve my essay, which is partially completed, for separate publication.

THE EVIDENCE OF THE DESCENT OF MAN FROM SOME LOWER FORM.

He who wishes to decide whether man is the modified descendant of some pre-existing form, would probably first enquire whether man varies, however slightly, in bodily structure and in mental faculties; and if so, whether the variations are transmitted to his offspring in accordance with the laws which prevail with the lower animals; such as that of the transmission of characters to the same age or sex. Again, are the variations the result, as far as our ignorance permits us to judge, of the same general causes, and are they governed by the same general laws, as in the case of other organisms; for instance by correlation, the inherited effects of use and disuse, &c.? Is man subject to similar malconformations, the result of arrested development, of reduplication of parts, &c., and does he display in any of his anomalies reversion to some former and ancient type of structure? It might also naturally be enquired whether man, like so many other animals, has given rise to varieties and sub-races, differing but slightly from each other, or to races differing so much that they must be classed as doubtful species? How are such races distributed over the world; and how, when crossed, do they react on each other, both in the first and succeeding generations? And so with many other points.

The enquirer would next come to the important point, whether man tends to increase at so rapid a rate, as to lead to occasional severe struggles for existence, and consequently to beneficial variations, whether in body or mind, being preserved, and injurious ones eliminated. Do the races or species of men, whichever term may be applied, encroach on and replace each other, so that some finally become extinct? We shall see that all these questions, as indeed is obvious in respect to most of them, must be answered in the affirmative, in the same manner as with the lower animals. But the several considerations just referred to may be conveniently deferred for a time; and we will first see how far the bodily structure of man shows traces, more or less plain, of his descent from some lower form. In the two succeeding chapters the mental powers of man, in comparison with those of the lower animals, will be considered.

• • •

The bearing of the three great classes of facts now given is unmistakeable. But it would be superfluous here fully to recapitulate the line of argument given in detail in my 'Origin of Species.' The homological construction of the whole frame in the members of the same class is intelligible, if we admit their descent from a common progenitor, together with their subsequent adaptation to diversified conditions. On any other view the similarity of pattern between the hand of a man or monkey, the foot of a horse, the flipper of a seal, the wing of a bat, &c., is utterly inexplicable. It is no scientific explanation to assert that they have all been formed on the same ideal plan. With respect to development, we can clearly understand, on the principle of variations supervening at a rather late embryonic period, and being inherited at a corresponding period, how it is that the embryos of wonderfully different forms should still retain, more or less perfectly, the structure of their common progenitor. No other explanation has ever been given of the marvellous fact that the embryo of a man, dog, seal, bat, reptile, &c., can at first hardly be distinguished from each other. In order to understand the existence of rudimentary organs, we have only to suppose that a former progenitor possessed the parts in question in a perfect state, and that under changed habits of life they became greatly reduced, either from simple disuse, or through the natural selection of those individuals which were least encumbered with a superfluous part, aided by the other means previously indicated.

Thus we can understand how it has come to pass that man and all other vertebrate animals have been constructed on the same general model, why they pass through the same early stages of development, and why they retain certain rudiments in common. Consequently we ought frankly to admit their community of descent: to take any other view, is to admit that our own structure and that of all the animals around us, is a mere snare laid to entrap our judgment. This conclusion is greatly strengthened, if we look to the members of the whole animal series, and consider the evidence derived from their affinities or classification, their geographical distribution and geological succession. It is only our natural prejudice, and that arrogance which made our forefathers declare that they were descended from demi-gods, which leads us to demur to this conclusion. But the time will before long come when it will be thought wonderful, that naturalists, who were well acquainted with the comparative structure and development of man and other mammals, should have believed that each was the work of a separate act of creation.

• • •

BELIEF IN GOD—RELIGION.

There is no evidence that man was aboriginally endowed with the ennobling belief in the existence of an Omnipotent God. On the contrary there is ample evidence, derived not from hasty travellers, but from men who have long resided with savages, that numerous races have existed and still exist, who have no idea of one or more gods, and who have no words in their languages to express such an idea. The question is of course wholly distinct from that higher one, whether there exists a Creator and Ruler of the universe; and this has been answered in the affirmative by the highest intellects that have ever lived.

If, however, we include under the term "religion" the belief in unseen or spiritual agencies, the case is wholly different; for this belief seems to be almost universal with the less civilised races. Nor is it difficult to comprehend how it arose. As soon as the important faculties of the imagination, wonder, and curiosity, together with some power of reasoning, had become partially developed, man would naturally have craved to understand what was passing around him, and have vaguely speculated on his own existence. As Mr. M'Lennan has remarked, "Some explanation of the phenomena of life, a man must feign for himself; and to judge from the universality of it, the simplest hypothesis, and the first to occur to men, seems to have been that natural phenomena are ascribable to the presence in animals, plants, and things, and in the forces of nature, of such spirits prompting to action as men are conscious they themselves possess." It is probable, as Mr. Tylor has clearly shewn, that dreams may have first given rise to the notion of spirits; for savages do not readily distinguish between subjective and objective impressions. When a savage dreams, the figures which appear before him are believed to have come from a distance and to stand over him; or "the soul of the dreamer goes out on its travels, and comes home with a remembrance of what it has seen." But until the above-named faculties of imagination, curiosity, reason, &c., had been fairly well developed in the mind of man, his dreams would not have led him to believe in spirits, any more than in the case of a dog.

The tendency in savages to imagine that natural objects and agencies are animated by spiritual or living essences, is perhaps illustrated by a little fact which I once noticed: my dog, a full-grown and very sensible animal, was lying on the lawn during a hot and still day; but at a little distance a slight breeze occasionally moved an open parasol, which would have been wholly disregarded by the dog, had any one stood near it. As it was, every time that the parasol slightly moved, the dog growled fiercely and barked. He must, I think, have reasoned to himself in a rapid and unconscious manner, that movement without any apparent cause indicated the presence of some strange living agent, and no stranger had a right to be on his territory.

The belief in spiritual agencies would easily pass into the belief in the existence of one or more gods. For savages would naturally attribute to spirits the same passions, the same love of vengeance or simplest form of justice, and the same affections which they themselves experienced. The Fuegians appear to be in this respect in an intermediate condition, for when the surgeon on board the "Beagle" shot some young ducklings as specimens, York Minster declared in the most solemn manner, "Oh! Mr. Bynoe, much rain, much snow, blow much;" and this was evidently a retributive punishment for wasting human food. So again he related how, when his brother killed a "wild man," storms long raged, much rain and snow fell. Yet we could never discover that the Fuegians believed in what we should call a God, or practised any religious rites; and Jemmy Button, with justifiable pride, stoutly maintained that there was no devil in his land. This latter assertion is the more remarkable, as with savages the belief in bad spirits is far more common than the belief in good spirits.

The feeling of religious devotion is a highly complex one, consisting of love, complete submission to an exalted and mysterious superior, a strong sense of dependence, fear, reverence, gratitude, hope for the future, and perhaps other elements. No being could experience so complex an emotion until advanced in his intellectual and moral faculties to at least a moderately high level. Nevertheless we see some distant approach to this state of mind, in the deep love of a dog for his master, associated with complete submission, some fear, and perhaps other feelings. The behaviour of a dog when returning to his master after an absence, and, as I may add, of a monkey to his beloved keeper, is widely different from that towards their fellows. In the latter case the transports of joy appear to be somewhat less, and the sense of equality is shewn in every action. Professor Braubach goes so far as to maintain that a dog looks on his master as on a god.

The same high mental faculties which first led man to believe in unseen spiritual agencies, then in fetishism, polytheism, and ultimately in monotheism, would infallibly lead him, as long as his reasoning powers remained poorly developed, to various strange superstitions and customs. Many of these are terrible to think of—such as the sacrifice of human beings to a blood-loving god; the trial of innocent persons by the ordeal of poison or fire; witchcraft, &c.—yet it is well occasionally to reflect on these superstitions, for they shew us what an infinite debt of gratitude we owe to the improvement of our reason, to science, and our accumulated knowledge. As Sir J. Lubbock has well observed, "it is not too much to say that the horrible dread of unknown evil hangs like a thick cloud over savage life, and embitters every pleasure." These miserable and indirect consequences of our highest faculties may be compared with the incidental and occasional mistakes of the instincts of the lower animals.

• • •

SUMMARY OF THE TWO LAST CHAPTERS. [ON MENTAL POWERS AND MORAL SENSE]

There can be no doubt that the difference between the mind of the lowest man and that of the highest animal is immense. An anthropomorphous ape, if he could take a dispassionate view of his own case, would admit that though he could form an artful plan to plunder a garden—though he could use stones for fighting or for breaking open nuts, yet that the thought of fashioning a stone into a tool was quite beyond his scope. Still less, as he would admit, could he follow out a train of metaphysical reasoning, or solve a mathematical problem, or reflect on God, or admire a grand natural scene. Some apes, however, would probably declare that they could and did admire the beauty of the coloured skin and fur of their partners in marriage. They would admit, that though they could make other apes understand by cries some of their perceptions and simpler wants, the notion of expressing definite ideas by definite sounds had never crossed their minds. They might insist that they were ready to aid their fellow-apes of the same troop in many ways, to risk their lives for them, and to take charge of their orphans; but they would be forced to acknowledge that disinterested love for all living creatures, the most noble attribute of man, was quite beyond their comprehension.

Nevertheless the difference in mind between man and the higher animals, great as it is, is certainly one of degree and not of kind. We have seen that the senses and intuitions, the various emotions and faculties, such as love, memory, attention, curiosity, imitation, reason, &c., of which man boasts, may be found in an incipient, or even sometimes in a well-developed condition, in the lower animals. They are also capable of some inherited improvement, as we see in the domestic dog compared with the wolf or jackal. If it be maintained that certain powers, such as self-consciousness, abstraction, &c., are peculiar to man, it may well be that these are the incidental results of other highly-advanced intellectual faculties; and these again are mainly the result of the continued use of a highly developed language. At what age does the new-born infant possess the power of abstraction, or become self-conscious and reflect on its own existence? We cannot answer; nor can we answer in regard to the ascending organic scale. The half-art and half-instinct of language still bears the stamp of its gradual evolution. The ennobling belief in God is not universal with man; and the belief in active spiritual agencies naturally follows from his other mental powers. The moral sense perhaps affords the best and highest distinction between man and the lower animals; but I need not say anything on this head, as I have so lately endeavoured to shew that the social instincts,—the prime principle of man's moral constitution—with the aid of active intellectual powers and the effects of habit, naturally lead to the golden rule, "As ye would that men should do to you, do ye to them likewise;" and this lies at the foundation of morality.

In a future chapter I shall make some few remarks on the probable steps and means by which the several mental and moral faculties of man have been gradually evolved. That this at least is possible ought not to be denied, when we daily see their development in every infant;

and when we may trace a perfect gradation from the mind of an utter idiot, lower than that of the lowest animal, to the mind of a Newton.

• • •

ON THE MANNER OF DEVELOPMENT OF MAN FROM THE LOWER FORMS.

I have now endeavoured to shew that some of the most distinctive characters of man have in all probability been acquired, either directly, or more commonly indirectly, through natural selection. We should bear in mind that modifications in structure or constitution, which are of no service to an organism in adapting it to its habits of life, to the food which it consumes, or passively to the surrounding conditions, cannot have been thus acquired. We must not, however, be too confident in deciding what modifications are of service to each being: we should remember how little we know about the use of many parts, or what changes in the blood or tissues may serve to fit an organism for a new climate or some new kind of food. Nor must we forget the principle of correlation, by which, as Isidore Geoffroy has shewn in the case of man, many strange deviations of structure are tied together. Independently of correlation, a change in one part often leads through the increased or decreased use of other parts, to other changes of a quite unexpected nature. It is also well to reflect on such facts, as the wonderful growth of galls on plants caused by the poison of an insect, and on the remarkable changes of colour in the plumage of parrots when fed on certain fishes, or inoculated with the poison of toads; for we can thus see that the fluids of the system, if altered for some special purpose, might induce other strange changes. We should especially bear in mind that modifications acquired and continually used during past ages for some useful purpose would probably become firmly fixed and might be long inherited.

Thus a very large yet undefined extension may safely be given to the direct and indirect results of natural selection; but I now admit, after reading the essay by Nägeli on plants, and the remarks by various authors with respect to animals, more especially those recently made by Professor Broca, that in the earlier editions of my 'Origin of Species'

I probably attributed too much to the action of natural selection or the survival of the fittest. I have altered the fifth edition of the Origin so as to confine my remarks to adaptive changes of structure. I had not formerly sufficiently considered the existence of many structures which appear to be, as far as we can judge, neither beneficial nor injurious; and this I believe to be one of the greatest oversights as yet detected in my work. I may be permitted to say as some excuse, that I had two distinct objects in view, firstly, to shew that species had not been separately created, and secondly, that natural selection had been the chief agent of change, though largely aided by the inherited effects of habit, and slightly by the direct action of the surrounding conditions. Nevertheless I was not able to annul the influence of my former belief, then widely prevalent, that each species had been purposely created; and this led to my tacitly assuming that every detail of structure, excepting rudiments, was of some special, though unrecognised, service. Any one with this assumption in his mind would naturally extend the action of natural selection, either during past or present times, too far. Some of those who admit the principle of evolution, but reject natural selection, seem to forget, when criticising my book, that I had the above two objects in view; hence if I have erred in giving to natural selection great power, which I am far from admitting, or in having exaggerated its power, which is in itself probable, I have at least, as I hope, done good service in aiding to overthrow the dogma of separate creations.

That all organic beings, including man, present many modifications of structure which are of no service to them at present, nor have been formerly, is, as I can now see, probable. We know not what produces the numberless slight differences between the individuals of each species, for reversion only carries the problem a few steps backwards; but each peculiarity must have had its own efficient cause. If these causes, whatever they may be, were to act more uniformly and energetically during a lengthened period (and no reason can be assigned why this should not sometimes occur), the result would probably be not mere slight individual differences, but well-marked, constant modifications. Modifications which are in no way beneficial cannot have been kept uniform through natural selection, though any which were injurious would have been thus eliminated. Uniformity of character would, however, naturally follow from the assumed uniformity of the exciting causes, and likewise from the free intercrossing of many individuals. The same organism might acquire in this manner during successive periods successive modifications, and these would be transmitted in a nearly uniform state as long as the exciting causes remained the same and there was free intercrossing. With respect to the exciting causes we can only say, as when speaking of so-called spontaneous variations, that they relate much more closely to the constitution of the varying organism, than to the nature of the conditions to which it has been subjected.

• • •

ON THE AFFINITIES AND GENEALOGY OF MAN.

Conclusion. The best definition of advancement or progress in the organic scale ever given, is that by Von Baer; and this rests on the amount of differentiation and specialisation of the several parts of the same being, when arrived, as I should be inclined to add, at maturity. Now as organisms have become slowly adapted by means of natural selection for diversified lines of life, their parts will have become, from the advantage gained by the division of physiological labour, more and more differentiated and specialised for various functions. The same part appears often to have been modified first for one purpose, and then long afterwards for some other and quite distinct purpose; and thus all the parts are rendered more and more complex. But each organism will still retain the general type of structure of the progenitor from which it was aboriginally derived. In accordance with this view it seems, if we turn to geological evidence, that organisation on the whole has advanced throughout the world by slow and interrupted steps. In the great kingdom of the Vertebrata it has culminated in man. It must not, however, be supposed that groups of organic beings are always supplanted and disappear as soon as they have given birth to other and more perfect groups. The latter, though victorious over their predecessors, may not have become better adapted for all places in the economy of nature. Some old forms appear to have survived from inhabiting protected sites, where they have not been exposed to very severe competition; and these often aid us in constructing our genealogies, by giving us a fair idea of former and lost populations. But we must not fall into the error of looking at the existing members of any lowly-organised group as perfect representatives of their ancient predecessors.

The most ancient progenitors in the kingdom of the Vertebrata, at which we are able to obtain an obscure glance, apparently consisted of a group of marine animals, resembling the larvæ of existing Ascidians. These animals probably gave rise to a group of fishes, as lowly organised as the lancelet; and from these the Ganoids, and other fishes like the Lepidosiren, must have been developed. From such fish a very small advance would carry us on to the amphibians. We have seen that birds and reptiles were once intimately connected together; and the Monotremata now, in a slight degree, connect mammals with reptiles. But no one can at present say by what line of descent the three higher and related classes, namely, mammals, birds, and reptiles, were derived from either of the two lower vertebrate classes, namely amphibians and fishes. In the class of mammals the steps are not difficult to conceive which led from the ancient Monotremata to the ancient Marsupials; and from these to the early progenitors of the placental mammals. We may thus ascend to the Lemuridæ; and the interval is not wide from these to the Simiadæ. The Simiadæ then branched off into two great stems, the New World and Old World monkeys; and from the latter, at a remote period, Man, the wonder and glory of the Universe, proceeded.

Thus we have given to man a pedigree of prodigious length, but not, it may be said, of noble quality. The world, it has often been remarked, appears as if it had long been preparing for the advent of man; and this, in one sense is strictly true, for he owes his birth to a long line of progenitors. If any single link in this chain had never existed, man would not have been exactly what he now is. Unless we wilfully close our eyes, we may, with our present knowledge, approximately recognise our parentage; nor need we feel ashamed of it. The most humble organism is something much higher than the inorganic dust under our feet; and no one with an unbiassed mind can study any living creature, however humble, without being struck with enthusiasm at its marvellous structure and properties.

• • •

ON THE RACES OF MAN.

The question whether mankind consists of one or several species has of late years been much agitated by anthropologists, who are divided into two schools of monogenists and polygenists. Those who do not admit the principle of evolution, must look at species either as separate creations or as in some manner distinct entities; and they must decide what forms to rank as species by the analogy of other organic beings which are commonly thus received. But it is a hopeless endeavour to decide this point on sound grounds, until some definition of the term "species" is generally accepted; and the definition must not include an element which cannot possibly be ascertained, such as an act of creation. We might as well attempt without any definition to decide whether a certain number of houses should be called a village, or town, or city. We have a practical illustration of the difficulty in the never-ending doubts whether many closely-allied mammals, birds, insects, and plants, which represent each other in North America and Europe, should be ranked species or geographical races; and so it is with the productions of many islands situated at some little distance from the nearest continent.

Those naturalists, on the other hand, who admit the principle of evolution, and this is now admitted by the greater number of rising men, will feel no doubt that all the races of man are descended from a single primitive stock; whether or not they think fit to designate them as distinct species, for the sake of expressing their amount of difference. With our domestic animals the question whether the various races have arisen from one or more species is different. Although all such races, as well as all the natural species within the same genus, have undoubtedly sprung from the same primitive stock, yet it is a fit subject for discussion, whether, for instance, all the domestic races of the dog have acquired their present differences since some one species was first domesticated and bred by man; or whether they owe some of their characters to inheritance from distinct species, which had already been modified in a state of nature. With mankind no such question can arise, for he cannot be said to have been domesticated at any particular period.

When the races of man diverged at an extremely remote epoch from their common progenitor, they will have differed but little from each other, and been few in number; consequently they will then, as far as their distinguishing characters are concerned, have had less claim to rank as distinct species, than the existing so-called races. Nevertheless such early races would perhaps have been ranked by some naturalists as distinct species, so arbitrary is the term, if their differences, although extremely slight, had been more constant than at present, and had not graduated into each other.

It is, however, possible, though far from probable, that the early progenitors of man might at first have diverged much in character, until they became more unlike each other than are any existing races; but that subsequently, as suggested by Vogt, they converged in character. When man selects for the same object the offspring of two distinct species, he sometimes induces, as far as general appearance is concerned, a considerable amount of convergence. This is the case, as shewn by Von Nathusius, with the improved breeds of pigs, which are descended from two distinct species; and in a less well-marked manner with the improved breeds of cattle. A great anatomist, Gratiolet, maintains that the anthropomorphous apes do not form a natural sub-group; but that the orang is a highly developed gibbon or semnopithecus; the chimpanzee a highly developed macacus; and the gorilla a highly developed mandrill. If this conclusion; which rests almost exclusively on brain-characters, be admitted, we should have a case of convergence at least in external characters, for the anthropomorphous apes are certainly more like each other in many points than they are to other apes. All analogical resemblances, as of a whale to a fish, may indeed be said to be cases of convergence; but this term has never been applied to superficial and adaptive resemblances. It would be extremely rash in most cases to attribute to convergence close similarity in many points of structure in beings which had once been widely different. The form of a crystal is determined solely by the molecular forces, and it is not surprising that dissimilar substances should sometimes assume the same form; but with organic beings we should bear in mind that the form of each depends on an infinitude of complex relations, namely on the variations which have arisen, these being due to causes far too intricate to be followed out,—on the nature of the variations which have been preserved, and this depends on the surrounding physical conditions, and in a still higher degree on the surrounding organisms with which each has come into competition, —and lastly, on inheritance (in itself a fluctuating element) from innumerable progenitors, all of which have had their forms determined through equally complex relations. It appears utterly incredible that two organisms, if differing in a marked manner, should ever afterwards converge so closely as to lead to a near approach to identity throughout their whole organisation. In the case of the convergent pigs above referred to, evidence of their descent from two primitive stocks is still plainly retained, according to Von Nathusius, in certain bones of their skulls. If the races of man were descended, as supposed by some naturalists, from two or more distinct species, which had differed as much, or nearly as much, from each other, as the orang differs from the gorilla, it can hardly be doubted that marked differences in the structure of certain bones would still have been discoverable in man as he now exists.

Although the existing races of man differ in many respects, as in colour, hair, shape of skull, proportions of the body, &c., yet if their whole organisation be taken into consideration they are found to resemble each other closely in a multitude of points. Many of these points are of so unimportant or of so singular a nature, that it is extremely improbable that they should have been independently acquired by aboriginally distinct species or races. The same remark holds good with equal or greater force with respect to the numerous points of mental similarity between the most distinct races of man. The American aborigines, Negroes and Europeans differ as much from each other in mind as any three races that can be named; yet I was incessantly struck, whilst living with the Fuegians on board the "Beagle," with the many little traits of character, shewing how similar their minds were to ours; and so it was with a full-blooded negro with whom I happened once to be intimate.

He who will carefully read Mr. Tylor's and Sir J. Lubbock's interesting works can hardly fail to be deeply impressed with the close similarity between the men of all races in tastes, dispositions and habits. This is shewn by the pleasure which they all take in dancing, rude music, acting, painting, tattooing, and otherwise decorating themselves,—in their mutual comprehension of gesture-language—and, as I shall be able to shew in a future essay, by the same expression in their features, and by the same inarticulate cries, when they are excited by various emotions. This similarity, or rather identity, is striking, when contrasted with the different expressions which may be observed in distinct species of monkeys. There is good evidence that the art of shooting with bows and arrows has not been handed down from any common progenitor of mankind, yet the stone arrow-heads, brought from the most distant parts of the world and manufactured at the most remote periods, are, as Nilsson has shewn, almost identical; and this fact can only be accounted for by the various races having similar inventive or mental powers. The same observation has been made by archæologists with respect to certain widely-prevalent ornaments, such as zigzags, &c.; and with respect to various simple beliefs and customs, such as the burying of the dead under megalithic structures. I remember observing in South America that there, as in so many other parts of the world, man has generally chosen the summits of lofty hills, on which to throw up piles of stones, either for the sake of recording some remarkable event, or for burying his dead.

Now when naturalists observe a close agreement in numerous small details of habits, tastes and dispositions between two or more domestic races, or between nearly-allied natural forms, they use this fact as an argument that all are descended from a common progenitor who was thus endowed; and consequently that all should be classed under the same species. The same argument may be applied with much force to the races of man.

As it is improbable that the numerous and unimportant points of resemblance between the several races of man in bodily structure and mental faculties (I do not here refer to similar customs) should all have been independently acquired, they must have been inherited from progenitors who were thus characterised. We thus gain some insight into the early state of man, before he had spread step by step over the face of the earth. The spreading of man to regions widely separated by the sea, no doubt, preceded any considerable amount of divergence of character in the several races; for otherwise we should sometimes meet with the same race in distinct continents; and this is never the case. Sir J. Lubbock, after comparing the arts now practised by savages in all parts of the world, specifies those which man could not have known, when he first wandered from his original birth-place; for if once learnt they would never have been forgotten. He thus shews that "the spear, which is but a development of the knife-point, and the club, which is but a long hammer, are the only things left." He admits, however, that the art of making fire probably had already been discovered, for it is common to all the races now existing, and was known to the ancient cave-inhabitants of Europe. Perhaps the art of making rude canoes or rafts was likewise known; but as man existed at a remote epoch, when the land in many places stood at a very different level, he would have been able, without the aid of canoes, to have spread widely. Sir J. Lubbock further remarks how improbable it is that our earliest ancestors could have "counted as high as ten, considering that so many races now in existence cannot get beyond four." Nevertheless, at this early period, the intellectual and social faculties of man could hardly have been inferior in any extreme degree to those now possessed by the lowest savages; otherwise primeval man could not have been so eminently successful in the struggle for life, as proved by his early and wide diffusion.

From the fundamental differences between certain languages, some philologists have inferred that when man first became widely diffused he was not a speaking animal; but it may be suspected that languages, far less perfect than any now spoken, aided by gestures, might have been used, and yet have left no traces on subsequent and more highly-developed tongues. Without the use of some language, however imperfect, it appears doubtful whether man's intellect could have risen to the standard implied by his dominant position at an early period.

Whether primeval man, when he possessed very few arts of the rudest kind, and when his power of language was extremely imperfect, would have deserved to be called man, must depend on the definition which we employ. In a series of forms graduating insensibly from some ape-like creature to man as he now exists, it would be impossible to fix on any definite point when the term "man" ought to be used. But this is a matter of very little importance. So again it is almost a matter of indifference whether the so-called races of man are thus designated, or are ranked as species or sub-species; but the latter term appears the most appropriate. Finally, we may conclude that when the principles of evolution are generally accepted, as they surely will be before long, the dispute between the monogenists and the polygenists will die a silent and unobserved death.

• • •

We have now seen that the characteristic differences between the races of man cannot be accounted for in a satisfactory manner by the direct action of the conditions of life, nor by the effects of the continued use of parts, nor through the principle of correlation. We are therefore led to inquire whether slight individual differences, to which man is eminently liable, may not have been preserved and augmented during a long series of generations through natural selection. But here we are at once met by the objection that beneficial variations alone can be thus preserved; and as far as we are enabled to judge (although always liable to error on this head) not one of the external differences between the races of man are of any direct or special service to him. The intellectual and moral or social faculties must of course be excepted from this remark; but differences in these faculties can have had little or no influence on external characters. The variability of all the characteristic differences between the races, before referred to, likewise indicates that these differences cannot be of much importance; for, had they been important, they would long ago have been either fixed and preserved, or eliminated. In this respect man resembles those forms, called by naturalists protean or polymorphic, which have remained extremely variable, owing, as it seems, to their variations being of an indifferent nature, and consequently to their having escaped the action of natural selection.

We have thus far been baffled in all our attempts to account for the differences between the races of man; but there remains one important agency, namely Sexual Selection, which appears to have acted as powerfully on man, as on many other animals. I do not intend to assert that sexual selection will account for all the differences between the races. An unexplained residuum is left, about which we can in our ignorance only say, that as individuals are continually born with, for instance, heads a little rounder or narrower, and with noses a little longer or shorter, such slight differences might become fixed and uniform, if the unknown agencies which induced them were to act in a more constant manner, aided by long-continued intercrossing. Such modifications come under the provisional class, alluded to in our fourth chapter, which for the want of a better term have been called spontaneous variations. Nor do I pretend that the effects of sexual selection can be indicated with scientific precision; but it can be shewn that it would be an inexplicable fact if man had not been modified by this agency, which has acted so powerfully on innumerable animals, both high and low in the scale. It can further be shewn that the differences between the races of man, as in colour, hairiness, form of features, &c., are of the nature which it might have been expected would have been acted on by sexual selection. But in order to treat this subject in a fitting manner, I have found it necessary to pass the whole animal kingdom in review; I have therefore devoted to it the Second Part of this work. At the close I shall return to man, and, after attempting to shew how far he has been modified through sexual selection, will give a brief summary of the chapters in this First Part.

PRINCIPLES OF SEXUAL SELECTION.

With animals which have their sexes separated, the males necessarily differ from the females in their organs of reproduction; and these afford the primary sexual characters. But the sexes often differ in what Hunter has called secondary sexual characters, which are not directly connected with the act of reproduction; for instance, in the male possessing certain organs of sense or locomotion, of which the female is quite destitute, or in having them more highly-developed, in order that he may readily find or reach her; or again, in the male having special organs of prehension so as to hold her securely. These latter organs of infinitely diversified kinds graduate into, and in some cases can hardly be distinguished from, those which are commonly ranked as primary, such as the complex appendages at the apex of the abdomen in male insects. Unless indeed we confine the term "primary" to the reproductive glands, it is scarcely possible to decide, as far as the organs of prehension are concerned, which ought to be called primary and which secondary.

The female often differs from the male in having organs for the nourishment or protection of her young, as the mammary glands of mammals, and the abdominal sacks of the marsupials. The male, also, in some few cases differs from the female in possessing analogous organs, as the receptacles for the ova possessed by the males of certain fishes, and those temporarily developed in certain male frogs. Female bees have a special apparatus for collecting and carrying pollen, and their ovipositor is modified into a sting for the defence of their larvæ and the community. In the females of many insects the ovipositor is modified in the most complex manner for the safe placing of the eggs. Numerous similar cases could be given, but they do not here concern us. There are, however, other sexual differences quite disconnected with the primary organs with which we are more especially concerned—such as the greater size, strength, and pugnacity of the male, his weapons of offence or means of defence against rivals, his gaudy colouring and various ornaments, his power of song, and other such characters.

Besides the foregoing primary and secondary sexual differences, the male and female sometimes differ in structures connected with different habits of life, and not at all, or only indirectly, related to the reproductive functions. Thus the females of certain flies (Culicidæ and Tabanidæ) are blood-suckers, whilst the males live on flowers and have their mouths destitute of mandibles. The males alone of certain moths and of some crustaceans (*e.g.* Tanais) have imperfect, closed mouths, and cannot feed. The Complemental males of certain cirripedes live like epiphytic plants either on the female or hermaphrodite form, and are destitute of a mouth and prehensile limbs. In these cases it is the male which has been modified and has lost certain important organs, which the other members of the same group possess. In other cases it is the female which has lost such parts; for instance, the female glow-worm is destitute of wings, as are many female moths, some of which never leave their cocoons. Many female parasitic crustaceans have lost their natatory legs. In some weevil-beetles (Curculionidæ) there is a great difference between the male and female in the length of the rostrum or snout; but the meaning of this and of many analogous differences, is not at all understood. Differences of structure between the two sexes in relation to different habits of life are generally confined to the lower animals; but with some few birds the beak of the male differs from that of the female. No doubt in most, but apparently not in all these cases, the differences are indirectly connected with the propagation of the species: thus a female which has to nourish a multitude of ova will require more food than the male, and consequently will require special means for procuring it. A male animal which lived for a very short time might without detriment lose through disuse its organs for procuring food; but he would retain his locomotive organs in a perfect state, so that he might reach the female. The female, on the other hand, might safely lose her organs for flying, swimming, or walking, if she gradually acquired habits which rendered such powers useless.

We are, however, here concerned only with that kind of selection, which I have called sexual selection. This depends on the advantage which certain individuals have over other individuals of the same sex and species, in exclusive relation to reproduction. When the two sexes differ in structure in relation to different habits of life, as in the cases above mentioned, they have no doubt been modified through natural selection, accompanied by inheritance limited to one and the same sex. So again the primary sexual organs, and those for nourishing or protecting the young, come under this same head; for those individuals which generated or nourished their offspring best, would leave, *cæteris paribus*, the greatest number to inherit their superiority; whilst those which generated or nourished their offspring badly, would leave but few to inherit their weaker powers. As the male has to search for the female, he requires for this purpose organs of sense and locomotion, but if these organs are necessary for the other purposes of life, as is generally the case, they will have been developed through natural selection. When the male has found the female he sometimes absolutely requires prehensile organs to hold her; thus Dr. Wallace informs me that the males of certain moths cannot unite with the females if their tarsi or feet are broken. The males of many oceanic crustaceans have their legs and antennæ modified in an extraordinary manner for the prehension of the female; hence we may suspect that owing to these animals being washed about by the waves of the open sea, they absolutely require these organs in order to propagate their kind, and if so their development will have been the result of ordinary or natural selection.

When the two sexes follow exactly the same habits of life, and the male has more highly developed sense or locomotive organs than the female, it may be that these in their perfected state are indispensable to the male for finding the female; but in the vast majority of cases, they serve only to give one male an advantage over another, for the less well-endowed males, if time were allowed them, would succeed in pairing with the females; and they would in all other respects, judging from the structure of the female, be equally well adapted for their ordinary habits of life. In such cases sexual selection must have come into action, for the males have acquired their present structure, not from being better fitted to survive in the struggle for existence, but from having gained an advantage over other males, and from having transmitted this advantage to their male offspring alone. It was the importance of this distinction which led me to designate this form of selection as sexual selection. So again, if the chief service rendered to the male by his prehensile organs is to prevent the escape of the female before the arrival of other males, or when assaulted by them, these organs will have been perfected through sexual selection, that is by the advantage acquired by certain males over their rivals. But in most cases it is scarcely possible to distinguish between the effects of natural and sexual selection. Whole chapters could easily be filled with details on the differences between the sexes in their sensory, locomotive, and prehensile organs. As, however, these structures are not more interesting than others adapted for the ordinary purposes of life, I shall almost pass them over, giving only a few instances under each class.

There are many other structures and instincts which must have been developed through sexual selection—such as the weapons of offence and the means of defence possessed by the males for fighting with and driving away their rivals—their courage and pugnacity—their ornaments of many kinds—their organs for producing vocal or instrumental music—and their glands for emitting odours; most of these latter structures serving only to allure or excite the female. That these characters are the result of sexual and not of ordinary selection is clear, as unarmed, unornamented, or unattractive males would succeed equally well in the battle for life and in leaving a numerous progeny, if better endowed males were not present. We may infer that this would be the case, for the females, which are unarmed and unornamented, are able to survive and procreate their kind. Secondary sexual characters of the kind just referred to, will be fully discussed in the following chapters, as they are in many respects interesting, but more especially as they depend on the will, choice, and rivalry of the individuals of either sex. When we behold two males fighting for the possession of the female, or several male birds displaying their gorgeous plumage, and performing the strangest antics before an assembled body of females, we cannot doubt that, though led by instinct, they know what they are about, and consciously exert their mental and bodily powers.

In the same manner as man can improve the breed of his game-cocks by the selection of those birds which are victorious in the cock-pit, so it appears that the strongest and most vigorous males, or those provided with the best weapons, have prevailed under nature, and have led to the improvement of the natural breed or species. Through repeated deadly contests, a slight degree of variability, if it led to some advantage, however slight, would suffice for the work of sexual selection; and it is certain that secondary sexual characters are eminently variable. In the same manner as man can give beauty, according to his standard of taste, to his male poultry—can give to the Sebright bantam a new and elegant plumage, an erect and peculiar carriage—so it appears that in a state of nature female birds, by having long selected the more attractive males, have added to their beauty. No doubt this implies powers of discrimination and taste on the part of the female which will at first appear extremely improbable; but I hope hereafter to shew that this is not the case.

From our ignorance on several points, the precise manner in which sexual selection acts is to a certain extent uncertain. Nevertheless if those naturalists who already believe in the mutability of species, will read the following chapters, they will, I think, agree with me that sexual selection has played an important part in the history of the organic world. It is certain that with almost all animals there is a struggle between the males for the possession of the female. This fact is so notorious that it would be superfluous to give instances. Hence the females, supposing that their mental capacity sufficed for the exertion of a choice, could select one out of several males. But in numerous cases it appears as if it had been specially arranged that there should be a struggle between many males. Thus with migratory birds, the males generally arrive before the females at their place of breeding, so that many males are ready to contend for each female. The bird-catchers assert that this is invariably the case with the nightingale and blackcap, as I am informed by Mr. Jenner Weir, who confirms the statement with respect to the latter species.

Mr. Swaysland of Brighton, who has been in the habit, during the last forty years, of catching our migratory birds on their first arrival, writes to me that he has never known the females of any species to arrive before their males. During one spring he shot thirty-nine males of Ray's wagtail (*Budytes Raii*) before he saw a single female. Mr. Gould has ascertained by dissection, as he informs me, that male snipes arrive in this country before the females. In the case of fish, at the period when the salmon ascend our rivers, the males in large numbers are ready to breed before the females. So it apparently is with frogs and toads. Throughout the great class of insects the males almost always emerge from the pupal state before the other sex, so that they generally swarm for a time before any females can be seen. The cause of this difference between the males and females in their periods of arrival and maturity is sufficiently obvious. Those males which annually first migrated into any country, or which in the spring were first ready to breed, or were the most eager, would leave the largest number of offspring; and these would tend to inherit similar instincts and constitutions. On the whole there can be no doubt that with almost all animals, in which the sexes are separate, there is a constantly recurrent struggle between the males for the possession of the females.

Our difficulty in regard to sexual selection lies in understanding how it is that the males which conquer other males, or those which prove the most attractive to the females, leave a greater number of offspring to inherit their superiority than the beaten and less attractive males. Unless this result followed, the characters which gave to certain males an advantage over others, could not be perfected and augmented through sexual selection. When the sexes exist in exactly equal numbers, the worst-endowed males will ultimately find females (excepting where polygamy prevails), and leave as many offspring, equally well fitted for their general habits of life, as the best-endowed males. From various facts and considerations, I formerly inferred that with most animals, in which secondary sexual characters were well developed, the males considerably exceeded the females in number; and this does hold good in some few cases. If the males were to the females as two to one, or as three to two, or even in a somewhat lower ratio, the whole affair would be simple; for the better-armed or more attractive males would leave the largest number of offspring. But after investigating, as far as possible, the numerical proportions of the sexes, I do not believe that any great inequality in number commonly exists. In most cases sexual selection appears to have been effective in the following manner.

Let us take any species, a bird for instance, and divide the females inhabiting a district into two equal bodies: the one consisting of the more vigorous and better-nourished individuals, and the other of the less vigorous and healthy. The former, there can be little doubt, would be ready to breed in the spring before the others; and this is the opinion of Mr. Jenner Weir, who has during many years carefully attended to the habits of birds. There can also be no doubt that the most vigorous, healthy, and best-nourished females would on an average succeed in rearing the largest number of offspring. The males, as we have seen, are generally ready to breed before the females; of the males the strongest, and with some species the best armed, drive away the weaker males; and the former would then unite with the more vigorous and best-nourished females, as these are the first to breed. Such vigorous pairs would surely rear a larger number of offspring than the retarded females, which would be compelled, supposing the sexes to be numerically equal, to unite with the conquered and less powerful males; and this is all that is wanted to add, in the course of successive generations, to the size, strength, and courage of the males, or to improve their weapons.

But in a multitude of cases the males which conquer other males, do not obtain possession of the females, independently of choice on the part of the latter. The courtship of animals is by no means so simple and short an affair as might be thought. The females are most excited by, or prefer pairing with, the more ornamented males, or those which are the best songsters, or play the best antics; but it is obviously probable, as has been actually observed in some cases, that they would at the same time prefer the more vigorous and lively males. Thus the more vigorous females, which are the first to breed, will have the choice of many males; and though they may not always select the strongest or best armed, they will select those which are vigorous and well armed, and in other respects the most attractive. Such early pairs would have the same advantage in rearing offspring on the female side as above explained, and nearly the same advantage on the male side. And this apparently has sufficed during a long course of generations to add not only to the strength and fighting-powers of the males, but likewise to their various ornaments or other attractions.

In the converse and much rarer case of the males selecting particular females, it is plain that those which were the most vigorous and had conquered others, would have the freest choice; and it is almost certain that they would select vigorous as well as attractive females. Such pairs would have an advantage in rearing offspring, more especially if the male had the power to defend the female during the pairing-season, as occurs with some of the higher animals, or aided in providing for the young. The same principles would apply if both sexes mutually preferred and selected certain individuals of the opposite sex; supposing that they selected not only the more attractive, but likewise the more vigorous individuals.

• • •

GENERAL SUMMARY AND CONCLUSION.

A brief summary will here be sufficient to recall to the reader's mind the more salient points in this work. Many of the views which have been advanced are highly speculative, and some no doubt will prove erroneous; but I have in every case given the reasons which have led me to one view rather than to another. It seemed worth while to try how far the principle of evolution would throw light on some of the more complex problems in the natural history of man. False facts are highly injurious to the progress of science, for they often long endure; but false views, if supported by some evidence, do little harm, as every one takes a salutary pleasure in proving their falseness; and when this is done, one path towards error is closed and the road to truth is often at the same time opened.

The main conclusion arrived at in this work, and now held by many naturalists who are well competent to form a sound judgment, is that man is descended from some less highly organised form. The grounds upon which this conclusion rests will never be shaken, for the close similarity between man and the lower animals in embryonic development, as well as in innumerable points of structure and constitution, both of high and of the most trifling importance,—the rudiments which he retains, and the abnormal reversions to which he is occasionally liable,—are facts which cannot be disputed. They have long been known, but until recently they told us nothing with respect to the origin of man. Now when viewed by the light of our knowledge of the whole organic world, their meaning is unmistakeable. The great principle of evolution stands up clear and firm, when these groups of facts are considered in connection with others, such as the mutual affinities of the members of the same group, their geographical distribution in past and present times, and their geological succession. It is incredible that all these facts should speak falsely. He who is not content to look, like a savage, at the phenomena of nature as disconnected, cannot any longer believe that man is the work of a separate act of creation. He will be forced to admit that the close resemblance of the embryo of man to that, for instance, of a dog—the construction of his skull, limbs, and whole frame, independently of the uses to which the parts may be put, on the same plan with that of other mammals—the occasional reappearance of various structures, for instance of several distinct muscles, which man does not normally possess, but which are common to the Quadrumana—and a crowd of analogous facts—all point in the plainest manner to the conclusion that man is the co-descendant with other mammals of a common progenitor.

We have seen that man incessantly presents individual differences in all parts of his body and in his mental faculties. These differences or variations seem to be induced by the same general causes, and to obey the same laws as with the lower animals. In both cases similar laws of inheritance prevail. Man tends to increase at a greater rate than his means of subsistence; consequently he is occasionally subjected to a severe struggle for existence, and natural selection will have effected whatever lies within its scope. A succession of strongly-marked variations of a similar nature are by no means requisite; slight fluctuating differences in the individual suffice for the work of natural selection. We may feel assured that the inherited effects of the long-continued use or disuse of parts will have done much in the same direction with natural selection. Modifications formerly of importance, though no longer of any special use, will be long inherited. When one part is modified, other parts will change through the principle of correlation, of which we have instances in many curious cases of correlated monstrosities. Something may be attributed to the direct and definite action of the surrounding conditions of life, such as abundant food, heat, or moisture; and lastly, many characters of slight physiological importance, some indeed of considerable importance, have been gained through sexual selection.

No doubt man, as well as every other animal, presents structures, which as far as we can judge with our little knowledge, are not now of any service to him, nor have been so during any former period of his existence, either in relation to his general conditions of life, or of one sex to the other. Such structures cannot be accounted for by any form of selection, or by the inherited effects of the use and disuse of parts. We know, however, that many strange and strongly-marked peculiarities of structure occasionally appear in our domesticated productions, and if the unknown causes which produce them were to act more uniformly, they would probably become common to all the individuals of the species. We may hope hereafter to understand something about the causes of such occasional modifications, especially through the study of monstrosities: hence the labours of experimentalists, such as those of M. Camille Dareste, are full of promise for the future. In the greater number of cases we can only say that the cause of each slight variation and of each monstrosity lies much more in the nature or constitution of the organism, than in the nature of the surrounding conditions; though new and changed conditions certainly play an important part in exciting organic changes of all kinds.

Through the means just specified, aided perhaps by others as yet undiscovered, man has been raised to his present state. But since he attained to the rank of manhood, he has diverged into distinct races, or as they may be more appropriately called sub-species. Some of these, for instance the Negro and European, are so distinct that, if specimens had been brought to a naturalist without any further information, they would undoubtedly have been considered by him as good and true species. Nevertheless all the races agree in so many unimportant details of structure and in so many mental peculiarities, that these can be accounted for only through inheritance from a common progenitor; and a progenitor thus characterised would probably have deserved to rank as man.

It must not be supposed that the divergence of each race from the other races, and of all the races from a common stock, can be traced back to any one pair of progenitors. On the contrary, at every stage in the process of modification, all the individuals which were in any way best fitted for their conditions of life, though in different degrees, would have survived in greater numbers than the less well fitted. The process would have been like that followed by man, when he does not intentionally select particular individuals, but breeds from all the superior and neglects all the inferior individuals. He thus slowly but surely modifies his stock, and unconsciously forms a new strain. So with respect to modifications, acquired independently of selection, and due to variations arising from the nature of the organism and the action of the surrounding conditions, or from changed habits of life, no single pair will have been modified in a much greater degree than the other pairs which inhabit the same country, for all will have been continually blended through free intercrossing.

By considering the embryological structure of man—the homologies which he presents with the lower animals,—the rudiments which he retains, and the reversions to which he is liable,—we can partly recall in imagination the former condition of our early progenitors; and can approximately place them in their proper position in the zoological series. We thus learn that man is descended from a hairy quadruped, furnished with a tail and pointed ears, probably arboreal in its habits, and an inhabitant of the Old World. This creature, if its whole structure had been examined by a naturalist, would have been classed amongst the Quadrumana, as surely as would the common and still more ancient progenitor of the Old and New World monkeys. The Quadrumana and all the higher mammals are probably derived from an ancient marsupial animal, and this through a long line of diversified forms, either from some reptile-like or some amphibian-like creature, and this again from some fish-like animal. In the dim obscurity of the past we can see that the early progenitor of all the Vertebrata must have been an aquatic animal, provided with branchiæ, with the two sexes united in the same individual, and with the most important organs of the body (such as the brain and heart) imperfectly developed. This animal seems to have been more like the larvæ of our existing marine Ascidians than any other known form.

The greatest difficulty which presents itself, when we are driven to the above conclusion on the origin of man, is the high standard of intellectual power and of moral disposition which he has attained. But every one who admits the general principle of evolution, must see that the mental powers of the higher animals, which are the same in kind with those of mankind, though so different in degree, are capable of advancement. Thus the interval between the mental powers of one of the higher apes and of a fish, or between those of an ant and scale-insect, is immense. The development of these powers in animals does not offer any special difficulty; for with our domesticated animals, the mental faculties are certainly variable, and the variations are inherited. No one doubts that these faculties are of the utmost importance to animals in a state of nature. Therefore the conditions are favourable for their development through natural selection. The same conclusion may be extended to man; the intellect must have been all-important to him, even at a very remote period, enabling him to use language, to invent and make weapons, tools, traps, &c.; by which means, in combination with his social habits, he long ago became the most dominant of all living creatures.

A great stride in the development of the intellect will have followed, as soon as, through a previous considerable advance, the half-art and half-instinct of language came into use; for the continued use of language will have reacted on the brain, and produced an inherited effect; and this again will have reacted on the improvement of language. The large size of the brain in man, in comparison with that of the lower animals, relatively to the size of their bodies, may be attributed in chief part, as Mr. Chauncey Wright has well remarked, to the early use of some simple form of language,—that wonderful engine which affixes signs to all sorts of objects and qualities, and excites trains of thought which would never arise from the mere impression of the senses, and if they did arise could not be followed out. The higher intellectual powers of man, such as those of ratiocination, abstraction, self-consciousness, &c., will have followed from the continued improvement of other mental faculties; but without considerable culture of the mind, both in the race and in the individual, it is doubtful whether these high powers would be exercised, and thus fully attained.

The development of the moral qualities is a more interesting and difficult problem. Their foundation lies in the social instincts, including in this term the family ties. These instincts are of a highly complex nature, and in the case of the lower animals give special tendencies towards certain definite actions; but the more important elements for us are love, and the distinct emotion of sympathy. Animals endowed with the social instincts take pleasure in each other's company, warn each other of danger, defend and aid each other in many ways. These instincts are not extended to all the individuals of the species, but only to those of the same community. As they are highly beneficial to the species, they have in all probability been acquired through natural selection.

A moral being is one who is capable of comparing his past and future actions and motives,—of approving of some and disapproving of others; and the fact that man is the one being who with certainty can be thus designated makes the greatest of all distinctions between him and the lower animals. But in our third chapter I have endeavoured to show that the moral sense follows, firstly, from the enduring and always present nature of the social instincts, in which respect man agrees with the lower animals; and secondly, from his mental faculties being highly active and his impressions of past events extremely vivid, in which respects he differs from the lower animals. Owing to this condition of mind, man cannot avoid looking backwards and comparing the impressions of past events and actions. He also continually looks forward. Hence after some temporary desire or passion has mastered his social instincts, he will reflect and compare the now weakened impression of such past impulses, with the ever present social instinct; and he will then feel that sense of dissatisfaction which all unsatisfied instincts leave behind them. Consequently he resolves to act differently for the future—and this is conscience. Any instinct which is permanently stronger or more enduring than another, gives rise to a feeling which we express by saying that it ought to be obeyed. A pointer dog, if able to reflect on his past conduct, would say to himself, I ought (as indeed we say of him) to have pointed at that hare and not have yielded to the passing temptation of hunting it.

Social animals are partly impelled by a wish to aid the members of the same community in a general manner, but more commonly to perform certain definite actions. Man is impelled by the same general wish to aid his fellows, but has few or no special instincts. He differs also from the lower animals in being able to express his desires by words, which thus become the guide to the aid required and bestowed. The motive to give aid is likewise somewhat modified in man: it no longer consists solely of a blind instinctive impulse, but is largely influenced by the praise or blame of his fellow men. Both the appreciation and the bestowal of praise and blame rest on sympathy; and this emotion, as we have seen, is one of the most important elements of the social instincts. Sympathy, though gained as an instinct, is also much strengthened by exercise or habit. As all men desire their own happiness, praise or blame is bestowed on actions and motives, according as they lead to this end; and as happiness is an essential part of the general good, the greatest-happiness principle indirectly serves as a nearly safe standard of right and wrong. As the reasoning powers advance and experience is gained, the more remote effects of certain lines of conduct on the character of the individual, and on the general good, are perceived; and then the self-regarding virtues, from coming within the scope of public opinion, receive praise, and their opposites receive blame. But with the less civilised nations reason often errs, and many bad customs and base superstitions come within the same scope, and consequently are esteemed as high virtues, and their breach as heavy crimes.

The moral faculties are generally esteemed, and with justice, as of higher value than the intellectual powers. But we should always bear in mind that the activity of the mind in vividly recalling past impressions is one of the fundamental though secondary bases of conscience. This fact affords the strongest argument for educating and stimulating in all possible ways the intellectual faculties of every human being. No doubt a man with a torpid mind, if his social affections and sympathies are well developed, will be led to good actions, and may have a fairly sensitive conscience. But whatever renders the imagination of men more vivid and strengthens the habit of recalling and comparing past impressions, will make the conscience more sensitive, and may even compensate to a certain extent for weak social affections and sympathies.

The moral nature of man has reached the highest standard as yet attained, partly through the advancement of the reasoning powers and consequently of a just public opinion, but especially through the sympathies being rendered more tender and widely diffused through the effects of habit, example, instruction, and reflection. It is not improbable that virtuous tendencies may through long practice be inherited. With the more civilised races, the conviction of the existence of an all-seeing Deity has had a potent influence on the advancement of morality. Ultimately man no longer accepts the praise or blame of his fellows as his chief guide, though few escape this influence, but his habitual convictions controlled by reason afford him the safest rule. His conscience then becomes his supreme judge and monitor. Nevertheless the first foundation or origin of the moral sense lies in the social instincts, including sympathy; and these instincts no doubt were primarily gained, as in the case of the lower animals, through natural selection.

The belief in God has often been advanced as not only the greatest, but the most complete of all the distinctions between man and the lower animals. It is however impossible, as we have seen, to maintain that this belief is innate or instinctive in man. On the other hand a belief in all-pervading spiritual agencies seems to be universal; and apparently follows from a considerable advance in the reasoning powers of man, and from a still greater advance in his faculties of imagination, curiosity and wonder. I am aware that the assumed instinctive belief in God has been used by many persons as an argument for His existence. But this is a rash argument, as we should thus be compelled to believe in the existence of many cruel and malignant spirits, possessing only a little more power than man; for the belief in them is far more general than of a beneficent Deity. The idea of a universal and beneficent Creator of the universe does not seem to arise in the mind of man, until he has been elevated by long-continued culture.

He who believes in the advancement of man from some lowly-organised form, will naturally ask how does this bear on the belief in the immortality of the soul. The barbarous races of man, as Sir J. Lubbock has shewn, possess no clear belief of this kind; but arguments derived from the primeval beliefs of savages are, as we have just seen, of little or no avail. Few persons feel any anxiety from the impossibility of determining at what precise period in the development of the individual, from the first trace of the minute germinal vesicle to the child either before or after birth, man becomes an immortal being; and there is no greater cause for anxiety because the period in the gradually ascending organic scale cannot possibly be determined.

I am aware that the conclusions arrived at in this work will be denounced by some as highly irreligious; but he who thus denounces them is bound to shew why it is more irreligious to explain the origin of man as a distinct species by descent from some lower form, through the laws of variation and natural selection, than to explain the birth of the individual through the laws of ordinary reproduction. The birth both of the species and of the individual are equally parts of that grand sequence of events, which our minds refuse to accept as the result of blind chance. The understanding revolts at such a conclusion, whether or not we are able to believe that every slight variation of structure,—the union of each pair in marriage,—the dissemination of each seed,—and other such events, have all been ordained for some special purpose.

Sexual selection has been treated at great length in these volumes; for, as I have attempted to shew, it has played an important part in the history of the organic world. As summaries have been given to each chapter, it would be superfluous here to add a detailed summary. I am aware that much remains doubtful, but I have endeavoured to give a fair view of the whole case. In the lower divisions of the animal kingdom, sexual selection seems to have done nothing: such animals are often affixed for life to the same spot, or have the two sexes combined in the same individual, or what is still more important, their perceptive and intellectual faculties are not sufficiently advanced to allow of the feelings of love and jealousy, or of the exertion of choice. When, however, we come to the Arthropoda and Vertebrata, even to the lowest classes in these two great Sub-Kingdoms, sexual selection has effected much; and it deserves notice that we here find the intellectual faculties developed, but in two very distinct lines, to the highest standard, namely in the Hymenoptera (ants, bees, &c.) amongst the Arthropoda, and in the Mammalia, including man, amongst the Vertebrata.

In the most distinct classes of the animal kingdom, with mammals, birds, reptiles, fishes, insects, and even crustaceans, the differences between the sexes follow almost exactly the same rules. The males are almost always the wooers; and they alone are armed with special weapons for fighting with their rivals. They are generally stronger and larger than the females, and are endowed with the requisite qualities of courage and pugnacity. They are provided, either exclusively or in a much higher degree than the females, with organs for producing vocal or instrumental music, and with odoriferous glands. They are ornamented with infinitely diversified appendages, and with the most brilliant or conspicuous colours, often arranged in elegant patterns, whilst the females are left unadorned. When the sexes differ in more important structures, it is the male which is provided with special sense-organs for discovering the female, with locomotive organs for reaching her, and often with prehensile organs for holding her. These various structures for securing or charming the female are often developed in the male during only part of the year, namely the breeding season. They have in many cases been transferred in a greater or less degree to the females; and in the latter case they appear in her as mere rudiments. They are lost by the males after emasculation. Generally they are not developed in the male during early youth, but appear a short time before the age for reproduction. Hence in most cases the young of both sexes resemble each other; and the female resembles her young offspring throughout life. In almost every great class a few anomalous cases occur in which there has been an almost complete transposition of the characters proper to the two sexes; the females assuming characters which properly belong to the males. This surprising uniformity in the laws regulating the differences between the sexes in so many and such widely separated classes, is intelligible if we admit the action throughout all the higher divisions of the animal kingdom of one common cause, namely sexual selection.

Sexual selection depends on the success of certain individuals over others of the same sex in relation to the propagation of the species; whilst natural selection depends on the success of both sexes, at all ages, in relation to the general conditions of life. The sexual struggle is of two kinds; in the one it is between the individuals of the same sex, generally the male sex, in order to drive away or kill their rivals, the females remaining passive; whilst in the other, the struggle is likewise between the individuals of the same sex, in order to excite or charm those of the opposite sex, generally the females, which no longer remain passive, but select the more agreeable partners. This latter kind of selection is closely analogous to that which man unintentionally, yet effectually, brings to bear on his domesticated productions, when he continues for a long time choosing the most pleasing or useful individuals, without any wish to modify the breed.

The laws of inheritance determine whether characters gained through sexual selection by either sex shall be transmitted to the same sex, or to both sexes; as well as the age at which they shall be developed. It appears that variations which arise late in life are commonly transmitted to one and the same sex. Variability is the necessary basis for the action of selection, and is wholly independent of it. It follows from this, that variations of the same general nature have often been taken advantage of and accumulated through sexual selection in relation to the propagation of the species, and through natural selection in relation to the general purposes of life. Hence secondary sexual characters, when equally transmitted to both sexes can be distinguished from ordinary specific characters only by the light of analogy. The modifications acquired through sexual selection are often so strongly pronounced that the two sexes have frequently been ranked as distinct species, or even as distinct genera. Such strongly marked differences must be in some manner highly important; and we know that they have been acquired in some instances at the cost not only of inconvenience, but of exposure to actual danger.

The belief in the power of sexual selection rests chiefly on the following considerations. The characters which we have the best reason for supposing to have been thus acquired are confined to one sex; and this alone renders it probable that they are in some way connected with the act of reproduction. These characters in innumerable instances are fully developed only at maturity; and often during only a part of the year, which is always the breeding-season. The males (passing over a few exceptional cases) are the most active in courtship; they are the best armed, and are rendered the most attractive in various ways. It is to be especially observed that the males display their attractions with elaborate care in the presence of the females; and that they rarely or never display them excepting during the season of love. It is incredible that all this display should be purposeless. Lastly we have distinct evidence with some quadrupeds and birds that the individuals of the one sex are capable of feeling a strong antipathy or preference for certain individuals of the opposite sex.

Bearing these facts in mind, and not forgetting the marked results of man's unconscious selection, it seems to me almost certain that if the individuals of one sex were during a long series of generations to prefer pairing with certain individuals of the other sex, characterised in some peculiar manner, the offspring would slowly but surely become modified in this same manner. I have not attempted to conceal that, excepting when the males are more numerous than the females, or when polygamy prevails, it is doubtful how the more attractive males succeed in leaving a larger number of offspring to inherit their superiority in ornaments or other charms than the less attractive males; but I have shewn that this would probably follow from the females, —especially the more vigorous females which would be the first to breed, preferring not only the more attractive but at the same time the more vigorous and victorious males.

Although we have some positive evidence that birds appreciate bright and beautiful objects, as with the Bower-birds of Australia, and although they certainly appreciate the power of song, yet I fully admit that it is an astonishing fact that the females of many birds and some mammals should be endowed with sufficient taste for what has apparently been effected through sexual selection; and this is even more astonishing in the case of reptiles, fish, and insects. But we really know very little about the minds of the lower animals. It cannot be supposed that male Birds of Paradise or Peacocks, for instance, should take so much pains in erecting, spreading, and vibrating their beautiful plumes before the females for no purpose. We should remember the fact given on excellent authority in a former chapter, namely that several peahens, when debarred from an admired male, remained widows during a whole season rather than pair with another bird.

Nevertheless I know of no fact in natural history more wonderful than that the female Argus pheasant should be able to appreciate the exquisite shading of the ball-and-socket ornaments and the elegant patterns on the wing-feathers of the male. He who thinks that the male was created as he now exists must admit that the great plumes, which prevent the wings from being used for flight, and which, as well as the primary feathers, are displayed in a manner quite peculiar to this one species during the act of courtship, and at no other time, were given to him as an ornament. If so, he must likewise admit that the female was created and endowed with the capacity of appreciating such ornaments. I differ only in the conviction that the male Argus pheasant acquired his beauty gradually, through the females having preferred during many generations the more highly ornamented males; the æsthetic capacity of the females having been advanced through exercise or habit in the same manner as our own taste is gradually improved. In the male, through the fortunate chance of a few feathers not having been modified, we can distinctly see how simple spots with a little fulvous shading on one side might have been developed by small and graduated steps into the wonderful ball-and-socket ornaments; and it is probable that they were actually thus developed.

Everyone who admits the principle of evolution, and yet feels great difficulty in admitting that female mammals, birds, reptiles, and fish, could have acquired the high standard of taste which is implied by the beauty of the males, and which generally coincides with our own standard, should reflect that in each member of the vertebrate series the nerve-cells of the brain are the direct offshoots of those possessed by the common progenitor of the whole group. It thus becomes intelligible that the brain and mental faculties should be capable under similar conditions of nearly the same course of development, and consequently of performing nearly the same functions.

The reader who has taken the trouble to go through the several chapters devoted to sexual selection, will be able to judge how far the conclusions at which I have arrived are supported by sufficient evidence. If he accepts these conclusions, he may, I think, safely extend them to mankind; but it would be superfluous here to repeat what I have so lately said on the manner in which sexual selection has apparently acted on both the male and female side, causing the two sexes of man to differ in body and mind, and the several races to differ from each other in various characters, as well as from their ancient and lowly-organized progenitors.

He who admits the principle of sexual selection will be led to the remarkable conclusion that the cerebral system not only regulates most of the existing functions of the body, but has indirectly influenced the progressive development of various bodily structures and of certain mental qualities. Courage, pugnacity, perseverance, strength and size of body, weapons of all kinds, musical organs, both vocal and instrumental, bright colours, stripes and marks, and ornamental appendages, have all been indirectly gained by the one sex or the other, through the influence of love and jealousy, through the appreciation of the beautiful in sound, colour or form, and through the exertion of a choice; and these powers of the mind manifestly depend on the development of the cerebral system.

Man scans with scrupulous care the character and pedigree of his horses, cattle, and dogs before he matches them; but when he comes to his own marriage he rarely, or never, takes any such care. He is impelled by nearly the same motives as are the lower animals when left to their own free choice, though he is in so far superior to them that he highly values mental charms and virtues. On the other hand he is strongly attracted by mere wealth or rank. Yet he might by selection do something not only for the bodily constitution and frame of his offspring, but for their intellectual and moral qualities. Both sexes ought to refrain from marriage if in any marked degree inferior in body or mind; but such hopes are Utopian and will never be even partially realised until the laws of inheritance are thoroughly known. All do good service who aid towards this end. When the principles of breeding and of inheritance are better understood, we shall not hear ignorant members of our legislature rejecting with scorn a plan for ascertaining by an easy method whether or not consanguineous marriages are injurious to man.

The advancement of the welfare of mankind is a most intricate problem: all ought to refrain from marriage who cannot avoid abject poverty for their children; for poverty is not only a great evil, but tends to its own increase by leading to recklessness in marriage. On the other hand, as Mr. Galton has remarked, if the prudent avoid marriage, whilst the reckless marry, the inferior members will tend to supplant the better members of society. Man, like every other animal, has no doubt advanced to his present high condition through a struggle for existence consequent on his rapid multiplication; and if he is to advance still higher he must remain subject to a severe struggle. Otherwise he would soon sink into indolence, and the more highly-gifted men would not be more successful in the battle of life than the less gifted. Hence our natural rate of increase, though leading to many and obvious evils, must not be greatly diminished by any means. There should be open competition for all men; and the most able should not be prevented by laws or customs from succeeding best and rearing the largest number of offspring. Important as the struggle for existence has been and even still is, yet as far as the highest part of man's nature is concerned there are other agencies more important. For the moral qualities are advanced, either directly or indirectly, much more through the effects of habit, the reasoning powers, instruction, religion,

&c., than through natural selection; though to this latter agency the social instincts, which afforded the basis for the development of the moral sense, may be safely attributed.

The main conclusion arrived at in this work, namely that man is descended from some lowly-organised form, will, I regret to think, be highly distasteful to many persons. But there can hardly be a doubt that we are descended from barbarians. The astonishment which I felt on first seeing a party of Fuegians on a wild and broken shore will never be forgotten by me, for the reflection at once rushed into my mind—such were our ancestors. These men were absolutely naked and bedaubed with paint, their long hair was tangled, their mouths frothed with excitement, and their expression was wild, startled, and distrustful. They possessed hardly any arts, and like wild animals lived on what they could catch; they had no government, and were merciless to every one not of their own small tribe. He who has seen a savage in his native land will not feel much shame, if forced to acknowledge that the blood of some more humble creature flows in his veins. For my own part I would as soon be descended from that heroic little monkey, who braved his dreaded enemy in order to save the life of his keeper; or from that old baboon, who, descending from the mountains, carried away in triumph his young comrade from a crowd of astonished dogs—as from a savage who delights to torture his enemies, offers up bloody sacrifices, practises infanticide without remorse, treats his wives like slaves, knows no decency, and is haunted by the grossest superstitions.

Man may be excused for feeling some pride at having risen, though not through his own exertions, to the very summit of the organic scale; and the fact of his having thus risen, instead of having been aboriginally placed there, may give him hopes for a still higher destiny in the distant future. But we are not here concerned with hopes or fears, only with the truth as far as our reason allows us to discover it. I have given the evidence to the best of my ability; and we must acknowledge, as it seems to me, that man with all his noble qualities, with sympathy which feels for the most debased, with benevolence which extends not only to other men but to the humblest living creature, with his god-like intellect which has penetrated into the movements and constitution of the solar system—with all these exalted powers—Man still bears in his bodily frame the indelible stamp of his lowly origin.

Pangenesis. [1871]

The "Provisional Hypothesis of Pangenesis," brought forth by Darwin in Variation *as an explanation for inheritance, continued to cause him problems. In the paper below, Darwin takes to task his second cousin Francis Galton, well known as a eugenicist and statistician, for assuming that the "gemmules" travel through an animal's body via its bloodstream. In fact Galton's inference as to how gemmules would move from somatic to reproductive tissue under Darwin's hypothesis is reasonable, and his experiment is ingenious.*

Henry Bence Jones, mentioned in Darwin's text, was his personal physician for many years. The "Mr. Paget" referred to is probably the surgeon James Paget, who supplied Darwin with information for Variation *and* The Expression of the Emotions. *It is interesting to note that the new journal* Nature *in which this paper appears had been founded, at least in part, to assist in spreading Darwinian views. To this day* Nature *remains the leading British journal of science.*

In a paper, read March 30, 1871, before the Royal Society, and just published in the Proceedings, Mr. Galton gives the results of his interesting experiments on the inter-transfusion of the blood of distinct varieties of rabbits. These experiments were undertaken to test whether there was any truth in my provisional hypothesis of Pangenesis. Mr. Galton, in recapitulating "the cardinal points," says that the gemmules are supposed "to swarm in the blood." He enlarges on this head, and remarks, "Under Mr. Darwin's theory, the gemmules in each individual must, therefore, be looked upon as entozoa of his blood," &c. Now, in the chapter on Pangenesis in my "Variation of Animals and Plants under Domestication," I have not said one word about the blood, or about any fluid proper to any circulating system. It is, indeed, obvious that the presence of gemmules in the blood can form no necessary part of my hypothesis; for I refer in illustration of it to the lowest animals, such as the Protozoa, which do not possess blood or any vessels; and I refer to plants in which the fluid, when present in the vessels, cannot be considered as true blood. The fundamental laws of growth, reproduction, inheritance, &c., are so closely similar throughout the whole organic kingdom, that the means by which the gemmules (assuming for the moment their existence) are diffused through the body, would probably be the same in all beings; therefore the means can hardly be diffusion through the blood. Nevertheless, when I first heard of Mr. Galton's experiments, I did not sufficiently reflect on the subject, and saw not the difficulty of believing in the presence of gemmules in the blood. I have said (*Variation*, &c., vol. ii., p. 379) that "the gemmules in each organism must be thoroughly diffused; nor does this seem improbable, considering their minuteness, and the steady circulation of fluids throughout the body." But when I used these latter words and other similar ones, I presume that I was thinking of the diffusion of the gemmules through the tissues, or from cell to cell, independently of the presence of vessels,—as in the remarkable experiments by Dr. Bence Jones, in which chemical elements absorbed by the stomach were detected in the course of some minutes in the crystalline lens of the eye; or again as in the repeated loss of colour and its recovery after a few days by the hair, in the singular case of a neuralgic lady recorded by Mr. Paget. Nor can it be objected that the gemmules could not pass through tissues or cell-walls, for the contents of each pollen-grain have to pass through the coats, both of the pollen-tube and embryonic sack. I may add, with respect to the passage of fluids through membrane, that they pass from cell to cell in the absorbing hairs of the roots of living plants at a rate, as I have myself observed under the microscope, which is truly surprising.

When, therefore, Mr. Galton concludes from the fact that rabbits of one variety, with large proportion of the blood of another variety in their veins, do not produce mongrelised offspring, that the hypothesis of Pangenesis is false, it seems to me that his conclusion is a little hasty. His words are, "I have now made experiments of transfusion and cross circulation on a large scale in rabbits, and have arrived at definite results, negativing, in my opinion, beyond all doubt the truth of the doctrine of Pangenesis." If Mr. Galton could have proved that the reproductive elements were contained in the blood of the higher animals, and were merely separated or collected by the reproductive glands, he would have made a most important physiological discovery. As it is, I think every one will admit that his experiments are extremely curious, and that he deserves the highest credit for his ingenuity and preservence. But it does not appear to me that Pangenesis has, as yet, received its death blow; though, from presenting so many vulnerable points, its life is always in jeopardy; and this is my excuse for having said a few words in its defence.

The Expression of the Emotions in Man and Animals. [1872]

DARWIN HAD EXPECTED TO include a chapter on human emotions in The Descent of Man. *When he gathered his notes, however, he realized that he had so much material that another book was in order. As was so often the case, Darwin had been interested in the subject for many years. When his first child, William, was born on December 27, 1839, Darwin had begun a notebook on the development of the infant's expressions. His notes became the source of "A Biographical Sketch of an Infant," reprinted later in this volume. Darwin had continued to accumulate material related to human and animal expressions over the years. In 1867 he circulated a sheet titled "Queries about Expressions" to a number of correspondents whom he considered likely sources of information on primitive peoples. Other researchers reprinted and distributed the sheet at least three times and requested that the answers be sent to Darwin. The sheet's sixteen questions appeared in the introduction to* The Expression of the Emotions in Man and Animals, *reprinted below. Darwin also listed the twenty-nine respondents to his questionnaire and alluded to their contributions throughout his text.*

In the Autobiography *Darwin claimed that booksellers subscribed to 5267 copies of* The Expression of the Emotions *on the day of publication, November 26, 1872. This figure would make* The Expression of the Emotions *his first-day best-seller. A posthumous second edition, edited by his son Francis and containing notes accumulated by Darwin, appeared in 1890. One of Darwin's most generally accessible books,* The Expression of the Emotions *contains interesting illustra-*

tions and a wide range of material. It succeeds in showing that humans share many behavioral traits with other animals, especially mammals. Darwin argues that even spiritual qualities are not unique to humans: "Every dog reveres its master" was an axiom he had learned as a British sportsman long before he became a scientist. The sole distinctive human response, he asserts, is blushing, "the most peculiar and the most human of all emotions." Accordingly, we include below the chapter on blushing to represent Darwin's seminal work in ethology.

INTRODUCTION.

. . .

Many writers consider the whole subject of Expression as inexplicable. Thus the illustrious physiologist Müller, says, "The completely different expression of the features in different passions shows that, according to the kind of feeling excited, entirely different groups of the fibres of the facial nerve are acted on. Of the cause of this we are quite ignorant."

No doubt as long as man and all other animals are viewed as independent creations, an effectual stop is put to our natural desire to investigate as far as possible the causes of Expression. By this doctrine, anything and everything can be equally well explained; and it has proved as pernicious with respect to Expression as to every other branch of natural history. With mankind some expressions, such as the bristling of the hair under the influence of extreme terror, or the uncovering of the teeth under that of furious rage, can hardly be understood, except on the belief that man once existed in a much lower and animal-like condition. The community of certain expressions in distinct though allied species, as in the movements of the same facial muscles during laughter by man and by various monkeys, is rendered somewhat more intelligible, if we believe in their descent from a common progenitor. He who admits on general grounds that the structure and habits of all animals have been gradually evolved, will look at the whole subject of Expression in a new and interesting light.

The study of Expression is difficult, owing to the movements being often extremely slight, and of a fleeting nature. A difference may be clearly perceived, and yet it may be impossible, at least I have found it so, to state in what the difference consists. When we witness any deep emotion, our sympathy is so strongly excited, that close observation is forgotten or rendered almost impossible; of which fact I have had many curious proofs. Our imagination is another and still more serious source of error; for if from the nature of the circumstances we expect to see any expression, we readily imagine its presence. Notwithstanding Dr. Duchenne's great experience, he for a long time fancied, as he states, that several muscles contracted under certain emotions, whereas he ultimately convinced himself that the movement was confined to a single muscle.

In order to acquire as good a foundation as possible, and to ascertain, independently of common opinion, how far particular movements of the features and gestures are really expressive of certain states of the mind, I have found the following means the most serviceable. In the first place, to observe infants; for they exhibit many emotions, as Sir C. Bell remarks, "with extraordinary force;" whereas, in after life, some of our expressions "cease to have the pure and simple source from which they spring in infancy."

In the second place, it occurred to me that the insane ought to be studied, as they are liable to the strongest passions, and give uncontrolled vent to them. I had, myself, no opportunity of doing this, so I applied to Dr. Maudsley and received from him an introduction to Dr. J. Crichton Browne, who has charge of an immense asylum near Wakefield, and who, as I found, had already attended to the subject. This excellent observer has with unwearied kindness sent me copious notes and descriptions, with valuable suggestions on many points; and I can hardly over-estimate the value of his assistance. I owe also, to the kindness of Mr. Patrick Nicol, of the Sussex Lunatic Asylum, interesting statements on two or three points.

Thirdly Dr. Duchenne galvanized, as we have already seen, certain muscles in the face of an old man, whose skin was little sensitive, and thus produced various expressions which were photographed on a large scale. It fortunately occurred to me to show several of the best plates, without a word of explanation, to above twenty educated persons of various ages and both sexes, asking them, in each case, by what emotion or feeling the old man was supposed to be agitated; and I recorded their answers in the words which they used. Several of the expressions were instantly recognised by almost everyone, though described in not exactly the same terms; and these may, I think, be relied on as truthful, and will hereafter be specified. On the other hand, the most widely different judgments were pronounced in regard to some of them. This exhibition was of use in another way, by convincing me how easily we may be misguided by our imagination; for when I first looked through

Dr. Duchenne's photographs, reading at the same time the text, and thus learning what was intended, I was struck with admiration at the truthfulness of all, with only a few exceptions. Nevertheless, if I had examined them without any explanation, no doubt I should have been as much perplexed, in some cases, as other persons have been.

Fourthly, I had hoped to derive much aid from the great masters in painting and sculpture, who are such close observers. Accordingly, I have looked at photographs and engravings of many well-known works; but, with a few exceptions, have not thus profited. The reason no doubt is, that in works of art, beauty is the chief object; and strongly contracted facial muscles destroy beauty. The story of the composition is generally told with wonderful force and truth by skilfully given accessories.

Fifthly, it seemed to me highly important to ascertain whether the same expressions and gestures prevail, as has often been asserted without much evidence, with all the races of mankind, especially with those who have associated but little with Europeans. Whenever the same movements of the features or body express the same emotions in several distinct races of man, we may infer with much probability, that such expressions are true ones,—that is, are innate or instinctive. Conventional expressions or gestures, acquired by the individual during early life, would probably have differed in the different races, in the same manner as do their languages. Accordingly I circulated, early in the year 1867, the following printed queries with a request, which has been fully responded to, that actual observations, and not memory, might be trusted. These queries were written after a considerable interval of time, during which my attention had been otherwise directed, and I can now see that they might have been greatly improved. To some of the later copies, I appended, in manuscript, a few additional remarks:—

1. Is astonishment expressed by the eyes and mouth being opened wide, and by the eyebrows being raised?
2. Does shame excite a blush when the color of the skin allows it to be visible? and especially how low down the body does the blush extend?
3. When a man is indignant or defiant does he frown, hold his body and head erect, square his shoulders and clench his fists?
4. When considering deeply on any subject, or trying to understand any puzzle, does he frown, or wrinkle the skin beneath the lower eyelids?

5. When in low spirits, are the corners of the mouth depressed, and the inner corner of the eyebrows raised by that muscle which the French call the "Grief muscle"? The eyebrow in this state becomes slightly oblique, with a little swelling at the inner end; and the forehead is transversely wrinkled in the middle part, but not across the whole breadth, as when the eyebrows are raised in surprise.
6. When in good spirits do the eyes sparkle, with the skin a little wrinkled round and under them, and with the mouth a little drawn back at the corners?
7. When a man sneers or snarls at another, is the corner of the upper lip over the canine or eye tooth raised on the side facing the man whom he addresses?
8. Can a dogged or obstinate expression be recognized, which is chiefly shown by the mouth being firmly closed, a lowering brow and a slight frown?
9. Is contempt expressed by a slight protrusion of the lips and by turning up the nose, and with a slight expiration?
10. Is disgust shown by the lower lip being turned down, the upper lip slightly raised, with a sudden expiration, something like incipient vomiting, or like something spit out of the mouth?
11. Is extreme fear expressed in the same general manner as with Europeans?
12. Is laughter ever carried to such an extreme as to bring tears into the eyes?
13. When a man wishes to show that he cannot prevent something being done, or cannot himself do something, does he shrug his shoulders, turn inwards his elbows, extend outwards his hands and open the palms; with the eyebrows raised?
14. Do the children when sulky, pout or greatly protrude the lips?
15. Can guilty, or sly, or jealous expressions be recognized? though I know not how these can be defined.
16. Is the head nodded vertically in affirmation, and shaken laterally in negation?

Observations on natives who have had little communication with Europeans would be of course the most valuable,

> though those made on any natives would be of much interest to me. General remarks on expression are of comparatively little value; and memory is so deceptive that I earnestly beg it may not be trusted. A definite description of the countenance under any emotion or frame of mind, with a statement of the circumstances under which it occurred, would possess much value.

To these queries I have received thirty-six answers from different observers, several of them missionaries or protectors of the aborigines, to all of whom I am deeply indebted for the great trouble which they have taken, and for the valuable aid thus received. I will specify their names, &c., towards the close of this chapter, so as not to interrupt my present remarks. The answers relate to several of the most distinct and savage races of man. In many instances, the circumstances have been recorded under which each expression was observed, and the expression itself described. In such cases, much confidence may be placed in the answers. When the answers have been simply yes or no, I have always received them with caution. It follows, from the information thus acquired, that the same state of mind is expressed throughout the world with remarkable uniformity; and this fact is in itself interesting as evidence of the close similarity in bodily structure and mental disposition of all the races of mankind.

Sixthly, and lastly, I have attended, as closely as I could, to the expression of the several passions in some of the commoner animals; and this I believe to be of paramount importance, not of course for deciding how far in man certain expressions are characteristic of certain states of mind, but as affording the safest basis for generalisation on the causes, or origin, of the various movements of Expression. In observing animals, we are not so likely to be biassed by our imagination; and we may feel safe that their expressions are not conventional.

From the reasons above assigned, namely, the fleeting nature of some expressions (the changes in the features being often extremely slight); our sympathy being easily aroused when we behold any strong emotion, and our attention thus distracted; our imagination deceiving us, from knowing in a vague manner what to expect, though certainly few of us know what the exact changes in the countenance are; and lastly, even our long familiarity with the subject,—from all these causes combined, the observation of Expression is by no means easy, as many persons, whom I have asked to observe certain points, have soon discovered. Hence it is difficult to determine, with certainty, what are the movements of the features and of the body, which commonly characterize certain states of the mind. Nevertheless, some of the doubts and difficulties have, as I hope, been cleared away by the observation of infants,—of the insane,—of the different races of man,—of works of art,—and lastly,—of the facial muscles under the action of galvanism, as effected by Dr. Duchenne.

But there remains the much greater difficulty of understanding the cause or origin of the several expressions, and of judging whether any theoretical explanation is trustworthy. Besides, judging as well as we can by our reason, without the aid of any rules, which of two or more explanations is the most satisfactory, or are quite unsatisfactory, I see only one way of testing our conclusions. This is to observe whether the same principle by which one expression can, as it appears, be explained, is applicable in other allied cases; and especially, whether the same general principles can be applied with satisfactory results, both to man and the lower animals. This latter method, I am inclined to think, is the most serviceable of all. The difficulty of judging of the truth of any theoretical explanation, and of testing it by some distinct line of investigation, is the great drawback to that interest which the study seems well fitted to excite.

Finally, with respect to my own observations, I may state that they were commenced in the year 1838; and from that time to the present day, I have occasionally attended to the subject. At the above date, I was already inclined to believe in the principle of evolution, or of the derivation of species from other and lower forms. Consequently, when I read Sir C. Bell's great work, his view, that man had been created with certain muscles specially adapted for the expression of his feelings, struck me as unsatisfactory. It seemed probable that the habit of expressing our feelings by certain movements, though now rendered innate, had been in some manner gradually acquired. But to discover how such habits had been acquired was perplexing in no small degree. The whole subject had to be viewed under a new aspect, and each expression demanded a rational explanation. This belief led me to attempt the present work, however imperfectly it may have been executed.

• • •

SELF-ATTENTION—SHAME—SHYNESS—MODESTY: BLUSHING.

Blushing is the most peculiar and the most human of all expressions. Monkeys redden from passion, but it would require an overwhelming amount of evidence to make us believe that any animal could blush. The reddening of the face from a blush is due to the relaxation of the muscular coats of the small arteries, by which the capillaries become filled with blood; and this depends on the proper vaso-motor centre being affected. No doubt if there be at the same time much mental agitation, the general circulation will be affected; but it is not due to the action of the heart that the network of minute vessels covering the face becomes under a sense of shame gorged with blood. We can cause laughing by tickling the skin, weeping or frowning by a blow, trembling from the fear of pain, and so forth; but we cannot cause a blush, as Dr. Burgess remarks, by any physical means,—that is by any action on the body. It is the mind which must be affected. Blushing is not only involuntary; but the wish to restrain it, by leading to self-attention actually increases the tendency.

The young blush much more freely than the old, but not during infancy, which is remarkable, as we know that infants at a very early age redden from passion. I have received authentic accounts of two little girls blushing at the ages of between two and three years; and of another sensitive child, a year older, blushing, when reproved for a fault. Many children, at a somewhat more advanced age blush in a strongly marked manner. It appears that the mental powers of infants are not as yet sufficiently developed to allow of their blushing. Hence, also, it is that idiots rarely blush. Dr. Crichton Browne observed for me those under his care, but never saw a genuine blush, though he has seen their faces flush, apparently from joy, when food was placed before them, and from anger. Nevertheless some, if not utterly degraded, are capable of blushing. A microcephalous idiot, for instance, thirteen years old, whose eyes brightened a little when he was pleased or amused, has been described by Dr. Behn, as blushing and turning to one side, when undressed for medical examination.

Women blush much more than men. It is rare to see an old man, but not nearly so rare to see an old woman blushing. The blind do not escape. Laura Bridgman, born in this condition, as well as completely deaf, blushes. The Rev. R. H. Blair, Principal of the Worcester College, informs me that three children born blind, out of seven or eight then in the Asylum, are great blushers. The blind are not at first conscious that they are observed, and it is a most important part of their education, as Mr. Blair informs me, to impress this knowledge on their minds; and the impression thus gained would greatly strengthen the tendency to blush, by increasing the habit of self-attention.

The tendency to blush is inherited. Dr. Burgess gives the case of a family consisting of a father, mother, and ten children, all of whom, without exception, were prone to blush to a most painful degree. The children were grown up; "and some of them were sent to travel in order to wear away this diseased sensibility, but nothing was of the slightest avail." Even peculiarities in blushing seem to be inherited. Sir James Paget, whilst examining the spine of a girl, was struck at her singular manner of blushing; a big splash of red appeared first on one cheek, and then other splashes, variously scattered over the face and neck. He subsequently asked the mother whether her daughter always blushed in this peculiar manner; and was answered, "Yes, she takes after me." Sir J. Paget then perceived that by asking this question he had caused the mother to blush; and she exhibited the same peculiarity as her daughter.

In most cases the face, ears and neck are the sole parts which redden; but many persons, whilst blushing intensely, feel that their whole bodies grow hot and tingle; and this shows that the entire surface must be in some manner affected. Blushes are said sometimes to commence on the forehead, but more commonly on the cheeks, afterwards spreading to the ears and neck. In two Albinos examined by Dr. Burgess, the blushes commenced by a small circumscribed spot on the cheeks, over the parotidean plexus of nerves, and then increased into a circle; between this blushing circle and the blush on the neck there was an evident line of demarcation; although both arose simultaneously. The retina, which is naturally red in the Albino, invariably increased at the same time in redness. Every one must have noticed how easily after one blush fresh blushes chase each other over the face. Blushing is preceded by a peculiar sensation in the skin. According to Dr. Burgess the reddening of the skin is generally succeeded by a slight pallor, which shows that the capillary vessels contract after dilating. In some rare cases paleness instead of redness is caused under conditions which would naturally induce a blush. For instance, a young lady told me that in a large and crowded party she caught her hair so firmly on the button of a passing servant, that it took some time before she could be extricated; from her sensations she imagined that she had blushed crimson; but was assured by a friend that she had turned extremely pale.

I was desirous to learn how far down the body blushes extend; and Sir J. Paget, who necessarily has frequent opportunities for observation, has kindly attended to this point for me during two or three years. He finds that with women who blush intensely on the face, ears, and nape of neck, the blush does not commonly extend any lower down the body. It is rare to see it as low down as the collar-bones and shoulder-blades; and he has never himself seen a single instance in which it extended below the upper part of the chest. He has also noticed that blushes sometimes die away downwards, not gradually and insensibly, but by irregular ruddy blotches. Dr. Langstaff has likewise observed for me several women whose bodies did not in the least redden while their faces were crimsoned with blushes. With the insane, some of whom appear to be particularly liable to blushing, Dr. J. Crichton Browne has several times seen the blush extend as far down as the collar-bones, and in two instances to the breasts. He gives me the case of a married woman, aged twenty-seven, who suffered from epilepsy. On the morning after her arrival in the Asylum, Dr. Browne, together with his assistants, visited her whilst she was in bed. The moment that he approached, she blushed deeply over her cheeks and temples; and the blush spread quickly to her ears. She was much agitated and tremulous. He unfastened the collar of her chemise in order to examine the state of her lungs; and then a brilliant blush rushed over her chest, in an arched line over the upper third of each breast, and extended downwards between the breasts nearly to the ensiform cartilage of the sternum. This case is interesting, as the blush did not thus extend downwards until it became intense by her attention being drawn to this part of her person. As the examination proceeded she became composed, and the blush disappeared; but on several subsequent occasions the same phenomena were observed.

The foregoing facts show that, as a general rule, with English women, blushing does not extend beneath the neck and upper part of the chest. Nevertheless Sir J. Paget informs me that he has lately heard of a case, on which he can fully rely, in which a little girl, shocked by what she imagined to be an act of indelicacy, blushed all over her abdomen and the upper parts of her legs. Moreau also relates, on the authority of a celebrated painter, that the chest, shoulders, arms, and whole body of a girl, who unwillingly consented to serve as a model, reddened when she was first divested of her clothes.

It is a rather curious question why, in most cases the face, ears, and neck alone redden, inasmuch as the whole surface of the body often tingles and grows hot. This seems to depend, chiefly, on the face and adjoining parts of the skin having been habitually exposed to the air, light, and alternations of temperature, by which the small arteries not only have acquired the habit of readily dilating and contracting, but appear to have become unusually developed in comparison with other parts of the surface. It is probably owing to this same cause, as M. Moreau and Dr. Burgess have remarked, that the face is so liable to redden under various circumstances, such as a fever-fit, ordinary heat, violent exertion, anger, a slight blow, &c.; and on the other hand that it is liable to grow pale from cold and fear, and to be discoloured during pregnancy. The face is also particularly liable to be affected by cutaneous complaints, by small-pox, erysipelas, &c. This view is likewise supported by the fact that the men of certain races, who habitually go nearly naked, often blush over their arms and chests and even down to their waists. A lady, who is a great blusher, informs Dr. Crichton Browne, that when she feels ashamed or is agitated, she blushes over her face, neck, wrists, and hands,—that is, over all the exposed portions of her skin. Nevertheless it may be doubted whether the habitual exposure of the skin of the face and neck, and its consequent power of reaction under stimulants of all kinds, is by itself sufficient to account for the much greater tendency in English women of these parts than of others to blush; for the hands are well supplied with nerves and small vessels, and have been as much exposed to the air as the face or neck, and yet the hands rarely blush. We shall presently see that the attention of the mind having been directed much more frequently and earnestly to the face than to any other part of the body, probably affords a sufficient explanation.

Blushing in the various races of man.—The small vessels of the face become filled with blood, from the emotion of shame, in almost all the races of man, though in the very dark races no distinct change of colour can be perceived. Blushing is evident in all the Aryan nations of Europe, and to a certain extent with those of India. But Mr. Erskine has never noticed that the necks of the Hindoos are decidedly affected. With the Lepchas of Sikhim, Mr. Scott has often observed a faint blush on the cheeks, base of the ears, and sides of the neck, accompanied by sunken eyes and lowered head. This has occurred when he has detected them in a falsehood, or has accused them of ingratitude. The pale, sallow complexions of these men render a blush much more conspicuous than in most of the other natives of India. With the latter, shame, or it may be in part fear, is expressed, according to Mr. Scott, much more plainly by the head being averted or bent down, with the eyes wavering or turned askant, than by any change of colour in the skin.

The Semitic races blush freely, as might have been expected, from their general similitude to the Aryans. Thus with the Jews, it is said in the Book of Jeremiah (chap. vi. 15), "Nay, they were not at all ashamed, neither could they blush." Mrs. Asa Gray saw an Arab managing his boat clumsily on the Nile, and when laughed at by his companions, "he blushed quite to the back of his neck." Lady Duff Gordon remarks that a young Arab blushed on coming into her presence.

Mr. Swinhoe has seen the Chinese blushing, but he thinks it is rare; yet they have the expression "to redden with shame." Mr. Geach informs me that the Chinese settled in Malacca and the native Malays of the interior both blush. Some of these people go nearly naked, and he particularly attended to the downward extension of the blush. Omitting the cases in which the face alone was seen to blush, Mr. Geach observed that the face, arms, and breast of a Chinaman, aged 24 years, reddened from shame; and with another Chinese, when asked why he had not done his work in better style, the whole body was similarly affected. In two Malays he saw the face, neck, breast, and arms blushing; and in a third Malay (a Bugis) the blush extended down to the waist.

The Polynesians blush freely. The Rev. Mr. Stack has seen hundreds of instances with the New Zealanders. The following case is worth giving, as it relates to an old man who was unusually dark-coloured and partly tattooed. After having let his land to an Englishman for a small yearly rental, a strong passion seized him to buy a gig, which had lately become the fashion with the Maoris. He consequently wished to draw all the rent for four years from his tenant, and consulted Mr. Stack whether he could do so. The man was old, clumsy, poor, and ragged, and the idea of his driving himself about in his carriage for display amused Mr. Stack so much that he could not help bursting out into a laugh; and then "the old man blushed up to the roots of his hair." Forster says that "you may easily distinguish a spreading blush" on the cheeks of the fairest women in Tahiti. The natives also of several of the other archipelagoes in the Pacific have been seen to blush.

Mr. Washington Matthews has often seen a blush on the faces of the young squaws belonging to various wild Indian tribes of North America. At the opposite extremity of the continent in Tierra del

Fuego, the natives, according to Mr. Bridges, "blush much, but chiefly in regard to women; but they certainly blush also at their own personal appearance." This latter statement agrees with what I remember of the Fuegian, Jemmy Button, who blushed when he was quizzed about the care which he took in polishing his shoes, and in otherwise adorning himself. With respect to the Aymara Indians on the lofty plateaus of Bolivia, Mr. Forbes says, that from the colour of their skins it is impossible that their blushes should be as clearly visible as in the white races; still under such circumstances as would raise a blush in us, "there can always be seen the same expression of modesty or confusion; and even in the dark, a rise of temperature of the skin of the face can be felt, exactly as occurs in the European." With the Indians who inhabit the hot, equable, and damp parts of South America, the skin apparently does not answer to mental excitement so readily as with the natives of the northern and southern parts of the continent, who have long been exposed to great vicissitudes of climate; for Humboldt quotes without a protest the sneer of the Spaniard, "How can those be trusted, who know not how to blush?" Von Spix and Martius, in speaking of the aborigines of Brazil, assert that they cannot properly be said to blush; "it was only after long intercourse with the whites, and after receiving some education, that we perceived in the Indians a change of colour expressive of the emotions of their minds." It is, however, incredible that the power of blushing could have thus originated; but the habit of self-attention, consequent on their education and new course of life, would have much increased any innate tendency to blush.

Several trustworthy observers have assured me that they have seen on the faces of negroes an appearance resembling a blush, under circumstances which would have excited one in us, though their skins were of an ebony-black tint. Some describe it as blushing brown, but most say that the blackness becomes more intense. An increased supply of blood in the skin seems in some manner to increase its blackness; thus certain exanthematous diseases cause the affected places in the negro to appear blacker, instead of, as with us, redder. The skin, perhaps, from being rendered more tense by the filling of the capillaries, would reflect a somewhat different tint to what it did before. That the capillaries of the face in the negro become filled with blood, under the emotion of shame, we may feel confident; because a perfectly characterized albino negress, described by Buffon, showed a faint tinge of crimson on her cheeks when she exhibited herself naked. Cicatrices of the skin remain for a long time white in the negro, and Dr. Burgess, who had frequent opportunities of observing a scar of this kind on the face of a negress, distinctly saw that it "invariably became red whenever she was abruptly spoken to, or charged with any trivial offence." The blush could be seen proceeding from the circumference of the scar towards the middle, but it did not reach the centre. Mulattoes are often great blushers, blush succeeding blush over their faces. From these facts there can be no doubt that negroes blush, although no redness is visible on the skin.

I am assured by Gaika and by Mrs. Barber that the Kafirs of South Africa never blush; but this may only mean that no change of colour is distinguishable. Gaika adds that under the circumstances which would make a European blush, his countrymen "look ashamed to keep their heads up."

It is asserted by four of my informants that the Australians, who are almost as black as negroes, never blush. A fifth answers doubtfully, remarking that only a very strong blush could be seen, on account of the dirty state of their skins. Three observers state that they do blush; Mr. S. Wilson adding that this is noticeable only under a strong emotion, and when the skin is not too dark from long exposure and want of cleanliness. Mr. Lang answers, "I have noticed that shame almost always excites a blush, which frequently extends as low as the neck." Shame is also shown, as he adds, "by the eyes being turned from side to side." As Mr. Lang was a teacher in a native school, it is probable that he chiefly observed children; and we know that they blush more than adults. Mr. G. Taplin has seen half-castes blushing, and he says that the aborigines have a word expressive of shame. Mr. Hagenauer, who is one of those who has never observed the Australians to blush, says that he has "seen them looking down to the ground on account of shame;" and the missionary, Mr. Bulmer, remarks that though "I have not been able to detect anything like shame in the adult aborigines, I have noticed that the eyes of the children, when ashamed, present a restless, watery appearance, as if they did not know where to look."

The facts now given are sufficient to show that blushing, whether or not there is any change of colour, is common to most, probably to all, of the races of man.

Movements and gestures which accompany Blushing.—Under a keen sense of shame there is a strong desire for concealment. We turn away the whole body, more especially the face, which we endeavour in some manner to hide. An ashamed person can hardly endure to meet the gaze of those present, so that he almost invariably casts down his eyes or looks askant. As there generally exists at the same time a strong wish to avoid the appearance of shame, a vain attempt is made to look direct at the person who causes this feeling; and the antagonism between these opposite tendencies leads to various restless movements in the eyes. I have noticed two ladies who, whilst blushing, to which they are very liable, have thus acquired, as it appears, the oddest trick of incessantly blinking their eyelids with extraordinary rapidity. An intense blush is sometimes accompanied by a slight effusion of tears; and this, I presume, is due to the lacrymal glands partaking of the increased supply of blood, which we know rushes into the capillaries of the adjoining parts, including the retina.

Many writers, ancient and modern, have noticed the foregoing movements; and it has already been shown that the aborigines in various parts of the world often exhibit their shame by looking downwards or askant, or by restless movements of their eyes. Ezra cries out (ch. ix. 6), "O, my God! I am ashamed, and blush to lift up my head to thee, my God." In Isaiah (ch. 1. 6) we meet with the words, "I hid not my face from shame." Seneca remarks (Epist. xi. 5) "that the Roman players hang down their heads, fix their eyes on the ground and keep them lowered, but are unable to blush in acting shame." According to Macrobius, who lived in the fifth century ('Saturnalia,' B. vii. C. 11), "Natural philosophers assert that nature being moved by shame spreads the blood before herself as a veil, as we see any one blushing often puts his hands before his face." Shakespeare makes Marcus ('Titus Andronicus,' act ii, sc. 5) say to his niece, "Ah! now thou turn'st away thy face for shame." A lady informs me that she found in the Lock Hospital a girl whom she had formerly known, and who had become a wretched castaway, and the poor creature, when approached, hid her face under the bed-clothes, and could not be persuaded to uncover it. We often see little children, when shy or ashamed, turn away, and still standing up, bury their faces in their mother's gown; or they throw themselves face downwards on her lap.

Confusion of mind.—Most persons, whilst blushing intensely, have their mental powers confused. This is recognized in such common expressions as "she was covered with confusion." Persons in this condition lose their presence of mind, and utter singularly inappropriate remarks. They are often much distressed, stammer, and make awkward movements or strange grimaces. In certain cases involuntary twitchings of some of the facial muscles may be observed. I have been informed by a young lady, who blushes excessively, that at such times she does not even know what she is saying. When it was suggested to her that this might be due to her distress from the consciousness that her blushing was noticed, she answered that this could not be the case, "as she had sometimes felt quite as stupid when blushing at a thought in her own room."

I will give an instance of the extreme disturbance of mind to which some sensitive men are liable. A gentleman, on whom I can rely, assured me that he had been an eye-witness of the following scene:—A small dinner-party was given in honour of an extremely shy man, who, when he rose to return thanks, rehearsed the speech, which he had evidently learnt by heart, in absolute silence, and did not utter a single word; but he acted as if he were speaking with much emphasis. His friends, perceiving how the case stood, loudly applauded the imaginary bursts of eloquence, whenever his gestures indicated a pause, and the man never discovered that he had remained the whole time completely silent. On the contrary, he afterwards remarked to my friend, with much satisfaction, that he thought he had succeeded uncommonly well.

When a person is much ashamed or very shy, and blushes intensely, his heart beats rapidly and his breathing is disturbed. This can hardly fail to affect the circulation of the blood within the brain, and perhaps the mental powers. It seems however doubtful, judging from the still more powerful influence of anger and fear on the circulation, whether we can thus satisfactorily account for the confused state of mind in persons whilst blushing intensely.

The true explanation apparently lies in the intimate sympathy which exists between the capillary circulation of the surface of the head and face, and that of the brain. On applying to Dr. J. Crichton Browne for information, he has given me various facts bearing on this subject. When the sympathetic nerve is divided on one side of the head, the capillaries on this side are relaxed and become filled with blood, causing the skin to redden and to grow hot, and at the same time the temperature within the cranium on the same side rises. Inflammation of the membranes of the brain leads to the engorgement of the face, ears, and eyes with blood. The first stage of an epileptic fit appears to be the contraction of the vessels of the brain, and the first outward manifestation is an extreme pallor of countenance. Erysipelas of the head commonly induces delirium. Even the relief given to a severe headache by burning the skin with strong lotion, depends, I presume, on the same principle.

Dr. Browne has often administered to his patients the vapour of the nitrate of amyl, which has the singular property of causing vivid redness of the face in from thirty to sixty seconds. This flushing resembles blushing in almost every detail: it begins at several distinct points on the face, and spreads till it involves the whole surface of the head, neck, and front of the chest; but has been observed to extend only in one case to the abdomen. The arteries in the retina become enlarged; the eyes glisten, and in one instance there was a slight effusion of tears. The patients are at first pleasantly stimulated, but, as the flushing increases, they become confused and bewildered. One woman to whom the vapour had often been administered asserted that, as soon as she grew hot, she grew *muddled.* With persons just commencing to blush it appears, judging from their bright eyes and lively behaviour, that their mental powers are somewhat stimulated. It is only when the blushing is excessive that the mind grows confused. Therefore it would seem that the capillaries of the face are affected, both during the inhalation of the nitrite of amyl and during blushing, before that part of the brain is affected on which the mental powers depend.

Conversely when the brain is primarily affected, the circulation of the skin is so in a secondary manner. Dr. Browne has frequently observed, as he informs me, scattered red blotches and mottlings on the chests of epileptic patients. In these cases, when the skin on the thorax or abdomen is gently rubbed with a pencil or other object, or, in strongly-marked cases, is merely touched by the finger, the surface becomes suffused in less than half a minute with bright red marks, which spread to some distance on each side of the touched point, and persist for several minutes. These are the *cerebral maculæ* of Trousseau; and they indicate, as Dr. Browne remarks, a highly modified condition of the cutaneous vascular system. If, then, there exists, as cannot be doubted, an intimate sympathy between the capillary circulation in that part of the brain on which our mental powers depend, and in the skin of the face, it is not surprising that the moral causes which induce intense blushing should likewise induce, independently of their own disturbing influence, much confusion of mind.

The Nature of the Mental States which induce Blushing.—These consist of shyness, shame, and modesty; the essential element in all being self-attention. Many reasons can be assigned for believing that originally self-attention directed to personal appearance, in relation to the opinion of others, was the exciting cause; the same effect being subsequently produced, through the force of association, by self-attention in relation to moral conduct. It is not the simple act of reflecting on our own appearance, but the thinking what others think of us, which excites a blush. In absolute solitude the most sensitive person would be quite indifferent about his appearance. We feel blame or disapprobation more acutely than approbation; and consequently depreciatory remarks or ridicule, whether of our appearance or conduct, causes us to blush much more readily than does praise. But undoubtedly praise and admiration are highly efficient: a pretty girl blushes when a man gazes intently at her, though she may know perfectly well that he is not depreciating her. Many children, as well as old and sensitive persons blush, when they are much praised. Hereafter the question will be discussed, how it has arisen that the consciousness that others are attending to our personal appearance should have led to the capillaries, especially those of the face, instantly becoming filled with blood.

My reasons for believing that attention directed to personal appearance, and not to moral conduct, has been the fundamental element in the acquirement of the habit of blushing, will now be given. They are separately light, but combined possess, as it appears to me, considerable weight. It is notorious that nothing makes a shy person blush so much as any remark, however slight, on his personal appearance. One cannot notice even the dress of a woman much given to blushing, without causing her face to crimson. It is sufficient to stare hard at some persons to make them, as Coleridge remarks, blush,—"account for that he who can."

With the two albinos observed by Dr. Burgess, "the slightest attempt to examine their peculiarities invariably" caused them to blush deeply. Women are much more sensitive about their personal appearance than men are, especially elderly women in comparison with elderly men, and they blush much more freely. The young of both sexes are much more sensitive on this same head than the old, and they also blush much more freely than the old. Children at a very early age do not blush; nor do they show those other signs of self-consciousness which generally accompany blushing; and it is one of their chief charms that they think nothing about what others think of them. At this early age they will stare at a stranger with a fixed gaze and unblinking eyes, as on an inanimate object, in a manner which we elders cannot imitate.

It is plain to every one that young men and women are highly sensitive to the opinion of each other with reference to their personal appearance; and they blush incomparably more in the presence of the opposite sex than in that of their own. A young man, not very liable to blush, will blush intensely at any slight ridicule of his appearance from a girl whose judgment on any important subject he would disregard. No happy pair of young lovers, valuing each other's admiration and love more than anything else in the world, probably ever courted each other without many a blush. Even the barbarians of Tierra del Fuego, according to Mr. Bridges, blush "chiefly in regard to women, but certainly also at their own personal appearance."

Of all parts of the body, the face is most considered and regarded, as is natural from its being the chief seat of expression and the source of the voice. It is also the chief seat of beauty and of ugliness, and throughout the world is the most ornamented. The face, therefore, will have been subjected during many generations to much closer and more earnest self-attention than any other part of the body; and in accordance with the principle here advanced we can understand why it should be the most liable to blush. Although exposure to alternations of temperature, &c., has probably much increased the power of dilatation and contraction in the capillaries of the face and adjoining parts, yet this by itself will hardly account for these parts blushing much more than the rest of the body; for it does not explain the fact of the hands rarely blushing. With Europeans the whole body tingles slightly when the face blushes intensely; and with the races of men who habitually go nearly naked, the blushes extend over a much larger surface than with us. These facts are, to a certain extent, intelligible, as the self-attention of primeval man, as well as of the existing races which still go naked, will not have been so exclusively confined to their faces, as is the case with the people who now go clothed.

We have seen that in all parts of the world persons who feel shame for some moral delinquency, are apt to avert, bend down, or hide their faces, independently of any thought about their personal appearance. The object can hardly be to conceal their blushes, for the face is thus averted or hidden under circumstances which exclude any desire to conceal shame, as when guilt is fully confessed and repented of. It is, however, probable that primeval man before he had acquired much moral sensitiveness would have been highly sensitive about his personal appearance, at least in reference to the other sex, and he would consequently have felt distress at any depreciatory remarks about his appearance; and this is one form of shame. And as the face is the part of the body which is most regarded, it is intelligible that anyone ashamed of his personal appearance would desire to conceal this part of his body. The habit having been thus acquired, would naturally be carried on when shame from strictly moral causes was felt; and it is not easy otherwise to see why under these circumstances there should be a desire to hide the face more than any other part of the body.

The habit, so general with every one who feels ashamed, of turning away, or lowering his eyes, or restlessly moving them from side to side, probably follows from each glance directed towards those present, bringing home the conviction that he is intently regarded; and he endeavours, by not looking at those present, and especially not at their eyes, momentarily to escape from this painful conviction.

Shyness.—This odd state of mind, often called shamefacedness, or false shame, or *mauvaise honte*, appears to be one of the most efficient of all the causes of blushing. Shyness is, indeed, chiefly recognized by the face reddening, by the eyes being averted or cast down, and by awkward, nervous movements of the body. Many a woman blushes from this cause, a hundred, perhaps a thousand times, to once that she blushes from having done anything deserving blame, and of which she is truly ashamed. Shyness seems to depend on sensitiveness to the opinion, whether good or bad, of others, more especially with respect to external appearance. Strangers neither know nor care anything about our conduct or character, but they may, and often do, criticize our appearance: hence shy persons are particularly apt to be shy and to blush in the presence of strangers. The consciousness of anything peculiar, or even new, in the dress, or any slight blemish on the person, and more especially on the face—points which are likely to attract the attention of strangers—makes the shy intolerably shy. On the other hand, in those cases in which conduct and not personal appearance is concerned, we are much more apt to be shy in the presence of acquaintances, whose judgment we in some degree value, than in that of strangers. A physician told me that a young man, a wealthy duke, with whom he had travelled as medical attendant, blushed like a girl, when he paid him his fee; yet this young man probably would not have blushed and been shy, had he been paying a bill to a tradesman. Some persons, however, are so sensitive, that the mere act of speaking to almost any one is sufficient to rouse their self-consciousness, and a slight blush is the result.

Disapprobation or ridicule, from our sensitiveness on this head, causes shyness and blushing much more readily than does approbation; though the latter with some persons is highly efficient. The conceited are rarely shy; for they value themselves much too highly to expect depreciation. Why a proud man is often shy, as appears to be the case, is not so obvious, unless it be that, with all his self-reliance, he really thinks much about the opinion of others, although in a disdainful spirit. Persons who are exceedingly shy are rarely shy in the presence of those with whom they are quite familiar, and of whose good opinion and sympathy they are perfectly assured,—for instance, a girl in the presence of her mother. I neglected to inquire in my printed paper whether shyness can be detected in the different races of man; but a Hindoo gentleman assured Mr. Erskine that it is recognizable in his countrymen.

Shyness, as the derivation of the word indicates in several languages, is closely related to fear; yet it is distinct from fear in the ordinary sense. A shy man no doubt dreads the notice of strangers, but can hardly be said to be afraid of them; he may be as bold as a hero in battle, and yet have no self-confidence about trifles in the presence of strangers. Almost every one is extremely nervous when first addressing a public assembly, and most men remain so throughout their lives; but this appears to depend on the consciousness of a great coming exertion, with its associated effects on the system, rather than on shyness; although a timid or shy man no doubt suffers on such occasions infinitely more than another. With very young children it is difficult to distinguish between fear and shyness; but this latter feeling with them has often seemed to me to partake of the character of the wildness of an untamed animal. Shyness comes on at a very early age. In one of my own children, when two years and three months old, I saw a trace of what certainly appeared to be shyness, directed towards myself after an absence from home of only a week. This was shown not by a blush, but by the eyes being for a few minutes slightly averted from me. I have noticed on other occasions that shyness or shamefacedness and real shame are exhibited in the eyes of young children before they have acquired the power of blushing.

As shyness apparently depends on self-attention, we can perceive how right are those who maintain that reprehending children for shyness, instead of doing them any good, does much harm, as it calls their attention still more closely to themselves. It has been well urged that "nothing hurts young people more than to be watched continually about their feelings, to have their countenances scrutinized, and the degrees of their sensibility measured by the surveying eye of the unmerciful spectator. Under the constraint of such examinations they can think of nothing but that they are looked at, and feel nothing but shame or apprehension."

Moral causes: guilt.—With respect to blushing from strictly moral causes, we meet with the same fundamental principle as before, namely, regard for the opinion of others. It is not the conscience which raises a blush, for a man may sincerely regret some slight fault committed in solitude, or he may suffer the deepest remorse for an undetected crime, but he will not blush. "I blush," says Dr. Burgess, "in the presence of my accusers." It is not the sense of guilt, but the thought that others think or know us to be guilty which crimsons the face. A man may feel thoroughly ashamed at having told a small falsehood, without blushing; but if he even suspects that he is detected he will instantly blush, especially if detected by one whom he reveres.

On the other hand, a man may be convinced that God witnesses all his actions, and he may feel deeply conscious of some fault and pray for forgiveness; but this will not, as a lady who is a great blusher believes, ever excite a blush. The explanation of this difference between the knowledge by God and man of our actions lies, I presume, in man's disapprobation of immoral conduct being somewhat akin in nature to his depreciation of our personal appearance, so that through association both lead to similar results; whereas the disapprobation of God brings up no such association.

Many a person has blushed intensely when accused of some crime, though completely innocent of it. Even the thought, as the lady before referred to has observed to me, that others think that we have made an unkind or stupid remark, is amply sufficient to cause a blush, although we know all the time that we have been completely misunderstood. An action may be meritorious or of an indifferent nature, but a sensitive person, if he suspects that others take a different view of it, will blush. For instance, a lady by herself may give money to a beggar without a trace of a blush, but if others are present, and she doubts whether they approve, or suspects that they think her influenced by display, she will blush. So it will be, if she offers to relieve the distress of a decayed gentlewoman, more particularly of one whom she had previously known under better circumstances, as she cannot then feel sure how her conduct will be viewed. But such cases as these blend into shyness.

Breaches of etiquette.—The rules of *etiquette* always refer to conduct in the presence of, or towards others. They have no necessary connection with the moral sense, and are often meaningless. Nevertheless as they depend on the fixed custom of our equals and superiors, whose opinion we highly regard, they are considered almost as binding as are the laws of honour to a gentleman. Consequently the breach of the laws of etiquette, that is, any impoliteness or *gaucherie*, any impropriety, or an inappropriate remark, though quite accidental, will cause the most intense blushing of which a man is capable. Even the recollection of such an act, after an interval of many years, will make the whole body to tingle. So strong, also, is the power of sympathy that a sensitive person, as a lady has assured me, will sometimes blush at a flagrant breach of etiquette by a perfect stranger, though the act may in no way concern her.

Modesty.—This is another powerful agent in exciting blushes; but the word modesty includes very different states of the mind. It implies humility, and we often judge of this by persons being greatly pleased and blushing at slight praise, or by being annoyed at praise which seems to them too high according to their own humble standard of themselves. Blushing here has the usual signification of regard for the opinion of others. But modesty frequently relates to acts of indelicacy; and indelicacy is an affair of etiquette, as we clearly see with the nations that go altogether or nearly naked. He who is modest, and blushes easily at acts of this nature, does so because they are breaches of a firmly and wisely established etiquette. This is indeed shown by the derivation of the word *modest* from *modus*, a measure or standard of behaviour. A blush due to this form of modesty is, moreover, apt to be intense, because it generally relates to the opposite sex; and we have seen how in all cases our liability to blush is thus increased. We apply the term 'modest,' as it would appear, to those who have a humble opinion of themselves, and to those who are extremely sensitive about an indelicate word or deed, simply because in both cases blushes are readily excited, for these two frames of mind have nothing else in common. Shyness also, from this same cause, is often mistaken for modesty in the sense of humility.

Some persons flush up, as I have observed and have been assured, at any sudden and disagreeable recollection. The commonest cause seems to be the sudden remembrance of not having done something for another person which had been promised. In this case it may be that the thought passes half unconsciously through the mind, "What will he think of me?" and then the flush would partake of the nature of a true blush. But whether such flushes are in most cases due to the capillary circulation being affected, is very doubtful; for we must remember that almost every strong emotion, such as anger or great joy, acts on the heart, and causes the face to redden.

The fact that blushes may be excited in absolute solitude seems opposed to the view here taken, namely that the habit originally arose from thinking about what others think of us. Several ladies, who are great blushers, are unanimous in regard to solitude; and some of them believe that they have blushed in the dark. From what Mr. Forbes has stated with respect to the Aymaras, and from my own sensations, I have no doubt that this latter statement is correct. Shakespeare, therefore, erred when he made Juliet, who was not even by herself, say to Romeo (act ii. sc. 2):—

> *Thou know'st the mask of night is on my face;*
> *Else would a maiden blush bepaint my cheek,*
> *For that which thou hast heard me speak to-night.*

But when a blush is excited in solitude, the cause almost always relates to the thoughts of others about us—to acts done in their presence, or suspected by them; or again when we reflect what others would have thought of us had they known of the act. Nevertheless one or two of my informants believe that they have blushed from shame at acts in no way relating to others. If this be so, we must attribute the result to the force of inveterate habit and association, under a state of mind closely analogous to that which ordinarily excites a blush; nor need we feel surprise at this, as even sympathy with another person who commits a flagrant breach of etiquette is believed, as we have just seen, sometimes to cause a blush.

Finally, then, I conclude that blushing,—whether due to shyness—to shame for a real crime—to shame from a breach of the laws of etiquette—to modesty from humility—to modesty from an indelicacy—depends in all cases on the same principle; this principle being a sensitive regard for the opinion, more particularly for the depreciation of others, primarily in relation to our personal appearance, especially of our faces; and secondarily, through the force of association and habit, in relation to the opinion of others on our conduct.

Theory of Blushing.—We have now to consider, why should the thought that others are thinking about us affect our capillary circulation? Sir C. Bell insists that blushing "is a provision for expression, as may be inferred from the colour extending only to the surface of the face, neck, and breast, the parts most exposed. It is not acquired; it is from the beginning." Dr. Burgess believes that it was designed by the Creator in "order that the soul might have sovereign power of displaying in the cheeks the various internal emotions of the moral feelings;" so as to serve as a check on ourselves, and as a sign to others, that we were violating rules which ought to be held sacred. Gratiolet merely remarks,—"Or, comme il est dans l'ordre de la nature que l'être social le plus intelligent soit aussi le plus intelligible, cette faculté de rougeur et de pâleur qui distingue l'homme, est un signe naturel de sa haute perfection." ["Therefore, since within the natural order the most intelligent social being must also be the most intelligible, this facility for blushing and paling that distinguishes man is a natural sign of his great perfection."—editors]

The belief that blushing was *specially* designed by the Creator is opposed to the general theory of evolution, which is now so largely accepted; but it forms no part of my duty here to argue on the general question. Those who believe in design, will find it difficult to account for shyness being the most frequent and efficient of all the causes of blushing, as it makes the blusher to suffer and the beholder uncomfortable, without being of the least service to either of them. They will also find it difficult to account for negroes and other dark-coloured races blushing, in whom a change of colour in the skin is scarcely or not at all visible.

No doubt a slight blush adds to the beauty of a maiden's face; and the Circassian women who are capable of blushing, invariably fetch a higher price in the seraglio of the Sultan than less susceptible women. But the firmest believer in the efficacy of sexual selection will hardly suppose that blushing was acquired as a sexual ornament. This view would also be opposed to what has just been said about the dark-coloured races blushing in an invisible manner.

The hypothesis which appears to me the most probable, though it may at first seem rash, is that attention closely directed to any part of the body tends to interfere with the ordinary and tonic contraction of the small arteries of that part. These vessels, in consequence, become at such times more or less relaxed, and are instantly filled with arterial blood. This tendency will have been much strengthened, if frequent attention has been paid during many generations to the same part, owing to nerve-force readily flowing along accustomed channels, and by the power of inheritance. Whenever we believe that others are depreciating or even considering our personal appearance, our attention is vividly directed to the outer and visible parts of our bodies; and of all such parts we are most sensitive about our faces, as no doubt has been the case during many past generations. Therefore, assuming for the moment that the capillary vessels can be acted on by close attention, those of the face will have become eminently susceptible. Through the force of association, the same effects will tend to follow whenever we think that others are considering or censuring our actions or character.

As the basis of this theory rests on mental attention having some power to influence the capillary circulation, it will be necessary to give a considerable body of details, bearing more or less directly on this subject. Several observers, who from their wide experience and knowledge are eminently capable of forming a sound judgment, are convinced that attention or consciousness (which latter term Sir H. Holland thinks the more explicit) concentrated on almost any part of the body produces some direct physical effect on it. This applies to the movements of the involuntary muscles, and of the voluntary muscles when acting involuntarily,—to the secretion of the glands,—to the activity of the senses and sensations,—and even to the nutrition of parts.

It is known that the involuntary movements of the heart are affected if close attention be paid to them. Gratiolet gives the case of a man, who by continually watching and counting his own pulse, at last caused one beat out of every six to intermit. On the other hand, my father told me of a careful observer, who certainly had heart-disease and died from it, and who positively stated that his pulse was habitually irregular to an extreme degree; yet to his great disappointment it invariably became regular as soon as my father entered the room. Sir H. Holland remarks, that "the effect upon the circulation of a part from the consciousness suddenly directed and fixed upon it, is often obvious and immediate." Professor Laycock, who has particularly attended to phenomena of this nature, insists that "when the attention is directed to any portion of the body, innervation and circulation are excited locally, and the functional activity of that portion developed."

It is generally believed that the peristaltic movements of the intestines are influenced by attention being paid to them at fixed recurrent periods; and these movements depend on the contraction of unstriped and involuntary muscles. The abnormal action of the voluntary muscles in epilepsy, chorea, and hysteria is known to be influenced by the expectation of an attack, and by the sight of other patients similarly affected. So it is with the involuntary acts of yawning and laughing.

Certain glands are much influenced by thinking of them, or of the conditions under which they have been habitually excited. This is familiar to every one in the increased flow of saliva, when the thought, for instance, of intensely acid fruit is kept before the mind. It was shown in our sixth chapter, that an earnest and long-continued desire either to repress, or to increase, the action of the lacrymal glands is effectual. Some curious cases have been recorded in thc casc of women, of the power of the mind on the mammary glands; and still more remarkable ones in relation to the uterine functions.

When we direct our whole attention to any one sense, its acuteness is increased; and the continued habit of close attention, as with blind people to that of hearing, and with the blind and deaf to that of touch, appears to improve the sense in question permanently. There is, also, some reason to believe, judging from the capacities of different races of man, that the effects are inherited. Turning to ordinary sensations, it is well known that pain is increased by attending to it; and Sir B. Brodie goes so far as to believe that pain may be felt in any part of the body to which attention is closely drawn. Sir H. Holland also remarks that we become not only conscious of the existence of a part subjected to concentrated attention, but we experience in it various odd sensations, as of weight, heat, cold, tingling, or itching.

Lastly, some physiologists maintain that the mind can influence the nutrition of parts. Sir J. Paget has given a curious instance of the power, not indeed of the mind, but of the nervous system, on the hair. A lady "who is subject to attacks of what is called nervous headache, always finds in the morning after such an one, that some patches of her hair are white, as if powdered with starch. The change is effected in a night, and in a few days after, the hairs gradually regain their dark brownish colour."

We thus see that close attention certainly affects various parts and organs, which are not properly under the control of the will. By what means attention—perhaps the most wonderful of all the wondrous powers of the mind—is effected, is an extremely obscure subject. According to Müller, the process by which the sensory cells of the brain are rendered, through the will, susceptible of receiving more intense and distinct impressions, is closely analogous to that by which the motor cells are excited to send nerve-force to the voluntary muscles. There are many points of analogy in the action of the sensory and motor nerve-cells; for instance, the familiar fact that close attention to any one sense causes fatigue, like the prolonged exertion of any one muscle. When therefore we voluntarily concentrate our attention on any part of the body, the cells of the brain which receive impressions or sensations from that part are, it is probable, in some unknown manner stimulated into activity. This may account, without any local change in the part to which our attention is earnestly directed, for pain or odd sensations being there felt or increased.

If, however, the part is furnished with muscles, we cannot feel sure, as Mr. Michael Foster has remarked to me, that some slight impulse may not be unconsciously sent to such muscles; and this would probably cause an obscure sensation in the part.

In a large number of cases, as with the salivary and lacrymal glands, intestinal canal, &c., the power of attention seems to rest, either chiefly, or as some physiologists think, exclusively, on the vaso-motor system being affected in such a manner that more blood is allowed to flow into the capillaries of the part in question. This increased action of the capillaries may in some cases be combined with the simultaneously increased activity of the sensorium.

The manner in which the mind affects the vaso-motor system may be conceived in the following manner. When we actually taste sour fruit, an impression is sent through the gustatory nerves to a certain part of the sensorium; this transmits nerve-force to the vaso-motor centre, which consequently allows the muscular coats of the small arteries that permeate the salivary glands to relax. Hence more blood flows into these glands, and they secrete a copious supply of saliva. Now it does not seem an improbable assumption, that, when we reflect intently on a sensation, the same part of the sensorium, or a closely connected part of it, is brought into a state of activity, in the same manner as when we actually perceive the sensation. If so, the same cells in the brain will be excited, though, perhaps, in a less degree, by vividly thinking about a sour taste, as by perceiving it; and they will transmit in the one case, as in the other, nerve-force to the vaso-motor centre with the same results.

To give another, and, in some respects, more appropriate illustration. If a man stands before a hot fire, his face reddens. This appears to be due, as Mr. Michael Foster informs me, in part to the local action of the heat, and in part to a reflex action from the vaso-motor centres. In this latter case, the heat affects the nerves of the face; these transmit an impression to the sensory cells of the brain, which act on the vaso-motor centre, and this reacts on the small arteries of the face, relaxing them and allowing them to become filled with blood. Here, again, it seems not improbable that if we were repeatedly to concentrate with great earnestness our attention on the recollection of our heated faces, the same part of the sensorium which gives us the consciousness of actual heat would be in some slight degree stimulated, and would in consequence tend to transmit some nerve-force to the vaso-motor centres, so as to relax the capillaries of the face. Now as men during endless generations have had their attention often and earnestly directed to their personal appearance, and especially to their faces, any incipient tendency in the facial capillaries to be thus affected will have become in the course of time greatly strengthened through the principles just referred to, namely, nerve-force passing readily along accustomed channels, and inherited habit. Thus, as it appears to me, a plausible explanation is afforded of the leading phenomena connected with the act of blushing.

Recapitulation.—Men and women, and especially the young, have always valued, in a high degree, their personal appearance; and have likewise regarded the appearance of others. The face has been the chief object of attention, though, when man aboriginally went naked, the whole surface of his body would have been attended to. Our self-attention is excited almost exclusively by the opinion of others, for no person living in absolute solitude would care about his appearance. Every one feels blame more acutely than praise. Now, whenever we know, or suppose, that others are depreciating our personal appearance, our attention is strongly drawn towards ourselves, more especially to our faces. The probable effect of this will be, as has just been explained, to excite into activity that part of the sensorium which receives the sensory nerves of the face; and this will react through the vaso-motor system on the facial capillaries. By frequent reiteration during numberless generations, the process will have become so habitual, in association with the belief that others are thinking of us, that even a suspicion of their depreciation suffices to relax the capillaries, without any conscious thought about our faces. With some sensitive persons it is enough even to notice their dress to produce the same effect. Through the force, also, of association and inheritance our capillaries are relaxed, whenever we know, or imagine, that any one is blaming, though in silence, our actions, thoughts, or character; and, again, when we are highly praised.

On this hypothesis we can understand how it is that the face blushes much more than any other part of the body, though the whole surface is somewhat affected, more especially with the races which still go nearly naked. It is not at all surprising that the dark-coloured races should blush, though no change of colour is visible in their skins. From the principle of inheritance it is not surprising that persons born blind should blush. We can understand why the young are much more affected than the old, and women more than men; and why the opposite sexes especially excite each other's blushes. It becomes obvious why personal remarks should be particularly liable to cause blushing, and why the most powerful of all the causes is shyness; for shyness relates to the presence and opinion of others, and the shy are always more or less self-conscious. With respect to real shame from moral delinquencies, we can perceive why it is not guilt, but the thought that others think us guilty, which raises a blush. A man reflecting on a crime committed in solitude, and stung by his conscience, does not blush; yet he will blush under the vivid recollection of a detected fault, or of one committed in the presence of others, the degree of blushing being closely related to the feeling of regard for those who have detected, witnessed, or suspected his fault. Breaches of conventional rules of conduct, if they are rigidly insisted on by our equals or superiors, often cause more intense blushes even than a detected crime; and an act which is really criminal, if not blamed by our equals, hardly raises a tinge of colour on our cheeks. Modesty from humility, or from an indelicacy, excites a vivid blush, as both relate to the judgment or fixed customs of others.

From the intimate sympathy which exists between the capillary circulation of the surface of the head and of the brain, whenever there is intense blushing, there will be some, and often great, confusion of mind. This is frequently accompanied by awkward movements, and sometimes by the involuntary twitching of certain muscles.

As blushing, according to this hypothesis, is an indirect result of attention, originally directed to our personal appearance, that is to the surface of the body, and more especially to the face, we can understand the meaning of the gestures which accompany blushing throughout the world. These consist in hiding the face, or turning it towards the ground, or to one side. The eyes are generally averted or are restless, for to look at the man who causes us to feel shame or shyness, immediately brings home in an intolerable manner the consciousness that his gaze is directed on us. Through the principle of associated habit, the same movements of the face and eyes are practised, and can, indeed, hardly be avoided, whenever we know or believe that others are blaming, or too strongly praising, our moral conduct.

Perception in the Lower Animals.
[1873]

THIS LITTLE PAPER on animal perception, published in Nature *in March 1873, typifies the ever-active Darwinian mind, always observing and appraising particular phenomena.*

As several persons seem interested in Mr. Wallace's suggestion that animals find their way home by recognizing the odour of the places which they have passed whilst shut up, you may perhaps think the following little fact worth giving. Many years ago I was on a mail-coach, and as soon as we came to a public-house, the coachman pulled up for the fraction of a second. He did so when we came to a second public-house, and I then asked him the reason. He pointed to the off-hand wheeler, and said that she had been long completely blind, and she would stop at every place on the road at which she had before stopped. He had found by experience that less time was wasted by pulling up his team than by trying to drive her past the place, for she was contented with a momentary stop. After this I watched her, and it was evident that she knew exactly, before the coachman began to pull up the other horses, every public-house on the road, for she had at some time stopped at all. I think there can be little doubt that this mare recognized all these houses by her sense of smell. With respect to cats, so many cases have been recorded of their returning from a considerable distance to their homes, after having been carried away shut up in baskets, that I can hardly disbelieve them, though these stories are disbelieved by some persons. Now, as far as I have observed, cats do not possess a very acute sense of smell, and they seem to discover their prey by eyesight and by hearing. This leads me to mention another trifling fact: I sent a riding-horse by railway from Kent *via* Yarmouth, to Freshwater Bay, in the Isle of Wight. On the first day that I rode eastward, my horse, when I turned to go home, was very unwilling to return towards his stable, and he several times turned round. This led me to make repeated trials, and every time that I slackened the reins, he turned sharply round and began to trot to the eastward by a little north, which was nearly in the direction of his home in Kent. I had ridden this horse daily for several years, and he had never before behaved in this manner. My impression was that he somehow knew the direction whence he had been brought. I should state that the last stage from Yarmouth to Freshwater is almost due south, and along this road he had been ridden by my groom; but he never once showed any wish to return in this direction. I had purchased this horse several years before from a gentleman in my own neighbourhood, who had possessed him for a considerable time. Nevertheless it is possible, though far from probable, that the horse may have been born in the Isle of Wight. Even if we grant to animals a sense of the points of the compass, of which there is no evidence, how can we account, for instance, for the turtles which formerly congregated in multitudes, only at one season of the year, on the shores of the Isle of Ascension, finding their way to that speck of land in the midst of the great Atlantic Ocean?

Flowers of the Primrose Destroyed by Birds. [1874]

Another engaging account of specific observations made and the questions those observations raise, the following paper on primroses appeared in Nature, *April 23, 1874. Darwin had been especially attentive to primrose flowers since he began studying their mechanisms of pollination in 1861. His classic paper "On the two forms, or dimorphic condition, in the species of* Primula, *and their remarkable sexual relations" had been published in the* Journal of the Proceedings of the Linnean Society (Botany) *in 1862. In this paper Darwin had explained that the primrose's possession of two lengths of styles and placements of anthers is an adaptation promoting cross-pollination.*

For above twenty years I have observed every spring in my shrubberies and in the neighbouring woods, that a large number of the flowers of the primrose are cut off, and lie strewn on the ground close round the plants. So it is sometimes with the flowers of the cowslip and polyanthus, when they are borne on short stalks. This year the devastation has been greater than ever; and in a little wood not far from my house many hundred flowers have been destroyed, and some clumps have been completely denuded. For reasons presently to be given, I have no doubt that this is done by birds; and as I once saw some green-finches flying away from some primroses, I suspect that this is the enemy. The object of the birds in thus cutting off the flowers long perplexed me. As we have little water hereabouts, I at one time thought that it was done in order to squeeze the juice out of the stalks; but I have since observed that they are as frequently cut during very rainy, as during dry weather. One of my sons then suggested that the object was to get the nectar of the flowers; and I have no doubt that this is the right explanation. On a hasty glance it appears as if the foot-stalk had been cut through; but on close inspection, it will invariably be found that the extreme base of the calyx and the young ovary are left attached to the foot-stalk. And if the cut-off ends of the flowers be examined, it will be seen that they do not fit the narrow cut-off ends of the calyx, which remains attached to the stalk. A piece of the calyx between one and two-tenths of an inch in length, has generally been cut clean away; and these little bits of the calyx can often be found on the ground; but sometimes they remain hanging by a few fibres to the upper part of the calyx of the detached flowers. Now no animal that I can think of, except a bird, could make two almost parallel clean cuts, transversely across the calyx of a flower. The part which is cut off contains within the narrow tube of the corolla the nectar; and the pressure of the bird's beak would force this out at both the cut-off ends. I have never heard of any bird in Europe feeding on nectar; though there are many that do so in the tropical parts of the New and Old Worlds, and which are believed to aid in the cross-fertilisation of the species. In such cases both the bird and the plant would profit. But with the primrose it is an unmitigated evil, and might well lead to its extermination; for in the wood above alluded to many hundred flowers have been destroyed this season, and cannot produce a single seed. My object in this communication to *Nature* is to ask your correspondents in England and abroad to observe whether the primroses there suffer, and to state the result, whether negative or affirmative, adding whether primroses are abundant in each district. I cannot remember having formerly seen anything of the kind in the midland counties of England. If the habit of cutting off the flowers should prove, as seems probable, to be general, we must look at it as inherited or instinctive; for it is unlikely that each bird should have discovered during its individual life-time the exact spot where the nectar lies concealed within the tube of the corolla, and should have learnt to bite off the flowers so skilfully that a minute portion of the calyx is always left attached to the foot-stalk. If, on the other hand, the evil is confined to this part of Kent, it will be a curious case of a new habit or instinct arising in this primrose-decked land.

Insectivorous Plants. [1875]

*In the summer and fall of 1860, Darwin's daughter Henrietta was ill. The family took her first to Hartfield, the Sussex home of Emma's sister Sarah Elizabeth Wedgwood. While walking there and later on the south coastal downs at Eastbourne, where they took Henrietta for the sea air, Darwin noticed how common sundews (*Drosera rotundifolia*) capture insects on their sticky-haired leaves. He took some specimens home and watched their responses to insects placed on their leaves. Then he began to experiment with minute samples of other materials, such as sand, glass, hair, and meat. "Fortunately," Darwin states in the* Autobiography, *"a crucial test occurred to me, that of placing a large number of leaves in various nitrogenous and non-nitrogenous fluids of equal density, and as soon as I found that the former alone excited energetic movements, it was obvious that here was a fine new field for investigation." Thus was born yet another of his long-term studies of plant adaptation. Darwin performed a great many such experiments on insectivorous plants in the early 1860s and returned to them in the 1870s. While Darwin's experiments were proceeding, Emma wryly wrote to Lady Lyell, "At present he is treating Drosera just like a living creature, and I suppose hopes to end in proving it to be an animal."*

The resulting book, Insectivorous Plants, *was published in 1875. A posthumous second edition incorporating Darwin's marginal comments from his copy of the first edition appeared in 1888, its editor and annotator being Francis Darwin, who had trained in botany at Cambridge, received his Ph.D. in Germany, and assisted his father in his later botanical studies. Despite its title,* Insectivorous Plants *is not principally a book about how insects are captured. It chiefly chronicles the experiments on how the plants reacted to various substances placed on their leaves, and thus has more to do with physiology than with morphology or behavior. We offer Darwin's descriptions of how insects are captured by the sundew, Venus's flytrap, and bladderwort. Not much has been added to what Darwin learned about sundews and flytraps; however, we now understand that the bladderwort's means of entrapment is less passive than Darwin describes it as being. When the hairs on a bladder are pressed, they cause the bladder to open and to suck the insect inside, where its proteins are digested and absorbed.*

DROSERA ROTUNDIFOLIA, OR THE COMMON SUNDEW.

During the summer of 1860, I was surprised by finding how large a number of insects were caught by the leaves of the common sun-dew (*Drosera rotundifolia*) on a heath in Sussex. I had heard that insects were thus caught, but knew nothing further on the subject. I gathered by chance a dozen plants, bearing fifty-six fully expanded leaves, and on thirty-one of these dead insects or remnants of them adhered; and, no doubt, many more would have been caught afterwards by these same leaves, and still more by those as yet not expanded. On one plant all six leaves had caught their prey; and on several plants very many leaves had caught more than a single insect. On one large leaf I found the remains of thirteen distinct insects. Flies (Diptera) are captured much oftener than other insects. The largest kind which I have seen caught was a small butterfly (*Cænonympha pamphilus*); but the Rev. H. M. Wilkinson informs me that he found a large living dragon-fly with its body firmly held by two leaves. As this plant is extremely common in some districts, the number of insects thus annually slaughtered must be prodigious. Many plants cause the death of insects, for instance the sticky buds of the horse-chestnut (*Æsculus hippocastanum*), without thereby receiving, as far as we can perceive, any advantage; but it was soon evident that Drosera was excellently adapted

FIG. 1.*
(*Drosera rotundifolia.*)

Leaf viewed from above; enlarged four times.

for the special purpose of catching insects, so that the subject seemed well worthy of investigation.

The results have proved highly remarkable; the more important ones being—firstly, the extraordinary sensitiveness of the glands to slight pressure and to minute doses of certain nitrogenous fluids, as shown by the movements of the so-called hairs or tentacles; secondly, the power possessed by the leaves of rendering soluble or digesting nitrogenous substances, and of afterwards absorbing them; thirdly, the changes which take place within the cells of the tentacles, when the glands are excited in various ways.

* The drawings of Drosera and Dionæa, given in this work, were made for me by my son George Darwin: those of Aldrovanda, and of the several species of Utricularia, by my son Francis. They have been excellently reproduced on wood by Mr. Cooper, 188 Strand.

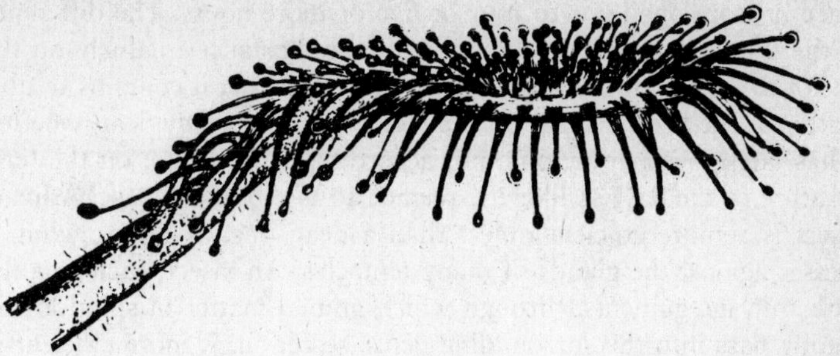

FIG. 2.
(*Drosera rotundifolia.*)
Old leaf viewed laterally; enlarged about five times.

It is necessary, in the first place, to describe briefly the plant. It bears from two or three to five or six leaves, generally extended more or less horizontally, but sometimes standing vertically upwards. The shape and general appearance of a leaf is shown, as seen from above, in fig. 1, and as seen laterally, in fig. 2. The leaves are commonly a little broader than long, but this was not the case in the one here figured. The whole upper surface is covered with gland-bearing filaments, or tentacles, as I shall call them, from their manner of acting. The glands were counted on thirty-one leaves, but many of these were of unusually large size, and the average number was 192; the greatest number being 260, and the least 130. The glands are each surrounded by large drops of extremely viscid secretion, which, glittering in the sun, have given rise to the plant's poetical name of the sun-dew.

• • •

Preliminary Sketch of the Action of the several Parts, and of the Manner in which Insects are Captured. If a small organic or inorganic object be placed on the glands in the centre of a leaf, these transmit a motor impulse to the marginal tentacles. The nearer ones are first affected and slowly bend towards the centre, and then those farther off, until at last all become closely inflected over the object. This takes place in from one hour to four or five or more hours. The difference in the time required depends on many circumstances; namely on the size of the object and on its nature, that is, whether it contains soluble matter of the proper kind; on the vigour and age of the leaf; whether it has lately been in action; and, according to Nitschke, on the temperature of the day, as likewise seemed to me to be the case. A living insect is a more efficient object than a dead one, as in struggling it presses against the glands of many tentacles. An insect, such as a fly, with thin integuments, through which animal matter in solution can readily pass into the surrounding dense secretion, is more efficient in causing prolonged inflection than an insect with a thick coat, such as a beetle. The inflection of the tentacles takes place indifferently in the light and darkness; and the plant is not subject to any nocturnal movement of so-called sleep.

If the glands on the disc are repeatedly touched or brushed, although no object is left on them, the marginal tentacles curve inwards. So again, if drops of various fluids, for instance of saliva or of a solution of any salt of ammonia, are placed on the central glands, the same result quickly follows, sometimes in under half an hour.

The tentacles in the act of inflection sweep through a wide space; thus a marginal tentacle, extended in the same plane with the blade, moves through an angle of 180°; and I have seen the much reflected tentacles of a leaf which stood upright move through an angle of not less than 270°. The bending part is almost confined to a short space near the base; but a rather larger portion of the elongated exterior tentacles becomes slightly incurved; the distal half in all cases remaining straight. The short tentacles in the centre of the disc when directly excited, do not become inflected; but they are capable of inflection if excited by a motor impulse received from other glands at a distance. Thus, if a leaf is immersed in an infusion of raw meat, or in a weak solution of ammonia (if the solution is at all strong, the leaf is paralysed), all the exterior tentacles bend inwards excepting those near the centre, which remain upright; but these bend towards any exciting object placed on one side of the disc. The glands may be seen to form a dark ring round the centre; and this follows from the exterior tentacles increasing in length in due proportion, as they stand nearer to the circumference.

The kind of inflection which the tentacles undergo is best shown when the gland of one of the long exterior tentacles is in any way excited; for the surrounding ones remain unaffected. . . . A gland may

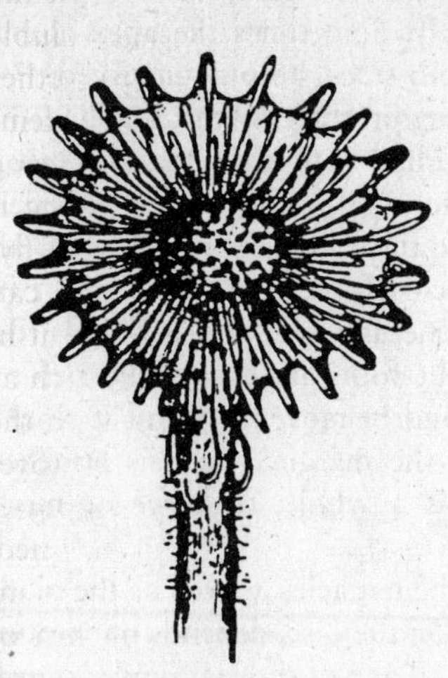

FIG. 4.
(*Drosera rotundifolia.*)
Leaf (enlarged) with all the tentacles closely inflected, from immersion in a solution of phosphate of ammonia (one part to 87,500 of water).

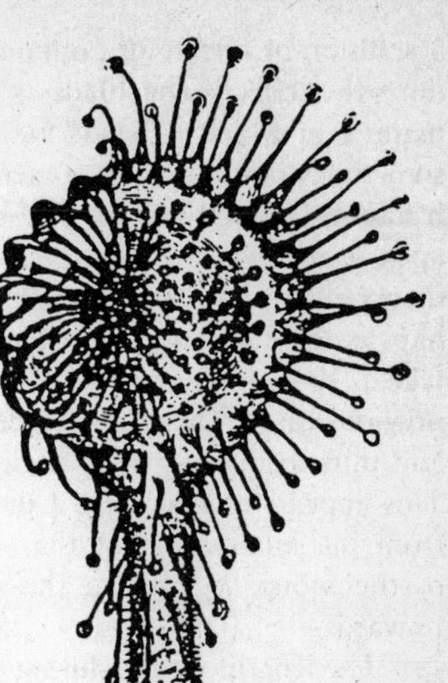

FIG. 5.
(*Drosera rotundifolia.*)
Leaf (enlarged) with the tentacles on one side inflected over a bit of meat placed on the disc.

be excited by being simply touched three or four times, or by prolonged contact with organic or inorganic objects, and various fluids. I have distinctly seen, through a lens, a tentacle beginning to bend in ten seconds, after an object had been placed on its gland; and I have often seen strongly pronounced inflection in under one minute. It is surprising how minute a particle of any substance, such as a bit of thread or hair or splinter of glass, if placed in actual contact with the surface of a gland, suffices to cause the tentacle to bend. If the object, which has been carried by this movement to the centre, be not very small, or if it contains soluble nitrogenous matter, it acts on the central glands; and these transmit a motor impulse to the exterior tentacles, causing them to bend inwards.

Not only the tentacles, but the blade of the leaf often, but by no means always, becomes much incurved, when any strongly exciting substance or fluid is placed on the disc. Drops of milk and of a solution of nitrate of ammonia or soda are particularly apt to produce this effect. The blade is thus converted into a little cup. The manner in which it bends varies greatly. Sometimes the apex alone, sometimes one side, and sometimes both sides, become incurved. For instance, I placed bits of hard-boiled egg on three leaves; one had the apex bent towards the base; the second had both distal margins much incurved, so that it became almost triangular in outline, and this perhaps is the commonest case; whilst the third blade was not at all affected, though the tentacles were as closely inflected as in the two previous cases. The whole blade also generally rises or bends upwards, and thus forms a smaller angle with the footstalk than it did before. This appears at first sight a distinct kind of movement, but it results from the incurvation of that part of the margin which is attached to the footstalk, causing the blade, as a whole, to curve or move upwards.

The length of time during which the tentacles as well as the blade remain inflected over an object placed on the disc, depends on various circumstances; namely on the vigour and age of the leaf, and, according to Dr. Nitschke, on the temperature, for during cold weather when the leaves are inactive, they re-expand at an earlier period than when the weather is warm. But the nature of the object is by far the most important circumstance; I have repeatedly found that the tentacles remain clasped for a much longer average time over objects which yield soluble nitrogenous matter than over those, whether organic or inorganic, which yield no such matter. After a period varying from one to seven days, the tentacles and blade re-expand, and are then ready to act again. I have seen the same leaf inflected three successive times over insects placed on the disc; and it would probably have acted a greater number of times.

The secretion from the glands is extremely viscid, so that it can be drawn out into long threads. It appears colourless, but stains little balls of paper pale pink. An object of any kind placed on a gland always causes it, as I believe, to secrete more freely; but the mere presence of the object renders this difficult to ascertain. In some cases, however, the effect was strongly marked, as when particles of sugar were added; but the result in this case is probably due merely to exosmose. Particles of carbonate and phosphate of ammonia and of some other salts, for instance sulphate of zinc, likewise increase the secretion. Immersion in a solution of one part of chloride of gold, or of some other salts, to 437 of water, excites the glands to largely increased secretion; on the other hand, tartrate of antimony produces no such effect. Immersion in many acids (of the strength of one part to 437 of water) likewise causes a wonderful amount of secretion, so that when the leaves are lifted out, long ropes of extremely viscid fluid hang from them. Some acids, on the other hand, do not act in this manner. Increased secretion is not necessarily dependent on the inflection of the tentacle, for particles of sugar and of sulphate of zinc cause no movement.

It is a much more remarkable fact that when an object, such as a bit of meat or an insect, is placed on the disc of a leaf, as soon as the surrounding tentacles become considerably inflected, their glands pour forth an increased amount of secretion. I ascertained this by selecting leaves with equal-sized drops on the two sides, and by placing bits of meat on one side of the disc; and as soon as the tentacles on this side became much inflected, but before the glands touched the meat, the drops of secretion became larger. This was repeatedly observed, but a record was kept of only thirteen cases, in nine of which increased secretion was plainly observed; the four failures being due either to the leaves being rather torpid, or to the bits of meat being too small to cause much inflection. We must therefore conclude that the central glands, when strongly excited, transmit some influence to the glands of the circumferential tentacles, causing them to secrete more copiously.

It is a still more important fact (as we shall see more fully when we treat of the digestive power of the secretion) that when the tentacles become inflected, owing to the central glands having been stimulated mechanically, or by contact with animal matter, the secretion not only increases in quantity, but changes its nature and becomes acid; and this occurs before the glands have touched the object on the centre of the leaf. This acid is of a different nature from that contained in the tissue of the leaves. As long as the tentacles remain closely inflected, the glands continue to secrete, and the secretion is acid; so that, if neutralised by carbonate of soda, it again becomes acid after a few hours. I have observed the same leaf with the tentacles closely inflected over rather indigestible substances, such as chemically prepared casein, pouring forth acid secretion for eight successive days, and over bits of bone for ten successive days.

The secretion seems to possess, like the gastric juice of the higher animals, some antiseptic power. During very warm weather I placed close together two equal-sized bits of raw meat, one on a leaf of the Drosera, and the other surrounded by wet moss. They were thus left for 48 hrs., and then examined. The bit on the moss swarmed with infusoria, and was so much decayed that the transverse striæ on the muscular fibres could no longer be clearly distinguished; whilst the bit on the leaf, which was bathed by the secretion, was free from infusoria, and its striæ were perfectly distinct in the central and undissolved portion. In like manner small cubes of albumen and cheese placed on wet moss became threaded with filaments of mould, and had their surfaces slightly discoloured and disintegrated; whilst those on the leaves of Drosera remained clean, the albumen being changed into transparent fluid.

As soon as tentacles, which have remained closely inflected during several days over an object, begin to re-expand, their glands secrete less freely, or cease to secrete, and are left dry. In this state they are covered with a film of whitish, semi-fibrous matter, which was held in solution by the secretion. The drying of the glands during the act of re-expansion is of some little service to the plant; for I have often observed that objects adhering to the leaves could then be blown away by a breath of air; the leaves being thus left unencumbered and free for future action. Nevertheless, it often happens that all the glands do not become completely dry; and in this case delicate objects, such as fragile insects, are sometimes torn by the re-expansion of the tentacles into fragments, which remain scattered all over the leaf. After the re-expansion is complete, the glands quickly begin to re-secrete, and as soon as full-sized drops are formed, the tentacles are ready to clasp a new object.

When an insect alights on the central disc, it is instantly entangled by the viscid secretion, and the surrounding tentacles after a time begin to bend, and ultimately clasp it on all sides. Insects are generally killed, according to Dr. Nitschke, in about a quarter of an hour, owing to their tracheæ being closed by the secretion. If an insect adheres to only a few of the glands of the exterior tentacles, these soon become inflected and carry their prey to the tentacles next succeeding them inwards; these then bend inwards, and so onwards, until the insect is ultimately carried by a curious sort of rolling movement to the centre of the leaf. Then, after an interval, the tentacles on all sides become inflected and bathe their prey with their secretion, in the same manner as if the insect had first alighted on the central disc. It is surprising how minute an insect suffices to cause this action: for instance, I have seen one of the smallest species of gnats (Culex), which had just settled with its excessively delicate feet on the glands of the outermost tentacles, and these were already beginning to curve inwards, though not a single gland had as yet touched the body of the insect. Had I not interfered, this minute gnat would assuredly have been carried to the centre of the leaf and been securely clasped on all sides. We shall hereafter see what excessively small doses of certain organic fluids and saline solutions cause strongly marked inflection.

Whether insects alight on the leaves by mere chance, as a resting-place, or are attracted by the odour of the secretion, I know not. I suspect from the number of insects caught by the English species of Drosera, and from what I have observed with some exotic species kept in my greenhouse, that the odour is attractive. In this latter case the leaves may be compared with a baited trap; in the former case with a trap laid in a run frequented by game, but without any bait.

That the glands possess the power of absorption, is shown by their almost instantaneously becoming dark-coloured when given a minute quantity of carbonate of ammonia; the change of colour being chiefly or exclusively due to the rapid aggregation of their contents. When certain other fluids are added, they become pale-coloured. Their power of absorption is, however, best shown by the widely different results which follow, from placing drops of various nitrogenous and non-nitrogenous fluids of the same density on the glands of the disc, or on a single marginal gland; and likewise by the very different lengths of time during which the tentacles remain inflected over objects, which yield or do not yield soluble nitrogenous matter. This same conclusion might indeed have been inferred from the structure and movements of the leaves, which are so admirably adapted for capturing insects.

The absorption of animal matter from captured insects explains how Drosera can flourish in extremely poor peaty soil,—in some cases where nothing but sphagnum moss grows, and mosses depend altogether on the atmosphere for their nourishment. Although the leaves at a hasty glance do not appear green, owing to the purple colour of the tentacles, yet the upper and lower surfaces of the blade, the pedicels of the central tentacles, and the petioles contain chlorophyll, so that, no doubt, the plant obtains and assimilates carbonic acid from the air. Nevertheless, considering the nature of the soil where it grows, the supply of nitrogen would be extremely limited, or quite deficient, unless the plant had the power of obtaining this important element from captured insects. We can thus understand how it is that the roots are so poorly developed. These usually consist of only two or three slightly divided branches, from half to one inch in length, furnished with absorbent hairs. It appears, therefore, that the roots serve only to imbibe water; though, no doubt, they would absorb nutritious matter if present in the soil; for as we shall hereafter see, they absorb a weak solution of carbonate of ammonia. A plant of Drosera, with the edges of its leaves curled inwards, so as to form a temporary stomach, with the glands of the closely inflected tentacles pouring forth their acid secretion, which dissolves animal matter, afterwards to be absorbed, may be said to feed like an animal. But, differently from an animal, it drinks by means of its roots; and it must drink largely, so as to retain many drops of viscid fluid round the glands, sometimes as many as 260, exposed during the whole day to a glaring sun.

DIONEA MUSCIPULA.

This plant, commonly called Venus' fly-trap, from the rapidity and force of its movements, is one of the most wonderful in the world. It is a member of the small family of the Droseraceæ, and is found only in the eastern part of North Carolina, growing in damp situations. The roots are small; those of a moderately fine plant which I examined consisted of two branches about 1 inch in length, springing from a bulbous enlargement. They probably serve, as in the case of Drosera, solely for the absorption of water; for a gardener, who has been very successful in the cultivation of this plant, grows it, like an epiphytic orchid, in well-drained damp moss without any soil. The form of the bilobed leaf, with its foliaceous foot-stalk, is shown in the accompanying drawing. The two lobes stand at rather less than a right angle to each other. Three minute pointed processes or filaments, placed triangularly, project from the upper surfaces of both; but I have seen two leaves with four filaments on each side, and another with only two. These filaments are remarkable from their extreme sensitiveness to a touch, as shown not by their own movement, but by that of the lobes. The margins of the leaf are prolonged into sharp rigid projections which I will call spikes, into each of which a bundle of spiral vessels enters. The spikes stand in such a position that, when the lobes close, they inter-lock like the teeth of a rat-trap. The midrib of the leaf, on the lower side, is strongly developed and prominent.

The upper surface of the leaf is thickly covered, excepting towards

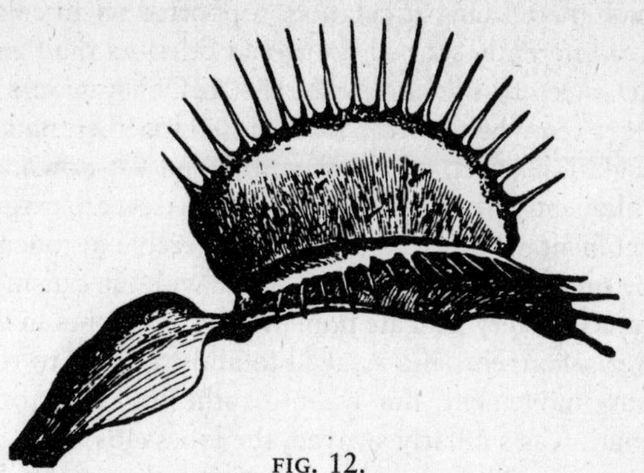

FIG. 12.
(*Dionæa muscipula.*)
Leaf viewed laterally in its expanded state.

the margins, with minute glands of a reddish or purplish colour, the rest of the leaf being green. There are no glands on the spikes, or on the foliaceous footstalk. The glands are formed of from twenty to thirty polygonal cells, filled with purple fluid. Their upper surface is convex. They stand on very short pedicels, into which spiral vessels do not enter, in which respect they differ from the tentacles of Drosera. They secrete, but only when excited by the absorption of certain matters; and they have the power of absorption. Minute projections, formed of eight divergent arms of a reddish-brown or orange colour, and appearing under the microscope like elegant little flowers, are scattered in considerable numbers over the foot-stalk, the backs of the leaves, and the spikes, with a few on the upper surface of the lobes. These octofid projections are no doubt homologous with the papillae on the leaves of *Drosera rotundifolia.* There are also a few very minute, simple, pointed hairs, about $\frac{7}{12000}$ (.0148 mm.) of an inch in length on the backs of the leaves.

The sensitive filaments are formed of several rows of elongated cells, filled with purplish fluid. They are a little above the $\frac{1}{20}$ of an inch in length; are thin and delicate, and taper to a point. I examined the bases of several, making sections of them, but no trace of the entrance of any vessel could be seen. The apex is sometimes bifid or even trifid, owing to a slight separation between the terminal pointed cells. Towards the base there is constriction, formed of broader cells, beneath which there is an articulation, supported on an enlarged base, consisting of differently shaped polygonal cells. As the filaments project at right angles to the surface of the leaf, they would have been liable to be broken whenever the lobes closed together, had it not been for the articulation which allows them to bend flat down.

These filaments, from their tips to their bases, are exquisitely sensitive to a momentary touch. It is scarcely possible to touch them ever so lightly or quickly with any hard object without causing the lobes to close. A piece of very delicate human hair, 2½ inches in length, held dangling over a filament, and swayed to and fro so as to touch it, did not excite any movement. But when a rather thick cotton thread of the same length was similarly swayed, the lobes closed. Pinches of fine wheaten flour, dropped from a height, produced no effect. The above-mentioned hair was then fixed into a handle, and cut off so that 1 inch projected; this length being sufficiently rigid to support itself in a nearly horizontal line. The extremity was then brought by a slow movement laterally into contact with the tip of a filament, and the leaf instantly closed. On another occasion two or three touches of the same kind were necessary before any movement ensued. When we consider how flexible a fine hair is, we may form some idea how slight must be the touch given by the extremity of a piece, 1 inch in length, moved slowly.

Although these filaments are so sensitive to a momentary and delicate touch, they are far less sensitive than the glands of Drosera to prolonged pressure. Several times I succeeded in placing on the tip of a filament, by the aid of a needle moved with extreme slowness, bits of rather thick human hair, and these did not excite movement, although they were more than ten times as long as those which caused the tentacles of Drosera to bend; and although in this latter case they were largely supported by the dense secretion. On the other hand, the glands of Drosera may be struck with a needle or any hard object, once, twice, or even thrice, with considerable force, and no movement ensues. This singular difference in the nature of the sensitiveness of the filaments of Dionæa and of the glands of Drosera evidently stands in relation to the habits of the two plants. If a minute insect alights with its delicate feet on the glands of Drosera, it is caught by the viscid secretion, and the slight, though prolonged pressure, gives notice of the presence of prey, which is secured by the slow bending of the tentacles. On the other hand, the sensitive filaments of Dionæa are not viscid, and the capture of insects can be assured only by their sensitiveness to a momentary touch, followed by the rapid closure of the lobes.

• • •

Digestive Power of the Secretion.—When a leaf closes over any object, it may be said to form itself into a temporary stomach; and if the object yields ever so little animal matter, this serves, to use Schiff's expression, as a peptogene, and the glands on the surface pour forth their acid secretion, which acts like the gastric juice of animals. As so many experiments were tried on the digestive power of Drosera, only a few were made with Dionæa, but they were amply sufficient to prove that it digests. This plant, moreover, is not so well fitted as Drosera for observation, as the process goes on within the closed lobes. Insects, even beetles, after being subjected to the secretion for several days, are surprisingly softened, though their chitinous coats are not corroded.

• • •

On the Manner in which Insects are caught.—We will now consider the action of the leaves when insects happen to touch one of the sensitive filaments. This often occurred in my greenhouse, but I do not know whether insects are attracted in any special way by the leaves. They are caught in large numbers by the plant in its native country. As soon as a filament is touched, both lobes close with astonishing quickness; and as they stand at less than a right angle to each other, they have a good chance of catching any intruder. The angle between the blade and footstalk does not change when the lobes close. The chief seat of movement is near the midrib, but is not confined to this part; for, as the lobes come together, each curves inwards across its whole breadth; the marginal spikes however, not becoming curved. This movement of the whole lobe was well seen in a leaf to which a large fly had been given, and from which a large portion had been cut off the end of one lobe; so that the opposite lobe, meeting with no resistance in this part, went on curving inwards much beyond the medial line. The whole of the lobe, from which a portion had been cut, was afterwards removed, and the opposite lobe now curled completely over, passing through an angle of from 120° to 130°, so as to occupy a position almost at right angles to that which it would have held had the opposite lobe been present.

From the curving inwards of the two lobes, as they move towards each other, the straight marginal spikes intercross by their tips at first, and ultimately by their bases. The leaf is then completely shut and encloses a shallow cavity. If it has been made to shut merely by one of the sensitive filaments having been touched, or if it includes an object not yielding soluble nitrogenous matter, the two lobes retain their inwardly concave form until they re-expand. The re-expansion under these circumstances—that is when no organic matter is enclosed—was observed in ten cases. In all of these, the leaves re-expanded to about two-thirds of the full extent in 24 hrs. from the time of closure. Even the leaf from which a portion of one lobe had been cut off opened to a slight degree within this same time. In one case a leaf re-expanded to about two-thirds of the full extent in 7 hrs., and completely in 32 hrs.; but one of its filaments had been touched merely with a hair just enough to cause the leaf to close. Of these ten leaves only a few re-expanded completely in less than two days, and two or three required even a little longer time. Before, however, they fully re-expand, they are ready to close instantly if their sensitive filaments are touched. How many times a leaf is capable of shutting and opening if no animal matter is left enclosed, I do not know; but one leaf was made to close four times, reopening afterwards, within six days. On the last occasion it caught a fly, and then remained closed for many days.

This power of reopening quickly after the filaments have been accidentally touched by blades of grass, or by objects blown on the leaf by the wind, as occasionally happens in its native place, must be of some importance to the plant; for as long as a leaf remains closed, it cannot of course capture an insect.

When the filaments are irritated and a leaf is made to shut over an insect, a bit of meat, albumen, gelatine, casein, and, no doubt, any other substance containing soluble nitrogenous matter, the lobes, instead of remaining concave, thus including a concavity, slowly press closely together throughout their whole breadth. As this takes place, the margins gradually become a little everted, so that the spikes, which at first intercrossed, at last project in two parallel rows. The lobes press against each other with such force that I have seen a cube of albumen much flattened, with distinct impressions of the little prominent glands; but this latter circumstance may have been partly caused by the corroding action of the secretion. So firmly do they become pressed together that, if any large insect or other object has been caught, a corresponding projection on the outside of the leaf is distinctly visible. When the two lobes are thus completely shut, they resist being opened, as by a thin wedge driven between them, with astonishing force, and are generally ruptured rather than yield. If not ruptured, they close again, as Dr. Canby informs me in a letter, "with quite a loud flap." But if the end of a leaf is held firmly between the thumb and finger, or by a clip, so that the lobes cannot begin to close, they exert, whilst in this position, very little force.

I thought at first that the gradual pressing together of the lobes was caused exclusively by captured insects crawling over and repeatedly irritating the sensitive filaments; and this view seemed the more probable when I learnt from Dr. Burdon Sanderson that whenever the filaments of a closed leaf are irritated, the normal electric current is disturbed. Nevertheless, such irritation is by no means necessary, for a dead insect, or a bit of meat, or of albumen, all act equally well; proving that in these cases it is the absorption of animal matter which excites the lobes slowly to press close together. We have seen that the absorption of an extremely small quantity of such matter also causes a fully expanded leaf to close slowly; and this movement is clearly analogous to the slow pressing together of the concave lobes. This latter action is of high functional importance to the plant, for the glands on both sides are thus brought into contact with a captured insect, and consequently secrete. The secretion with animal matter in solution is then drawn by capillary attraction over the whole surface of the leaf, causing all the glands to secrete and allowing them to absorb the diffused animal matter. The movement, excited by the absorption of such matter, though slow, suffices for its final purpose, whilst the movement excited by one of the sensitive filaments being touched is rapid, and this is indispensable for the capturing of insects. These two movements, excited by two such widely different means, are thus both well adapted, like all the other functions of the plant, for the purposes which they subserve.

There is another wide difference in the action of leaves which enclose objects, such as bits of wood, cork, balls of paper, or which have had their filaments merely touched, and those which enclose organic bodies yielding soluble nitrogenous matter. In the former case the leaves, as we have seen, open in under 24 hrs. and are then ready, even before being fully expanded, to shut again. But if they have closed over nitrogen-yielding bodies, they remain closely shut for many days; and after re-expanding are torpid, and never act again, or only after a considerable interval of time. In four instances, leaves after catching insects never reopened, but began to wither, remaining closed—in one case for fifteen days over a fly; in a second, for twenty-four days, though the fly was small; in a third for twenty-four days over a woodlouse; and in a fourth, for thirty-five days over a large Tipula. In two other cases leaves remained closed for at least nine days over flies, and for how many more I do not know. It should, however, be added that in two instances in which very small insects had been naturally caught the leaf opened as quickly as if nothing had been caught; and I suppose that this was due to such small insects not having been crushed or not having excreted any animal matter, so that the glands were not excited. Small angular bits of albumen and gelatine were placed at both ends of three leaves, two of which remained closed for thirteen and the other for twelve days. Two other leaves remained closed over bits of meat for eleven days, a third leaf for eight days, and a fourth (but this had been cracked and injured) for only six days. Bits of cheese, or casein, were placed at one end and albumen at the other end of three leaves; and the ends with the former opened after six, eight, and nine days, whilst the opposite ends opened a little later. None of the above bits of meat, albumen, &c., exceeded a cube of 1/10 of an inch (2.54 mm.) in size, and were sometimes smaller; yet these small portions sufficed to keep the leaves closed for many days. Dr. Canby informs me that leaves remain shut for a longer time over insects than over meat; and from what I have seen, I can well believe that this is the case, especially if the insects are large.

In all the above cases, and in many others in which leaves remained closed for a long but unknown period over insects naturally caught, they were more or less torpid when they reopened. Generally they were so torpid during many succeeding days that no excitement of the filaments caused the least movement. In one instance, however, on the day after a leaf opened which had clasped a fly, it closed with extreme slowness when one of its filaments was touched; and although no object was left enclosed, it was so torpid that it did not re-open for the second time until 44 hrs. had elapsed. In a second case, a leaf which had expanded after remaining closed for at least nine days over a fly, when greatly irritated, moved one alone of its two lobes, and retained this unusual position for the next two days. A third case offers the strongest exception which I have observed; a leaf, after remaining clasped for an unknown time over a fly, opened, and when one of its filaments was touched, closed, though rather slowly. Dr. Canby, who observed in the United States a large number of plants which, although not in their native site, were probably more vigorous than my plants, informs me that he has "several times known vigorous leaves to devour their prey several times; but ordinarily twice, or, quite often, once was enough to render them unserviceable." Mrs. Treat, who cultivated many plants in New Jersey, also informs me that "several leaves caught successively three insects each, but most of them were not able to digest the third fly, but died in the attempt. Five leaves, however, digested each three flies, and closed over the fourth, but died soon after the fourth capture. Many leaves did not digest even one large insect." It thus appears that the power of digestion is somewhat limited, and it is certain that leaves always remain clasped for many days over an insect, and do not recover their power of closing again for many subsequent days. In this respect Dionæa differs from Drosera, which catches and digests many insects after shorter intervals of time.

We are now prepared to understand the use of the marginal spikes, which form so conspicuous a feature in the appearance of the plant, and which at first seemed to me in my ignorance useless appendages. From the inward curvature of the lobes as they approach each other, the tips of the marginal spikes first intercross, and ultimately their bases. Until the edges of the lobes come into contact, elongated spaces between the spikes, varying from the 1/15 to the 1/10 of an inch (1.693 to 2.54 mm.) in breadth, according to the size of the leaf, are left open. Thus an insect, if its body is not thicker than these measurements, can easily escape between the crossed spikes, when disturbed by the closing lobes and increasing darkness; and one of my sons actually saw a small insect thus escaping. A moderately large insect, on the other hand, if it tries to escape between the bars will surely be pushed back again into its horrid prison with closing walls, for the spikes continue to cross more and more until the edges of the lobes come into contact. A very strong insect, however, would be able to free itself, and Mrs. Treat saw this effected by a rose-chafer (*Macrodactylus subspinosus*) in the United States. Now it would manifestly be a great disadvantage to the plant to waste many days in remaining clasped over a minute insect, and several additional days or weeks in afterwards recovering its sensibility; inasmuch as a minute insect would afford but little nutriment. It would be far better for the plant to wait for a time until a moderately large insect was captured, and to allow all the little ones to escape; and this advantage is secured by the slowly intercrossing marginal spikes, which act like the large meshes of a fishing-net, allowing the small and useless fry to escape.

As I was anxious to know whether this view was correct—and as it seems a good illustration of how cautious we ought to be in assuming, as I had done with respect to the marginal spikes, that any fully developed structure is useless—I applied to Dr. Canby. He visited the native site of the plant, early in the season, before the leaves had grown to their full size, and sent me fourteen leaves, containing naturally captured insects. Four of these had caught rather small insects, viz. three of them ants, and the fourth a rather small fly, but the other ten had all caught large insects, namely, five elaters, two chrysomelas, a curculio, a thick and broad spider, and a scolopendra. Out of these ten insects, no less than eight were beetles, and out of the whole fourteen there was only one, viz. a dipterous insect, which could readily take flight. Drosera, on the other hand, lives chiefly on insects which are good flyers, especially Diptera, caught by the aid of its viscid secretion. But what most concerns us is the size of the ten larger insects. Their average length from head to tail was .256 of an inch, the lobes of the leaves being on an average .53 of an inch in length, so that the insects were very nearly half as long as the leaves within which they were enclosed. Only a few of these leaves, therefore, had wasted their powers by capturing small prey, though it is probable that many small insects had crawled over them and been caught, but had then escaped through the bars.

• • •

CONCLUDING REMARKS ON THE DROSERACEÆ.

Little light can be thrown on the gradual acquirement of the third remarkable character possessed by the more highly developed genera of the Droseraceæ, namely the power of movement when excited. It should, however, be borne in mind that leaves and their homologues, as well as flower-peduncles, have gained this power, in innumerable instances, independently of inheritance from any common parent form; for instance, in tendril-bearers and leaf-climbers (i.e. plants with their leaves, petioles, and flower-peduncles, &c., modified for prehension) belonging to a large number of the most widely distinct orders, —in the leaves of the many plants which go to sleep at night, or move when shaken,—and in the irritable stamens and pistils of not a few species. We may therefore infer that the power of movement can be by some means readily acquired. Such movements imply irritability or sensitiveness, but, as Cohn has remarked, the tissues of the plants thus endowed do not differ in any uniform manner from those of ordinary plants; it is therefore probable that all leaves are to a slight degree irritable. Even if an insect alights on a leaf, a slight molecular change is probably transmitted to some distance across its tissue, with the sole difference that no perceptible effect is produced. We have some evidence in favour of this belief, for we know that a single touch on the glands of Drosera does not excite inflection; yet it must produce some effect, for if the glands have been immersed in a solution of camphor, inflection follows within a shorter time than would have followed from the effects of camphor alone. So again with Dionæa, the blades in their ordinary state may be roughly touched without their closing; yet some effect must be thus caused and transmitted across the whole leaf, for if the glands have recently absorbed animal matter, even a delicate touch causes them to close instantly. On the whole we may conclude that the acquirement of a high degree of sensitiveness and of the power of movement by certain genera of the Droseraceæ presents no greater difficulty than that presented by the similar but feebler powers of a multitude of other plants.

The specialised nature of the sensitiveness possessed by Drosera and Dionæa, and by certain other plants, well deserves attention. A gland of Drosera may be forcibly hit once, twice, or even thrice, without any effect being produced, whilst the continued pressure of an extremely minute particle excites movement. On the other hand, a particle many times heavier may be gently laid on one of the filaments of Dionæa with no effect; but if touched only once by the slow movement of a delicate hair, the lobes close; and this difference in the nature of the sensitiveness of these two plants stands in manifest adaptation to their manner of capturing insects. So does the fact, that when the central glands of Drosera absorb nitrogenous matter, they transmit a motor impulse to the exterior tentacles much more quickly than when they are mechanically irritated; whilst with Dionæa the absorption of nitrogeneous matter causes the lobes to press together with extreme slowness, whilst a touch excites rapid movement. Somewhat analogous cases may be observed, as I have shown in another work, with the tendrils of various plants; some being most excited by contact with fine fibres, others by contact with bristles, others with a flat or a creviced surface. The sensitive organs of Drosera and Dionæa are also specialised, so as not to be uselessly affected by the weight or impact of drops of rain, or by blasts of air. This may be accounted for by supposing that these plants and their progenitors have grown accustomed to the repeated action of rain and wind, so that no molecular change is thus induced; whilst they have been rendered more sensitive by means of natural selection to the rarer impact or pressure of solid bodies. Although the absorption by the glands of Drosera of various fluids excites movement, there is a great difference in the action of allied fluids; for instance, between certain vegetable acids, and between citrate and phosphate of ammonia. The specialised nature and perfection of the sensitiveness in these two plants is all the more astonishing as no one supposes that they possess nerves; and by testing Drosera with several substances which act powerfully on the nervous system of animals, it does not appear that they include any diffused matter analogous to nerve-tissue.

Although the cells of Drosera and Dionæa are quite as sensitive to certain stimulants as are the tissues which surround the terminations of the nerves in the higher animals, yet these plants are inferior even to animals low down in the scale, in not being affected except by stimulants in contact with their sensitive parts. They would, however, probably be affected by radiant heat; for warm water excites energetic movement. When a gland of Drosera, or one of the filaments of Dionæa, is excited, the motor impulse radiates in all directions, and is not, as in the case of animals, directed towards special points or organs. This holds good even in the case of Drosera when some exciting substance has been placed at two points on the disc, and when the tentacles all round are inflected with marvellous precision towards the two points. The rate at which the motor impulse is transmitted, though rapid in Dionæa, is much slower than in most or all animals. This fact, as well as that of the motor impulse not being specially directed to certain points, are both no doubt due to the absence of nerves. Nevertheless we perhaps see the prefigurement of the formation of nerves in animals in the transmission of the motor impulse being so much more rapid down the confined space within the tentacles of Drosera than elsewhere, and somewhat more rapid in a longitudinal than in a transverse direction across the disc. These plants exhibit still more plainly their inferiority to animals in the absence of any reflex action, except in so far as the glands of Drosera, when excited from a distance, send back some influence which causes the contents of the cells to become aggregated down to the bases of the tentacles. But the greatest inferiority of all is the absence of a central organ, able to receive impressions from all points. to transmit their effects in any definite direction, to store them up and reproduce them.

UTRICULARIA.

I was led to investigate the habits and structure of the species of this genus partly from their belonging to the same natural family as Pinguicula, but more especially by Mr. Holland's statement, that "water insects are often found imprisoned in the bladders," which he suspects "are destined for the plant to feed on." The plants which I first received as *Utricularia vulgaris* from the New Forest in Hampshire and from Cornwall, and which I have chiefly worked on, have been determined by Dr. Hooker to be a very rare British species, the *Utricularia neglecta* of Lehm. I subsequently received the true *Utricularia vulgaris* from Yorkshire. Since drawing up the following description from my own observations and those of my son, Francis Darwin, an important memoir by Prof. Cohn on *Utricularia vulgaris* has appeared; and it has been no small satisfaction to me to find that my account agrees almost completely with that of this distinguished observer. I will publish my description as it stood before reading that by Prof. Cohn, adding occasionally some statements on his authority.

• • •

The Uses of the several Parts.—After the above long but necessary description of the parts, we will turn to their uses. The bladders have been supposed by some authors to serve as floats; but branches which bore no bladders, and others from which they had been removed, floated perfectly, owing to the air in the intercellular spaces. Bladders containing dead and captured animals usually include bubbles of air, but these cannot have been generated solely by the process of decay, as I have often seen air in young, clean, and empty bladders; and some old bladders with much decaying matter had no bubbles.

The real use of the bladders is to capture small aquatic animals, and this they do on a large scale. In the first lot of plants, which I received from the New Forest early in July, a large proportion of the fully grown bladders contained prey; in a second lot, received in the beginning of August, most of the bladders were empty, but plants had been selected which had grown in unusually pure water. In the first lot, my son examined seventeen bladders, including prey of some kind, and eight of these contained entomostracan crustaceans, three larvæ of insects, one being still alive, and six remnants of animals so much decayed that their nature could not be distinguished. I picked out five bladders which seemed very full, and found in them four, five, eight, and ten crustaceans, and in the fifth a single much elongated larva. In five other bladders, selected from containing remains, but not appearing very full, there were one, two, four, two, and five crustaceans. A plant of *Utricularia vulgaris*, which had been kept in almost pure water, was placed by Cohn one evening into water swarming with crustaceans, and by the next morning most of the bladders contained these animals entrapped and swimming round and round their prisons. They remained alive for several days; but at last perished, asphyxiated, as I suppose, by the oxygen in the water having been all consumed. Fresh-water worms were also found by Cohn in some bladders. In all cases the bladders with decayed remains swarmed with living Algæ of many kinds, Infusoria, and other low organisms, which evidently lived as intruders.

Animals enter the bladders by bending inwards the posterior free edge of the valve, which from being highly elastic shuts again instantly. As the edge is extremely thin, and fits closely against the edge of the collar, both projecting into the bladder, it would evidently be very difficult for any animal to get out when once imprisoned, and apparently they never do escape. To show how closely the edge fits, I may mention that my son found a Daphnia which had inserted one of its antennæ into the slit, and it was thus held fast during a whole day. On three or four occasions I have seen long narrow larvae, both dead and alive, wedged between the corner of the valve and collar, with half their bodies within the bladder and half out.

As I felt much difficulty in understanding how such minute and weak animals, as are often captured, could force their way into the bladders, I tried many experiments to ascertain how this was effected. The free margin of the valve bends so easily that no resistance is felt when a needle or thin bristle is inserted. A thin human hair, fixed to a handle, and cut off so as to project barely 1/4 of an inch, entered with some difficulty; a longer piece yielded instead of entering. On three occasions minute particles of blue glass (so as to be easily distinguished) were placed on valves whilst under water; and on trying gently to move them with a needle, they disappeared so suddenly that, not seeing what had happened, I thought that I had flirted them off; but on examining the bladders, they were found safely enclosed. The same thing occurred to my son, who placed little cubes of green box-wood (about 1/60 of an inch, .423 mm.) on some valves; and thrice in the act of placing them on, or whilst gently moving them to another spot, the valve suddenly opened and they were engulfed. He then placed similar bits of wood on other valves, and moved them about for some time, but they did not enter. Again, particles of blue glass were placed by me on three valves, and extremely minute shavings of lead on two other valves; after 1 or 2 hrs. none had entered, but in from 2 to 5 hrs. all five were enclosed. One of the particles of glass was a long splinter, of which one end rested obliquely on the valve, and after a few hours it was found fixed, half within the bladder and half projecting out, with the edge of the valve fitting closely all round, except at one angle, where a small open space was left. It was so firmly fixed, like the above mentioned larvæ, that the bladder was torn from the branch and shaken, and yet the splinter did not fall out. My son also placed little cubes (about 1/65 of an inch, .391 mm.) of green box-wood, which were just heavy enough to sink in water, on three valves. These were examined after 19 hrs. 30 m., and were still lying on the valves; but after 22 hrs. 30 m. one was found enclosed. I may here mention that I found in a bladder on a naturally growing plant a grain of sand, and in another bladder three grains; these must have fallen by some accident on the valves, and then entered like the particles of glass.

The slow bending of the valve from the weight of particles of glass and even of box-wood, though largely supported by the water, is, I suppose, analogous to the slow bending of colloid substances. For instance, particles of glass were placed on various points of narrow strips of moistened gelatine, and these yielded and became bent with extreme slowness. It is much more difficult to understand how gently moving a particle from one part of a valve to another causes it suddenly to open. To ascertain whether the valves were endowed with irritability, the surfaces of several were scratched with a needle or brushed with a fine camel-hair brush, so as to imitate the crawling movement of small crustaceans, but the valve did not open. Some bladders, before being brushed, were left for a time in water at temperatures between 80° and 130° F. (26.6°–54.4° Cent.), as, judging from a widespread analogy, this would have rendered them more sensitive to irritation, or would by itself have excited movement; but no effect was produced. We may, therefore, conclude that animals enter merely by forcing their way through the slit-like orifice; their heads serving as a wedge. But I am surprised that such small and weak creatures as are often captured (for instance, the nauplius of a crustacean, and a tardigrade) should be strong enough to act in this manner, seeing that it was difficult to push in one end of a bit of a hair 1/4 of an inch in length. Nevertheless, it is certain that weak and small creatures do enter, and Mrs. Treat, of New Jersey, has been more successful than any other observer, and has often witnessed in the case of *Utricularia clandestina* the whole process. She saw a tardigrade slowly walking round a bladder, as if reconnoitring; at last it crawled into the depression where the valve lies, and then easily entered. She also witnessed the entrapment of various minute crustaceans. Cypris "was quite wary, but nevertheless was often caught. Coming to the entrance of a bladder, it would sometimes pause a moment, and then dash away; at other times it would come close up, and even venture part of the way into the entrance and back out as if afraid. Another, more heedless, would open the door and walk in; but it was no sooner in than it manifested alarm, drew in its feet and antennæ, and closed its shell." Larvæ, apparently of gnats, when "feeding near the entrance, are pretty certain to run their heads into the net, whence there is no retreat. A large larva is sometimes three or four hours in being swallowed, the process bringing to mind what I have witnessed when a small snake makes a large frog its victim." But as the valve does not appear to be in the least irritable, the slow swallowing process must be the effect of the onward movement of the larva.

It is difficult to conjecture what can attract so many creatures, animal- and vegetable-feeding crustaceans, worms, tardigrades, and various larvæ, to enter the bladders. Mrs. Treat says that the larvæ just referred to are vegetable-feeders, and seem to have a special liking for the long bristles round the valve, but this taste will not account for the entrance of animal-feeding crustaceans. Perhaps small aquatic animals habitually try to enter every small crevice, like that between the valve and collar, in search of food or protection. It is not probable that the remarkable transparency of the valve is an accidental circumstance, and the spot of light thus formed may serve as a guide. The long bristles round the entrance apparently serve for the same purpose. I believe that this is the case, because the bladders of some epiphytic and marsh species of Utricularia which live embedded either in entangled vegetation or in mud, have no bristles round the entrance, and these under such conditions would be of no service as a guide. Nevertheless, with these epiphytic and marsh species, two pairs of bristles project from the surface of the valve, as in the aquatic species; and their use probably is to prevent too large animals from trying to force an entrance into the bladder, thus rupturing the orifice.

• • •

Conclusion. It has now been shown that many species of Utricularia and of two closely allied genera, inhabiting the most distant parts of the world—Europe, Africa, India, the Malay Archipelago, Australia, North and South America—are admirably adapted for capturing by two methods small aquatic or terrestrial animals, and that they absorb the products of their decay.

Ordinary plants of the higher classes procure the requisite inorganic elements from the soil by means of their roots, and absorb carbonic acid from the atmosphere by means of their leaves and stems. But we have seen in a previous part of this work that there is a class of plants which digest and afterwards absorb animal matter, namely, all the Droseraceæ, Pinguicula, and, as discovered by Dr. Hooker, Nepenthes, and to this class other species will almost certainly soon be added. These plants can dissolve matter out of certain vegetable substances, such as pollen, seeds, and bits of leaves. No doubt their glands likewise absorb the salts of ammonia brought to them by the rain. It has also been shown that some other plants can absorb ammonia by their glandular hairs; and these will profit by that brought to them by the rain. There is a second class of plants which, as we have just seen, cannot digest, but absorb the products of the decay of the animals which they capture, namely, Utricularia and its close allies; and from the excellent observations of Dr. Mellichamp and Dr. Canby, there can scarcely be a doubt that Sarracenia and Darlingtonia may be added to this class, though the fact can hardly be considered as yet fully proved. There is a third class of plants which feed, as is now generally admitted, on the products of the decay of vegetable matter, such as the bird's-nest orchis (Neottia), &c. Lastly, there is the well-known fourth class of parasites (such as the mistletoe), which are nourished by the juices of living plants. Most, however, of the plants belonging to these four classes obtain part of their carbon, like ordinary species, from the atmosphere. Such are the diversified means, as far as at present known, by which higher plants gain their subsistence.

The Effects of Cross and Self-Fertilisation in the Vegetable Kingdom. [1876]

*C*ROSS AND *S*ELF-*F*ERTILISATION *is another book that can be traced back to Darwin's earliest evolutionary interests. In its introduction he claims to have worked on this project since 1839, and indeed his early notebooks contain comments on pollination and ideas for experiments to demonstrate how selection works. In the* Autobiography *Darwin states that the actual pollination experiments whose reports occupy much of* Cross and Self-Fertilisation *took place over an eleven-year period. The book, which he considered a complement to his study of orchids, appeared in November 1876. A second edition—little altered, uncharacteristically—was published in 1878.*

Darwin found that the overwhelming majority of the flowering plants he studied were cross-pollinating. Self-pollination apparently has evolved as a secondary phenomenon, and self-pollinating species are considered likelier candidates for extinction than are crossing species. Curiously, however, self-pollination may be advantageous in the short term for plants, such as weeds, that colonize new habitats. Thanks to self-pollination, colonizing plants can reproduce more rapidly—and of course self-pollination is the only means by which a single specimen arriving in a new habitat can perpetuate itself. On the other hand, self-pollination affords less variation in the population; and when environmental changes take place, a self-pollinated population is much less likely to evolve new adaptations than is one characterized by cross-pollination. The result can be extinction for the self-pollinator.

While Darwin was conducting his experiments in England, the

Moravian monk Gregor Mendel was engaged in his own crossing of garden peas in Brno. Darwin's experiments showed him that close inbreeding was deleterious, that it results in the formation of fewer seeds than when the same plant is outcrossed with another plant. Mendel's experiments yielded more momentous results: his laws of genetics. These laws are segregation (the principle that parental traits of the same character, seed-coat color, for example, somehow separate so that a sperm cell or egg cell passes on the trait of one or another parent but not both) and independent assortment (the principle that distinct characteristics, such as seed-coat color and seed-coat surface, are inherited independently of one another). Why didn't Darwin discover Mendel's laws? Probably because Mendel was a mathematician and Darwin was not. Unlike Darwin, Mendel could recognize such significant mathematical relationships as the 3:1 ratio in inheritance of one pair of traits, or the 9:3:3:1 distribution of two pairs of inherited traits. In addition, Mendel was studying the inheritance of distinct characteristics such as the color and surface characteristics of pea seeds, matters of single gene differences. Darwin, on the other hand, studied the heredity of features which we now know to be governed by a number of genes and more difficult to quantify. Finally, Mendel began from the premise that discrete particles determined single hereditary characteristics. Darwin did not: his theory of pangenesis involved particles from each cell and tissue migrating to the gametes and intermingling before fertilization to determine the characteristics of offspring.

Even though Darwin proceeded in ignorance of how genetic inheritance operates, his introductory remarks in Cross and Self-Fertilisation *offer a fine example of how to set up an experimental research project and interpret its results. His discussion of the adaptations of flowers and their pollinators, chiefly insects, effectively demonstrates by what means cross-pollination takes place.*

INTRODUCTORY REMARKS.

There is weighty and abundant evidence that the flowers of most kinds of plants are constructed so as to be occasionally or habitually cross-fertilised by pollen from another flower, produced either by the same plant, or generally, as we shall hereafter see reason to believe, by a distinct plant. Cross-fertilisation is sometimes ensured by the sexes being separated, and in a large number of cases by the pollen and stigma of the same flower being matured at different times. Such plants are called dichogamous, and have been divided into two sub-classes: proterandrous species, in which the pollen is mature before the stigma, and proterogynous species, in which the reverse occurs; this latter form of dichogamy not being nearly so common as the other. Cross-fertilisation is also ensured, in many cases, by mechanical contrivances of wonderful beauty, preventing the impregnation of the flowers by their own pollen. There is a small class of plants, which I have called dimorphic and trimorphic, but to which Hildebrand has given the more appropriate name of heterostyled; this class consists of plants presenting two or three distinct forms, adapted for reciprocal fertilisation, so that, like plants with separate sexes, they can hardly fail to be intercrossed in each generation. The male and female organs of some flowers are irritable, and the insects which touch them get dusted with pollen, which is thus transported to other flowers. Again, there is a class, in which the ovules absolutely refuse to be fertilised by pollen from the same plant, but can be fertilised by pollen from any other individual of the same species. There are also very many species which are partially sterile with their own pollen. Lastly, there is a large class in which the flowers present no apparent obstacle of any kind to self-fertilisation, nevertheless these plants are frequently intercrossed, owing to the prepotency of pollen from another individual or variety over the plant's own pollen.

As plants are adapted by such diversified and effective means for cross-fertilisation, it might have been inferred from this fact alone that they derived some great advantage from the process; and it is the object of the present work to show the nature and importance of the benefits thus derived. There are, however, some exceptions to the rule of plants being constructed so as to allow of or to favour cross-fertilisation, for some few plants seem to be invariably self-fertilised; yet even these retain traces of having been formerly adapted for cross-fertilisation. These exceptions need not make us doubt the truth of the above rule, any more than the existence of some few plants which produce flowers, and yet never set seed, should make us doubt that flowers are adapted for the production of seed and the propagation of the species.

We should always keep in mind the obvious fact that the production of seed is the chief end of the act of fertilisation; and that this end can be gained by hermaphrodite plants with incomparably greater certainty by self-fertilisation, than by the union of the sexual elements belonging to two distinct flowers or plants. Yet it is as unmistakably plain that innumerable flowers are adapted for cross-fertilisation, as that the teeth and talons of a carnivorous animal are adapted for catching prey; or that the plumes, wings, and hooks of a seed are adapted for its dissemination. Flowers, therefore, are constructed so as to gain two objects which are, to a certain extent, antagonistic, and this explains many apparent anomalies in their structure. The close proximity of the anthers to the stigma in a multitude of species favours, and often leads to, self-fertilisation; but this end could have been gained far more safely if the flowers had been completely closed, for then the pollen would not have been injured by the rain or devoured by insects, as often happens. Moreover, in this case, a very small quantity of pollen would have been sufficient for fertilisation, instead of millions of grains being produced. But the openness of the flower and the production of a great and apparently wasteful amount of pollen are necessary for cross-fertilisation. These remarks are well illustrated by the plants called cleistogene, which bear on the same stock two kinds of flowers. The flowers of the one kind are minute and completely closed, so that they cannot possibly be crossed; but they are abundantly fertile, although producing an extremely small quantity of pollen. The flowers of the other kind produce much pollen and are open; and these can be, and often are, cross-fertilised. Hermann Müller has also made the remarkable discovery that there are some plants which exist under two forms; that is, produce on distinct stocks two kinds of hermaphrodite flowers. The one form bears small flowers constructed for self-fertilisation; whilst the other bears larger and much more conspicuous flowers plainly constructed for self-fertilisation by the aid of insects; and without their aid these produce no seed.

The adaptation of flowers for cross-fertilisation is a subject which has interested me for the last thirty-seven years, and I have collected a large mass of observations, but these are now rendered superfluous by the many excellent works which have been lately published. In the year 1857 I wrote a short paper on the fertilisation of the kidney bean; and in 1862 my work 'On the Contrivances by which British and Foreign Orchids are Fertilised by Insects' appeared. It seemed to me a better plan to work out one group of plants as carefully as I could, rather than to publish many miscellaneous and imperfect observations. My present work is the complement of that on Orchids, in which it was shown how admirably these plants are constructed so as to permit of, or to favour, or to necessitate cross-fertilisation. The adaptations for cross-fertilisation are perhaps more obvious in the Orchideæ than in any other group of plants, but it is an error to speak of them, as some authors have done, as an exceptional case. The lever-like action of the stamens of Salvia (described by Hildebrand, Dr. W. Ogle, and others), by which the anthers are depressed and rubbed on the backs of bees, shows as perfect a structure as can be found in any orchid. Papilionaceous flowers, as described by various authors—for instance, by Mr. T. H. Farrer—offer innumerable curious adaptations for cross-fertilisation. The case of *Posoqueria fragrans* (one of the Rubiaceæ), is as wonderful as that of the most wonderful orchid. The stamens, according to Fritz Müller, are irritable, so that as soon as a moth visits a flower, the anthers explode and cover the insect with pollen; one of the filaments which is broader than the others then moves and closes the flower for about twelve hours, after which time it resumes its original position. Thus the stigma cannot be fertilised by pollen from the same flower, but only by that brought by a moth from some other flower. Endless other beautiful contrivances for this same purpose could be specified.

Long before I had attended to the fertilisation of flowers, a remarkable book appeared in 1793 in Germany, 'Das Entdeckte Geheimniss der Natur,' by C. K. Sprengel, in which he clearly proved by innumerable observations, how essential a part insects play in the fertilisation of many plants. But he was in advance of his age, and his discoveries were for a long time neglected. Since the appearance of my book on Orchids, many excellent works on the fertilisation of flowers, such as those by Hildebrand, Delpino, Axell, and Hermann Müller, and numerous shorter papers, have been published. A list would occupy several pages, and this is not the proper place to give their titles, as we are not here concerned with the means, but with the results of cross-fertilisation. No one who feels interest in the mechanism by which nature effects her ends, can read these books and memoirs without the most lively interest.

From my own observations on plants, guided to a certain extent by the experience of the breeders of animals, I became convinced many years ago that it is a general law of nature that flowers are adapted to be crossed, at least occasionally, by pollen from a distinct plant. Sprengel at times foresaw this law, but only partially, for it does not appear that he was aware that there was any difference in power between pollen from the same plant and from a distinct plant. In the introduction to his book (p. 4) he says, as the sexes are separated in so many flowers, and as so many other flowers are dichogamous, "it appears that nature has not willed that any one flower should be fertilised by its own pollen." Nevertheless, he was far from keeping this conclusion always before his mind, or he did not see its full importance, as may be perceived by anyone who will read his observations carefully; and he consequently mistook the meaning of various structures. But his discoveries are so numerous and his work so excellent, that he can well afford to bear a small amount of blame. A most capable judge, H. Müller, likewise says: "It is remarkable in how very many cases Sprengel rightly perceived that pollen is necessarily transported to the stigmas of other flowers of the same species by the insects which visit them, and yet did not imagine that this transportation was of any service to the plants themselves."

Andrew Knight saw the truth much more clearly, for he remarks, "Nature intended that a sexual intercourse should take place between neighbouring plants of the same species." After alluding to the various means by which pollen is transported from flower to flower, as far as was then imperfectly known, he adds, "Nature has something more in view than that its own proper males should fecundate each blossom." In 1811 Kölreuter plainly hinted at the same law, as did afterwards another famous hybridiser of plants, Herbert. But none of these distinguished observers appear to have been sufficiently impressed with the truth and generality of the law, so as to insist on it and impress their belief on others.

In 1862 I summed up my observations on Orchids by saying that nature "abhors perpetual self-fertilisation." If the word perpetual had been omitted, the aphorism would have been false. As it stands, I believe that it is true, though perhaps rather too strongly expressed; and I should have added the self-evident proposition that the propagation of the species, whether by self-fertilisation or by cross-fertilisation, or asexually by buds, stolons, &c. is of paramount importance. Hermann Müller has done excellent service by insisting repeatedly on this latter point.

It often occurred to me that it would be advisable to try whether seedlings from cross-fertilised flowers were in any way superior to those from self-fertilised flowers. But as no instance was known with animals of any evil appearing in a single generation from the closest possible interbreeding, that is between brothers and sisters, I thought that the same rule would hold good with plants; and that it would be necessary at the sacrifice of too much time to self-fertilise and intercross plants during several successive generations, in order to arrive at any result. I ought to have reflected that such elaborate provisions favouring cross-fertilisation, as we see in innumerable plants, would not have been acquired for the sake of gaining a distant and slight advantage, or of avoiding a distant and slight evil. Moreover, the fertilisation of a flower by its own pollen corresponds to a closer form of interbreeding than is possible with ordinary bi-sexual animals; so that an earlier result might have been expected.

I was at last led to make the experiments recorded in the present volume from the following circumstance. For the sake of determining certain points with respect to inheritance, and without any thought of the effects of close interbreeding, I raised close together two large beds of self-fertilised and crossed seedlings from the same plant of *Linaria vulgaris*. To my surprise, the crossed plants when fully grown were plainly taller and more vigorous than the self-fertilised ones. Bees incessantly visit the flowers of this Linaria and carry pollen from one to the other; and if insects are excluded, the flowers produce extremely few seeds; so that the wild plants from which my seedlings were raised must have been intercrossed during all previous generations. It seemed therefore quite incredible that the difference between the two beds of seedlings could have been due to a single act of self-fertilisation; and I attributed the result to the self-fertilised seeds not having been well ripened, improbable as it was that all should have been in this state, or to some other accidental and inexplicable cause. During the next year, I raised for the same purpose as before two large beds close together of self-fertilised and crossed seedlings from the carnation, *Dianthus caryophyllus*. This plant, like the Linaria, is almost sterile if insects are excluded; and we may draw the same inference as before, namely, that the parent-plants must have been intercrossed during every or almost every previous generation. Nevertheless, the self-fertilised seedlings were plainly inferior in height and vigour to the crossed.

My attention was now thoroughly aroused, for I could hardly doubt that the difference between the two beds was due to the one set being the offspring of crossed, and the other of self-fertilised flowers. Accordingly I selected almost by hazard two other plants, which happened to be in flower in the greenhouse, namely, *Mimulus luteus* and *Ipomœa purpurea*, both of which, unlike the Linaria and Dianthus, are highly self-fertile if insects are excluded. Some flowers on a single plant of both species were fertilised with their own pollen, and others were crossed with pollen from a distinct individual; both plants being protected by a net from insects. The crossed and self-fertilised seeds thus produced were sown on opposite sides of the same pots, and treated in all respects alike; and the plants when fully grown were measured and compared. With both species, as in the cases of the Linaria and Dianthus, the crossed seedlings were conspicuously superior in height and in other ways to the self-fertilised. I therefore determined to begin a long series of experiments with various plants, and these were continued for the following eleven years; and we shall see that in a large majority of cases the crossed beat the self-fertilised plants. Several of the exceptional cases, moreover, in which the crossed plants were not victorious, can be explained.

It should be observed that I have spoken for the sake of brevity, and shall continue to do so, of crossed and self-fertilised seeds, seedlings, or plants; these terms implying that they are the product of crossed or self-fertilised flowers. Cross-fertilisation always means a cross between distinct plants which were raised from seeds and not from cuttings or buds. Self-fertilisation always implies that the flowers in question were impregnated with their own pollen.

My experiments were tried in the following manner. A single plant, if it produced a sufficiency of flowers, or two or three plants were placed under a net stretched on a frame, and large enough to cover the plant (together with the pot, when one was used) without touching it. This latter point is important, for if the flowers touch the net they may be cross-fertilised by bees, as I have known to happen; and when the net is wet the pollen may be injured. I used at first "white cotton net," with very fine meshes, but afterwards a kind of net with meshes one-tenth of an inch in diameter; and this I found by experience effectually excluded all insects excepting Thrips, which no net will exclude. On the plants thus protected several flowers were marked, and were fertilised with their own pollen; and an equal number on the same plants, marked in a different manner, were at the same time crossed with pollen from a distinct plant. The crossed flowers were never castrated, in order to make the experiments as like as possible to what occurs under nature with plants fertilised by the aid of insects. Therefore, some of the flowers which were crossed may have failed to be thus fertilised, and afterwards have been self-fertilised. But this and some other sources of error will presently be discussed. In some few cases of spontaneously self-fertile species, the flowers were allowed to fertilise themselves under the net; and in still fewer cases uncovered plants were allowed to be freely crossed by the insects which incessantly visited them. There are some great advantages and some disadvantages in my having occasionally varied my method of proceeding; but when there was any difference in the treatment, it is always so stated under the head of each species.

Care was taken that the seeds were thoroughly ripened before being gathered. Afterwards the crossed and self-fertilised seeds were in most cases placed on damp sand on opposite sides of a glass tumbler covered by a glass plate, with a partition between the two lots; and the glass was placed on the chimney-piece in a warm room. I could thus observe the germination of the seeds. Sometimes a few would germinate on one side before any on the other, and these were thrown away. But as often as a pair germinated at the same time, they were planted on opposite sides of a pot, with a superficial partition between the two; and I thus proceeded until from half-a-dozen to a score or more seedlings of exactly the same age were planted on the opposite sides of several pots. If one of the young seedlings became sickly or was in any way injured, it was pulled up and thrown away, as well as its antagonist on the opposite side of the same pot.

As a large number of seeds were placed on the sand to germinate, many remained after the pairs had been selected, some of which were in a state of germination and others not so; and these were sown crowded together on the opposite sides of one or two rather larger pots, or sometimes in two long rows out of doors. In these cases there was the most severe struggle for life among the crossed seedlings on one side of the pot, and the self-fertilised seedlings on the other side, and between the two lots which grew in competition in the same pot. A vast number soon perished, and the tallest of the survivors on both sides when fully grown were measured. Plants treated in this manner, were subjected to nearly the same conditions as those growing in a state of nature, which have to struggle to maturity in the midst of a host of competitors.

On other occasions, from the want of time, the seeds, instead of being allowed to germinate on damp sand, were sown on the opposite sides of pots, and the fully grown plants measured. But this plan is less accurate, as the seeds sometimes germinated more quickly on one side than on the other. It was however necessary to act in this manner with some few species, as certain kinds of seeds would not germinate well when exposed to the light; though the glasses containing them were kept on the chimney-piece on one side of a room, and some way from the two windows which faced the N.E.

The soil in the pots in which the seedlings were planted, or the seeds sown, was well mixed, so as to be uniform in composition. The plants on the two sides were always watered at the same time and as equally as possible; and even if this had not been done, the water would have spread almost equally to both sides, as the pots were not large. The crossed and self-fertilised plants were separated by a superficial partition, which was always kept directed towards the chief source of the light, so that the plants on both sides were equally illuminated. I do not believe it possible that two sets of plants could have been subjected to more closely similar conditions, than were my crossed and self-fertilised seedlings, as grown in the above described manner.

In comparing the two sets, the eye alone was never trusted. Generally the height of every plant on both sides was carefully measured, often more than once, viz., whilst young, sometimes again when older, and finally when fully or almost fully grown. But in some cases, which are always specified, owing to the want of time, only one or two of the tallest plants on each side were measured. This plan, which is not a good one, was never followed (except with the crowded plants raised from the seeds remaining after the pairs had been planted) unless the tallest plants on each side seemed fairly to represent the average difference between those on both sides. It has, however, some great advantages, as sickly or accidentally injured plants, or the offspring of ill-ripened seeds, are thus eliminated. When the tallest plants alone on each side were measured, their average height of course exceeds that of all the plants on the same side taken together. But in the case of the much crowded plants raised from the remaining seeds, the average height of the tallest plants was less than that of the plants in pairs, owing to the unfavourable conditions to which they were subjected from being greatly crowded. For our purpose, however, of the comparison of the crossed and self-fertilised plants, their absolute height signifies little.

As the plants were measured by an ordinary English standard divided into inches and eighths of an inch, I have not thought it worth while to change the fractions into decimals. The average or mean heights were calculated in the ordinary rough method by adding up the measurements of all, and dividing the product by the number of plants measured; the result being here given in inches and decimals. As the different species grow to various heights, I have always for the sake of easy comparison given in addition the average height of the crossed plants of each species taken as 100, and have calculated the average height of the self-fertilised plant in relation to this standard. With respect to the crowded plants raised from the seeds remaining after the pairs had been planted, and of which only some of the tallest on each side were measured, I have not thought it worth while to complicate the results by giving separate averages for them and for the pairs, but have added up all their heights, and thus obtained a single average.

I long doubted whether it was worth while to give the measurements of each separate plant, but have decided to do so, in order that it may be seen that the superiority of the crossed plants over the self-fertilised, does not commonly depend on the presence of two or three extra fine plants on the one side, or of a few very poor plants on the other side. Although several observers have insisted in general terms on the offspring from intercrossed varieties being superior to either parent-form, no precise measurements have been given; and I have met with no observations on the effects of crossing and self-fertilising the individuals of the same variety. Moreover, experiments of this kind require so much time—mine having been continued during eleven years—that they are not likely soon to be repeated.

• • •

When I began my experiments I did not intend to raise crossed and self-fertilised plants for more than a single generation; but as soon as the plants of the first generation were in flower I thought that I would raise one more generation, and acted in the following manner. Several flowers on one or more of the self-fertilised plants were again self-fertilised; and several flowers on one or more of the crossed plants were fertilised with pollen from another crossed plant of the same lot. Having thus once begun, the same method was followed for as many as ten successive generations with some of the species. The seeds and seedlings were always treated in exactly the same manner as already described. The self-fertilised plants, whether originally descended from one or two mother-plants, were thus in each generation as closely interbred as was possible; and I could not have improved on my plan. But instead of crossing one of the crossed plants with another crossed plant, I ought to have crossed the self-fertilised plants of each generation with pollen taken from a non-related plant—that is, one belonging to a distinct family or stock of the same species and variety. This was done in several cases as an additional experiment, and gave very striking results. But the plan usually followed was to put into competition and compare intercrossed plants, which were almost always the offspring of more or less closely related plants, with the self-fertilised plants of each succeeding generation;—all having been grown under closely similar conditions. I have, however, learnt more by this method of proceeding, which was begun by an oversight and then necessarily followed, than if I had always crossed the self-fertilised plants of each succeeding generation with pollen from a fresh stock.

I have said that the crossed plants of the successive generations were almost always inter-related. When the flowers on an hermaphrodite plant are crossed with pollen taken from a distinct plant, the seedlings thus raised may be considered as hermaphrodite brothers or sisters; those raised from the same capsule being as close as twins or animals of the same litter. But in one sense the flowers on the same plant are distinct individuals, and as several flowers on the mother-plant were crossed by pollen taken from several flowers on the father-plant, such seedlings would be in one sense half-brothers or sisters, but more closely related than are the half-brothers and sisters of ordinary animals. The flowers on the mother-plant were, however, commonly crossed by pollen taken from two or more distinct plants; and in these cases the seedlings might be called with more truth half-brothers or sisters. When two or three mother-plants were crossed, as often happened, by pollen taken from two or three father-plants (the seeds being all intermingled), some of the seedlings of the first generation would be in no way related, whilst many others would be whole or half-brothers and sisters. In the second generation a large number of the seedlings would be what may be called whole or half first-cousins, mingled with whole and half-brothers and sisters, and with some plants not at all related. So it would be in the succeeding generations, but there would also be many cousins of the second and more remote degrees. The relationship will thus have become more and more inextricably complex in the later generations; with most of the plants in some degree and many of them closely related.

I have only one other point to notice, but this is one of the highest importance; namely, that the crossed and self-fertilised plants were subjected in the same generation to as nearly similar and uniform conditions as was possible. In the successive generations they were exposed to slightly different conditions as the seasons varied, and they were raised at different periods. But in other respects all were treated alike, being grown in pots in the same artificially prepared soil, being watered at the same time, and kept close together in the same greenhouse or hothouse. They were therefore not exposed during successive years to such great vicissitudes of climate as are plants growing out of doors.

On some apparent and real Causes of Error in my Experiments.—It has been objected to such experiments as mine, that covering plants with a net, although only for a short time whilst in flower, may affect their health and fertility. I have seen no such effect except in one instance with a Myosotis, and the covering may not then have been the real cause of injury. But even if the net were slightly injurious, and certainly it was not so in any high degree, as I could judge by the appearance of the plants and by comparing their fertility with that of neighbouring uncovered plants, it would not have vitiated my experiments; for in all the more important cases the flowers were crossed as well as self-fertilised under a net, so that they were treated in this respect exactly alike.

As it is impossible to exclude such minute pollen-carrying insects as Thrips, flowers which it was intended to fertilise with their own pollen may sometimes have been afterwards crossed with pollen brought by these insects from another flower on the same plant; but as we shall hereafter see, a cross of this kind does not produce any effect, or at most only a slight one. When two or more plants were placed near one another under the same net, as was often done, there is some real though not great danger of the flowers which were believed to be self-fertilised being afterwards crossed with pollen brought by Thrips from a distinct plant. I have said that the danger is not great because I have often found that plants which are self-sterile, unless aided by insects, remained sterile when several plants of the same species were placed under the same net. If, however, the flowers which had been presumably self-fertilised by me were in any case afterwards crossed by Thrips with pollen brought from a distinct plant, crossed seedlings would have been included amongst the self-fertilised; but it should be especially observed that this occurrence would tend to diminish and not to increase any superiority in average height, fertility, &c., of the crossed over the self-fertilised plants.

As the flowers which were crossed were never castrated, it is probable or even almost certain that I sometimes failed to cross-fertilise them effectually, and that they were afterwards spontaneously self-fertilised. This would have been most likely to occur with dichogamous species, for without much care it is not easy to perceive whether their stigmas are ready to be fertilised when the anthers open. But in all cases, as the flowers were protected from wind, rain, and the access of insects, any pollen placed by me on the stigmatic surface whilst it was immature, would generally have remained there until the stigma was mature; and the flowers would then have been crossed as was intended. Nevertheless, it is highly probable that self-fertilised seedlings have sometimes by this means got included amongst the crossed seedlings. The effect would be, as in the former case, not to exaggerate but to diminish any average superiority of the crossed over the self-fertilised plants.

Errors arising from the two causes just named, and from others, —such as some of the seeds not having been thoroughly ripened, though care was taken to avoid this error—the sickness or unperceived injury of any of the plants,—will have been to a large extent eliminated, in those cases in which many crossed and self-fertilised plants were measured and an average struck. Some of these causes of error will also have been eliminated by the seeds having been allowed to germinate on bare damp sand, and being planted in pairs; for it is not likely that ill-matured and well-matured, or diseased and healthy seeds, would germinate at exactly the same time. The same result will have been gained in the several cases in which only a few of the tallest, finest, and healthiest plants on each side of the pots were measured.

Kölreuter and Gärtner have proved that with some plants several, even as many as from fifty to sixty, pollen-grains are necessary for the fertilisation of all the ovules in the ovarium. Naudin also found in the case of Mirabilis that if only one or two of its very large pollen-grains were placed on the stigma, the plants raised from such seeds were dwarfed. I was therefore careful to give an amply sufficient supply of pollen, and generally covered the stigma with it; but I did not take any special pains to place exactly the same amount on the stigmas of the self-fertilised and crossed flowers. After having acted in this manner during two seasons, I remembered that Gärtner thought, though without any direct evidence, that an excess of pollen was perhaps injurious; and it has been proved by Spallanzani, Quatrefages, and Newport, that with various animals an excess of the seminal fluid entirely prevents fertilisation. It was therefore necessary to ascertain whether the fertility of the flowers was affected by applying a rather small and an extremely large quantity of pollen to the stigma. Accordingly a very small mass of pollen-grains was placed on one side of the large stigma in sixty-four flowers of *Ipomœa purpurea*, and a great mass of pollen over the whole surface of the stigma in sixty-four other flowers. In order to vary the experiment, half the flowers of both lots were on plants produced from self-fertilised seeds, and the other half on plants from crossed seeds. The sixty-four flowers with an excess of pollen yielded sixty-one capsules; and excluding four capsules, each of which contained only a single poor seed, the remainder contained on an average 5.07 seeds per capsule. The sixty-four flowers with only a little pollen placed on one side of the stigma yielded sixty-three capsules, and excluding one from the same cause as before, the remainder contained on an average 5.129 seeds. So that the flowers fertilised with little pollen yielded rather more capsules and seeds than did those fertilised with an excess; but the difference is too slight to be of any significance. On the other hand, the seeds produced by the flowers with an excess of pollen were a little heavier of the two; for 170 of them weighed 79.67 grains, whilst 170 seeds from the flowers with very little pollen weighed 79.20 grains. Both lots of seeds having been placed on damp sand presented no difference in their rate of germination. We may therefore conclude that my experiments were not affected by any slight difference in the amount of pollen used; a sufficiency having been employed in all cases.

The order in which our subject will be treated in the present volume is as follows. A long series of experiments will first be given in Chapters II. to VI. Tables will afterwards be appended, showing in a condensed form the relative heights, weights, and fertility of the offspring of the various crossed and self-fertilised species. Another table exhibits the striking results from fertilising plants, which during several generations had either been self-fertilised or had been crossed with plants kept all the time under closely similar conditions, with pollen taken from plants of a distinct stock and which had been exposed to different conditions. In the concluding chapters various related points and questions of general interest will be discussed.

Anyone not specially interested in the subject need not attempt to read all the details; though they possess, I think, some value, and cannot be all summarised. But I would suggest to the reader to take as an example the experiments on Ipomœa in Chapter II; to which may be added those on Digitalis, Origanum, Viola, or the common cabbage, as in all these cases the crossed plants are superior to the self-fertilised in a marked degree, but not in quite the same manner. As instances of self-fertilised plants being equal or superior to the crossed, the experiments on Bartonia, Canna, and the common pea ought to be read; but in the last case, and probably in that of Canna, the want of any superiority in the crossed plants can be explained.

Species were selected for experiment belonging to widely distinct families, inhabiting various countries. In some few cases several genera belonging to the same family were tried, and these are grouped together; but the families themselves have been arranged not in any natural order, but in that which was the most convenient for my purpose. The experiments have been fully given, as the results appear to me of sufficient value to justify the details. Plants bearing hermaphrodite flowers can be interbred more closely than is possible with bisexual animals, and are therefore well-fitted to throw light on the nature and extent of the good effects of crossing, and on the evil effects of close interbreeding or self-fertilisation. The most important conclusion at which I have arrived is that the mere act of crossing by itself does no good. The good depends on the individuals which are crossed differing slightly in constitution, owing to their progenitors having been subjected during several generations to slightly different conditions, or to what we call in our ignorance spontaneous variation. This conclusion, as we shall hereafter see, is closely connected with various important physiological problems, such as the benefit derived from slight changes in the conditions of life, and this stands in the closest connection with life itself. It throws light on the origin of the two sexes and on their separation or union in the same individual, and lastly on the whole subject of hybridism, which is one of the greatest obstacles to the general acceptance and progress of the great principle of evolution.

In order to avoid misapprehension, I beg leave to repeat that throughout this volume a crossed plant, seedling, or seed, means one of crossed *parentage*, that is, one derived from a flower fertilised with pollen from a distinct plant of the same species. And that a self-fertilised plant, seedling, or seed, means one of self-fertilised *parentage*, that is, one derived from a flower fertilised with pollen from the same flower, or sometimes, when thus stated, from another flower on the same plant.

THE HABITS OF INSECTS IN RELATION TO THE FERTILISATION OF FLOWERS.

Bees and various other insects must be directed by instinct to search flowers for nectar and pollen, as they act in this manner without instruction as soon as they emerge from the pupa state. Their instincts, however, are not of a specialised nature, for they visit many exotic flowers as readily as the endemic kinds, and they often search for nectar in flowers which do not secrete any; and they may be seen attempting to suck it out of nectaries of such length that it cannot be reached by them. All kinds of bees and certain other insects usually visit the flowers of the same species as long as they can, before going to another species. This fact was observed by Aristotle with respect to the hive-bee more than 2000 years ago, and was noticed by Dobbs in a paper published in 1736 in the Philosophical Transactions. It may be observed by any one, both with hive and humble-bees, in every flower-garden; not that the habit is invariably followed. Mr. Bennett watched for several hours many plants of *Lamium album*, *L. purpureum*, and another Labiate plant, *Nepeta glechoma*, all growing mingled together on a bank near some hives; and he found that each bee confined its visits to the same species. The pollen of these three plants differs in colour, so that he was able to test his observations by examining that which adhered to the bodies of the captured bees, and he found one kind on each bee.

Humble and hive-bees are good botanists, for they know that varieties may differ widely in the colour of their flowers and yet belong to the same species. I have repeatedly seen humble-bees flying straight from a plant of the ordinary red *Dictamnus fraxinella* to a white variety; from one to another very differently coloured variety of *Delphinium consolida* and of *Primula veris*; from a dark purple to a bright yellow variety of *Viola tricolor*; and with two species of Papaver, from one variety to another which differed much in colour; but in this latter case some of the bees flew indifferently to either species, although passing by other genera, and thus acted as if the two species were merely varieties. H. Müller also has seen hive-bees flying from flower to flower of *Ranunculus bulbosus* and *arvensis*, and of *Trifolium fragiferum* and *repens*; and even from blue hyacinths to blue violets.

Some species of Diptera or flies keep to the flowers of the same species with almost as much regularity as do bees; and when captured they are found covered with pollen. I have seen *Rhingia rostrata* acting in this manner with the flowers of *Lychnis dioica, Ajuga reptans*, and *Vicia sepium. Volucella plumosa* and *Empis cheiroptera* flew straight from flower to flower of *Myosotis sylvatica. Dolichopus nigripennis* behaved in the same manner with *Potentilla tormentilla*; and other Diptera with *Stellaria holostea, Helianthemum vulgare, Bellis perennis, Veronica hederæfolia* and *chamœdrys*; but some flies visited indifferently the flowers of these two latter species. I have seen more than once a minute Thrips, with pollen adhering to its body, fly from one flower to another of the same kind; and one was observed by me crawling about within a convolvulus with four grains of pollen adhering to its head, which were deposited on the stigma.

Fabricius and Sprengel state that when flies have once entered the flowers of Aristolochia they never escape,—a statement which I could not believe, as in this case the insects would not aid in the cross-fertilisation of the plant; and this statement has now been shown by Hildebrand to be erroneous. As the spathes of *Arum maculatum* are furnished with filaments apparently adapted to prevent the exit of insects, they resemble in this respect the flowers of Aristolochia; and on examining several spathes, from thirty to sixty minute Diptera belonging to three species were found in some of them; and many of these insects were found in some of them; and many of these insects were lying dead at the bottom, as if they had been permanently entrapped. In order to discover whether the living ones could escape and carry pollen to another plant, I tied in the spring of 1842 a fine muslin bag tightly round a spathe; and on returning in an hour's time several little flies were crawling about on the inner surface of the bag. I then gathered a spathe and breathed hard into it; several flies soon crawled out, and all without exception were dusted with arum pollen. These flies quickly flew away, and I distinctly saw three of them fly to another plant about a yard off; they alighted on the inner or concave surface of the spathe, and suddenly flew down into the flower. I then opened this flower, and although not a single anther had burst, several grains of pollen were lying at the bottom, which must have been brought from another plant by one of these flies or by some other insect. In another flower little flies were crawling about, and I saw them leave pollen on the stigmas.

I do not know whether Lepidoptera generally keep to the flowers of the same species; but I once observed many minute moths (I believe *Lampronia* (Tinea) *calthella*) apparently eating the pollen of *Mercurialis annua*, and they had the whole front of their bodies covered with pollen. I then went to a female plant some yards off, and saw in the course of fifteen minutes three of these moths alight on the stigmas. Lepidoptera are probably often induced to frequent the flowers of the same species, whenever these are provided with a long and narrow nectary, as in this case other insects cannot suck the nectar, which will thus be preserved for those having an elongated proboscis. No doubt the Yucca moth visits only the flowers whence its name is derived, for a most wonderful instinct guides this moth to place pollen on the stigma, so that the ovules may be developed on which the larvæ feed. With respect to Coleoptera, I have seen Meligethes covered with pollen flying from flower to flower of the same species; and this must often occur, as, according to M. Brisout, "many of the species affect only one kind of plant."

It must not be supposed from these several statements that insects strictly confine their visits to the same species. They often visit other species when only a few plants of the same kind grow near together. In a flower-garden containing some plants of Œnothera, the pollen of which can easily be recognised, I found not only single grains but masses of it within many flowers of Mimulus, Digitalis, Antirrhinum, and Linaria. Other kinds of pollen were likewise detected in these same flowers. A large number of the stigmas of a plant of Thyme, in which the anthers were completely aborted, were examined; and these stigmas, though scarcely larger than a split needle, were covered not only with pollen of Thyme brought from other plants by the bees, but with several other kinds of pollen.

That insects should visit the flowers of the same species as long as they can, is of great importance to the plant, as it favours the cross-fertilisation of distinct individuals of the same species; but no one will suppose that insects act in this manner for the good of the plant. The cause probably lies in insects being thus enabled to work quicker; they have just learned how to stand in the best position on the flower, and how far and in what direction to insert their proboscides. They act on the same principle as does an artificer who has to make half-a-dozen engines, and who saves time by making consecutively each wheel and part for all of them. Insects, or at least bees, seem much influenced by habit in all their manifold operations; and we shall presently see that this holds good in their felonious practice of biting holes through the corolla.

It is a curious question how bees recognise the flowers of the same species. That the coloured corolla is the chief guide cannot be doubted. On a fine day, when hive-bees were incessantly visiting the little blue flowers of *Lobelia erinus*, I cut off all the petals of some, and only the lower striped petals of others, and these flowers were not once again sucked by the bees, although some actually crawled over them. The removal of the two little upper petals alone made no difference in their visits. Mr. J. Anderson likewise states that when he removed the corollas of the Calceolaria, bees never visited the flowers. On the other hand, in some large masses of *Geranium phæum* which had escaped out of a garden, I observed the unusual fact of the flowers continuing to secrete an abundance of nectar after all the petals had fallen off; and the flowers in this state were still visited by humble-bees. But the bees might have learned that these flowers with all their petals lost were still worth visiting, by finding nectar in those with only one or two lost. The colour alone of the corolla serves as an approximate guide: thus I watched for some time humble-bees which were visiting exclusively plants of the white-flowered *Spiranthes autumnalis*, growing on short turf at a considerable distance apart; and these bees often flew within a few inches of several other plants with white flowers, and then without further examination passed onwards in search of the *Spiranthes*. Again, many hive-bees which confined their visits to the common ling (*Calluna vulgaris*), repeatedly flew towards *Erica tetralix*, evidently attracted by the nearly similar tint of their flowers, and then instantly passed on in search of the Calluna.

That the colour of the flower is not the sole guide, is clearly shown by the six cases above given of bees which repeatedly passed in a direct line from one variety to another of the same species, although they bore very differently coloured flowers. I observed also bees flying in a straight line from one clump of a yellow-flowered Œnothera to every other clump of the same plant in the garden, without turning an inch from their course to plants of Eschscholtzia and others with yellow flowers which lay only a foot or two on either side. In these cases the bees knew the position of each plant in the garden perfectly well, as we may infer by the directness of their flight; so that they were guided by experience and memory. But how did they discover at first that the above varieties with differently coloured flowers belonged to the same species? Improbable as it may appear, they seem, at least sometimes, to recognise plants even from a distance by their general aspect, in the same manner as we should do. On three occasions I observed humble-bees flying in a perfectly straight line from a tall larkspur (Delphinium) which was in full flower to another plant of the same species at the distance of fifteen yards which had not as yet a single flower open, and on which the buds showed only a faint tinge of blue. Here neither odour nor the memory of former visits could have come into play, and the tinge of blue was so faint that it could hardly have served as a guide.

The conspicuousness of the corolla does not suffice to induce repeated visits from insects, unless nectar is at the same time secreted, together perhaps with some odour emitted. I watched for a fortnight many times daily a wall covered with *Linaria cymbaluria* in full flower, and never saw a bee even looking at one. There was then a very hot day, and suddenly many bees were industriously at work on the flowers. It appears that a certain degree of heat is necessary for the secretion of nectar; for I observed with *Lobelia erinus* that if the sun ceased to shine for only half an hour, the visits of the bees slackened and soon ceased. An analogous fact with respect to the sweet excretion from the stipules of *Vicia sativa* has been already given. As in the case of the Linaria, so with *Pedicularis sylvatica, Polygala vulgaris, Viola tricolor*, and some species of Trifolium, I have watched the flowers day after day without seeing a bee at work, and then suddenly all the flowers were visited by many bees. Now how did so many bees discover at once that the flowers were secreting nectar? I presume that it must have been by their odour; and that as soon as a few bees began to suck the flowers, others of the same and of different kinds observed the fact and profited by it. We shall presently see, when we treat of the perforation of the corolla, that bees are fully capable of profiting by the labour of other species. Memory also comes into play, for, as already remarked, bees know the position of each clump of flowers in a garden. I have repeatedly seen them passing round a corner, but otherwise in as straight a line as possible, from one plant of Fraxinella and of Linaria to another and distant one of the same species; although, owing to the intervention of other plants, the two were not in sight of each other.

It would appear that either the taste or the odour of the nectar of certain flowers is unattractive to hive or to humble-bees, or to both; for there seems no other reason why certain open flowers which secrete nectar are not visited by them. The small quantity of nectar secreted by some of these flowers can hardly be the cause of their neglect, as hive-bees search eagerly for the minute drops on the glands on the leaves of the *Prunus laurocerasus*. Even the bees from different hives sometimes visit different kinds of flowers, as is said to be the case by Mr. Grant with respect to the Polyanthus and *Viola tricolor*. I have known humble-bees to visit the flowers of *Lobelia fulgens* in one garden and not in another at the distance of only a few miles. The cupful of nectar in the labellum of *Epipactis latifolia* is never touched by hive or humble-bees, although I have seen them flying close by; and yet the nectar has a pleasant taste to us, and is habitually consumed by the common wasp. As far as I have seen, wasps seek for nectar in this country only from the flowers of this Epipactis, *Scrophularia aquatica, Symphoricarpus racemosa*, and Tritoma; the two former plants being endemic, and the two latter exotic. As wasps are so fond of sugar and of any sweet fluid, and as they do not disdain the minute drops on the glands of *Prunus laurocerasus*, it is a strange fact that they do not suck the nectar of many open flowers, which they could do without the aid of a proboscis. Hive-bees visit the flowers of the Symphoricarpus and Tritoma, and this makes it all the stranger that they do not visit the flowers of the Epipactis, or, as far as I have seen, those of the *Scrophularia aquatica*; although they do visit the flowers of *Scrophularia nodosa*, at least in North America.

The extraordinary industry of bees and the number of flowers which they visit within a short time, so that each flower is visited repeatedly, must greatly increase the chance of each receiving pollen from a distinct plant. When the nectar is in any way hidden, bees cannot tell without inserting their proboscides whether it has lately been exhausted by other bees, and this, as remarked in a former chapter, forces them to visit many more flowers than they otherwise would. But they endeavour to lose as little time as they can; thus in flowers having several nectaries, if they find one dry they do not try the others, but as I have often observed, pass on to another flower. They work so industriously and effectually, that even in the case of social plants, of which hundreds of thousands grow together, as with the several kinds of heath, every single flower is visited, of which evidence will presently be given. They lose no time and fly very quickly from plant to plant, but I do not know the rate at which hive-bees fly. Humble-bees fly at the rate of ten miles an hour, as I was able to ascertain in the case of the males from their curious habit of calling at certain fixed points, which made it easy to measure the time taken in passing from one place to another.

With respect to the number of flowers which bees visit in a given time, I observed that in exactly one minute a humble-bee visited twenty-four of the closed flowers of the *Linaria cymbalaria*; another bee visited in the same time twenty-two flowers of the *Symphoricarpus racemosa*; and another seventeen flowers on two plants of a Delphinium. In the course of fifteen minutes a single flower on the summit of a plant of Œnothera was visited eight times by several humble-bees, and I followed the last of these bees, whilst it visited in the course of a few additional minutes every plant of the same species in a large flower-garden. In nineteen minutes every flower on a small plant of *Nemophila insignis* was visited twice. In one minute six flowers of a Campanula were entered by a pollen-collecting hive-bee; and bees when thus employed work slower than when sucking nectar. Lastly, seven flower-stalks on a plant of *Dictamnus fraxinella* were observed on the 15th of June 1841 during ten minutes; they were visited by thirteen humble-bees each of which entered many flowers. On the 22nd the same flower-stalks were visited within the same time by eleven humble-bees. This plant bore altogether 280 flowers, and from the above data, taking into consideration how late in the evening humble-bees work, each flower must have been visited at least thirty times daily, and the same flower keeps open during several days. The frequency of the visits of bees is also sometimes shown by the manner in which the petals are scratched by their hooked tarsi; I have seen large beds of Mimulus, Stachys, and Lathyrus with the beauty of their flowers thus sadly defaced.

Perforation of the Corolla by Bees.—I have already alluded to bees biting holes in flowers for the sake of obtaining the nectar. They often act in this manner, both with endemic and exotic species, in many parts of Europe, in the United States, and in the Himalaya; and therefore probably in all parts of the world. The plants, the fertilisation of which actually depends on insects entering the flowers, will fail to produce seed when their nectar is stolen from the outside; and even with those species which are capable of fertilising themselves without any aid, there can be no cross-fertilisation, and this, as we know, is a serious evil in most cases. The extent to which humble-bees carry on the practice of biting holes is surprising: a remarkable case was observed by me near Bournemouth, where there were formerly extensive heaths. I took a long walk, and every now and then gathered a twig of *Erica tetralix*, and when I had got a handful all the flowers were examined through a lens. This process was repeated many times; but though many hundreds were examined, I did not succeed in finding a single flower which had not been perforated. Humble-bees were at the time sucking the flowers through these perforations. On the following day a large number of flowers were examined on another heath with the same result, but here hive-bees were sucking through the holes. This case is all the more remarkable, as the innumerable holes had been made within a fortnight, for before that time I saw the bees everywhere sucking in the proper manner at the mouths of the corolla. In an extensive flower-garden some large beds of *Salvia grahami, Stachys coccinea*, and *Pentstemon argutus* (?) had every flower perforated, and many scores were examined. I have seen whole fields of red clover (*Trifolium pratense*) in the same state. Dr. Ogle found that 90 per cent. of the flowers of *Salvia glutinosa* had been bitten. In the United States Mr. Bailey says it is difficult to find a blossom of the native *Gerardia pedicularia* without a hole in it; and Mr. Gentry, in speaking of the introduced *Wistaria sinensis*, says "that nearly every flower had been perforated."

As far as I have seen, it is always humble-bees which first bite the holes, and they are well fitted for the work by possessing powerful mandibles; but hive-bees afterwards profit by the holes thus made. Dr. H. Müller, however, writes to me that hive-bees sometimes bite holes through the flowers of *Erica tetralix*. No insects except bees, with the single exception of wasps in the case of Tritoma, have sense enough, as far as I have observed, to profit by the holes already made. Even humble-bees do not always discover that it would be advantageous to them to perforate certain flowers. There is an abundant supply of nectar in the nectary of *Tropæolum tricolor*, yet I have found this plant untouched in more than one garden, while the flowers of other plants had been extensively perforated; but a few years ago Sir J. Lubbock's gardener assured me that he had seen humble-bees boring through the nectary of this Tropæolum. Müller has observed humble-bees trying to suck at the mouths of the flowers of *Primula elatior* and of an Aquilegia, and, failing in their attempts, they made holes through the corolla; but they often bite holes, although they could with very little more trouble obtain the nectar in a legitimate manner by the mouth of the corolla.

Dr. W. Ogle has communicated to me a curious case. He gathered in Switzerland 100 flower-stems of the common blue variety of the monkshood (*Aconitum napellus*), and not a single flower was perforated; he then gathered 100 stems of a white variety growing close by, and every one of the open flowers had been perforated. This surprising difference in the state of the flowers may be attributed with much probability to the blue variety being distasteful to bees, from the presence of the acrid matter which is so general in the Ranunculaceæ, and to its absence in the white variety in correlation with the loss of the blue tint. According to Sprengel, this plant is strongly proterandrous; it would therefore be more or less sterile unless bees carried pollen from the younger to the older flowers. Consequently the white variety, the flowers of which were always bitten instead of being properly entered by the bees, would fail to yield the full number of seeds and would be a comparatively rare plant, as Dr. Ogle informs me was the case.

Bees show much skill in their manner of working, for they always make their holes from the outside close to the spot where the nectar lies hidden within the corolla. All the flowers in a large bed of *Stachys coccinea* had either one or two slits made on the upper side of the corolla near the base. The flowers of a Mirabilis and of *Salvia coccinea* were perforated in the same manner; whilst those of *Salvia grahami*, in which the calyx is much elongated, had both the calyx and the corolla invariably perforated. The flowers of *Pentstemon argutus* are broader than those of the plants just named, and two holes alongside each other had here always been made just above the calyx. In these several cases the perforations were on the upper side, but in *Antirrhi-*

num majus one or two holes had been made on the lower side, close to the little protuberance which represents the nectary, and therefore directly in front of and close to the spot where the nectar is secreted.

But the most remarkable case of skill and judgment known to me, is that of the perforation of the flowers of *Lathyrus sylvestris*, as described by my son Francis. The nectar in this plant is enclosed within a tube, formed by the united stamens, which surround the pistil so closely that a bee is forced to insert its proboscis outside the tube; but two natural rounded passages or orifices are left in the tube near the base, in order that the nectar may be reached by the bees. Now my son found in sixteen out of twenty-four flowers on this plant, and in eleven out of sixteen of those on the cultivated everlasting pea, which is either a variety of the same species or a closely allied one, that the left passage was larger than the right one. And here comes the remarkable point,—the humble-bees bite holes through the standard petal, and they always operated on the left side over the passage, which is generally the larger of the two. My son remarks: "It is difficult to say how the bees could have acquired this habit. Whether they discovered the inequality in the size of the nectar-holes in sucking the flowers in the proper way, and then utilised this knowledge in determining where to gnaw the hole; or whether they found out the best situation by biting through the standard at various points, and afterwards remembered its situation in visiting other flowers. But in either case they show a remarkable power of making use of what they have learnt by experience." It seems probable that bees owe their skill in biting holes through flowers of all kinds to their having long practised the instinct of moulding cells and pots of wax, or of enlarging their old cocoons with tubes of wax; for they are thus compelled to work on the inside and outside of the same object.

In the early part of the summer of 1857 I was led to observe during some weeks several rows of the scarlet kidney-bean (*Phaseolus multiflorus*), whilst attending to the fertilisation of this plant, and daily saw humble- and hive-bees sucking at the mouths of the flowers. But one day I found several humble-bees employed in cutting holes in flower after flower; and on the next day every single hive-bee, without exception, instead of alighting on the left wing-petal and sucking the flower in the proper manner, flew straight without the least hesitation to the calyx, and sucked through the holes which had been made only the day before by the humble-bees; and they continued this habit for many following days. Mr. Belt has communicated to me (July 28th, 1874) a similar case, with the sole difference that less than half of the flowers had been perforated by the humble-bees; nevertheless, all the hive-bees gave up sucking at the mouths of the flowers and visited exclusively the bitten ones. Now how did the hive-bees find out so quickly that holes had been made? Instinct seems to be out of the question, as the plant is an exotic. The holes cannot be seen by bees whilst standing on the wing-petals, where they had always previously alighted. From the ease with which bees were deceived when the petals of *Lobelia erinus* were cut off, it was clear that in this case they were not guided to the nectar by its smell; and it may be doubted whether they were attracted to the holes in the flowers of the Phaseolus by the odour emitted from them. Did they perceive the holes by the sense of touch in their proboscides, whilst sucking the flowers in the proper manner, and then reason that it would save them time to alight on the outside of the flowers and use the holes? This seems almost too abstruse an act of reason for bees; and it is more probable that they saw the humble-bees at work, and understanding what they were about, imitated them and took advantage of the shorter path to the nectar. Even with animals high in the scale, such as monkeys, we should be surprised at hearing that all the individuals of one species within the space of twenty-four hours understood an act performed by a distinct species, and profited by it.

I have repeatedly observed with various kinds of flowers that all the hive and humble-bees which were sucking through the perforations, flew to them, whether on the upper or under side of the corolla, without the least hesitation; and this shows how quickly all the individuals within the district had acquired the same knowledge. Yet habit comes into play to a certain extent, as in so many of the other operations of bees. Dr. Ogle, Messrs. Farrer and Belt have observed in the case of *Phaseolus multiflorus* that certain individuals went exclusively to the perforations, while others of the same species visited only the mouths of the flowers. I noticed in 1861 exactly the same fact with *Trifolium pratense*. So persistent is the force of habit, that when a bee which is visiting perforated flowers comes to one which has not been bitten, it does not go to the mouth, but instantly flies away in search of another bitten flower. Nevertheless, I once saw a humble-bee visiting the hybrid *Rhododendron azaloides*, and it entered the mouths of some flowers and cut holes into the others. Dr. H. Müller informs me that in the same district he has seen some individuals of *Bombus mastrucatus* boring through the calyx and corolla of *Rhinanthus alecterolophus*, and others through the corolla alone. Different species of bees may, however, sometimes be observed acting differently at the same time on the same plant. I have seen hive-bees sucking at the mouths of the flowers of the common bean; humble-bees of one kind sucking through holes bitten in the calyx, and humble-bees of another kind sucking the little drops of fluid excreted by the stipules. Mr. Beal of Michigan informs me that the flowers of the Missouri currant (*Ribes aureum*) abound with nectar, so that children often suck them; and he saw hive-bees sucking through holes made by a bird, the oriole, and at the same time humble-bees sucking in the proper manner at the mouths of the flowers. This statement about the oriole calls to mind what I have before said of certain species of humming-birds boring holes through the flowers of the Brugmansia, whilst other species entered by the mouth.

The motive which impels bees to gnaw holes through the corolla seems to be the saving of time, for they lose much time in climbing into and out of large flowers, and in forcing their heads into closed ones. They were able to visit nearly twice as many flowers, as far as I could judge, of a Stachys and Pentstemon by alighting on the upper surface of the corolla and sucking through the cut holes, than by entering in the proper way. Nevertheless each bee before it has had much practice, must lose some time in making each new perforation, especially when the perforation has to be made through both calyx and corolla. This action therefore implies foresight, of which faculty we have abundant evidence in their building operations; and may we not further believe that some trace of their social instinct, that is, of working for the good of other members of the community, may here likewise play a part?

Many years ago I was struck with the fact that humble-bees as a general rule perforate flowers only when these grow in large numbers near together. In a garden where there were some very large beds of *Stachys coccinea* and of *Pentstemon argutus*, every single flower was perforated, but I found two plants of the former species growing quite separate with their petals much scratched, showing that they had been frequently visited by bees, and yet not a single flower was perforated. I found also a separate plant of the Pentstemon, and saw bees entering the mouth of the corolla, and not a single flower had been perforated. In the following year (1842) I visited the same garden several times: on the 19th of July humble-bees were sucking the flowers of *Stachys coccinea* and *Salvia grahami* in the proper manner, and none of the corollas were perforated. On the 7th of August all the flowers were perforated, even those on some few plants of the Salvia which grew at a little distance from the great bed. On the 21st of August only a few flowers on the summits of the spikes of both species remained fresh, and not one of these was now bored. Again, in my own garden every plant in several rows of the common bean had many flowers perforated; but I found three plants in separate parts of the garden which had sprung up accidentally, and these had not a single flower perforated. General Strachey formerly saw many perforated flowers in a garden in the Himalaya, and he wrote to the owner to inquire whether this relation between the plants growing crowded and their perforation by the bees there held good, and was answered in the affirmative. Hence it follows that the red clover (*Trifolium pratense*) and the common bean when cultivated in large numbers on heaths,—rows of the scarlet kidney-bean in the kitchen-garden,—and masses of any species in the flower-garden,—are all eminently liable to be perforated.

The explanation of this fact is not difficult. Flowers growing in large numbers afford a rich booty to the bees, and are conspicuous from a distance. They are consequently visited by crowds of these insects, and I once counted between twenty and thirty bees flying about a bed of Pentstemon. They are thus stimulated to work quickly by rivalry, and, what is much more important, they find a large proportion of the flowers, as suggested by my son, with their nectaries sucked dry. They thus waste much time in searching many empty flowers, and are led to bite the holes, so as to find out as quickly as possible whether there is any nectar present, and if so, to obtain it.

Flowers which are partially or wholly sterile unless visited by insects in the proper manner, such as those of most species of Salvia, of *Trifolium pratense, Phaseolus multiflorus*, &c., will fail more or less completely to produce seeds if the bees confine their visits to the perforations. The perforated flowers of those species, which are capable of fertilising themselves, will yield only self-fertilised seeds, and the seedlings will in consequence be less vigorous. Therefore all plants must suffer in some degree when bees obtain their nectar in a felonious manner by biting holes through the corolla; and many species, it might be thought, would thus be exterminated. But here, as is so general throughout nature, there is a tendency towards a restored equilibrium. If a plant suffers from being perforated, fewer individuals will be reared, and if its nectar is highly important to the bees, these in their turn will suffer and decrease in number; but, what is much more effective, as soon as the plant becomes somewhat rare so as not to grow in crowded masses, the bees will no longer be stimulated to gnaw holes in the flowers, but will enter them in a legitimate manner. More seed will then be produced, and the seedlings being the product of cross-fertilisation will be vigorous, so that the species will tend to increase in number, to be again checked, as soon as the plant again grows in crowded masses.

Sexual Selection in Relation to Monkeys. [1876]

THIS ADDENDUM TO The Descent of Man *appeared in* Nature *on November 2, 1876. Darwin's having read the German zoologist Johann von Fischer's paper on the expression of simian emotion led to several letters between the two scientists. Darwin's choice of subject here strains against Victorian sensibilities: even a scientific examination of the loves of primates, as opposed to plants, could be considered unseemly. The eminent man of letters and art historian John Ruskin, who met Darwin in 1879, remarked uneasily on Darwin's "deep & tender interest" in the "brightly coloured hinder half of certain monkeys"— and interestingly, Darwin's paper does not translate the quotation from von Fischer's German original.*

In the discussion on Sexual Selection in my "Descent of Man," no case interested and perplexed me so much as the brightly-coloured hinder ends and adjoining parts of certain monkeys. As these parts are more brightly coloured in one sex than the other, and as they become more brilliant during the season of love, I concluded that the colours had been gained as sexual attraction. I was well aware that I thus laid myself open to ridicule; though in fact it is not more surprising that a monkey should display his bright-red hinder end than that a peacock should display his magnificent tail. I had, however, at that time no evidence of monkeys exhibiting this part of their bodies during their courtship; and such display in the case of birds affords the best evidence that the ornaments of the males are of service to them by attracting or exciting the females. I have lately read an article by Joh. von Fischer, of Gotha, published in *Der Zoologische Garten*, April, 1876, on the expression of monkeys under various emotions, which is well worthy of study by any one interested in the subject, and which shows that the author is a careful and acute observer. In this article there is an account of the behaviour of a young male mandrill when he first beheld himself in a looking-glass, and it is added, that after a time he turned round and presented his red hinder end to the glass. Accordingly I wrote to Herr J. von Fischer to ask what he supposed was the meaning of this strange action, and he has sent me two long letters full of new and curious details, which will, I hope, be hereafter published. He says that he was himself at first perplexed by the above action, and was thus led carefully to observe several individuals of various other species of monkeys, which he has long kept in his house. He finds that not only the mandrill (*Cynocephalus mormon*) but the drill (*C. leucophaeus*) and three other kinds of baboons (*C. hamadryas, sphinx*, and *babouin*) also *Cynopithecus niger*, and *Macacus rhesus* and *nemestrinus*, turn this part of their bodies, which in all these species is more or less brightly coloured, to him when they are pleased, and to other persons as a sort of greeting. He took pains to cure a *Macacus rhesus*, which he had kept for five years, of this indecorous habit, and at last succeeded. These monkeys are particularly apt to act in this manner, grinning at the same time, when first introduced to a new monkey, but often also to their old monkey friends; and after this mutual display they begin to play together. The young mandrill ceased spontaneously after a time to act in this manner towards his master, von Fischer, but continued to do so towards persons who were strangers and to new monkeys. A young *Cynopithecus niger* never acted, excepting on one occasion, in this way towards his master, but frequently towards strangers, and continues to do so up to the present time. From these facts von Fischer concludes that the monkeys which behaved in this manner before a looking-glass (viz., the mandrill, drill, *Cynopithecus niger, Macascus rhesus*, and *nemestrinus*) acted as if their reflection were a new acquaintance. The mandrill and drill, which have their hinder ends especially ornamented, display it even whilst quite young, more frequently and more ostentatiously than do the other kinds. Next in order comes *Cynocephalus hamadryas*, whilst the other species act in this manner seldomer. The individuals, however, of the same species, vary in this respect, and some which were very shy never displayed their hinder ends. It deserves especial attention that von Fischer has never seen any species purposely exhibit the hinder part of its body, if not at all coloured. This remark applies to many individuals of *Macacus cynomolgus* and *Cercocebus radiatus* (which is closely allied to *M. rhesus*), to three species of Cercopithecus and several American monkeys. The habit of turning the hinder ends as a greeting to an old friend or new acquaintance, which seems to us so odd, is not really more so than the habits of many savages, for instance that of rubbing their bellies with their hands, or rubbing noses together. The habit with the mandrill and drill seems to be instinctive or inherited, as it was followed by very young animals; but it is modified or guided, like so many other instincts, by observation, for von Fischer says that they take pains to make their display fully, and if made before two observers, they turn to him who seems to pay the most attention.

With respect to the origin of the habit, von Fischer remarks that his monkeys like to have their naked hinder ends patted or stroked, and that they then grunt with pleasure. They often also turn this part of their bodies to other monkeys to have bits of dirt picked off, and so no doubt it would be with respect to thorns. But the habit with adult animals is connected to a certain extent with sexual feelings, for von Fischer watched through a glass door a female *Cynopithecus niger*, and she during several days "umdrehte und dem Männchen mit gurgelnden Tönen die stark geröthete Sitzfläche zeigte, was ich früher nie an diesem Thier bemerkt hatte. Beim Anblick dieses Gegenstandes erregte sich das Männchen sichtlich, denn es polterte heftig an den Stäben, ebenfalls gurgelnde Laute ausstossend" ["turned round and with gurgling sounds presented her reddened behind to the male, something I had never before observed in this animal. At the sight of it, the male became visibly aroused, for he barked furiously at the bars and gave out gurgling sounds"—editors]. As all the monkeys which have the hinder parts of their bodies more or less bright coloured live, according to von Fischer, in open rocky places, he thinks that these colours serve to render one sex conspicuous at a distance to the other; but as monkeys are such gregarious animals, I should have thought that there was no need for the sexes to recognise each other at a distance. It seems to me more probable that the bright colours, whether on the face or hinder end, or, as in the mandrill, on both, serve as a sexual ornament and attraction. Anyhow, as we now know that monkeys have the habit of turning their hinder ends towards other monkeys, it ceases to be at all surprising that it should have been this part of their bodies which has been more or less decorated. The fact that it is only the monkeys thus characterized which, as far as at present known, act in this manner as a greeting towards other monkeys, renders it doubtful whether the habit was first acquired from some independent cause, and that afterwards the parts in question were coloured as a sexual ornament; or whether the colouring and the habit of turning round were first acquired through variation and sexual selection, and that afterwards the habit was retained as a sign of pleasure or as a greeting, through the principle of inherited association. This principle apparently comes into play on many occasions: thus it is generally admitted that the songs of birds serve mainly as an attraction during the season of love, and that the *leks*, or great congregations of the black grouse, are connected with their courtship; but the habit of singing has been retained by some birds when they feel happy, for instance by the common robin, and the habit of congregating has been retained by the black grouse, during other seasons of the year.

I beg leave to refer to one other point in relation to sexual selection. It has been objected that this form of selection, as far as the ornaments of the males are concerned, implies that all the females within the same district must possess and exercise exactly the same taste. It should, however, be observed in the first place, that although the range of variation of a species may be very large, it is by no means indefinite. I have elsewhere given a good instance of this fact in the pigeon, of which there are at least a hundred varieties differing widely in their colours, and at least a score of varieties of the fowl differing in the same manner; but the range of colour in these two species is extremely distinct. Therefore the females of natural species cannot have an unlimited scope for their taste. In the second place, I presume that no supporter of the principle of sexual selection believes that the females select particular points of beauty in the males; they are merely excited or attracted in a greater degree by one male than by another, and this seems often to depend, especially with birds, on brilliant colouring. Even man, excepting perhaps an artist, does not analyse the slight differences in the features of the woman whom he may admire, on which her beauty depends. The male mandrill has not only the hinder end of his body, but his face gorgeously coloured and marked with oblique ridges, a yellow beard, and other ornaments. We may infer from what we see of the variation of animals under domestication, that the above several ornaments of the mandrill were gradually acquired by one individual varying a little in one way, and another individual in another way. The males which were the handsomest or the most attractive in any manner to the females would pair oftenest, and would leave rather more offspring than other males. The offspring of the former, although variously intercrossed, would either inherit the peculiarities of their fathers, or transmit an increased tendency to vary in the same manner. Consequently the whole body of males inhabiting the same country, would tend from the effects of constant intercrossing to become modified almost uniformly, but sometimes a little more in one character and sometimes in another, though at an extremely slow rate; all ultimately being thus rendered more attractive to the females. The process is like that which I have called unconscious selection by man, and of which I have given several instances. In one country the inhabitants value a fleet or light dog or horse, and in another country a heavier and more powerful one; in neither country is there any selection of the individual animals with lighter or stronger bodies and limbs; nevertheless after a considerable lapse of time the individuals are found to have been modified in the desired manner almost uniformly, though differently in each country. In two absolutely distinct countries inhabited by the same species, the individuals of which can never during long ages have intermigrated and intercrossed, and where, moreover, the variations will probably not have been identically the same, sexual selection might cause the males to differ. Nor does the belief appear to me altogether fanciful that two sets of females, surrounded by a very different environment, would be apt to acquire somewhat different tastes with respect to form, sound, or colour. However this may be, I have given in my "Descent of Man" instances of closely-allied birds inhabiting distinct countries, of which the young and the females cannot be distinguished, whilst the adult males differ considerably, and this may be attributed with much probability to the action of sexual selection.

The Different Forms of Flowers on Plants of the Same Species. [1877]

Falling in with the general research program that resulted in Orchids *and* Cross and Self-Fertilisation, The Different Forms of Flowers on Plants of the Same Species *updates and considerably amplifies a series of papers Darwin had published in the* Journal of the Proceedings of the Linnean Society *between 1862 and 1866. The book appeared in July 1877. A second edition in July 1880 reviewed the recent literature in its preface, but the text remained almost unchanged from the first edition. Only about two thousand copies of* The Different Forms of Flowers *sold in Darwin's lifetime; however, in the* Autobiography *Darwin stated that "no little discovery of mine ever gave me so much pleasure as the making out the meaning of heterostyled flowers."*

Early in the 1860s Darwin had become interested in the strangely variable stamen and style placement in the flowers of English primroses. Continuous gradations from an extremely long stamen or style to an extremely short one were not discernible, as is the case for many other physical characteristics. Instead, there were either long or short stamens or styles: a primrose flower with long stamens would have a short style, and vice versa. Furthermore, when Darwin sampled local populations of one hundred plants or so, he found that the two forms occurred in a roughly one-to-one ratio. Darwin reasoned that these alternative forms, which he called long-styled and short-styled, were adaptations for cross-pollination, a phenomenon he considered extremely important for the action of natural selection.

Today, Darwin's writings on distyly, which he explained, and tri-

styly, which he discovered, still form the basis of much research into plant sexuality. The kinds of breeding experiments Darwin describes in his book are still commonly undertaken by botanists. We have included the introduction detailing the various, sometimes bizarre, types of floral configurations in a species: hermaphrodite, dimorphic, trimorphic, cleistogamous, monoecious, dioecious, polygamous, polygamodioecious, polygamomonoecious, gynodioecious, andromonoecious, and gynomonoecious, all of which are explained in Darwin's text below. The only floral type not observed by Darwin is androdioecious (individuals of the species are either male or hermaphrodite), not discovered until 1990. Also included are Darwin's remarks on his experimental discovery that Linum *is heterostylous and his concluding comments on heterostyly.*

INTRODUCTION.

The subject of the present volume, namely the differently formed flowers normally produced by certain kinds of plants, either on the same stock or on distinct stocks, ought to have been treated by a professed botanist, to which distinction I can lay no claim. As far as the sexual relations of flowers are concerned, Linnæus long ago divided them into hermaphrodite, monœcious, diœcious, and polygamous species. This fundamental distinction, with the aid of several subdivisions in each of the four classes, will serve my purpose; but the classification is artificial, and the groups often pass into one another.

The hermaphrodite class contains two interesting sub-groups, namely, heterostyled and cleistogamic plants; but there are several other less important subdivisions, presently to be given, in which flowers differing in various ways from one another are produced by the same species.

Some plants were described by me several years ago, in a series of papers read before the Linnean Society, the individuals of which exist under two or three forms, differing in the length of their pistils and stamens and in other respects. They were called by me dimorphic and trimorphic, but have since been better named by Hildebrand, heterostyled. As I have many still unpublished observations with respect to these plants, it has seemed to me advisable to republish my former papers in a connected and corrected form, together with the new matter. It will be shown that these heterostyled plants are adapted for reciprocal fertilisation; so that the two or three forms, though all are hermaphrodites, are related to one another almost like the males and females of ordinary unisexual animals. I will also give a full abstract of such observations as have been published since the appearance of my papers; but only those cases will be noticed, with respect to which the evidence seems fairly satisfactory. Some plants have been supposed to be heterostyled merely from their pistils and stamens varying greatly in length, and I have been myself more than once thus deceived. With some species the pistil continues growing for a long time, so that if old and young flowers are compared they might be thought to be heterostyled. Again, a species tending to become diœcious, with the stamens reduced in some individuals and with the pistils in others, often presents a deceptive appearance. Unless it be proved that one form is fully fertile only when it is fertilised with pollen from another form, we have not complete evidence that the species is heterostyled. But when the pistils and stamens differ in length in two or three sets of individuals, and this is accompanied by a difference in the size of the pollen-grains or in the state of the stigma, we may infer with much safety that the species is heterostyled. I have, however, occasionally trusted to a difference between the two forms in the length of the pistil alone, or in the length of the stigma together with its more or less papillose condition; and in one instance differences of this kind have been proved by trials made on the fertility of the two forms, to be sufficient evidence.

The second sub-group above referred to consists of hermaphrodite plants, which bear two kinds of flowers—the one perfect and fully expanded—the other minute, completely closed, with the petals rudimentary, often with some of the anthers aborted, and the remaining ones together with the stigmas much reduced in size; yet these flowers are perfectly fertile. They have been called by Dr. Kuhn cleistogamic, and they will be described in the last chapter of this volume. They are manifestly adapted for self-fertilisation, which is effected at the cost of a wonderfully small expenditure of pollen; whilst the perfect flowers produced by the same plant are capable of cross-fertilisation. Certain aquatic species, when they flower beneath the water, keep their corollas closed, apparently to protect their pollen; they might therefore be called cleistogamic, but for reasons assigned in the proper place are not included in the present sub-group. Several cleistogamic species, as we shall hereafter see, bury their ovaries or young capsules in the ground. Some few plants produce subterranean flowers, as well as ordinary ones; and these might have been formed into a small separate subdivision.

Another interesting subdivision consists of certain plants, discovered by H. Müller, some individuals of which bear conspicuous flowers adapted for cross-fertilisation by the aid of insects, and others much smaller and less conspicuous flowers, which have often been slightly modified so as to ensure self-fertilisation. *Lysimachia vulgaris, Euphrasia officinalis, Rhinanthus crista-galli,* and *Viola tricolor* come under this head. The smaller and less conspicuous flowers are not closed, but as far as the purpose which they serve is concerned, namely, the assured propagation of the species, they approach in nature cleistogamic flowers; but they differ from them by the two kinds being produced on distinct plants.

With many plants, the flowers towards the outside of the inflorescence are much larger and more conspicuous than the central ones. As I shall not have occasion to refer to plants of this kind in the following chapters, I will here give a few details respecting them. It is familiar to every one that the ray-florets of the Compositæ often differ remarkably from the others; and so it is with the outer flowers of many Umbelliferæ, some Cruciferæ and a few other families. Several species of Hydrangea and Viburnum offer striking instances of the same fact. The Rubiaceous genus Mussænda presents a very curious appearance from some of the flowers having the tip of one of the sepals developed into a large petal-like expansion, coloured either white or purple. The outer flowers in several Acantaceous genera are large and conspicuous, but sterile; the next in order are smaller, open, moderately fertile and capable of cross-fertilisation; whilst the central ones are cleistogamic, being still smaller, closed and highly fertile; so that here the inflorescence consists of three kinds of flowers. From what we know in other cases of the use of the corolla, coloured bracteæ, &c., and from what H. Müller has observed on the frequency of the visits of insects to the flower-heads of the Umbelliferæ and Compositæ being largely determined by their conspicuousness, there can be no doubt that the increased size of the corolla of the outer flowers, the inner ones being in all the above cases small, serves to attract insects. The result is that cross-fertilisation is thus favoured. Most flowers wither soon after being fertilised, but Hildebrand states that the ray-florets of the Compositæ last for a long time, until all those on the disc are impregnated; and this clearly shows the use of the former. The ray-florets, however, are of service in another and very different manner, namely, by folding inwards at night and during cold rainy weather, so as to protect the florets of the disc. Moreover they often contain matter which is excessively poisonous to insects, as may be seen in the use of flea-powder, and in the case of Pyrethrum, M. Belhomme has shown that the ray-florets are more poisonous than the disc-florets in the ratio of about three to two. We may therefore believe that the ray-florets are useful in protecting the flowers from being gnawed by insects.

It is a well-known yet remarkable fact that the circumferential flowers of many of the foregoing plants have both their male and female reproductive organs aborted, as with the Hydrangea, Viburnum and certain Compositæ; or the male organs alone are aborted, as in many Compositæ. Between the sexless, female, and hermaphrodite states of these latter flowers, the finest gradations may be traced, as Hildebrand has shown. He also shows that there is a close relation between the size of the corolla in the ray-florets and the degree of abortion in their reproductive organs. As we have good reason to believe that these florets are highly serviceable to the plants which possess them, more especially by rendering the flower-heads conspicuous to insects, it is a natural inference that their corollas have been increased in size for this special purpose; and that their development has subsequently led, through the principle of compensation or balancement, to the more or less complete reduction of the reproductive organs. But an opposite view may be maintained, namely, that the reproductive organs first began to fail, as often happens under cultivation, and, as a consequence, the corolla became, through compensation, more highly developed. This view, however, is not probable, for when hermaphrodite plants become diœcious or gyno-diœcious—that is, are converted into hermaphrodites and females—the corolla of the female seems to be almost invariably reduced in size in consequence of the abortion of the male organs. The difference in the result in these two classes of cases may perhaps be accounted for by the matter saved through the abortion of the male organs in the females of gyno-diœcious and diœcious plants being directed (as we shall see in a future chapter) to the formation of an increased supply of seeds; whilst in the case of the exterior florets and flowers of the plants which we are here considering, such matter is expended in the development of a conspicuous corolla. Whether in the present class of cases the corolla was first affected, as seems to me the more probable view, or the reproductive organs first failed, their states of development are now firmly correlated. We see this well illustrated in Hydrangea and Viburnum; for when these plants are cultivated, the corollas of both the interior and exterior flowers become largely developed, and their reproductive organs are aborted.

There is a closely analogous subdivision of plants, including the genus Muscari (or Feather Hyacinth) and the allied Bellevalia, which bear both perfect flowers and closed bud-like bodies that never expand. The latter resemble in this respect cleistogamic flowers, but differ widely from them in being sterile and conspicuous. Not only the aborted flower-buds and their peduncles (which are elongated apparently through the principle of compensation) are brightly coloured, but so is the upper part of the spike—all, no doubt, for the sake of guiding insects to the inconspicuous perfect flowers. From such cases as these we may pass on to certain Labiatæ, for instance, *Salvia Horminum*, in which (as I hear from Mr. Thiselton Dyer) the upper bracts are enlarged and brightly coloured, no doubt for the same purpose as before, with the flowers suppressed.

In the Carrot and some allied Umbelliferæ, the central flower has its petals somewhat enlarged, and these are of a dark purplish-red tint; but it cannot be supposed that this one small flower makes the large white umbel at all more conspicuous to insects. The central flowers are said to be neuter or sterile, but I obtained by artificial fertilisation a seed (fruit) apparently perfect from one such flower. Occasionally two or three of the flowers next to the central one are similarly characterised; and according to Vaucher "cette singulière dégénération s'étend quelquefois à l'ombelle entière" ["this peculiar degeneration sometimes extends to the entire umbel"—editors]. That the modified central flower is of no functional importance to the plant is almost certain. It may perhaps be a remnant of a former and ancient condition of the species, when one flower alone, the central one, was female and yielded seeds, as in the umbelliferous genus Echinophora. There is nothing surprising in the central flower tending to retain its former condition longer than the others; for when irregular flowers become regular or peloric, they are apt to be central; and such peloric flowers apparently owe their origin either to arrested development—that is, to the preservation of an early stage of development—or to reversion. Central and perfectly developed flowers in not a few plants in their normal condition (for instance, the common Rue and Adoxa) differ slightly in structure, as in the number of the parts, from the other flowers on the same plant. All such cases seem connected with the fact of the bud which stands at the end of the shoot being better nourished than the others, as it receives the most sap.

The cases hitherto mentioned relate to hermaphrodite species which bear differently constructed flowers; but there are some plants that produce differently formed seeds, of which Dr. Kuhn has given a list. With the Umbelliferæ and Compositæ, the flowers that produce these seeds likewise differ, and the differences in the structure of the seeds are of a very important nature. The causes which have led to differences in the seeds on the same plant are not known; and it is very doubtful whether they subserve any special end.

We now come to our second Class, that of monœcious species, or those which have their sexes separated but borne on the same plant. The flowers necessarily differ, but when those of one sex include rudiments of the other sex, the difference between the two kinds is usually not great. When the difference is great, as we see in catkin-bearing plants, this depends largely on many of the species in this, as well as in the next or diœcious class, being fertilised by the aid of the wind; for the male flowers have in this case to produce a surprising amount of incoherent pollen. Some few monœcious plants consist of two bodies of individuals, with their flowers differing in function, though not in structure; for certain individuals mature their pollen before the female flowers on the same plant are ready for fertilisation, and are called proterandrous; whilst conversely other individuals, called proterogynous, have their stigmas mature before their pollen is ready. The purpose of this curious functional difference obviously is to favour the cross-fertilisation of distinct plants. A case of this kind was first observed by Delpino in the Walnut (*Juglans regia*), and has since been observed with the common Nut (*Corylus avellana*). According to H. Müller the individual plants of a few hermaphrodite species differ in a like manner; some being proterandrous and others proterogynous. On cultivated trees of the Walnut and Mulberry, the male flowers have been observed to abort on certain individuals, which have thus been converted into females; but whether there are any species in a state of nature which co-exist as monœcious and female individuals, I do not know.

The third Class consists of diœcious species, and the remarks made under the last class with respect to the amount of difference between the male and female flowers are here applicable. It is at present an inexplicable fact that with some diœcious plants, of which the Restiaceæ of Australia and the Cape of Good Hope offer the most striking instance, the differentiation of the sexes has affected the whole plant to such an extent (as I hear from Mr. Thiselton Dyer) that Mr. Bentham and Professor Oliver have often found it impossible to match the male and female specimens of the same species. In my seventh chapter some observations will be given on the gradual conversion of heterostyled and of ordinary hermaphrodite plants into diœcious or sub-diœcious species.

The fourth and last Class consists of the plants which were called polygamous by Linnæus; but it appears to me that it would be convenient to confine this term to the species which co-exist as hermaphrodites, males, and females; and to give new names to several other combinations of the sexes—a plan which I shall here follow. Polygamous plants, in this confined sense of the term, may be divided into two sub-groups, according as the three sexual forms are found on the same individual or on distinct individuals. Of this latter or trioicous sub-group, the common Ash (*Fraxinus excelsior*) offers a good instance: thus, I examined during the spring and autumn fifteen trees growing in the same field; and of these, eight produced male flowers alone, and in the autumn not a single seed; four produced only female flowers, which set an abundance of seeds; three were hermaphrodites, which had a different aspect from the other trees whilst in flower, and two of them produced nearly as many seeds as the female trees, whilst the third produced none, so that it was in function a male. The separation of the sexes, however, is not complete in the Ash; for the female flowers include stamens, which drop off at an early period, and their anthers, which never open or dehisce, generally contain pulpy matter instead of pollen. On some female trees, however, I found a few anthers containing pollen-grains apparently sound. On the male trees most of the flowers include pistils, but these likewise drop off at an early period; and the ovules, which ultimately abort, are very small compared with those in female flowers of the same age.

Of the other or monœcious sub-group of polygamous plants, or those which bear hermaphrodite, male and female flowers on the same individual, the common Maple (*Acer campestre*) offers a good instance; but Lecoq states that some trees are truly diœcious, and this shows how easily one state passes into another.

A considerable number of plants generally ranked as polygamous exist under only two forms, namely, as hermaphrodites and females; and these may be called gyno-diœcious, of which the common Thyme offers a good example. In my seventh chapter I shall give some observations on plants of this nature. Other species, for instance several kinds of Atriplex, bear on the same plant hermaphrodite and female flowers; and these might be called gyno-monœcious, if a name were desirable for them.

Again there are plants which produce hermaphrodite and male

flowers on the same individual, for instance, some species of Galium, Veratrum, &c.; and these might be called andro-monœcious. If there exist plants, the individuals of which consist of hermaphrodites and males, these might be distinguished as andro-diœcious. But, after making inquiries from several botanists, I can hear of no such cases. Lecoq, however, states, but without entering into full details, that some plants of *Caltha palustris* produce only male flowers, and that these live mingled with the hermaphrodites. The rarity of such cases as this last one is remarkable, as the presence of hermaphrodite and male flowers on the same individual is not an unusual occurrence; it would appear as if Nature did not think it worth while to devote a distinct individual to the production of pollen, excepting when this was indispensably necessary, as in the case of diœcious species.

I have now finished my brief sketch of the several cases, as far as known to me, in which flowers differing in structure or in function are produced by the same species of plant. Full details will be given in the following chapters with respect to many of these plants. I will begin with the heterostyled, then pass on to certain diœcious, sub-diœcious, and polygamous species, and end with the cleistogamic. For the convenience of the reader, and to save space, the less important cases and details have been printed in smaller type.

I cannot close this Introduction without expressing my warm thanks to Dr. Hooker for supplying me with specimens and for other aid; and to Mr. Thiselton Dyer and Professor Oliver for giving me much information and other assistance. Professor Asa Gray, also, has uniformly aided me in many ways. To Fritz Müller of St. Catharina, in Brazil, I am indebted for many dried flowers of heterostyled plants, often accompanied with valuable notes.

HETEROSTYLED DIMORPHIC PLANTS—CONTINUED.

It has long been known that several species of Linum present two forms, and having observed this fact in *L. flavum* more than thirty years ago, I was led, after ascertaining the nature of heterostylism in Primula, to examine the first species of Linum which I met with, namely, the beautiful *L. grandiflorum.* This plant exists under two forms, occurring in about equal numbers, which differ little in struc-

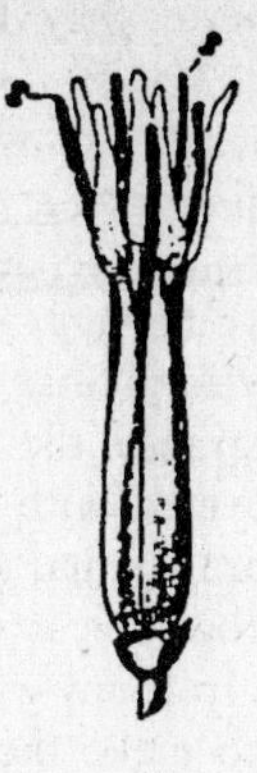

FIG. 4.

Long-styled form. Short-styled form.

s s stigmas.

LINUM GRANDIFLORUM.

ture, but greatly in function. The foliage, corolla, stamens, and pollen-grains (the latter examined both distended with water and dry) are alike in the two forms (Fig. 4). The difference is confined to the pistil; in the short-styled form the styles and the stigmas are only about half the length of those in the long-styled. A more important distinction is, that the five stigmas in the short-styled form diverge greatly from one another, and pass out between the filaments of the stamens, and thus lie within the tube of the corolla. In the long-styled form the elongated stigmas stand nearly upright, and alternate with the anthers. In this latter form the length of the stigmas varies considerably, their upper extremities projecting even a little above the anthers, or reaching up only to about their middle. Nevertheless, there is never the slightest difficulty in distinguishing between the two forms; for, besides the difference in the divergence of the stigmas, those of the short-styled form never reach even to the bases of the anthers. In this form the papillæ on the stigmatic surfaces are shorter, darker coloured, and more crowded together than in the long-styled form; but these differences seem due merely to the shortening of the stigma, for in the varieties of the long-styled form with shorter stigmas, the papillæ are more crowded and darker-coloured than in those with the longer stigmas. Considering the slight and variable differences between the two forms of this Linum, it is not surprising that hitherto they have been overlooked.

In 1861 I had eleven plants in my garden, eight of which were long-styled, and three short-styled. Two very fine long-styled plants grew in a bed a hundred yards off all the others, and separated from them by a screen of evergreens. I marked twelve flowers, and placed on their stigmas a little pollen from the short-styled plants. The pollen of the two forms is, as stated, identical in appearance; the stigmas of the long-styled flowers were already thickly covered with their own pollen—so thickly that I could not find one bare stigma, and it was late in the season, namely, September 15th. Altogether, it seemed almost childish to expect any result. Nevertheless from my experiments on Primula, I had faith, and did not hesitate to make the trial, but certainly did not anticipate the full result which was obtained. The germens of these twelve flowers all swelled, and ultimately six fine capsules (the seed of which germinated on the following year) and two poor capsules were produced; only four capsules shanking off. These same two long-styled plants produced, in the course of the summer, a vast number of flowers, the stigmas of which were covered with their own pollen; but they all proved absolutely barren, and their germens did not even swell.

The nine other plants, six long-styled and three short-styled, grew not very far apart in my flower-garden. Four of these long-styled plants produced no seed-capsules; the fifth produced two; and the remaining one grew so close to a short-styled plant that their branches touched, and this produced twelve capsules, but they were poor ones. The case was different with the short-styled plants. The one which grew close to the long-styled plant produced ninety-four imperfectly fertilised capsules containing a multitude of bad seeds, with a moderate number of good ones. The two other short-styled plants growing together were small, being partly smothered by other plants; they did not stand very close to any long-styled plants, yet they yielded together nineteen capsules. These facts seem to show that the short-styled plants are more fertile with their own pollen than are the long-styled, and we shall immediately see that this probably is the case. But I suspect that the difference in fertility between the two forms was in this instance in part due to a distinct cause. I repeatedly watched the flowers, and only once saw a humble-bee momentarily alight on one, and then fly away. If bees had visited the several plants, there cannot be a doubt that the four long-styled plants, which did not produce a single capsule, would have borne an abundance. But several times I saw small diptera sucking the flowers; and these insects, though not visiting the flowers with anything like the regularity of bees, would carry a little pollen from one form to the other, especially when growing near together; and the stigmas of the short-styled plants, diverging within the tube of the corolla, would be more likely than the upright stigmas of the long-styled plants, to receive a small quantity of pollen if brought to them by small insects. Moreover from the greater number of the long-styled than of the short-styled plants in the garden, the latter would be more likely to receive pollen from the long-styled, than the long-styled from the short-styled.

In 1862 I raised thirty-four plants of this Linum in a hot-bed; and these consisted of seventeen long-styled and seventeen short-styled forms. Seed sown later in the flower-garden yielded seventeen long-styled and twelve short-styled forms. These facts justify the statement that the two forms are produced in about equal numbers. The thirty-four plants of the first lot were kept under a net which excluded all insects, except such minute ones as Thrips. I fertilised fourteen long-styled flowers legitimately, with pollen from the short-styled, and got eleven fine seed-capsules, which contained on an average 8.6 seeds per capsule, but only 5.6 appeared to be good. It may be well to state that ten seeds is the maximum production for a capsule, and that our climate cannot be very favourable to this North-African plant. On three occasions the stigmas of nearly a hundred flowers were fertilised illegitimately with their own-form pollen, taken from separate plants, so as to prevent any possible ill effects from close inter-breeding. Many other flowers were also produced, which, as before stated, must have received plenty of their own pollen; yet from all these flowers, borne by the seventeen long-styled plants, only three capsules were produced. One of these included no seed, and the other two together gave only five good seeds. It is probable that this miserable product of two half-fertile capsules from the seventeen plants, each of which must have produced at least fifty or sixty flowers, resulted from their fertilisation with pollen from the short-styled plants by the aid of Thrips; for I made a great mistake in keeping the two forms under the same net, with their branches often interlocking; and it is surprising that a greater number of flowers were not accidentally fertilised.

Twelve short-styled flowers were in this instance castrated, and afterwards fertilised legitimately with pollen from the long-styled form; and they produced seven fine capsules. These included on an average 7.6 seeds, but of apparently good seed only 4.3 per capsule. At three separate times nearly a hundred flowers were fertilised illegitimately with their own-form pollen, taken from separate plants; and numerous other flowers were produced, many of which must have received their own pollen. From all these flowers on the seventeen short-styled plants only fifteen capsules were produced, of which only eleven contained any good seed, on an average 4.2 per capsule. As remarked in the case of the long-styled plants, some even of these capsules were perhaps the product of a little pollen accidentally fallen from the adjoining flowers of the other form on to the stigmas, or transported by Thrips. Nevertheless the short-styled plants seem to be slightly more fertile with their own pollen than the long-styled, in the proportion of fifteen capsules to three; nor can this difference be accounted for by the short-styled stigmas being more liable to receive their own pollen than the long-styled, for the reverse is the case. The greater self-fertility of the short-styled flowers was likewise shown in 1861 by the plants in my flower-garden, which were left to themselves, and were but sparingly visited by insects.

On account of the probability of some of the flowers on the plants of both forms, which were covered under the same net, having been legitimately fertilised in an accidental manner, the relative fertility of the two legitimate and two illegitimate unions cannot be compared with certainty; but judging from the number of good seeds per capsule, the difference was at least in the ratio of 100 to 7, and probably much greater.

Hildebrand tested my results, but only on a single short-styled plant, by fertilising many flowers with their own-form pollen; and these did not produce any seed. This confirms my suspicion that some of the few capsules produced by the foregoing seventeen short-styled plants were the product of accidental legitimate fertilisation. Other flowers on the same plant were fertilised by Hildebrand with pollen from the long-styled form, and all produced fruit.

The absolute sterility (judging from the experiments of 1861) of the long-styled plants with their own-form pollen led me to examine into its apparent cause; and the results are so curious that they are worth giving in detail. The experiments were tried on plants grown in pots and brought successively into the house.

First. Pollen from a short-styled plant was placed on the five stigmas of a long-styled flower, and these, after thirty hours, were found deeply penetrated by a multitude of pollen-tubes, far too numerous to be counted; the stigmas had also become discoloured and twisted. I repeated this experiment on another flower, and in eighteen hours the stigmas were penetrated by a multitude of long pollen-tubes. This is what might have been expected, as the union is a legitimate one. The converse experiment was likewise tried, and pollen from a long-styled flower was placed on the stigmas of a short-styled flower, and in twenty-four hours the stigmas were discoloured, twisted, and penetrated by numerous pollen-tubes; and this, again, is what might have been expected, as the union was a legitimate one.

Secondly. Pollen from a long-styled flower was placed on all five stigmas of a long-styled flower on a separate plant: after nineteen hours the stigmas were dissected, and only a single pollen-grain had emitted a tube, and this was a very short one. To make sure that the pollen was good, I took in this case, and in most of the other cases, pollen either from the same anther or from the same flower, and proved it to be good by placing it on the stigma of a short-styled plant, and found numerous pollen-tubes emitted.

Thirdly. Repeated last experiment, and placed own-form pollen on all five stigmas of a long-styled flower; after nineteen hours and a half, not one single grain had emitted its tube.

Fourthly. Repeated the experiment, with the same result after twenty-four hours.

Fifthly. Repeated last experiment, and, after leaving pollen on for nineteen hours, put on an additional quantity of own-form pollen on all five stigmas. After an interval of three days, the stigmas were examined, and, instead of being discoloured and twisted, they were straight and fresh-coloured. Only one grain had emitted a quite short tube, which was drawn out of the stigmatic tissue without being ruptured.

The following experiments are more striking:—

Sixthly. I placed own-form pollen on three of the stigmas of the long-styled flower, and pollen from a short-styled flower on the other two stigmas. After twenty-two hours these two stigmas were discoloured, slightly twisted, and penetrated by the tubes of numerous pollen-grains: the other three stigmas, covered with their own-form pollen, were fresh, and all the pollen-grains were loose; but I did not dissect the whole stigma.

Seventhly. Experiment repeated in the same manner, with the same result.

Eighthly. Experiment repeated, but the stigmas were carefully examined after an interval of only five hours and a half. The two stigmas with pollen from a short-styled flower were penetrated by innumerable tubes, which were as yet short, and the stigmas themselves were not at all discoloured. The three stigmas covered with their own-form pollen were not penetrated by a single pollen-tube.

Ninthly. Put pollen of a short-styled flower on a single long-styled stigma, and own-form pollen on the other four stigmas; after twenty-four hours the one stigma was somewhat discoloured and twisted, and penetrated by many long tubes: the other four stigmas were quite straight and fresh; but on dissecting them I found that three pollen-grains had protruded very short tubes into the tissue.

Tenthly. Repeated the experiment, with the same result after twenty-four hours, excepting that only two own-form grains had penetrated the stigmatic tissue with their tubes to a very short depth. The one stigma, which was deeply penetrated by a multitude of tubes from the short-styled pollen, presented a conspicuous difference in being much curled, half-shrivelled, and discoloured, in comparison with the other four straight and bright pink stigmas.

I could add other experiments: but those now given amply suffice to show that the pollen-grains of a short-styled flower placed on the stigma of a long-styled flower emit a multitude of tubes after an interval of from five to six hours, and penetrate the tissue ultimately to a great depth; and that after twenty-four hours the stigmas thus penetrated change colour, become twisted, and appear half-withered. On the other hand, pollen-grains from a long-styled flower placed on its own stigmas, do not emit their tubes after an interval of a day, or even three days; or at most only three or four grains out of a multitude emit their tubes, and these apparently never penetrate the stigmatic tissue deeply, and the stigmas themselves do not soon become discoloured and twisted.

This seems to me a remarkable physiological fact. The pollen-grains of the two forms are undistinguishable under the microscope; the stigmas differ only in length, degree of divergence, and in the size, shade of colour, and approximation of their papillæ, these latter differences being variable and apparently due merely to the degree of elongation of the stigma. Yet we plainly see that the two kinds of pollen and the two stigmas are widely dissimilar in their mutual reaction—the stigmas of each form being almost powerless on their own pollen, but causing, through some mysterious influence, apparently by simple contact (for I could detect no viscid secretion), the pollen-grains of the opposite form to protrude their tubes. It may be said that the two pollens and the two stigmas mutually recognise each other by some means. Taking fertility as the criterion of distinctness, it is no exaggeration to say that the pollen of the long-styled *Linum grandiflorum* (and conversely that of the other form) has been brought to a degree of differentiation, with respect to its action on the stigma of the same form, corresponding with that existing between the pollen and stigma of species belonging to distinct genera.

• • •

Final remarks. The existence of plants which have been rendered heterostyled is a highly remarkable phenomenon, as the two or three forms of the same undoubted species differ not only in important points of structure, but in the nature of their reproductive powers. As far as structure is concerned, the two sexes of many animals and of some plants differ to an extreme degree; and in both kingdoms the same species may consist of males, females, and hermaphrodites. Certain hermaphrodite cirripedes are aided in their reproduction by a whole cluster of what I have called complemental males, which differ wonderfully from the ordinary hermaphrodite form. With ants we have males and females, and two or three castes of sterile females or workers. With Termites there are, as Fritz Müller has shown, both winged and wingless males and females, besides the workers. But in none of these cases is there any reason to believe that the several males or several females of the same species differ in their sexual powers, except in the atrophied condition of the reproductive organs in the workers of social insects. Many hermaphrodite animals must unite for reproduction, but the necessity of such union apparently depends solely on their structure. On the other hand, with heterostyled dimorphic species there are two females and two sets of males, and with trimorphic species three females and three sets of males, which differ essentially in their sexual powers. We shall, perhaps, best perceive the complex and extraordinary nature of the marriage arrangements of a trimorphic plant by the following illustration. Let us suppose that the individuals of the same species of ant always lived in triple communities; and that in one of these, a large-sized female (differing also in other characters) lived with six middle-sized and six small-sized males; in the second community a middle-sized female lived with six large- and six small-sized males; and in the third, a small-sized female lived with six large- and six middle-sized males. Each of these three females, though enabled to unite with any male, would be nearly sterile with her own two sets of males, and likewise with two other sets of males of the same size with her own which lived in the other two communities; but she would be fully fertile when paired with a male of her own size. Hence the thirty-six males, distributed by half-dozens in the three communities, would be divided into three sets of a dozen each; and these sets, as well as the three females, would differ from one another in their reproductive powers in exactly the same manner as do the distinct species of the same genus. But it is a still more remarkable fact that young ants raised from any one of the three female ants, illegitimately fertilised by a male of a different size, would resemble in a whole series of relations the hybrid offspring from a cross between two distinct species of ants. They would be dwarfed in stature, and more or less, or even utterly barren. Naturalists are so much accustomed to behold great diversities of structure associated with the two sexes, that they feel no surprise at almost any amount of difference; but differences in sexual nature have been thought to be the very touchstone of specific distinction. We now see that such sexual differences—the greater or less power of fertilising and being fertilised—may characterise the co-existing individuals of the same species, in the same manner as they characterise and have kept separate those groups of individuals, produced during the lapse of ages, which we rank and denominate as distinct species.

A Biographical Sketch of an Infant. [1877]

"My first child was born on December 27th, 1839, and I at once commenced to make notes on the first dawn of the various expressions which he exhibited, for I felt convinced, even at this early period, that the most complex and fine shades of expression must all have had a gradual and natural origin." Thus Darwin wrote in the Autobiography. *A delightful and edifying case study in developmental psychology, the diary of young William Darwin's evolving expressions remained in manuscript for almost forty years. Published in the second number of the first psychology journal,* Mind: Quarterly Review of Psychology and Philosophy, *in July 1877, "The Biographical Sketch of an Infant" served as Darwin's response to "Taine on the Acquisition of Language by Children," a translation of part of a paper by the French philosopher, critic, and historian Hippolyte Taine, which had appeared in the first number of* Mind.

The Descent of Man *and* The Expression of the Emotions *established Darwin as a pioneer psychologist. Publication of "A Biographical Sketch of an Infant" demonstrated that in yet another of the nineteenth century's emerging scientific fields Darwin had been an acute observer and profound theorizer since the early days of his career. As he had done with William, Darwin recorded observations on his other children, but these comments were not published until 1988, when they appeared in volume 4 of* The Correspondence of Charles Darwin.

M. Taine's very interesting account of the mental development of an infant, translated in the last number of *Mind* (p. 252), has led me to look over a diary which I kept thirty-seven years ago with respect to one of my own infants. I had excellent opportunities for close observation, and wrote down at once whatever was observed. My chief object was expression, and my notes were used in my book on this subject; but as I attended to some other points, my observations may possibly possess some little interest in comparison with those by M. Taine, and with others which hereafter no doubt will be made. I feel sure, from what I have seen with my own infants, that the period of development of the several faculties will be found to differ considerably in different infants.

During the first seven days various reflex actions, namely sneezing, hickuping, yawning, stretching, and of course sucking and screaming, were well performed by my infant. On the seventh day, I touched the naked sole of his foot with a bit of paper, and he jerked it away, curling at the same time his toes, like a much older child when tickled. The perfection of these reflex movements shows that the extreme imperfection of the voluntary ones is not due to the state of the muscles or of the coordinating centres, but to that of the seat of the will. At this time, though so early, it seemed clear to me that a warm soft hand applied to his face excited a wish to suck. This must be considered as a reflex or an instinctive action, for it is impossible to believe that experience and association with the touch of his mother's breast could so soon have come into play. During the first fortnight he often started on hearing any sudden sound, and blinked his eyes. The same fact was observed with some of my other infants within the first fortnight. Once, when he was 66 days old, I happened to sneeze, and he started violently, frowned, looked frightened, and cried rather badly: for an hour afterwards he was in a state which would be called nervous in an older person, for every slight noise made him start. A few days before this same date, he first started at an object suddenly seen; but for a long time afterwards sounds made him start and wink his eyes much more frequently than did sight; thus when 114 days old, I shook a paste-board box with comfits in it near his face and he started, whilst the same box when empty or any other object shaken as near or much nearer to his face produced no effect. We may infer from these several facts that the winking of the eyes, which manifestly serves to protect them, had not been acquired through experience. Although so sensitive to sound in a general way, he was not able even when 124 days old easily to recognize whence a sound proceeded, so as to direct his eyes to the source.

With respect to vision,—his eyes were fixed on a candle as early as the 9th day, and up to the 45th day nothing else seemed thus to fix them; but on the 49th day his attention was attracted by a bright-coloured tassel, as was shown by his eyes becoming fixed and the movements of his arms ceasing. It was surprising how slowly he acquired the power of following with his eyes an object if swinging at all rapidly; for he could not do this well when seven and a half months old. At the age of 32 days he perceived his mother's bosom when three or four inches from it, as was shown by the protrusion of his lips and his eyes becoming fixed; but I much doubt whether this had any connection with vision; he certainly had not touched the bosom. Whether he was guided through smell or the sensation of warmth or through association with the position in which he was held, I do not at all know.

The movements of his limbs and body were for a long time vague and purposeless, and usually performed in a jerking manner; but there was one exception to this rule, namely, that from a very early period, certainly long before he was 40 days old, he could move his hands to his own mouth. When 77 days old, he took the sucking bottle (with which he was partly fed) in his right hand, whether he was held on the left or right arm of his nurse, and he would not take it in his left hand until a week later although I tried to make him do so; so that the right hand was a week in advance of the left. Yet this infant afterwards proved to be left-handed, the tendency being no doubt inherited—his grandfather, mother, and a brother having been or being left-handed. When between 80 and 90 days old, he drew all sorts of objects into this mouth, and in two or three weeks' time could do this with some skill; but he often first touched his nose with the object and then dragged it down into his mouth. After grasping my finger and drawing it to his mouth, his own hand prevented him from sucking it; but on the 114th day, after acting in this manner, he slipped his own hand down so that he could get the end of my finger into his mouth. This action was repeated several times, and evidently was not a chance but a rational one. The intentional movements of the hands and arms were thus much in advance of those of the body and legs, though the purposeless movements of the latter were from a very early period usually alternate as in the act of walking. When four months old, he often looked intently at his own hands and other objects close to him, and in doing so the eyes were turned much inwards, so that he often squinted frightfully. In a fortnight after this time (*i.e.* 132 days old) I observed that if an object was brought as near to his face as his own hands were, he tried to seize it, but often failed; and he did not try to do so in regard to more distant objects. I think there can be little doubt that the convergence of his eyes gave him the clue and excited him to move his arms. Although this infant thus began to use his hands at an early period, he showed no special aptitude in this respect, for when he was 2 years and 4 months old, he held pencils, pens, and other objects far less neatly and efficiently than did his sister who was then only 14 months old, and who showed great inherent aptitude in handling anything.

Anger.—It was difficult to decide at how early an age anger was felt; on his eighth day he frowned and wrinkled the skin round his eyes before a crying fit, but this may have been due to pain or distress, and not to anger. When about ten weeks old, he was given some rather cold milk and he kept a slight frown on his forehead all the time that he was sucking, so that he looked like a grown-up person made cross from being compelled to do something which he did not like. When nearly four months old, and perhaps much earlier, there could be no doubt, from the manner in which the blood gushed into his whole face and scalp, that he easily got into a violent passion. A small cause sufficed; thus, when a little over seven months old, he screamed with rage because a lemon slipped away and he could not seize it with his hands. When eleven months old, if a wrong plaything was given him, he would push it away and beat it; I presume that the beating was an instinctive sign of anger, like the snapping of the jaws by a young crocodile just out of the egg, and not that he imagined he could hurt the plaything. When two years and three months old, he became a great adept at throwing books or sticks, &c., at anyone who offended him; and so it was with some of my other sons. On the other hand, I could never see a trace of such aptitude in my infant daughters; and this makes me think that a tendency to throw objects is inherited by boys.

Fear.—This feeling probably is one of the earliest which is experienced by infants, as shown by their starting at any sudden sound when only a few weeks old, followed by crying. Before the present one was 4½ months old I had been accustomed to make close to him many strange and loud noises, which were all taken as excellent jokes, but at this period I one day made a loud snoring noise which I had never done before; he instantly looked grave and then burst out crying. Two or three days afterwards, I made through forgetfulness the same noise with the same result. About the same time (*viz.* on the 137th day) I approached with my back towards him and then stood motionless: he looked very grave and much surprised, and would soon have cried, had I not turned round; then his face instantly relaxed into a smile. It is well known how intensely older children suffer from vague and undefined fears, as from the dark, or in passing an obscure corner in a large hall, &c. I may give as an instance that I took the child in question, when 2¼ years old, to the Zoological Gardens, and he enjoyed looking at all the animals which were like those that he knew, such as deer, antelopes &c., and all the birds, even the ostriches, but was much alarmed at the various larger animals in cages. He often said afterwards that he wished to go again, but not to see "beasts in houses"; and we could in no manner account for his fear. May we not suspect that the vague but very real fears of children, which are quite independent of experience, are the inherited effects of real dangers and abject superstitions during ancient savage times? It is quite conformable with what we know of the transmission of formerly well-developed characters, that they should appear at an early period of life, and afterwards disappear.

Pleasurable Sensations.—It may be presumed that infants feel pleasure whilst sucking, and the expression of their swimming eyes seems to show that this is the case. This infant smiled when 45 days, a second infant when 46 days old; and these were true smiles, indicative of pleasure, for their eyes brightened and eyelids slightly closed. The smiles arose chiefly when looking at their mother, and were therefore probably of mental origin; but this infant often smiled then, and for some time afterwards, from some inward pleasurable feeling, for nothing was happening which could have in any way excited or amused him. When 110 days old he was exceedingly amused by a pinafore being thrown over his face and then suddenly withdrawn; and so he was when I suddenly uncovered my own face and approached his. He then uttered a little noise which was an incipient laugh. Here surprise was the chief cause of the amusement, as is the case to a large extent with the wit of grown-up persons. I believe that for three or four weeks before the time when he was amused by a face being suddenly uncovered, he received a little pinch on his nose and cheeks as a good joke. I was at first surprised at humour being appreciated by an infant only a little above three months old, but we should remember how very early puppies and kittens begin to play. When four months old, he showed in an unmistakable manner that he liked to hear that pianoforte played; so that here apparently was the earliest sign of an aesthetic feeling, unless the attraction of bright colours, which was exhibited much earlier, may be so considered.

Affection.—This probably arose very early in life, if we may judge by his smiling at those who had charge of him when under two months old; though I had no distinct evidence of his distinguishing and recognizing anyone, until he was nearly four months old. When nearly five months old, he plainly showed his wish to go to his nurse. But he did not spontaneously exhibit affection by overt acts until a little above a year old, namely, by kissing several times his nurse who had been absent for a short time. With respect to the allied feeling of sympathy, this was clearly shown at 6 months and 11 days by his melancholy face, with the corners of his mouth well depressed, when his nurse pretended to cry. Jealousy was plainly exhibited when I fondled a large doll, and when I weighed his infant sister, he being 15½ months old. Seeing how strong a feeling jealousy is in dogs, it would probably be exhibited by infants at an earlier age than that just specified, if they were tried in a fitting manner.

Association of Ideas, Reason, &c.—The first action which exhibited, as far as I observed, a kind of practical reasoning, has already been noticed, namely, the slipping his hand down my finger so as to get the end of it into his mouth; and this happened on the 114th day. When four and a half months old, he repeatedly smiled at my image and his own in a mirror, and no doubt mistook them for real objects; but he showed sense in being evidently surprised at my voice coming from behind him. Like all infants he much enjoyed thus looking at himself, and in less than two months perfectly understood that it was an image; for if I made quite silently any odd grimace, he would suddenly turn round to look at me. He was, however, puzzled at the age of seven months, when being out of doors he saw me on the inside of a large plate-glass window, and seemed in doubt whether or not it was an image. Another of my infants, a little girl, when exactly a year old, was not nearly so acute, and seemed quite perplexed at the image of a person in a mirror approaching her from behind. The higher apes which I tried with a small looking-glass behaved differently; they placed their hands behind the glass, and in doing so showed their sense, but far from taking pleasure in looking at themselves they got angry and would look no more.

When five months old, associated ideas arising independently of any instruction became fixed in his mind; thus as soon as his hat and cloak were put on, he was very cross if he was not immediately taken out of doors. When exactly seven months old, he made the great step of associating his nurse with her name, so that if I called it out he would look round for her. Another infant used to amuse himself by shaking his head laterally: we praised and imitated him, saying "Shake your head"; and when he was seven months old, he would sometimes do so on being told without any other guide. During the next four months the former infant associated many things and actions with words; thus when asked for a kiss he would protrude his lips and keep still,—would shake his head and say in a scolding voice "Ah" to the coal-box or a little spilt water, &c., which he had been taught to consider as dirty. I may add that when a few days under nine months old he associated his own name with his image in the looking glass, and when called by name would turn towards the glass even when at some distance from it. When a few days over nine months, he learnt spontaneously that a hand or other object causing a shadow to fall on the wall in front of him was to be looked for behind. Whilst under a year old, it was sufficient to repeat two or three times at intervals any short sentence to fix firmly in his mind some associated idea. In the infant described by M. Taine (pp. 254–256) the age at which ideas readily became associated seems to have been considerably later, unless indeed the earlier cases were overlooked. The facility with which associated ideas due to instruction and others spontaneously arising were acquired, seemed to me by far the most strongly marked of all the distinctions between the mind of an infant and that of the cleverest full-grown dog that I have ever known. What a contrast does the mind of an infant present to that of the pike, described by Professor Möbius, who during three whole months dashed and stunned himself against a glass partition which separated him from some minnows; and when, after at least learning that he could not attack them with impunity, he was placed in the aquarium with these same minnows, then in a persistent and senseless manner he would not attack them!

Curiosity, as M. Taine remarks, is displayed at an early age by infants, and is highly important in the development of their minds; but I made no special observation on this head. Imitation likewise comes into play. When our infant was only four months old I thought that he tried to imitate sounds; but I may have deceived myself, for I was not thoroughly convinced that he did so until he was ten months old. At the age of 11½ months he could readily imitate all sorts of actions, such as shaking his head and saying "Ah" to any dirty object, or by carefully and slowly putting his forefinger in the middle of the palm of his other hand, to the childish rhyme of "Pat it and pat it and mark it with T". It was amusing to behold his pleased expression after successfully performing any such accomplishment.

I do not know whether it is worth mentioning, as showing something about the strength of memory in a young child, that this one when 3 years and 23 days old on being shown an engraving of his grandfather, whom he had not seen for exactly six months, instantly recognised him and mentioned a whole string of events which had occurred whilst visiting him, and which certainly had never been mentioned in the interval.

Moral Sense.—The first sign of moral sense was noticed at the age of nearly 13 months. I said "Doddy (his nickname) won't give poor papa a kiss,—naughty Doddy". These words, without doubt, made him feel slightly uncomfortable; and at last when I had returned to my chair, he protruded his lips as a sign that he was ready to kiss me; and he then shook his hand in an angry manner until I came and received his kiss. Nearly the same little scene recurred in a few days, and the reconciliation seemed to give him so much satisfaction, that several times afterwards he pretended to be angry and slapped me, and then insisted on giving me a kiss. So that here we have a touch of the dramatic art, which is so strongly pronounced in most young children. About this time it became easy to work on his feelings and make him do whatever was wanted. When 2 years and 3 months old, he gave his last bit of gingerbread to his little sister, and then cried out with high self-approbation "Oh kind Doddy, kind Doddy". Two months later, he became extremely sensitive to ridicule, and was so suspicious that he often thought people who were laughing and talking together were laughing at him. A little later (2 years and 7½ months old) I met him coming out of the dining room with his eyes unnaturally bright, and an odd unnatural or affected manner, so that I went into the room to see who was there, and found that he had been taking pounded sugar, which he had been told not to do. As he had never been in any way punished, his odd manner certainly was not due to fear, and I suppose it was pleasurable excitement struggling with conscience. A fortnight afterwards, I met him coming out of the same room, and he was eyeing his pinafore which he had carefully rolled up; and again his manner was so odd that I determined to see what was within his pinafore, notwithstanding that he said that there was nothing and repeatedly commanded me to "go away," and I found it stained with pickle-juice; so that here was carefully planned deceit. As this child was educated solely by working on his good feelings, he soon became as truthful, open, and tender, as anyone could desire.

Unconsciousness, Shyness.—No one can have attended to very young children without being struck at the unabashed manner in which they fixedly stare without blinking their eyes at a new face; an old person can look in this manner only at an animal or inanimate object. This, I believe, is the result of young children not thinking in the least about themselves, and therefore not being in the least shy, though they are sometimes afraid of strangers. I saw the first symptom of shyness in my child when nearly two years and three months old: this was shown towards myself, after an absence of ten days from home, chiefly by his eyes being kept slightly averted from mine; but he soon came and sat on my knee and kissed me, and all trace of shyness disappeared.

Means of Communication.—The noise of crying or rather of squalling, as no tears are shed for a long time, is of course uttered in an instinctive manner, but serves to show that there is suffering. After a time the sound differs according to the cause, such as hunger or pain. This was noticed when this infant was eleven weeks old, and I believe at an earlier age in another infant. Moreover, he appeared soon to learn to begin crying voluntarily, or to wrinkle his face in the manner proper to the occasion, so as to show that he wanted something. When 46 days old, he first made little noises without any meaning to please himself, and these soon became varied. An incipient laugh was observed on the 113th day, but much earlier in another infant. At this date I thought, as already remarked, that he began to try to imitate sounds, as he certainly did at a considerably later period. When five and a half months old, he uttered an articulate sound "da" but without any meaning attached to it. When a little over a year old, he used gestures to explain his wishes; to give a simple instance, he picked up a bit of paper and giving it to me pointed to the fire, as he had often seen and liked to see paper burnt. At exactly the age of a year, he made the great step of inventing a word for food, namely, *mum*, but what led him to it I did not discover. And now instead of beginning to cry when he was hungry, he used this word in a demonstrative manner or as a verb, implying "Give me food". This word therefore corresponds with *ham* as used by M. Taine's infant at the later age of 14 months. But he also used *mum* as a substantive of wide signification; thus he called sugar *shu-mum*, and a little later after he had learned the word "black," he called liquorice *black-shu-mum,*—black-sugar-food.

I was particularly struck with the fact that when asking for food by the word *mum* he gave to it (I will copy the words written down at the time) "a most strongly marked interrogatory sound at the end". He also gave to "Ah," which he chiefly used at first when recognising any person or his own image in a mirror, an exclamatory sound, such as we employ when surprised. I remark in my notes that the use of these intonations seemed to have arisen instinctively, and I regret that more observations were not made on this subject. I record, however, in my notes that at a rather later period, when between 18 and 21 months old, he modulated his voice in refusing peremptorily to do anything by a defiant whine, so as to express "That I won't"; and again his humph of assent expressed "Yes, to be sure". M. Taine also insists strongly on the highly expressive tones of the sounds made by his infant before she had learnt to speak. The interrogatory sound which my child gave to the word *mum* when asking for food is especially curious; for if anyone will use a single word or a short sentence in this manner, he will find that the musical pitch of his voice rises considerably at the close. I did not then see that this fact bears on the view which I have elsewhere maintained that before man used articulate language, he uttered notes in a true musical scale as does the anthropoid ape Hylobates.

Finally, the wants of an infant are at first made intelligible by instinctive cries, which after a time are modified in part unconsciously, and in part, as I believe, voluntarily as a means of communication,—by the unconscious expression of the features,—by gestures and in a marked manner by different intonations,—lastly by words of a general nature invented by himself, then of a more precise nature imitated from those which he hears; and these latter are acquired at a wonderfully quick rate. An infant understands to a certain extent, and as I believe at a very early period, the meaning or feelings of those who tend him, by the expression of their features. There can hardly be a doubt about this with respect to smiling; and it seemed to me that the infant whose biography I have here given understood a compassionate expression at a little over five months old. When 6 months and 11 days old he certainly showed sympathy with his nurse on her pretending to cry. When pleased after performing some new accomplishment, being then almost a year old, he evidently studied the expression of those around him. It was probably due to differences of expression and not merely of the form of the features that certain faces clearly pleased him much more than others, even at so early an age as a little over six months. Before he was a year old, he understood intonations and gestures, as well as several words and short sentences. He understood one word, namely, his nurse's name, exactly five months before he invented his first word *mum*; and this is what might have been expected, as we know that the lower animals easily learn to understand spoken words.

Erasmus Darwin. [1879]

IN FEBRUARY 1879 Ernst Krause, a German botanist, published a brief scientific biography of Erasmus Darwin to honor Charles Darwin's seventieth birthday. Darwin was so taken with this account of his grandfather that he had it translated into English and published that November with supplementary information by himself. The oddly proportioned result was a 127-page family biography by Darwin serving as "preliminary notice" to Krause's eighty-five pages on Erasmus Darwin's scientific work. Darwin's contribution offers a fascinating glimpse of previous generations of his family.

The English publication of Erasmus Darwin *caused Darwin and Emma some discomfort. Without Darwin's knowledge, Krause had added to his part of the book comments that deprecated attempts to resurrect Erasmus Darwin's ideas as "mental anachronism." This addition seemed to allude to Samuel Butler's* Evolution Old and New, *which argued that Darwin's* Origin *was a debasement of his grandfather's and other early evolutionists' work. Butler considered Krause's words an undercover attack from Darwin and expressed his outrage in the* Athenaeum. *Darwin, his honor deeply offended, drafted replies but did not print them; and his silence further infuriated Butler. It adds an interesting twist to this tangled affair of evolution, ancestors, and descendants to know that Butler, whose satirical novels* Erewhon *and* The Fair Haven *had, despite their anonymous publication, made him a well-known proponent of evolutionary theory, was the grandson of the Samuel Butler who had been headmaster of Shrewsbury School during Darwin's days there.*

PREFACE.

In the February number, 1879, of a well-known German scientific journal, 'Kosmos,' Dr. Ernst Krause published a sketch of the life of Erasmus Darwin, the author of the 'Zoonomia,' 'Botanic Garden,' and other works. This article bears the title of a 'Contribution to the history of the Descent-Theory;' and Dr. Krause has kindly allowed my brother Erasmus and myself to have a translation made of it for publication in this country.

As I have private materials for adding to the knowledge of Erasmus Darwin's character, I have written a preliminary notice. These materials consist of a large collection of letters written by him; of his common-place book in folio, in the possession of his grandson Reginald Darwin; of some notes made shortly after his death, by my father, Dr. Robert Darwin, together with what little I can clearly remember that my father said about him; also some statements by his daughter, Violetta Darwin, afterwards Mrs. Tertius Galton, written down at the time by her daughters; and various short published notices. To these must be added the 'Memoirs of the Life of Dr. Darwin,' by Miss Seward, which appeared in 1804; and a lecture by Dr. Dowson on 'Erasmus Darwin, Philosopher, Poet, and Physician,' published in 1861, which contains many useful references and remarks.

PRELIMINARY NOTICE.

Erasmus Darwin was descended from a Lincolnshire family, and the first of his ancestors of whom we know anything was William Darwin, who possessed a small estate at Cleatham. He was also yeoman of the armoury of Greenwich to James I. and Charles I. This office was probably almost a sinecure, and certainly of very small value. He died in 1644, and we have reason to believe from gout. It is, therefore, probable that Erasmus, as well as many other members of the family, inherited from this William, or some of his predecessors, their strong tendency to gout; and it was an early attack of gout which made Erasmus a vehement advocate for temperance throughout his whole life.

The second William Darwin (born 1620) served as Captain-Lieutenant in Sir W. Pelham's troop of horse, and fought for the king. His estate was sequestrated by the Parliament, but he was afterwards pardoned on payment of a heavy fine. In a petition to Charles II. he speaks of his almost utter ruin from having adhered to the royal cause, and it appears that he had become a barrister. This circumstance probably led to his marrying the daughter of Erasmus Earle, Serjeant-at-law; and hence Erasmus Darwin derived his Christian name.

The eldest son from this marriage, William (born 1655), married the heiress of Robert Waring, of Wilsford, in the county of Nottingham. This lady also inherited the manor of Elston, which has remained ever since in the family.

This third William Darwin had two sons—William, and Robert who was educated as a barrister, and who was the father of Erasmus. I suppose that the Cleatham and the Waring properties were left to William, who seems to have followed no profession, and the Elston estate to Robert; for when the latter married, he gave up his profession and lived ever afterwards at Elston. There is a portrait of him at Elston Hall, and he looks, with his great wig and bands, like a dignified doctor of divinity. He seems to have had some taste for science, for he was an early member of the well-known Spalding Club; and the celebrated antiquary, Dr. Stukeley, in 'An account of the almost entire Sceleton of a large animal,' &c., published in the 'Philosophical Transactions,' April and May 1719, begins his paper as follows:—"Having an account from my friend, Robert Darwin, Esq., of Lincoln's Inn, a Person of Curiosity, of a human Sceleton impressed in Stone, found lately by the Rector of Elston," &c. Stukeley then speaks of it as a great rarity, "the like whereof has not been observed before in this island, to my knowledge." Judging from a sort of litany written by Robert, and handed down in the family, he was a strong advocate of temperance, which his son ever afterwards so strongly advocated:—

From a morning that doth shine,
From a boy that drinketh wine,
From a wife that talketh Latine,
Good Lord deliver me.

It is suspected that the third line may be accounted for by his wife, the mother of Erasmus, having been a very learned lady.

The eldest son of Robert, christened Robert Waring, succeeded to the estate of Elston, and died there at the age of ninety-two, a bachelor. He had a strong taste for poetry, like his youngest brother Erasmus. Robert also cultivated botany, and when an oldish man, he published his 'Principia Botanica.' This book in MS. was beautifully written, and my father declared that he believed it was published because his old uncle could not endure that such fine calligraphy should be wasted. But this was hardly just, as the work contains many curious notes on biology—a subject wholly neglected in England in the last century. The public, moreover, appreciated the book, as the copy in my possession is the third edition.

Of the second son, William Alvey, I know nothing. A third son, John, became the rector of Elston, the living being in the gift of the family. The fourth son, and the youngest of the children, was Erasmus, the subject of the present memoir, who was born on the 12th Dec. 1731, at Elston Hall.

His elder brother, Robert, states, in a letter to my father (May 19, 1802), that Erasmus "was always fond of poetry. He was also always fond of mechanicks. I remember him when very young making an ingenious alarum for his watch (clock?); he used also to show little experiments in electricity with a rude apparatus he then invented with a bottle." The same tastes, therefore, appeared very early in life which prevailed to the day of his death. "He had always a dislike to much exercise and rural diversions, and it was with great difficulty that we could ever persuade him to accompany us."

When ten years old (1741), he was sent to Chesterfield School, where he remained for nine years.

• • •

Judging from two letters—the first written in 1749, to one of the under-masters during the holidays, and the other to the head-master, shortly after he went to Cambridge, in 1750—he seems to have felt a degree of respect, gratitude, and affection for the several masters unusual in a schoolboy. Both these letters were accompanied by an inevitable copy of verses, those addressed to the head-master being of considerable length, and in imitation of the 5th Satire of Persius. His two elder brothers accompanied him to St. John's College, Cambridge; and this seems to have been a severe strain on their father's income. They appear, in consequence, to have been thrifty and honourably economical; so much so that they mended their own clothes; and, many years afterwards, Erasmus boasted to his second wife that, if she cut the heel out of a stocking, he would put a new one in without missing a stitch. He won the Exeter Scholarship at St. John's, which was worth only £16 per annum. No doubt he studied the classics whilst at Cambridge, for he did so to the end of his life, as shown by the many quotations in his latest work, 'The Temple of Nature.' He must also have studied mathematics to a certain extent, for, when he took his Bachelor of Arts degree, in 1754, he was at the head of the Junior Optimes. Nor did he neglect medicine; and he left Cambridge during one term to attend Hunter's lectures in London. As a matter of course, he wrote poetry whilst at Cambridge, and a poem on 'The Death of Prince Frederick,' in 1751, was published many years afterwards, in 1795, in the European Magazine.

In the autumn of 1754 he went to Edinburgh to study medicine, and while there, seems to have been as rigidly economical as at Cambridge; for amongst his papers there is a receipt for his board from July 13th to October 13th, amounting to only £6 12*s*.

• • •

In 1755 he returned to Cambridge, and took his Bachelor of Medicine degree. He then again went to Edinburgh, and early in Sept. 1756, settled as a physician in Nottingham. Here, however, he remained for only two or three months, as he got no patients. Whilst in Nottingham he wrote several letters, some in Latin and some in English, to his friend, the son of the famous German philosopher, Reimarus. Mechanics and medicine were the bonds of union between them. Erasmus also dedicated a poem to young Reimarus, on his taking his degree at Leyden in 1754. Various subjects were discussed between them, including the wildest speculations by Erasmus on the resemblance between the action of the human soul and that of electricity, but the letters are not worth publishing.

• • •

In November 1756, Erasmus settled in Lichfield, and now his life may be said to have begun in earnest; for it was here, and in or near Derby, to which place he removed in 1781, that he published all his works. Owing to two or three very successful cases, he soon got into some practice at Lichfield as a physician, when twenty-five years old. A year afterwards (Dec. 1757) he married Miss Mary Howard, aged 17–18 years, who, judging from all that I have heard of her, and from some of her letters, must have been a superior and charming woman. She died after a long and suffering illness in 1770. They seem to have lived together most happily during the thirteen years of their married life, and she was tenderly nursed by her husband during her last illness.

Miss Seward gives, on second-hand authority, a long speech of hers, ending with the words, "he has prolonged my days, and he has blessed them." This is probably true, but everything which Miss Seward says must be received with caution; and it is scarcely possible that a speech of such length could have been reported with any accuracy.

• • •

In 1781, eleven years after the death of his first wife, he married the widow of Colonel Chandos Pole, of Radburn Hall. He had become acquainted with her in the Spring of 1778, when she had come to Lichfield in order that he might attend her children professionally. It is evident from the many MS. verses addressed to her before their marriage, that Dr. Darwin was passionately attached to her, even during the lifetime of her husband, who died in 1780. These verses are somewhat less artificial than his published ones. On his second marriage he left Lichfield, and after living two years at Radburn Hall, he removed into the town of Derby, and ultimately to Breadsall Priory, a few miles from the town, where he died in 1802.

There is little to relate about his life at either Lichfield or Derby, and, as I am not attempting a connected narrative, I will here give such impressions as I have formed of his intellect and character, and a few of his letters which are either interesting in themselves, or which throw light upon what he thought and felt.

His correspondence with many distinguished men was large; but most of the letters which I possess or have seen are uninteresting, and not worth publication. Medicine and mechanics alone roused him to write with any interest. He occasionally corresponded with Rousseau, with whom he became acquainted in an odd manner, but none of their letters have been preserved. Rousseau was living in 1766 at Mr. Davenport's house, Wootton Hall, and used to spend much of his time "in the well-known cave upon the terrace in melancholy contemplation." He disliked being interrupted, so Dr. Darwin, who was then a stranger to him, sauntered by the cave, and minutely examined a plant growing in front of it. This drew forth Rousseau, who was interested in botany, and they conversed together, and afterwards corresponded during several years.

• • •

But it is fair to state that from my father's conversation, I infer that Dr. Darwin had acted towards him in his youth rather harshly and imperiously, and not always justly; and though in after years he felt the greatest interest in his son's success, and frequently wrote to him with affection, in my opinion the early impression on my father's mind was never quite obliterated.

I have heard indirectly (through one of his stepsons) that he was not always kind to his son Erasmus, being often vexed at his retiring nature, and at his not more fully displaying his great talents. On the other hand his children by his second marriage seem to have entertained the warmest affection for him.

• • •

My father spoke of Dr. Darwin as having great powers of conversation. Lady Charleville, who had been accustomed to the most brilliant society in London, told him that Dr. Darwin was one of the most agreeable men whom she had ever met. He himself used to say "there were two sorts of agreeable persons in conversation parties—agreeable talkers and agreeable listeners."

He stammered greatly, and it is surprising that this defect did not spoil his powers of conversation. A young man once asked him in, as he thought, an offensive manner, whether he did not find stammering very inconvenient. He answered, "No, Sir, it gives me time for reflection, and saves me from asking impertinent questions." Miss Seward speaks of him as being extremely sarcastic, but of this I can find no evidence in his letters or elsewhere. It is a pity that Dr. Johnson in his visits to Lichfield rarely met Dr. Darwin; but they seem to have disliked each other cordially, and to have felt that if they met they would have quarrelled like two dogs. There can, I suppose, be little doubt that Johnson would have come off victorious. In a volume of MSS. by Dr. Darwin, in the possession of one of his granddaughters, there is the following stanza:

From Lichfield famed two giant critics come,
Tremble, ye Poets! hear them! "Fe, Fo, Fum!"
By Seward's arm the mangled Beaumont bled,
And Johnson grinds poor Shakespear's bones for bread.

He is evidently alluding to Mr. Seward's edition of 'Beaumont and Fletcher's Plays,' and to Johnson's edition of 'Shakespear' in 1765.

He possessed, according to my father, great facility in explaining any difficult subject; and he himself attributed this power to his habit of always talking about whatever he was studying, "turning and moulding the subject according to the capacity of his hearers." He compared himself to Gil Blas's uncle, who learned the grammar by teaching it to his nephew.

• • •

Dr. Darwin has been frequently called an atheist, whereas in every one of his works distinct expressions may be found showing that he fully believed in God as the Creator of the universe. For instance, in the 'Temple of Nature,' published posthumously, he writes:

> Perhaps all the productions of nature are in their progress to greater perfection! an idea countenanced by modern discoveries and deductions concerning the progressive formation of the solid parts of the terraqueous globe, and consonant to the dignity of the creator of all things.

He concludes one chapter in 'Zoonomia' with the words of the Psalmist: "*The heavens declare the Glory of God, and the firmament sheweth his handiwork.*"

He published an ode on the folly of atheism, with the motto "I am fearfully and wonderfully made," of which the first verse is as follows:—

> *Dull atheist, could a giddy dance*
> *Of atoms lawless hurl'd*
> *Construct so wonderful, so wise,*
> *So harmonised a world?*

With reference to morality he says:

> The famous sentence of Socrates, 'Know yourself,' . . . however wise it may be, seems to be rather of a selfish nature. . . . But the sacred maxims of the author of Christianity, 'Do as you would be done by,' and 'Love your neighbour as yourself,' include all our duties of benevolence and morality; and, if sincerely obeyed by all nations, would a thousandfold multiply the present happiness of mankind.

Although Dr. Darwin was certainly a theist in the ordinary acceptation of the term, he disbelieved in any revelation. Nor did he feel much respect for unitarianism, for he used to say that "unitarianism was a feather-bed to catch a falling Christian."

• • •

Judging from his published works, letters, and all that I have been able to gather about him, the vividness of his imagination seems to have been one of his pre-eminent characteristics. This led to his great originality of thought, his prophetic spirit both in science and in the mechanical arts, and to his overpowering tendency to theorise and generalise. Nevertheless, his remarks, hereafter to be given, on the value of experiments and the use of hypotheses show that he had the true spirit of a philosopher. That he possessed uncommon powers of observation must be admitted. The diversity of the subjects to which he attended is surprising. But of all his characteristics, the incessant activity or energy of his mind was, perhaps, the most remarkable. . . .

• • •

No man has ever inculcated more persistently and strongly the evil effects of intemperance than did Dr. Darwin; but chiefly on the grounds of ill-health, with its inherited consequences; and this perhaps is the most practical line of attack. It is positively asserted that he diminished to a sensible extent the practice of drinking amongst the gentry of the county. He himself during many years never touched alcohol under any form; but he was not a bigot on the subject, for in old age he informed my father that he had taken to drink daily two glasses of home-made wine with advantage. Why he chose home-made wine is not obvious; perhaps he fancied that he thus did not depart so widely from his long-continued rule. . . .

• • •

I have said that, as far as I can judge, he was remarkably free from vanity, conceit, or display; nor does he appear to have been ambitious for a higher position in society. Miss Fielding, a granddaughter of Lady Charlotte Finch, governess to Queen Charlotte's daughters, was taken to Dr. Darwin, at Derby, on account of her health, and was invited to stay some time at his house. George the Third heard of my grandfather's fame through Lady Charlotte, and said: "Why does not Dr. Darwin come to London? He shall be my physician if he comes"; and he repeated this over and over again in his usual manner. But Dr. Darwin and his wife agreed that they disliked the thoughts of a London life so much, that the hint was not acted on. Others have expressed surprise that he never migrated to London.

• • •

As it is interesting to see how far Erasmus Darwin transmitted his characteristic qualities of mind to his descendants, I will give a short account of his children. He had three sons by his first wife (besides two who died in infancy), and four sons and three daughters by his second wife. His eldest son, Charles (born September 3, 1758), was a young man of extraordinary promise, but died (May 15, 1778) before he was twenty-one years old from the effects of a wound received whilst dissecting the brain of a child. He inherited from his father a strong taste for various branches of science, for writing verses, and for mechanics. "Tools were his playthings," and making "machines was one of the first efforts of his ingenuity, and one of the first sources of his amusement."

He also inherited stammering. With the hope of curing him, his father sent him to France when about eight years old (1766–67), with a private tutor, thinking that if he was not allowed to speak English for a time, the habit of stammering might be lost; and it is a curious fact that in after years when speaking French he never stammered. At a very early age he collected specimens of all kinds. When sixteen years old he was sent for a year to Oxford, but he did not like the place, and "thought (in the words of his father) that the vigor of the mind languished in the pursuit of classical elegance, like Hercules at the distaff, and sighed to be removed to the robuster exercise of the medical school of Edinburgh." He stayed three years at Edinburgh, working hard at his medical studies, and attending "with diligence all the sick poor of the parish of Waterleith, and supplying them with the necessary medicines." The Æsculapian Society awarded him its first gold medal for an experimental enquiry on pus and mucus. Notices of him appeared in various journals; and all the writers agree about his uncommon energy and abilities. He seems, like his father, to have excited the warm affection of his friends. Professor Andrew Duncan, in whose family vault Charles was buried, cut a lock of hair from the corpse, and took it to a jeweller, whose apprentice, afterwards the famous Sir H. Raeburn, set it in a locket for a memorial. The venerable professor spoke to me about him with the warmest affection forty-seven years after his death, when I was a young medical student in Edinburgh. The inscription on his tomb, written by his father, says, with more truth than is usual on such occasions: "Possessed of uncommon abilities and activity, he had acquired knowledge in every department of medical and philosophical science, much beyond his years."

Dr. Darwin was able to reach Edinburgh before Charles died, and had at first hopes of his recovery; but these hopes, as he informed my father, "with anguish," soon disappeared. Two days afterwards he wrote to Wedgwood to the same effect, ending his letter with the words, "God bless you, my dear friend, may your children succeed better." Two and a half years afterwards he again wrote to Wedgwood, "I am rather in a situation to demand than to administer consolation."

About the character of his second son, Erasmus (born 1759), I have little to say, for, though he wrote poetry, he seems to have had none of the other tastes of his father. He had, however, his own peculiar tastes, viz. genealogy, the collecting of coins, and statistics. When a boy he counted all the houses in the city of Lichfield, and found out the number of inhabitants in as many as he could; he thus made a census, and when a real one was first made, his estimate was found to be nearly accurate. His disposition was quiet and retiring. My father had a very high opinion of his abilities, and this was probably just, for he would not otherwise have been invited to travel with, and pay long visits to, men so distinguished in different ways as Boulton the engineer, and Day, the moralist and novelist. He was certainly very ingenious. He detected by a singularly subtle plan the author of a long series of anonymous letters, which had caused, during six or seven years, extreme annoyance and even misery to many of the inhabitants of the county. The author was found to be a county gentleman of not inconsiderable standing. He was a successful solicitor in Lichfield, but his death, Dec. 30, 1799, was a sad one, as I have already mentioned.

The third son, Robert Waring Darwin (my father, born 1766), did not inherit any aptitude for poetry or mechanics, nor did he possess, as I think, a scientific mind. He published, in Vol. lxxvi. of the 'Philosophical Transactions,' a paper on Ocular Spectra, which Wheatstone told me was a remarkable production for the period; but I believe that he was largely aided in writing it by his father. He was elected a Fellow of the Royal Society in 1788. I cannot tell why my father's mind did not appear to me fitted for advancing science; for he was fond of theorising, and was incomparably the most acute observer whom I ever knew. But his powers in this direction were exercised almost wholly in the practice of medicine, and in the observation of human character. He intuitively recognised the disposition or character, and even read the thoughts, of those with whom he came into contact with extraordinary acuteness. This skill partly accounts for his great success as a physician, for it impressed his patients with belief in him; and my father used to say that the art of gaining confidence was the chief element in a doctor's worldly success.

Erasmus brought him to Shrewsbury before he was twenty-one years old, and left him £20, saying, "Let me know when you want more, and I will send it you." His uncle, the rector of Elston, afterwards also sent him £20, and this was the sole pecuniary aid which he ever received. I have heard him say that his practice during the first year allowed him to keep two horses and a man-servant. Erasmus tells Mr. Edgeworth that his son Robert, after being settled in Shrewsbury for only six months, "already had between forty and fifty patients." By the second year he was in considerable, and ever afterwards in very large, practice. His success was the more remarkable, as he for some time detested the profession, and declared that if he had been sure of gaining £100 a year in any other way he would never have practised as a doctor.

He had an extraordinary memory for the dates of certain events, so that he knew the day of the birth, marriage, and death of most of the gentlemen of Shropshire. This power, however, far from giving him any pleasure, annoyed him, for he told me that his memory for dates reminded him of painful events, and so added to his regret for the death of old friends. His spirits were generally high, and he was a great talker. He was of an extremely sensitive nature, so that whatever annoyed or pained him, did so to an extreme degree. He was also somewhat easily roused to anger. One of his golden rules was never to become the friend of any one whom you could not thoroughly respect, and I think he always acted on it. But of all his characteristic qualities, his sympathy was pre-eminent, and I believe it was this which made him for a time hate his profession, as it constantly brought suffering before his eyes. Sympathy with the joy of others is a much rarer endowment than sympathy with their pains, and it is no exaggeration to say that to give pleasure to others was to my father an intense pleasure. He died November 13th, 1849. A short notice of his life appeared in the 'Proceedings of the Royal Society.'

Of the children of Erasmus by his second marriage, one son became a cavalry officer, a second rector of Elston, and a third, Francis (born 1786, died 1859), a physician, who travelled far in countries rarely visited in those days. He showed his taste for Natural History by being fond of keeping a number of wild and curious animals. I may add that one of his sons, Captain Darwin, is a great sportsman, and has published a little book, the 'Gamekeeper's Manual' (4th ed. 1863), which shows keen observation and knowledge of the habits of various animals. The eldest daughter of Erasmus, Violetta, married S. Tertius Galton, and I feel sure that their son, Francis, will be willing to attribute the remarkable originality of his mind in large part to inheritance from his maternal grandfather.

The Power of Movement in Plants. [1880]

In 1858 Darwin received from his friend the Harvard botanist Asa Gray a copy of Gray's short paper on the movements of tendrils in some plants related to pumpkins. The ever-curious Darwin began to make his own observations on climbers. Several years later his study bore fruit in a book, On the Movement and Habits of Climbing Plants, *published in 1865, with a slightly expanded second edition in 1875. Assisted in his research by his son Francis, Darwin continued his experiments and examined movements in plants other than climbers. The result was* The Power of Movement in Plants, *published in 1880. Most of this book's drawings are by another Darwin son, George.*

Darwin hypothesized that plants follow certain adaptive patterns of movement. Controlled by the interplay of internal biological factors and external environmental factors, these movements are generally spiral in nature and are related to differences in the turgidity of certain cells. Darwin also hypothesized that a certain substance produced in the tips of shoots and roots controls their bending. His collaborative research with Francis provided the first experimental evidence for the existence of auxin, a plant growth hormone not to be discovered until the 1930s. In addition the Darwins found that different environmental stimuli produced different kinds of movements in the various parts of plants: stems, roots, leaves, and so on. Thus, in many ways, The Power of Movement in Plants *heralds the beginning of modern plant physiology. Like* Climbing Plants, *it has proved a specialists' book, not among Darwin's best-sellers.*

INTRODUCTION.

The chief object of the present work is to describe and connect together several large classes of movement, common to almost all plants. The most widely prevalent movement is essentially of the same nature as that of the stem of a climbing plant, which bends successively to all points of the compass, so that the tip revolves. This movement has been called by Sachs "revolving nutation;" but we have found it much more convenient to use the terms *circumnutation* and *circumnutate.* As we shall have to say much about this movement, it will be useful here briefly to describe its nature. If we observe a circumnutating stem, which happens at the time to be bent, we will say towards the north, it will be found gradually to bend more and more easterly, until it faces the east; and so onwards to the south, then to the west, and back again to the north. If the movement had been quite regular, the apex would have described a circle, or rather, as the stem is always growing upwards, a circular spiral. But it generally describes irregular elliptical or oval figures; for the apex, after pointing in any one direction, commonly moves back to the opposite side, not, however, returning along the same line. Afterwards other irregular ellipses or ovals are successively described, with their longer axes directed to different points of the compass. Whilst describing such figures, the apex often travels in a zigzag line, or makes small subordinate loops or triangles. In the case of leaves the ellipses are generally narrow.

Until recently the cause of all such bending movements was believed to be due to the increased growth of the side which becomes for a time convex; that this side does temporarily grow more quickly than the concave side has been well established; but De Vries has lately shown that such increased growth follows a previously increased state of turgescence on the convex side. In the case of parts provided with a so-called joint, cushion or pulvinus, which consists of an aggregate of small cells that have ceased to increase in size from a very early age, we meet with similar movements; and here, as Pfeffer has shown and as we shall see in the course of this work, the increased turgescence of the cells on opposite sides is not followed by increased growth. Wiesner denies in certain cases the accuracy of De Vries' conclusion about turgescence, and maintains that the increased extensibility of the cell-walls is the more important element. That such extensibility must accompany increased turgescence in order that the part may bend is manifest, and this has been insisted on by several botanists; but in the case of unicellular plants it can hardly fail to be the more important element. On the whole we may at present conclude that increased growth, first on one side and then on another, is a secondary effect, and that the increased turgescence of the cells, together with the extensibility of their walls, is the primary cause of the movement of circumnutation.

In the course of the present volume it will be shown that apparently every growing part of every plant is continually circumnutating, though often on a small scale. Even the stems of seedlings before they have broken through the ground, as well as their buried radicles, circumnutate, as far as the pressure of the surrounding earth permits. In this universally present movement we have the basis or groundwork for the acquirement, according to the requirements of the plant, of the most diversified movements. Thus, the great sweeps made by the stems of twining plants, and by the tendrils of other climbers, result from a mere increase in the amplitude of the ordinary movement of circumnutation. The position which young leaves and other organs ultimately assume is acquired by the circumnutating movement being increased in some one direction. The leaves of various plants are said to sleep at night, and it will be seen that their blades then assume a vertical position through modified circumnutation, in order to protect their upper surfaces from being chilled through radiation. The movements of various organs to the light, which are so general throughout the vegetable kingdom, and occasionally from the light, or transversely with respect to it, are all modified forms of circumnutation; as again are the equally prevalent movements of stems, &c., towards the zenith, and of roots towards the centre of the earth. In accordance with these conclusions, a considerable difficulty in the way of evolution is in part removed, for it might have been asked, how did all their diversified movements for the most different purposes first arise? As the case stands, we know that there is always movement in progress, and its amplitude, or direction, or both, have only to be modified for the good of the plant in relation with internal or external stimuli.

Besides describing the several modified forms of circumnutation, some other subjects will be discussed. The two which have interested us most are, firstly, the fact that with some seedling plants the uppermost part alone is sensitive to light, and transmits an influence to the lower part, causing it to bend. If therefore the upper part be wholly protected from light, the lower part may be exposed for hours to it, and yet does not become in the least bent, although this would have occurred quickly if the upper part had been excited by light. Secondly, with the radicles of seedlings, the tip is sensitive to various stimuli, especially to very slight pressure, and, when thus excited, transmits an influence to the upper part, causing it to bend from the pressed side. On the other hand, if the tip is subjected to the vapour of water proceeding from one side, the upper part of the radicle bends towards this side. Again it is the tip, as stated by Ciesielski, though denied by others, which is sensitive to the attraction of gravity, and by transmission causes the adjoining parts of the radicle to bend towards the centre of the earth. These several cases of the effects of contact, other irritants, vapour, light, and the attraction of gravity being transmitted from the excited part for some little distance along the organ in question, have an important bearing on the theory of all such movements.

Terminology.—A brief explanation of some terms which will be used, must here be given. With seedlings, the stem which supports the *cotyledons* (i.e. the organs which represent the first leaves) has been called by many botanists the hypocotyledonous stem, but for brevity sake we will speak of it merely as the *hypocotyl*: the stem immediately above the cotyledons will be called the *epicotyl* or *plumule*. The *radicle* can be distinguished from the hypocotyl only by the presence of root-hairs and the nature of its covering. The meaning of the word *circumnutation* has already been explained. Authors speak of positive and negative heliotropism,—that is, the bending of an organ to or from the light; but it is much more convenient to confine the word *heliotropism* to bending towards the light, and to designate as *apheliotropism* bending from the light. There is another reason for this change, for writers, as we have observed, occasionally drop the adjectives *positive* and *negative*, and thus introduce confusion into their discussions. *Diaheliotropism* may express a position more or less transverse to the light and induced by it. In like manner positive geotropism, or bending towards the centre of the earth, will be called by us *geotropism; apogeotropism* will mean bending in opposition to gravity or from the centre of the earth; and *diageotropism*, a position more or less transverse to the radius of the earth. The words heliotropism and geotropism properly mean the act of moving in relation to the light or the earth; but in the same manner as gravitation, though defined as "the act of tending to the centre," is often used to express the cause of a body falling, so it will be found convenient occasionally to employ heliotropism and geotropism, &c., as the cause of the movements in question.

The term *epinasty* is now often used in Germany, and implies that the upper surface of an organ grows more quickly than the lower surface, and thus causes it to bend downwards. *Hyponasty* is the reverse, and implies increased growth along the lower surface, causing the part to bend upwards.

Methods of Observation.—The movements, sometimes very small and sometimes considerable in extent, of the various organs observed by us, were traced in the manner which after many trials we found to be best, and which must be described. Plants growing in pots were protected wholly from the light, or had light admitted from above, or on one side as the case might require, and were covered above by a large horizontal sheet of glass, and with another vertical sheet on one side. A glass filament, not thicker than a horsehair, and from a quarter to three-quarters of an inch in length, was affixed to the part to be observed by means of shellac dissolved in alcohol. The solution was allowed to evaporate, until it became so thick that it set hard in two or three seconds, and it never injured the tissues, even the tips of tender radicles, to which it was applied. To the end of the glass filament an excessively minute bead of black sealing-wax was cemented, below or behind which a bit of card with a black dot was fixed to a stick driven into the ground. The weight of the filament was so slight that even small leaves were not perceptibly pressed down. Another method of observation, when much magnification of the movement was not required, will presently be described. The bead and the dot on the card were viewed through the horizontal or vertical glass-plate (according to the position of the object), and when one exactly covered the other, a dot was made on the glass-plate with a sharply pointed stick dipped in thick Indian-ink. Other dots were made at short intervals of time and these were afterwards joined by straight lines. The figures thus traced were therefore angular; but if dots had been made every 1 or 2 minutes, the lines would have been more curvilinear, as occurred when radicles were allowed to trace their own courses on smoked glass-plates. To make the dots accurately was the sole difficulty, and required some practice. Nor could this be done quite accurately, when the movement was much magnified, such as 30 times and upwards; yet even in this case the general course may be trusted. To test the accuracy of the above method of observation, a filament was fixed to an inanimate object which was made to slide along a straight edge and dots were repeatedly made on a glass-plate; when these were joined, the result ought to have been a perfectly straight line, and the line was very nearly straight. It may be added that when the dot on the card was placed half-an-inch below or behind the bead of sealing-wax, and when the glass-plate (supposing it to have been properly curved) stood at a distance of 7 inches in front (a common distance), then the tracing represented the movement of the bead magnified 15 times.

Whenever a great increase of the movement was not required, another, and in some respects better, method of observation was followed. This consisted in fixing two minute triangles of thin paper, about 1/20 inch in height, to the two ends of the attached glass filament; and when their tips were brought into a line so that they covered one another, dots were made as before on the glass-plate. If we suppose the glass-plate to stand at a distance of seven inches from the end of the shoot bearing the filament, the dots when joined, will give nearly the same figure as if a filament seven inches long, dipped in ink, had been fixed to the moving shoot, and had inscribed its own course on the plate. The movement is thus considerably magnified; for instance, if a shoot one inch in length were bending, and the glass-plate stood at the distance of seven inches, the movement would be magnified eight times. It would, however, have been very difficult to have ascertained in each case how great a length of the shoot was bending; and this is indispensable for ascertaining the degree to which the movement is magnified.

After dots had been made on the glass-plates by either of the above methods, they were copied on tracing paper and joined by ruled lines, with arrows showing the direction of the movement. The nocturnal courses are represented by straight broken lines. The first dot is always made larger than the others, so as to catch the eye, as may be seen in the diagrams. The figures on the glass-plates were often drawn on too large a scale to be reproduced on the pages of this volume, and the proportion in which they have been reduced is always given. Whenever it could be approximately told how much the movement had been magnified, this is stated. We have perhaps introduced a superfluous number of diagrams; but they take up less space than a full description of the movements. Almost all the sketches of plants asleep, &c., were carefully drawn for us by Mr. George Darwin.

As shoots, leaves, &c., in circumnutating bend more and more, first in one direction and then in another, they were necessarily viewed at different times more or less obliquely; and as the dots were made on a flat surface, the apparent amount of movement is exaggerated according to the degree of obliquity of the point of view. It would, therefore, have been a much better plan to have used hemispherical glasses, if we had possessed them of all sizes, and if the bending part of the shoot had been distinctly hinged and could have been placed so as to have formed one of the radii of the sphere. But even in this case it would have been necessary afterwards to have projected the figures on paper; so that complete accuracy could not have been attained. From the distortion of our figures, owing to the above causes, they are of no use to any one who wishes to know the exact amount of movement, or the exact course pursued; but they serve excellently for ascertaining whether or not the part moved at all, as well as the general character of the movement.

In the following chapters, the movements of a considerable number of plants are described; and the species have been arranged according to the system adopted by Hooker in Le Maout and Decaisne's 'Descriptive Botany.' No one who is not investigating the present subject need read all the details, which, however, we have thought it advisable to give. He may, if he thinks fit, read the last chapter first, as it includes a summary of the whole volume; and he will thus see what points interest him, and on which he requires the full evidence.

Finally, we must have the pleasure of returning our sincere thanks to Sir Joseph Hooker and to Mr. W. Thiselton Dyer for their great kindness, in not only sending us plants from Kew, but in procuring others from several sources when they were required for our observations; also, for naming many species, and giving us information on various points.

• • •

GENERAL CONSIDERATIONS ON THE MOVEMENTS AND GROWTH OF SEEDLING PLANTS.

In concluding this section of the present chapter it may be convenient to summarise, under the form of an illustration, the usual movements of the hypocotyls and epicotyls of seedlings, whilst breaking through the ground and immediately afterwards. We may suppose a man to be thrown down on his hands and knees, and at the same time to one side, by a load of hay falling on him. He would first endeavour to get his arched back upright, wriggling at the same time in all directions to free himself a little from the surrounding pressure; and this may represent the combined effects of apogeotropism and circumnutation, when a seed is so buried that the arched hypocotyl or epicotyl protrudes at first in a horizontal or inclined plane. The man, still wriggling, would then raise his arched back as high as he could; and this may represent the growth and continued circumnutation of an arched hypocotyl or epicotyl, before it has reached the surface of the ground. As soon as the man felt himself at all free, he would raise the upper part of his body, whilst still on his knees and still wriggling; and this may represent the bowing backwards of the basal leg of the arch, which in most cases aids in the withdrawal of the cotyledons from the buried and ruptured seed-coats, and the subsequent straightening of the whole hypocotyl or epicotyl—circumnutation still continuing.

• • •

MODIFIED CIRCUMNUTATION: SLEEP OR NYCTITROPIC MOVEMENTS, THEIR USE: SLEEP OF COTYLEDONS.

The so-called sleep of leaves is so conspicuous a phenomenon that it was observed as early as the time of Pliny, and since Linnæus published his famous Essay, 'Somnus Plantarum,' it has been the subject of several memoirs. Many flowers close at night, and these are likewise said to sleep; but we are not here concerned with their movements, for although effected by the same mechanism as in the case of young leaves, namely, unequal growth on the opposite sides (as first proved by Pfeffer), yet they differ essentially in being excited chiefly by changes of temperature instead of light; and in being effected, as far as we can judge, for a different purpose. Hardly any one supposes that there is any real analogy between the sleep of animals and that of plants, whether of leaves or flowers. It seems, therefore, advisable to give a distinct name to the so-called sleep-movements of plants. These have also generally been confounded, under the term "periodic," with the slight daily rise and fall of leaves, as described in the fourth chapter; and this makes it all the more desirable to give some distinct name to sleep-movements. Nyctitropism and nyctitropic, i.e. night-turning, may be applied both to leaves and flowers, and will be occasionally used by us; but it would be best to confine the term to leaves. The leaves of some few plants move either upwards or downwards when the sun shines intensely on them, and this movement has sometimes been called diurnal sleep; but we believe it to be of an essentially different nature from the nocturnal movement, and it will be briefly considered in a future chapter.

The sleep or nyctitropism of leaves is a large subject, and we think that the most convenient plan will be first to give a brief account of the position which leaves assume at night, and of the advantages apparently thus gained. Afterwards the more remarkable cases will be described in detail, with respect to cotyledons in the present chapter, and to leaves in the next chapter. Finally, it will be shown that these movements result from circumnutation, much modified and regulated by the alternations of day and night, or light and darkness; but that they are also to a certain extent inherited.

Leaves, when they go to sleep, move either upwards or downwards, or in the case of the leaflets of compound leaves, forwards, that is, towards the apex of the leaf, or backwards, that is, towards its base; or, again, they may rotate on their own axes without moving either upwards or downwards. But in almost every case the plane of the blade is so placed as to stand nearly or quite vertically at night. Therefore the apex, or the base, or either lateral edge, may be directed towards the zenith. Moreover, the upper surface of each leaf, and more especially of each leaflet, is often brought into close contact with that of the opposite one; and this is sometimes effected by singularly complicated movements. This fact suggests that the upper surface requires more protection than the lower one. For instance, the terminal leaflet in Trifolium, after turning up at night so as to stand vertically, often continues to bend over until the upper surface is directed downwards whilst the lower surface is fully exposed to the sky; and an arched roof is thus formed over the two lateral leaflets, which have their upper surfaces pressed closely together. Here we have the unusual case of one of the leaflets not standing vertically, or almost vertically, at night.

Considering that leaves in assuming their nyctitropic positions often move through an angle of 90°; that the movement is rapid in the evening; that in some cases, as we shall see in the next chapter, it is extraordinarily complicated; that with certain seedlings, old enough to bear true leaves, the cotyledons move vertically upwards at night, whilst at the same time the leaflets move vertically downwards; and that in the same genus the leaves or cotyledons of some species move upwards, whilst those of other species move downwards;—from these and other such facts, it is hardly possible to doubt that plants must derive some great advantage from such remarkable powers of movement.

The nyctitropic movements of leaves and cotyledons are effected in two ways, firstly, by means of pulvini which become, as Pfeffer has shown, alternately more turgescent on opposite sides; and secondly, by increased growth along one side of the petiole or midrib, and then on the opposite side, as was first proved by Batalin. But as it has been shown by De Vries that in these latter cases increased growth is preceded by the increased turgescence of the cells, the difference between the above two means of movement is much diminished, and consists chiefly in the turgescence of the cells of a fully developed pulvinus, not being followed by growth. When the movements of leaves or cotyledons, furnished with a pulvinus and destitute of one, are compared, they are seen to be closely similar, and are apparently effected for the same purpose. Therefore, with our object in view, it does not appear advisable to separate the above two sets of cases into two distinct classes. There is, however, one important distinction between them, namely, that movements effected by growth on the alternate sides, are confined to young growing leaves, whilst those effected by means of a pulvinus last for a long time. We have already seen well-marked instances of this latter fact with cotyledons, and so it is with leaves, as has been observed by Pfeffer and by ourselves. The long endurance of the nyctitropic movements when effected by the aid of pulvini indicates, in addition to the evidence already advanced, the functional importance of such movements to the plant. There is another difference between the two sets of cases, namely, that there is never, or very rarely, any torsion of the leaves, excepting when a pulvinus is present; but this statement applies only to periodic and nyctitropic movements, as may be inferred from other cases given by Frank.

The fact that the leaves of many plants place themselves at night in widely different positions from what they hold during the day, but with the one point in common, that their upper surfaces avoid facing the zenith, often with the additional fact that they come into close contact with opposite leaves or leaflets, clearly indicates, as it seems to us, that the object gained is the protection of the upper surfaces from being chilled at night by radiation. There is nothing improbable in the upper surface needing protection more than the lower, as the two differ in function and structure. All gardeners know that plants suffer from radiation. It is this and not cold winds which the peasants of Southern Europe fear for their olives. Seedlings are often protected from radiation by a very thin covering of straw; and fruit-trees on walls by a few fir-branches, or even by a fishing-net, suspended over them. There is a variety of the gooseberry, the flowers of which from being produced before the leaves, are not protected by them from radiation, and consequently often fail to yield fruit. An excellent observer has remarked that one variety of the cherry has the petals of its flowers much curled backwards, and after a severe frost all the stigmas were killed; whilst at the same time, in another variety with incurved petals, the stigmas were not in the least injured.

This view that the sleep of leaves saves them from being chilled at night by radiation, would no doubt have occurred to Linnæus, had the principle of radiation been then discovered; for he suggests in many parts of his 'Somnus Plantarum' that the position of the leaves at night protects the young stems and buds, and often the young inflorescence, against cold winds. We are far from doubting that an additional advantage may be thus gained; and we have observed with several plants, for instance, *Desmodium gyrans*, that whilst the blade of the leaf sinks vertically down at night, the petiole rises, so that the blade has to move through a greater angle in order to assume its vertical position than would otherwise have been necessary; but with the result that all the leaves on the same plant are crowded together as if for mutual protection.

We doubted at first whether radiation would affect in any important manner objects so thin as are many cotyledons and leaves, and more especially affect differently their upper and lower surfaces; for although the temperature of their upper surfaces would undoubtedly fall when freely exposed to a clear sky, yet we thought that they would so quickly acquire by conduction the temperature of the surrounding air, that it could hardly make any sensible difference to them, whether they stood horizontally and radiated into the open sky, or vertically and radiated chiefly in a lateral direction towards neighbouring plants and other objects. We endeavoured, therefore, to ascertain something on this head by preventing the leaves of several plants from going to sleep, and by exposing to a clear sky when the temperature was beneath the freezing point, these, as well as the other leaves on the same plants which had already assumed their nocturnal vertical position. Our experiments show that leaves thus compelled to remain horizontal at night, suffered much more injury from frost than those which were allowed to assume their normal vertical position. It may, however, be said that conclusions drawn from such observations are not applicable to sleeping plants, the inhabitants of countries where frosts do not occur. But in every country, and at all seasons, leaves must be exposed to nocturnal chills through radiation, which might be in some degree injurious to them, and which they would escape by assuming a vertical position.

In our experiments, leaves were prevented from assuming their nyctitropic position, generally by being fastened with the finest entomological pins (which did not sensibly injure them) to thin sheets of cork supported on sticks. But in some instances they were fastened down by narrow strips of card, and in others by their petioles being passed through slits in the cork. The leaves were at first fastened close to the cork, for as this is a bad conductor, and as the leaves were not exposed for long periods, we thought that the cork, which had been kept in the house, would very slightly warm them; so that if they were injured by the frost in a greater degree than the free vertical leaves, the evidence would be so much the stronger that the horizontal position was injurious. But we found that when there was any slight difference in the result, which could be detected only occasionally, the leaves which had been fastened closely down suffered rather more than those fastened with very long and thin pins, so as to stand from ½ to ¾ inch above the cork. This difference in the result, which is in itself curious as showing what a very slight difference in the conditions influences the amount of injury inflicted, may be attributed, as we believe, to the surrounding warmer air not circulating freely beneath the closely pinned leaves and thus slightly warming them. This conclusion is supported by some analogous facts hereafter to be given.

We will now describe in detail the experiments which were tried. These were troublesome from our not being able to predict how much cold the leaves of the several species could endure. Many plants had every leaf killed, both those which were secured in a horizontal position and those which were allowed to sleep—that is, to rise up or sink down vertically. Others again had not a single leaf in the least injured, and these had to be re-exposed either for a longer time or to a lower temperature.

• • •

Concluding Remarks on the Radiation from Leaves at Night.—We exposed on two occasions during the summer to a clear sky several pinned-open leaflets of *Trifolium pratense*, which naturally rise at night, and of *Oxalis purpurea*, which naturally sink at night (the plants growing out of doors), and looked at them early on several successive mornings, after they had assumed their diurnal positions. The difference in the amount of dew on the pinned-open leaflets and on those which had gone to sleep was generally conspicuous; the latter being sometimes absolutely dry, whilst the leaflets which had been horizontal were coated with large beads of dew. This shows how much cooler the leaflets fully exposed to the zenith must have become, than those which stood almost vertically, either upwards or downwards, during the night.

From the several cases above given, there can be no doubt that the position of the leaves at night affects their temperature through radiation to such a degree, that when exposed to a clear sky during a frost, it is a question of life and death. We may therefore admit as highly probable, seeing that their nocturnal position is so well adapted to lessen radiation, that the object gained by their often complicated sleep movements, is to lessen the degree to which they are chilled at night. It should be kept in mind that it is especially the upper surface which is thus protected, as it is never directed towards the zenith, and is often brought into close contact with the upper surface of an opposite leaf or leaflet.

We failed to obtain sufficient evidence, whether the better protection of the upper surface has been gained from its being more easily injured than the lower surface, or from its injury being a greater evil to the plant. That there is some difference in constitution between the two surfaces is shown by the following cases. *Cassia floribunda* was exposed to a clear sky on a sharp frosty night, and several leaflets which had assumed their nocturnal dependent position with their lower surfaces turned outwards so as to be exposed obliquely to the zenith, nevertheless had these lower surfaces less blackened than the upper surfaces which were turned inwards and were in close contact with those of the opposite leaflets. Again, a pot full of plants of *Trifolium resupinatum*, which had been kept in a warm room for three days, was turned out of doors (Sept. 21st) on a clear and almost frosty night. Next morning ten of the terminal leaflets were examined as opaque objects under the microscope. These leaflets, in going to sleep, either turn vertically upwards, or more commonly bend a little over the lateral leaflets, so that their lower surfaces are more exposed to the zenith than their upper surfaces. Nevertheless, six of these ten leaflets were distinctly yellower on the upper than on the lower and more exposed surface. In the remaining four, the result was not so plain, but certainly whatever difference there was leaned to the side of the upper surface having suffered most.

It has been stated that some of the leaflets experimented on were fastened close to the cork, and others at a height of from ½ to ¾ of an inch above it; and that whenever, after exposure to a frost, any difference could be detected in their states, the closely pinned ones had suffered most. We attributed this difference to the air, not cooled by radiation, having been prevented from circulating freely beneath the closely pinned leaflets. That there was really a difference in the temperature of leaves treated in these two different methods, was plainly shown on one occasion; for after the exposure of a pot with plants of *Melilotus dentata* for 2 h. to a clear sky (the temperature on the surrounding grass being — 2° C.), it was manifest that more dew had congealed into hoar-frost on the closely pinned leaflets, than on those which stood horizontally a little above the cork. Again, the tips of some few leaflets, which had been pinned close to the cork, projected a little beyond the edge, so that the air could circulate freely round them. This occurred with six leaflets of *Oxalis acetosella*, and their tips certainly suffered rather less than the rest of the same leaflets; for on the following morning they were still slightly green. The same result followed, even still more clearly, in two cases with leaflets of *Melilotus officinalis* which projected a little beyond the cork; and in two other cases some leaflets which were pinned close to the cork were injured, whilst other free leaflets on the same leaves, which had not space to rotate and assume their proper vertical position, were not at all injured.

Another analogous fact deserves notice: we observed on several occasions that a greater number of free leaves were injured on the branches which had been kept motionless by some of their leaves having been pinned to the corks, than on the other branches. This was conspicuously the case with those of *Melilotus petitpierreana*, but the injured leaves in this instance were not actually counted. With *Arachis hypogæa*, a young plant with 7 stems bore 22 free leaves, and of these 5 were injured by the frost, all of which were on two stems, bearing four leaves pinned to the cork-supports. With *Oxalis carnosa*, 7 free leaves were injured, and every one of them belonged to a cluster of leaves, some of which had been pinned to the cork. We could account for these cases only by supposing that the branches which were quite free had been slightly waved about by the wind, and that their leaves had thus been a little warmed by the surrounding warmer air. If we hold our hands motion less before a hot fire, and then wave them about, we immediately feel relief; and this is evidently an analogous, though reversed, case. These several facts—in relation to leaves pinned close to or a little above the cork-supports—to their tips projecting beyond it—and to the leaves on branches kept motionless—seem to us curious, as showing how a difference, apparently trifling, may determine the greater or less injury of the leaves. We may even infer as probable that the less or greater destruction during a frost of the leaves on a plant which does not sleep, may often depend on the greater or less degree of flexibility of their petioles and of the branches which bear them.

• • •

SUMMARY AND CONCLUDING REMARKS.

It has now been shown that the following important classes of movement all arise from modified circumnutation, which is omnipresent whilst growth lasts, and after growth has ceased, whenever pulvini are present. These classes of movement consist of those due to epinasty and hyponasty,—those proper to climbing plants, commonly called revolving nutation,—the nyctitropic or sleep movements of leaves and cotyledons,—and the two immense classes of movement excited by light and gravitation. When we speak of modified circumnutation we mean that light, or the alternations of light and darkness, gravitation, slight pressure or other irritants, and certain innate or constitutional states of the plant, do not directly cause the movement; they merely lead to a temporary increase or diminution of those spontaneous changes in the turgescence of the cells which are already in progress. In what manner, light, gravitation, &c., act on the cells is not known; and we will here only remark that, if any stimulus affected the cells in such a manner as to cause some slight tendency in the affected part to bend in a beneficial manner, this tendency might easily be increased through the preservation of the more sensitive individuals. But if such bending were injurious, the tendency would be eliminated unless it was overpoweringly strong; for we know how commonly all characters in all organisms vary. Nor can we see any reason to doubt, that after the complete elimination of a tendency to bend in some one direction under a certain stimulus, the power to bend in a directly opposite direction might gradually be acquired through natural selection.

Although so many movements have arisen through modified circumnutation, there are others which appear to have had a quite independent origin; but they do not form such large and important classes. When a leaf of a Mimosa is touched it suddenly assumes the same position as when asleep, but Brücke has shown that this movement results from a different state of turgescence in the cells from that which occurs during sleep; and as sleep-movements are certainly due to modified circumnutation, those from a touch can hardly be thus due. The back of a leaf of *Drosera rotundifolia* was cemented to the summit of a stick driven into the ground, so that it could not move in the least, and a tentacle was observed during many hours under the microscope; but it exhibited no circumnutating movement, yet after being momentarily touched with a bit of raw meat, its basal part began to curve in 23 seconds. This curving movement therefore could not have resulted from modified circumnutation. But when a small object, such as a fragment of a bristle, was placed on one side of the tip of a radicle, which we know is continually circumnutating, the induced curvature was so similar to the movement caused by geotropism, that we can hardly doubt that it is due to modified circumnutation. A flower of a Mahonia was cemented to a stick, and the stamens exhibited no signs of circumnutation under the microscope, yet when they were lightly touched they suddenly moved towards the pistil. Lastly, the curling of the extremity of a tendril when touched seems to be independent of its revolving or circumnutating movement. This is best shown by the part which is the most sensitive to contact, circumnutating much less than the lower parts, or apparently not at all.

Although in these cases we have no reason to believe that the movement depends on modified circumnutation, as with the several classes of movement described in this volume, yet the difference between the two sets of cases may not be so great as it at first appears. In the one set, an irritant causes an increase or diminution in the turgescence of the cells, which are already in a state of change; whilst in the other set, the irritant first starts a similar change in their state of turgescence. Why a touch, slight pressure or any other irritant, such as electricity, heat, or the absorption of animal matter, should modify the turgescence of the affected cells in such a manner as to cause movement, we do not know. But a touch acts in this manner so often, and on such widely distinct plants, that the tendency seems to be a very general one; and if beneficial, it might be increased to any extent. In other cases, a touch produces a very different effect, as with Nitella, in which the protoplasm may be seen to recede from the walls of the cell; in Lactuca, in which a milky fluid exudes; and in the tendrils of certain Vitaceæ, Cucurbitaceæ, and Bignoniaceæ, in which slight pressure causes a cellular outgrowth.

Finally, it is impossible not to be struck with the resemblance between the foregoing movements of plants and many of the actions performed unconsciously by the lower animals. With plants an astonishingly small stimulus suffices; and even with allied plants one may be highly sensitive to the slightest continued pressure, and another highly sensitive to a slight momentary touch. The habit of moving at certain periods is inherited both by plants and animals; and several other points of similitude have been specified. But the most striking resemblance is the localisation of their sensitiveness, and the transmission of an influence from the excited part to another which consequently moves. Yet plants do not of course possess nerves or a central nervous system; and we may infer that with animals such structures serve only for the more perfect transmission of impressions, and for the more complete intercommunication of the several parts.

We believe that there is no structure in plants more wonderful, as far as its functions are concerned, than the tip of the radicle. If the tip be lightly pressed or burnt or cut, it transmits an influence to the upper adjoining part, causing it to bend away from the affected side; and, what is more surprising, the tip can distinguish between a slightly harder and softer object, by which it is simultaneously pressed on opposite sides. If, however, the radicle is pressed by a similar object a little above the tip, the pressed part does not transmit any influence to the more distant parts, but bends abruptly towards the object. If the tip perceives the air to be moister on one side than on the other, it likewise transmits an influence to the upper adjoining part, which bends towards the source of moisture. When the tip is excited by light (though in the case of radicles this was ascertained in only a single instance) the adjoining part bends from the light; but when excited by gravitation the same part bends towards the centre of gravity. In almost every case we can clearly perceive the final purpose or advantage of the several movements. Two, or perhaps more, of the exciting causes often act simultaneously on the tip, and one conquers the other, no doubt in accordance with its importance for the life of the plant. The course pursued by the radicle in penetrating the ground must be determined by the tip; hence it has acquired such diverse kinds of sensitiveness. It is hardly an exaggeration to say that the tip of the radicle thus endowed, and having the power of directing the movements of the adjoining parts, acts like the brain of one of the lower animals; the brain being seated within the anterior end of the body, receiving impressions from the sense-organs, and directing the several movements.

A Letter to Frithiof Holmgren; Mr. Darwin on Vivisection. [1881]

A SCIENTIFIC SCANDAL *marred the British Medical Association's August 1874 meeting at Norwich, where a French physiologist badly botched a demonstration of how injecting absinthe into the bloodstreams of two dogs would induce epilepsy. The British audience was outraged. The physiologist escaped home across the Channel, but three Norwich physicians were tried for unwarranted cruelty to animals. Although the accused were found innocent, the incident led to the establishment of the Anti-vivisectionist movement and the proposal of legislation to control experimental use of living animals.*

On November 8, 1875, Darwin testified before the Royal Commission on the Practice of Subjecting Live Animals to Experiments for Scientific Purposes. He stated that he had not personally conducted physiological experiments on animals but that he had signed a British Association for the Advancement of Science memorandum that defined the conditions under which such experiments were justifiable. He also observed that cruelty "deserves detestation and abhorrence." The Report of the Royal Commission led to the Cruelty to Animals Act of 1876, which allowed experimentation on animals to continue under certain restrictions.

Early in 1881, when the Swedish parliament was discussing legislation to regulate physiological and medical experimentation, Frithiof Holmgren, professor of physiology at the University of Uppsala, wrote to ask a number of eminent scientists for their views on vivisection. Darwin's reply to Holmgren was printed in the London Times *on*

April 18, 1881. It provoked comments by Frances Power Cobbe, an Anglo-Irish journalist and leading anti-vivisectionist, on the nineteenth. Darwin's answer to Cobbe was published in the Times *on April 22. Both of Darwin's letters appear below.*

Dear Sir,

In answer to your courteous letter of April 7, I have no objection to express my opinion with respect to the right of experimenting on living animals. I use this latter expression as more correct and comprehensive than that of vivisection. You are at liberty to make any use of this letter which you may think fit, but if published I should wish the whole to appear. I have all my life been a strong advocate for humanity to animals, and have done what I could in my writings to enforce this duty. Several years ago, when the agitation against physiologists commenced in England, it was asserted that inhumanity was here practiced, and useless suffering caused to animals; and I was led to think that it might be advisable to have an Act of Parliament on the subject. I then took an active part in trying to get a Bill passed, such as would have removed all just cause of complaint, and at the same time have left physiologists free to pursue their researches,—a Bill very different from the Act which has since been passed. It is right to add that the investigation of the matter by a Royal Commission proved that the accusations made against our English physiologists were false. From all that I have heard, however, I fear that in some parts of Europe little regard is paid to the sufferings of animals, and if this be the case, I should be glad to hear of legislation against inhumanity in any such country. On the other hand, I know that physiology cannot possibly progress except by means of experiments on living animals, and I feel the deepest conviction that he who retards the progress of physiology commits a crime against mankind. Any one who remembers, as I can, the state of this science half a century ago, must admit that it has made immense progress, and it is now progressing at an ever-increasing rate. What improvements in medical practice may be directly attributed to physiological research is a question which can be properly discussed only by those physiologists and medical practitioners who have studied the history of their subjects; but, as far as I can learn, the benefits are already great. However this may be, no one, unless he is grossly ignorant of what science has done for mankind, can entertain any doubt of the incalculable benefits which will hereafter be derived from physiology, not only by man, but by the lower animals. Look for instance at Pasteur's results in modifying the germs of the most malignant diseases, from which, as it so happens, animals will in the first place receive more relief than man. Let it be remembered how many lives and what a fearful amount of suffering have been saved by the knowledge gained of parasitic worms through the experiments of Virchow and others on living animals. In the future every one will be astonished at the ingratitude shown, at least in England, to these benefactors of mankind. As for myself, permit me to assure you that I honour, and shall always honour, every one who advances the noble science of physiology.

Sir,

I do not wish to discuss the views expressed by Miss Cobbe in the letter which appeared in the *Times* of the 19th inst.; but as she asserts that I have "misinformed" my correspondent in Sweden in saying that "the investigation of the matter by a Royal Commission proved that the accusations made against our English physiologists were false," I will merely ask leave to refer to some other sentences from the Report of the Commission.

(1.) The sentence—"It is not to be doubted that inhumanity may be found in persons of very high position as physiologists," which Miss Cobbe quotes from page 17 of the report, and which, in her opinion, "can necessarily concern English physiologists alone and not foreigners," is immediately followed by the words "We have seen that it was so in Majendie." Majendie was a French physiologist who became notorious some half-century ago for his cruel experiments on living animals.

(2.) The Commissioners, after speaking of the "general sentiment of humanity" prevailing in this country, say (p. 10):—

"This principle is accepted generally by the very highly educated men whose lives are devoted either to scientific investigation and education or to the mitigation or the removal of the sufferings of their fellow-creatures; though differences of degree in regard to its practical application will be easily discernible by those who study the evidence as it has been laid before us."

Again, according to the Commissioners (p. 10):—

"The secretary of the Royal Society for the Prevention of Cruelty to Animals, when asked whether the general tendency of the scientific world in this country is at variance with humanity, says he believes it to be very different, indeed, from that of foreign physiologists; and while giving it as the opinion of the Society that experiments are performed which are in their nature beyond any legitimate province of science, and that the pain which they inflict is pain which is not justifiable to inflict even for the scientific object in view, he readily acknowledges that he does not know a single case of wanton cruelty, and that in general the English physiologists have used anaesthetics where they think they can do so with safety to the experiment."

The Formation of Vegetable Mould, through the Action of Worms, with Observations on Their Habits. [1881]

TOWARD THE END *of his life Darwin was still experimenting—and perhaps there is some graveyard humor in the fact that his subject was earthworms. Once again, though, his research reached well back into the past, to 1837, when he made his first observations on how worms move soil. This line of enquiry began when his uncle and soon to be father-in-law, Josiah Wedgwood of Maer Hall, pointed out to him how much soil the casts of worms brought to the surface. It ended forty-four years later with the publication of his last book,* The Formation of Vegetable Mould, through the Action of Worms, with Observations on their Habits. *In the time between, Darwin had published short papers on his observations of worms in 1837, 1844, and 1869. In 1877 he had the opportunity to observe a Roman villa at Abinger, Surrey, being excavated and to reflect on how worms had aided in burying it. He then devised and conducted a series of ingenious experiments that revealed the details of earthworms' behavior and tested their intelligence. His experiments and their results are discussed, with a rare combination of scientific precision and charm, in the book. As is shown in the first two chapters of a symposium volume commemorating the centennial of the publication of* The Formation of Vegetable Mould, *Darwin was correct in his interpretation of what worms do and have done in the past.*

After The Voyage of the *Beagle*, The Formation of Vegetable Mould *is perhaps the most accessible of Darwin's books for the general reader. It sold well. According to Darwin, 3500 copies were purchased in the first two months; and in the years just following its appearance, it did better than had any of Darwin's other books, including* The Origin of Species. *In the first sentence of the worm book's concluding chapter, Darwin observes that "Worms have played a more important part in the history of the world than most persons would at first suppose." These simple words offer the clearest possible evidence of his lifelong talent for brilliantly recognizing and patiently tracing the natural connections between small causes and vast consequences.*

INTRODUCTION.

The share which worms have taken in the formation of the layer of vegetable mould, which covers the whole surface of the land in every moderately humid country, is the subject of the present volume. This mould is generally of a blackish colour and a few inches in thickness. In different districts it differs but little in appearance, although it may rest on various subsoils. The uniform fineness of the particles of which it is composed is one of its chief characteristic features; and this may be well observed in any gravelly country, where a recently-ploughed field immediately adjoins one which has long remained undisturbed for pasture, and where the vegetable mould is exposed on the sides of a ditch or hole. The subject may appear an insignificant one, but we shall see that it possesses some interest; and the maxim "de minimis lex non curat," ["the law takes no account of trifles"—editors] does not apply to science. Even Elie de Beaumont, who generally undervalues small agencies and their accumulated effects, remarks: "la couche très-mince de la terre végétale est un monument d'une haute antiquité, et, par le fait de sa permanence, un objet digne d'occuper le géologue, et capable de lui fournir des remarques intéressantes." ["the thinnest layer of vegetable mould is a monument of the greatest antiquity and, because of its permanence, an object worthy to occupy the geologist and capable of furnishing him with many interesting notes."—editors]

Although the superficial layer of vegetable mould as a whole no doubt is of the highest antiquity, yet in regard to its permanence, we shall hereafter see reason to believe that its component particles are in most cases removed at not a very slow rate, and are replaced by others due to the disintegration of the underlying materials.

As I was led to keep in my study during many months worms in pots filled with earth, I became interested in them, and wished to learn how far they acted consciously, and how much mental power they displayed. I was the more desirous to learn something on this head, as few observations of this kind have been made, as far as I know, on animals so low in the scale of organization and so poorly provided with sense-organs, as are earth-worms.

In the year 1837, a short paper was read by me before the Geological Society of London, "On the Formation of Mould," in which it was shown that small fragments of burnt marl, cinders, &c., which had been thickly strewed over the surface of several meadows, were found after a few years lying at the depth of some inches beneath the turf, but still forming a layer. This apparent sinking of superficial bodies is due, as was first suggested to me by Mr. Wedgwood of Maer Hall in Staffordshire, to the large quantity of fine earth continually brought up to the surface by worms in the form of castings. These castings are sooner or later spread out and cover up any object left on the surface. I was thus led to conclude that all the vegetable mould over the whole country has passed many times through, and will again pass many times through, the intestinal canals of worms. Hence the term "animal mould" would be in some respects more appropriate than that commonly used of "vegetable mould."

Ten years after the publication of my paper, M. D'Archiac, evidently influenced by the doctrines of Elie de Beaumont, wrote about my "singulière théorie," and objected that it could apply only to "les prairies basses et humides;" and that "les terres labourées, les bois, les prairies élevées, n'apportent aucune preuve à l'appui de cette manière de voir." ["tilled soil, woods, and elevated meadows have not provided a single proof to support this observation"—editors] But M. D'Archiac must have thus argued from inner consciousness and not from observation, for worms abound to an extraordinary degree in kitchen gardens where the soil is continually worked, though in such loose soil they generally deposit their castings in any open cavities or within their old burrows instead of on the surface. Von Hensen estimates that there are about twice as many worms in gardens as in corn-fields. With respect to "prairies élevées," I do not know how it may be in France, but nowhere in England have I seen the ground so thickly covered with castings as on commons, at a height of several hundred feet above the sea. In woods again, if the loose leaves in autumn are removed, the whole surface will be found strewed with castings. Dr. King, the superintendent of the Botanic Garden in Calcutta, to whose kindness I am indebted for many observations on earth-worms, informs me that he found, near Nancy in France, the bottom of the State forests covered over many acres with a spongy layer, composed of dead leaves and innumerable worm-castings. He there heard the Professor of "Aménagement des Forêts" lecturing to his pupils, and pointing out this case as a "beautiful example of the natural cultivation of the soil; for year after year the thrown-up castings cover the dead leaves; the result being a rich humus of great thickness."

In the year 1869, Mr. Fish rejected my conclusions with respect to the part which worms have played in the formation of vegetable mould, merely on account of their assumed incapacity to do so much work. He remarks that "considering their weakness and their size, the work they are represented to have accomplished is stupendous." Here we have an instance of that inability to sum up the effects of a continually recurrent cause, which has often retarded the progress of science, as formerly in the case of geology, and more recently in that of the principle of evolution.

Although these several objections seemed to me to have no weight, yet I resolved to make more observations of the same kind as those published, and to attack the problem on another side; namely, to weigh all the castings thrown up within a given time in a measured space, instead of ascertaining the rate at which objects left on the surface were buried by worms. But some of my observations have been rendered almost superfluous by an admirable paper by Von Hensen, already alluded to, which appeared in 1877. Before entering on details with respect to the castings, it will be advisable to give some account of the habits of worms from my own observations and from those of other naturalists.

HABITS OF WORMS.

Earth-worms are distributed throughout the world under the form of a few genera, which externally are closely similar to one another. The British species of Lumbricus have never been carefully monographed; but we may judge of their probable number from those inhabiting neighbouring countries. In Scandinavia there are eight species, according to Eisen; but two of these rarely burrow in the ground, and one inhabits very wet places or even lives under the water. We are here concerned only with the kinds which bring up earth to the surface in the form of castings. Hoffmeister says that the species in Germany are not well known, but gives the same number as Eisen, together with some strongly marked varieties.

Earth-worms abound in England in many different stations. Their castings may be seen in extraordinary numbers on commons and chalk-downs, so as almost to cover the whole surface, where the soil is poor and the grass short and thin. But they are almost or quite as numerous in some of the London parks, where the grass grows well and the soil appears rich. Even on the same field worms are much more frequent in some places than in others, without any visible difference in the nature of the soil. They abound in paved court-yards close to houses; and an instance will be given in which they had burrowed through the floor of a very damp cellar. I have seen worms in black peat in a boggy field; but they are extremely rare, or quite absent in the drier, brown, fibrous peat, which is so much valued by gardeners. On dry, sandy or gravelly tracks, where heath with some gorse, ferns, coarse grass, moss and lichens alone grow, hardly any worms can be found. But in many parts of England, wherever a path crosses a heath, its surface becomes covered with a fine short sward. Whether this change of vegetation is due to the taller plants being killed by the occasional trampling of man and animals, or to the soil being occasionally manured by the droppings from animals, I do not know. On such grassy paths worm-castings may often be seen. On a heath in Surrey, which was carefully examined, there were only a few castings on these paths, where they were much inclined; but on the more level parts, where a bed of fine earth had been washed down from the steeper parts and had accumulated to a thickness of a few inches, worm-castings abounded. These spots seemed to be overstocked with worms, so that they had been compelled to spread to a distance of a few feet from the grassy paths, and here their castings had been thrown up among the heath; but beyond this limit, not a single casting could be found. A layer, though a thin one, of fine earth, which probably long retains some moisture, is in all cases, as I believe, necessary for their existence; and the mere compression of the soil appears to be in some degree favourable to them, for they often abound in old gravel walks, and in foot-paths across fields.

Beneath large trees few castings can be found during certain seasons of the year, and this is apparently due to the moisture having been sucked out of the ground by the innumerable roots of the trees; for such places may be seen covered with castings after the heavy autumnal rains. Although most coppices and woods support many worms, yet in a forest of tall and ancient beech-trees in Knole Park, where the ground beneath was bare of all vegetation, not a single casting could be found over wide spaces, even during the autumn. Nevertheless, castings were abundant on some grass-covered glades and indentations which penetrated this forest. On the mountains of North Wales and on the Alps, worms, as I have been informed, are in most places rare; and this may perhaps be due to the close proximity of the subjacent rocks, into which worms cannot burrow during the winter so as to escape being frozen. Dr. McIntosh, however, found worm-castings at a height of 1500 feet on Schiehallion in Scotland. They are numerous on some hills near Turin at from 2000 to 3000 feet above the sea, and at a great altitude on the Nilgiri Mountains in south India and on the Himalaya.

Earth-worms must be considered as terrestrial animals, though they are still in one sense semi-aquatic, like the other members of the great class of annelids to which they belong. M. Perrier found that their exposure to the dry air of a room for only a single night was fatal to them. On the other hand he kept several large worms alive for nearly four months, completely submerged in water. During the summer when the ground is dry, they penetrate to a considerable depth and cease to work, as they do during the winter when the ground is frozen. Worms are nocturnal in their habits, and at night may be seen crawling about in large numbers, but usually with their tails still inserted in their burrows. By the expansion of this part of their bodies, and with the help of the short, slightly reflexed bristles, with which their bodies are armed, they hold so fast that they can seldom be dragged out of the ground without being torn into pieces. During the day they remain in their burrows, except at the pairing season, when those which inhabit adjoining burrows expose the greater part of their bodies for an hour or two in the early morning. Sick individuals, which are generally affected by the parasitic larvæ of a fly, must also be excepted, as they wander about during the day and die on the surface. After heavy rain succeeding dry weather, an astonishing number of dead worms may sometimes be seen lying on the ground. Mr. Galton informs me that on one such occasion (March, 1881), the dead worms averaged one for every two and a half paces in length on a walk in Hyde Park, four paces in width. He counted no less than 45 dead worms in one place in a length of sixteen paces. From the facts above given, it is not probable that these worms could have been drowned, and if they had been drowned they would have perished in their burrows. I believe that they were already sick, and that their deaths were merely hastened by the ground being flooded.

It has often been said that under ordinary circumstances healthy worms never, or very rarely, completely leave their burrows at night; but this is an error, as White of Selborne long ago knew. In the morning, after there has been heavy rain, the film of mud or of very fine sand over gravel-walks is often plainly marked with their tracks. I have noticed this from August to May, both months included, and it probably occurs during the two remaining months of the year when they are wet. On these occasions, very few dead worms could anywhere be seen. On January 31, 1881, after a long-continued and unusually severe frost with much snow, as soon as a thaw set in, the walks were marked with innumerable tracks. On one occasion, five tracks were counted crossing a space of only an inch square. They could sometimes be traced either to or from the mouths of the burrows in the gravel-walks, for distances between 2 or 3 up to 15 yards. I have never seen two tracks leading to the same burrow; nor is it likely, from what we shall presently see of their sense-organs, that a worm could find its way back to its burrow after having once left it. They apparently leave their burrows on a voyage of discovery, and thus they find new sites to inhabit.

Morren states that worms often lie for hours almost motionless close beneath the mouths of their burrows. I have occasionally noticed the same fact with worms kept in pots in the house; so that by looking down into their burrows, their heads could just be seen. If the ejected earth or rubbish over the burrows be suddenly removed, the end of the worm's body may very often be seen rapidly retreating. This habit of lying near the surface leads to their destruction to an immense extent. Every morning during certain seasons of the year, the thrushes and blackbirds on all the lawns throughout the country draw out of their holes an astonishing number of worms; and this they could not do, unless they lay close to the surface. It is not probable that worms behave in this manner for the sake of breathing fresh air, for we have seen that they can live for a long time under water. I believe that they lie near the surface for the sake of warmth, especially in the morning; and we shall hereafter find that they often coat the mouths of their burrows with leaves, apparently to prevent their bodies from coming into close contact with the cold damp earth. It is said that they completely close their burrows during the winter.

• • •

Senses.—Worms are destitute of eyes, and at first I thought that they were quite insensible to light; for those kept in confinement were repeatedly observed by the aid of a candle, and others out of doors by the aid of a lantern, yet they were rarely alarmed, although extremely timid animals. Other persons have found no difficulty in observing worms at night by the same means.

Hoffmeister, however, states that worms, with the exception of a few individuals, are extremely sensitive to light; but he admits that in most cases a certain time is requisite for its action. These statements led me to watch on many successive nights worms kept in pots, which were protected from currents of air by means of glass plates. The pots were approached very gently, in order that no vibration of the floor should be caused. When under these circumstances worms were illuminated by a bull's-eye lantern having slides of dark red and blue glass, which intercepted so much light that they could be seen only with some difficulty, they were not at all affected by this amount of light, however long they were exposed to it. The light, as far as I could judge, was brighter than that from the full moon. Its colour apparently made no difference in the result. When they were illuminated by a candle, or even by a bright paraffin lamp, they were not usually affected at first. Nor were they when the light was alternately admitted and shut off. Sometimes, however, they behaved very differently, for as soon as the light fell on them, they withdrew into their burrows with almost instantaneous rapidity. This occurred perhaps once out of a dozen times. When they did not withdraw instantly, they often raised the anterior tapering ends of their bodies from the ground, as if their attention was aroused or as if surprise was felt; or they moved their bodies from side to side as if feeling for some object. They appeared distressed by the light; but I doubt whether this was really the case, for on two occasions after withdrawing slowly, they remained for a long time with their anterior extremities protruding a little from the mouths of their burrows, in which position they were ready for instant and complete withdrawal.

When the light from a candle was concentrated by means of a large lens on the anterior extremity, they generally withdrew instantly; but this concentrated light failed to act perhaps once out of half a dozen trials. The light was on one occasion concentrated on a worm lying beneath water in a saucer, and it instantly withdrew into its burrow. In all cases the duration of the light, unless extremely feeble, made a great difference in the result; for worms left exposed before a paraffin lamp or a candle invariably retreated into their burrows within from five to fifteen minutes; and if in the evening the pots were illuminated before the worms had come out of their burrows, they failed to appear.

From the foregoing facts it is evident that light affects worms by its intensity and by its duration. It is only the anterior extremity of the body, where the cerebral ganglia lie, which is affected by light, as Hoffmeister asserts, and as I observed on many occasions. If this part is shaded, other parts of the body may be fully illuminated, and no effect will be produced. As these animals have no eyes, we must suppose that the light passes through their skins, and in some manner excites their cerebral ganglia. It appeared at first probable that the different manner in which they were affected on different occasions might be explained, either by the degree of extension of their skin and its consequent transparency, or by some particular incidence of the light; but I could discover no such relation. One thing was manifest, namely, that when worms were employed in dragging leaves into their burrows or in eating them, and even during the short intervals whilst they rested from their work, they either did not perceive the light or were regardless of it; and this occurred even when the light was concentrated on them through a large lens. So, again, whilst they are paired, they will remain for an hour or two out of their burrows, fully exposed to the morning light; but it appears from what Hoffmeister says that a light will occasionally cause paired individuals to separate.

When a worm is suddenly illuminated and dashes like a rabbit into its burrow—to use the expression employed by a friend—we are at first led to look at the action as a reflex one. The irritation of the cerebral ganglia appears to cause certain muscles to contract in an inevitable manner, independently of the will or consciousness of the animal, as if it were an automaton. But the different effect which a light produced on different occasions, and especially the fact that a worm when in any way employed and in the intervals of such employment, whatever set of muscles and ganglia may then have been brought into play, is often regardless of light, are opposed to the view of the sudden withdrawal being a simple reflex action. With the higher animals, when close attention to some object leads to the disregard of the impressions which other objects must be producing on them, we attribute this to their attention being then absorbed; and attention implies the presence of a mind. Every sportsman knows that he can approach animals whilst they are grazing, fighting or courting, much more easily than at other times. The state, also, of the nervous system of the higher animals differs much at different times, for instance, a horse is much more readily startled at one time than at another. The comparison here implied between the actions of one of the higher animals and of one so low in the scale as an earth-worm, may appear farfetched; for we thus attribute to the worm attention and some mental power, nevertheless I can see no reason to doubt the justice of the comparison.

Although worms cannot be said to possess the power of vision, their sensitiveness to light enables them to distinguish between day and night; and they thus escape extreme danger from the many diurnal animals which prey on them. Their withdrawal into their burrows during the day appears, however, to have become an habitual action; for worms kept in pots covered by glass-plates, over which sheets of black paper were spread, and placed before a north-east window, remained during the day-time in their burrows and came out every night; and they continued thus to act for a week. No doubt a little light may have entered between the sheets of glass and the blackened paper; but we know from the trials with coloured glass, that worms are indifferent to a small amount of light.

Worms appear to be less sensitive to moderate radiant heat than to a bright light. I judge of this from having held at different times a poker heated to dull redness near some worms, at a distance which caused a very sensible degree of warmth in my hand. One of them took no notice; a second withdrew into its burrow, but not quickly; the third and fourth much more quickly, and the fifth as quickly as possible. The light from a candle, concentrated by a lens and passing through a sheet of glass which would intercept most of the heat-rays, generally caused a much more rapid retreat than did the heated poker. Worms are sensitive to a low temperature, as may be inferred from their not coming out of their burrows during a frost.

Worms do not possess any sense of hearing. They took not the least notice of the shrill notes from a metal whistle, which was repeatedly sounded near them; nor did they of the deepest and loudest tones of a bassoon. They were indifferent to shouts, if care was taken that the breath did not strike them. When placed on a table close to the keys of a piano, which was played as loudly as possible, they remained perfectly quiet.

Although they are indifferent to undulations in the air audible by us, they are extremely sensitive to vibrations in any solid object. When the pots containing two worms which had remained quite indifferent to the sound of the piano, were placed on this instrument, and the note C in the bass clef was struck, both instantly retreated into their burrows. After a time they emerged, and when G above the line in the treble clef was struck they again retreated. Under similar circumstances on another night one worm dashed into its burrow on a very high note being struck only once, and the other worm when C in the treble clef was struck. On these occasions the worms were not touching the sides of the pots, which stood in saucers; so that the vibrations, before reaching their bodies, had to pass from the sounding board of the piano, through the saucer, the bottom of the pot and the damp, not very compact earth on which they lay with their tails in their burrows. They often showed their sensitiveness when the pot in which they lived, or the table on which the pot stood, was accidentally and lightly struck; but they appeared less sensitive to such jars than to the vibrations of the piano; and their sensitiveness to jars varied much at different times. It has often been said that if the ground is beaten or otherwise made to tremble, worms believe that they are pursued by a mole and leave their burrows. I beat the ground in many places where worms abounded, but not one emerged. When, however, the ground is dug with a fork and is violently disturbed beneath a worm, it will often crawl quickly out of its burrow.

The whole body of a worm is sensitive to contact. A slight puff of air from the mouth causes an instant retreat. The glass plates placed over the pots did not fit closely, and blowing through the very narrow chinks thus left, often sufficed to cause a rapid retreat. They sometimes perceived the eddies in the air caused by quickly removing the glass plates. When a worm first comes out of its burrow, it generally moves the much extended anterior extremity of its body from side to side in all directions, apparently as an organ of touch; and there is some reason to believe, as we shall see in the next chapter, that they are thus enabled to gain a general notion of the form of an object. Of all their senses that of touch, including in this term the perception of a vibration, seems much the most highly developed.

In worms the sense of smell apparently is confined to the perception of certain odours, and is feeble. They were quite indifferent to my breath, as long as I breathed on them very gently. This was tried, because it appeared possible that they might thus be warned of the approach of an enemy. They exhibited the same indifference to my breath whilst I chewed some tobacco, and while a pellet of cotton-wool with a few drops of millefleurs perfume or of acetic acid was kept in my mouth. Pellets of cotton-wool soaked in tobacco juice, and in millefleurs perfume, and in paraffin, were held with pincers and were waved about within two or three inches of several worms, but they took no notice. On one or two occasions, however, when acetic acid had been placed on the pellets, the worms appeared a little uneasy, and this was probably due to the irritation of their skins. The perception of such unnatural odours would be of no service to worms; and as such timid creatures would almost certainly exhibit some signs of any new impression, we may conclude that they did not perceive these odours.

The result was different when cabbage-leaves and pieces of onion were employed, both of which are devoured with much relish by worms. Small square pieces of fresh and half-decayed cabbage-leaves and of onion bulbs were on nine occasions buried in my pots, beneath about ¼ of an inch of common garden soil; and they were always discovered by the worms. One bit of cabbage was discovered and removed in the course of two hours; three were removed by the next morning, that is, after a single night; two others after two nights; and the seventh bit after three nights. Two pieces of onion were discovered and removed after three nights. Bits of fresh raw meat, of which worms are very fond, were buried, and were not discovered within forty-eight hours, during which time they had not become putrid. The earth above the various buried objects was generally pressed down only slightly, so as not to prevent the emission of any odour. On two occasions, however, the surface was well watered, and was thus rendered somewhat compact. After the bits of cabbage and onion had been removed, I looked beneath them to see whether the worms had accidentally come up from below, but there was no sign of a burrow; and twice the buried objects were laid on pieces of tin-foil which were not in the least displaced. It is of course possible that the worms whilst moving about on the surface of the ground, with their tails affixed within their burrows, may have poked their heads into the places where the above objects were buried; but I have never seen worms acting in this manner. Some pieces of cabbage-leaf and of onion were twice buried beneath very fine ferruginous sand, which was slightly pressed down and well watered, so as to be rendered very compact, and these pieces were never discovered. On a third occasion the same kind of sand was neither pressed down nor watered, and the pieces of cabbage were discovered and removed after the second night. These several facts indicate that worms possess some power of smell; and that they discover by this means odoriferous and much-coveted kinds of food.

It may be presumed that all animals which feed on various substances possess the sense of taste, and this is certainly the case with worms. Cabbage-leaves are much liked by worms; and it appears that they can distinguish between different varieties; but this may perhaps be owing to differences in their texture. On eleven occasions pieces of the fresh leaves of a common green variety and of the red variety used for pickling were given them, and they preferred the green, the red being either wholly neglected or much less gnawed. On two other occasions, however, they seemed to prefer the red. Half-decayed leaves of the red variety and fresh leaves of the green were attacked about equally. When leaves of the cabbage, horse-radish (a favourite food) and of the onion were given together, the latter were always and manifestly preferred. Leaves of the cabbage, lime-tree, Ampelopis, parsnip (Pastinaca), and celery (Apium) were likewise given together; and those of the celery were first eaten. But when leaves of cabbage, turnip, beet, celery, wild cherry and carrots were given together, the two latter kinds, especially those of the carrot, were preferred to all the others, including those of celery. It was also manifest after many trials that wild cherry leaves were greatly preferred to those of the lime-tree and hazel (Corylus). According to Mr. Bridgman the half-decayed leaves of *Phlox verna* are particularly liked by worms.

Pieces of the leaves of cabbage, turnip, horse-radish, and onion were left on the pots during 22 days, and were all attacked and had to be renewed; but during the whole of this time leaves of an Artemisia and of the culinary sage, thyme and mint, mingled with the above leaves, were quite neglected excepting those of the mint, which were occasionally and very slightly nibbled. These latter four kinds of leaves do not differ in texture in a manner which could make them disagreeable to worms; they all have a strong taste, but so have the four first mentioned kinds of leaves; and the wide difference in the result must be attributed to a preference by the worms for one taste over another.

Mental Qualities.—There is little to be said on this head. We have seen that worms are timid. It may be doubted whether they suffer as much pain when injured, as they seem to express by their contortions. Judging by their eagerness for certain kinds of food, they must enjoy the pleasure of eating. Their sexual passion is strong enough to overcome for a time their dread of light. They perhaps have a trace of social feeling, for they are not disturbed by crawling over each other's bodies, and they sometimes lie in contact. According to Hoffmeister they pass the winter either singly or rolled up with others into a ball at the bottom of their burrows. Although worms are so remarkably deficient in the several sense-organs, this does not necessarily preclude intelligence, as we know from such cases as those of Laura Bridgman; and we have seen that when their attention is engaged, they neglect impressions to which they would otherwise have attended; and attention indicates the presence of a mind of some kind. They are also much more easily excited at certain times than at others.

• • •

Worms seize leaves and other objects, not only to serve as food, but for plugging up the mouths of their burrows; and this is one of their strongest instincts. Leaves and petioles of many kinds, some flower-peduncles, often decayed twigs of trees, bits of paper, feathers, tufts of wool and horse-hairs are dragged into their burrows for this purpose. I have seen as many as seventeen petioles of a Clematis projecting from the mouth of one burrow, and ten from the mouth of another. Some of these objects, such as the petioles just named, feathers, &c., are never gnawed by worms. In a gravel-walk in my garden I found many hundred leaves of a pine-tree (*P. austriaca* or *nigricans*) drawn by their bases into burrows. The surfaces by which these leaves are articulated to the branches are shaped in as peculiar a manner as is the joint between the leg-bones of a quadruped; and if these surfaces had been in the least gnawed, the fact would have been immediately visible, but there was no trace of gnawing. Of ordinary dicotyledonous leaves, all those which are dragged into burrows are not gnawed. I have seen as many as nine leaves of the lime-tree drawn into the same burrow, and not nearly all of them had been gnawed; but such leaves may serve as a store for future consumption. Where fallen leaves are abundant, many more are sometimes collected over the mouth of a burrow than can be used, so that a small pile of unused leaves is left like a roof over those which have been partly dragged in.

A leaf in being dragged a little way into a cylindrical burrow is necessarily much folded or crumpled. When another leaf is drawn in, this is done exteriorly to the first one, and so on with the succeeding leaves; and finally all become closely folded and pressed together. Sometimes the worm enlarges the mouth of its burrow, or makes a fresh one close by, so as to draw in a still larger number of leaves. They often or generally fill up the interstices between the drawn-in leaves with moist viscid earth ejected from their bodies; and thus the mouths of the burrows are securely plugged. Hundreds of such plugged burrows may be seen in many places, especially during the autumnal and early winter months. But, as will hereafter be shown, leaves are dragged into the burrows not only for plugging them up and for food, but for the sake of lining the upper part or mouth.

When worms cannot obtain leaves, petioles, sticks, &c., with which to plug up the mouths of their burrows, they often protect them by little heaps of stones; and such heaps of smooth rounded pebbles may frequently be seen on gravel-walks. Here there can be no question about food. A lady, who was interested in the habits of worms, removed the little heaps of stones from the mouths of several burrows and cleared the surface of the ground for some inches all round. She went out on the following night with a lantern, and saw the worms with their tails fixed in their burrows, dragging the stones inwards by the aid of their mouths, no doubt by suction. "After two nights some of the holes had 8 or 9 small stones over them; after four nights one had about 30, and another 34 stones." One stone which had been dragged over the gravel-walk to the mouth of a burrow weighed two ounces; and this proves how strong worms are. But they show greater strength in sometimes displacing stones in a well-trodden gravel-walk; that they do so, may be inferred from the cavities left by the displaced stones being exactly filled by those lying over the mouths of adjoining burrows, as I have myself observed.

Work of this kind is usually performed during the night; but I have occasionally known objects to be drawn into the burrows during the day. What advantage the worms derive from plugging up the mouths of their burrows with leaves, &c., or from piling stones over them, is doubtful. They do not act in this manner at the times when they eject much earth from their burrows; for their castings then serve to cover the mouth. When gardeners wish to kill worms on a lawn, it is necessary first to brush or rake away the castings from the surface, in order that the lime-water may enter the burrows. It might be inferred from this fact that the mouths are plugged up with leaves, &c., to prevent the entrance of water during heavy rain; but it may be urged against this view that a few, loose, well-rounded stones are ill-adapted to keep out water. I have moreover seen many burrows in the perpendicularly cut turf-edgings to gravel-walks, into which water could hardly flow, as well plugged as burrows on a level surface. Can the plugs or piles of stones aid in concealing the burrows from scolopenders, which, according to Hoffmeister, are the bitterest enemies of worms? Or may not worms when thus protected be able to remain with safety with their heads close to the mouths of their burrows, which we know that they like to do, but which costs so many of them their lives? Or may not the plugs check the free ingress, of the lowest stratum of air, when chilled by radiation at night, from the surrounding ground and herbage. I am inclined to believe in this latter view; firstly, because when worms were kept in pots in a room with a fire, in which case cold air could not enter the burrows, they plugged them up in a slovenly manner; and secondarily, because they often coat the upper part of their burrows with leaves, apparently to prevent their bodies from coming into close contact with the cold damp earth. But the plugging-up process may perhaps serve for all the above purposes.

Whatever the motive may be, it appears that worms much dislike leaving the mouths of their burrows open. Nevertheless they will reopen them at night, whether or not they can afterwards close them. Numerous open burrows may be seen on recently-dug ground, for in this case the worms eject their castings in cavities left in the ground, or in the old burrows, instead of piling them over the mouths of their burrows, and they cannot collect objects on the surface by which the mouths might be protected. So again on a recently disinterred pavement of a Roman villa at Abinger (hereafter to be described) the worms pertinaciously opened their burrows almost every night, when these had been closed by being trampled on, although they were rarely able to find a few minute stones wherewith to protect them.

Intelligence shown by worms in their manner of plugging up their burrows.—If a man had to plug up a small cylindrical hole, with such objects as leaves, petioles or twigs, he would drag or push them in by their pointed ends; but if these objects were very thin relatively to the size of the hole, he would probably insert some by their thicker or broader ends. The guide in his case would be intelligence. It seemed therefore worth while to observe carefully how worms dragged leaves into their burrows; whether by their tips or bases or middle parts. It seemed more especially desirable to do this in the case of plants not natives to our country; for although the habit of dragging leaves into their burrows is undoubtedly instinctive with worms, yet instinct could not tell them how to act in the case of leaves about which their progenitors knew nothing. If, moreover, worms acted solely through instinct or an unvarying inherited impulse, they would draw all kinds of leaves into their burrows in the same manner. If they have no such definite instinct, we might expect that chance would determine whether the tip, base or middle was seized. If both these alternatives are excluded, intelligence alone is left; unless the worm in each case first tries many different methods, and follows that alone which proves possible or the most easy; but to act in this manner and to try different methods makes a near approach to intelligence.

In the first place 227 withered leaves of various kinds, mostly of English plants, were pulled out of worm-burrows in several places. Of these, 181 had been drawn into the burrows by or near their tips, so that the foot-stalk projected nearly upright from the mouth of the burrow; 20 had been drawn in by their bases, and in this case the tips projected from the burrows; and 26 had been seized near the middle, so that these had been drawn in transversely and were much crumpled. Therefore 80 per cent. (always using the nearest whole number) had been drawn in by the tip, 9 per cent. by the base or foot-stalk, and 11 per cent. transversely or by the middle. This alone is almost sufficient to show that chance does not determine the manner in which leaves are dragged into the burrows.

Of the above 227 leaves, 70 consisted of the fallen leaves of the common lime-tree, which is almost certainly not a native of England. These leaves are much acuminated towards the tip, and are very broad at the base with a well-developed foot-stalk. They are thin and quite flexible when half-withered. Of the 70, 79 per cent. had been drawn in by or near the tip; 4 per cent. by or near the base; and 17 per cent. transversely or by the middle. These proportions agree very closely, as far as the tip is concerned, with those before given. But the percentage drawn in by the base is smaller, which may be attributed to the breadth of the basal part of the blade. We here, also, see that the presence of a foot-stalk, which it might have been expected would have tempted the worms as a convenient handle, has little or no influence in determining the manner in which lime leaves are dragged into the burrows. The considerable proportion, viz., 17 per cent., drawn in more or less transversely depends no doubt on the flexibility of these half-decayed leaves. The fact of so many having been drawn in by the middle, and of some few having been drawn in by the base, renders it improbable that the worms first tried to draw in most of the leaves by one or both of these methods, and that they afterwards drew in 79 per cent. by their tips; for it is clear that they would not have failed in drawing them in by the base or middle.

The leaves of a foreign plant were next searched for, the blades of which were not more pointed towards the apex than towards the base. This proved to be the case with those of a laburnum (a hybrid between *Cytisus alpinus* and *laburnum*) for on doubling the terminal over the basal half, they generally fitted exactly; and when there was any difference, the basal half was a little the narrower. It might, therefore, have been expected that an almost equal number of these leaves would have been drawn in by the tip and base, or a slight excess in favour of the latter. But of 73 leaves (not included in the first lot of 227) pulled out of worm burrows, 63 per cent. had been drawn in by the tip; 27 per cent. by the base, and 10 per cent. transversely. We here see that a far larger proportion, viz., 27 per cent. were drawn in by the base than in the case of lime leaves, the blades of which are very broad at the base, and of which only 4 per cent. had thus been drawn in. We may perhaps account for the fact of a still larger proportion of the laburnum leaves not having been drawn in by the base, by worms having acquired the habit of generally drawing in leaves by their tips and thus avoiding the foot-stalk. For the basal margin of the blade in many kinds of leaves forms a large angle with the foot-stalk; and if such a leaf were drawn in by the foot-stalk, the basal margin would come abruptly into contact with the ground on each side of the burrow, and would render the drawing in of the leaf very difficult.

Nevertheless worms break through their habit of avoiding the foot-stalk, if this part offers them the most convenient means for drawing leaves into their burrows. The leaves of the endless hybridised varieties of the Rhododendron vary much in shape; some are narrowest towards the base and others towards the apex. After they have fallen off, the blade on each side of the midrib often becomes curled up while drying, sometimes along the whole length, sometimes chiefly at the base, sometimes towards the apex. Out of 28 fallen leaves on one bed of peat in my garden, no less than 23 were narrower in the basal quarter than in the terminal quarter of their length; and this narrowness was chiefly due to the curling in of the margins. Out of 36 fallen leaves on another bed, in which different varieties of the Rhododendron grew, only 17 were narrower towards the base than towards the apex. My son William, who first called my attention to this case, picked up 237 fallen leaves in his garden (where the Rhododendron grows in the natural soil) and of these 65 per cent. could have been drawn by worms into their burrows more easily by the base or foot-stalk than by the tip; and this was partly due to the shape of the leaf and in a less degree to the curling in of the margins: 27 per cent. could have been drawn in more easily by the tip than by the base: and 8 per cent. with about equal ease by either end. The shape of a fallen leaf ought to be judged of before one end has been drawn into a burrow, for after this has happened, the free end, whether it be the base or apex, will dry more quickly than the end embedded in the damp ground; and the exposed margins of the free end will consequently tend to become more curled inwards than they were when the leaf was first seized by the worm. My son found 91 leaves which had been dragged by worms into their burrows, though not to a great depth; of these 66 per cent. had been drawn in by the base or foot-stalk; and 34 per cent. by the tip. In this case, therefore, the worms judged with a considerable degree of correctness how best to draw the withered leaves of this foreign plant into their burrows; notwithstanding that they had to depart from their usual habit of avoiding the foot-stalk.

On the gravel-walks in my garden a very large number of leaves of three species of Pinus (*P. austriaca, nigricans* and *sylvestris*) are regularly drawn into the mouths of worm-burrows. These leaves consist of two needles, which are of considerable length in the two first and short in the last named species, and are united to a common base; and it is by this part that they are almost invariably drawn into the burrows. I have seen only two or at most three exceptions to this rule with worms in a state of nature. As the sharply pointed needles diverge a little, and as several leaves are drawn into the same burrow, each tuft forms a perfect *chevaux de frise*. On two occasions many of these tufts were pulled up in the evening, but by the following morning fresh leaves had been pulled in, and the burrows were again well protected. These leaves could not be dragged into the burrows to any depth, except by their bases, as a worm cannot seize hold of the two needles at the same time, and if one alone were seized by the apex, the other would be pressed against the ground and would resist the entry of the seized one. This was manifest in the above mentioned two or three exceptional cases. In order, therefore, that worms should do their work well, they must drag pine-leaves into their burrows by their bases, where the two needles are conjoined. But how they are guided in this work is a perplexing question.

This difficulty led my son Francis and myself to observe worms in confinement during several nights by the aid of a dim light, while they dragged the leaves of the above named pines into their burrows. They moved the anterior extremities of their bodies about the leaves, and on several occasions when they touched the sharp end of a needle they withdrew suddenly as if pricked. But I doubt whether they were hurt, for they are indifferent to very sharp objects, and will swallow even rose-thorns and small splinters of glass. It may also be doubted, whether the sharp ends of the needles serve to tell them that this is the wrong end to seize; for the points were cut off many leaves for a length of about one inch, and fifty-seven of them thus treated were drawn into the burrows by their bases, and not one by the cut-off ends. The worms in confinement often seized the needles near the middle and drew them towards the mouths of their burrows; and one worm tried in a senseless manner to drag them into the burrow by bending them. They sometimes collected many more leaves over the mouths of their burrows (as in the case formerly mentioned of lime-leaves) than could enter them. On other occasions, however, they behaved very differently; for as soon as they touched the base of a pine-leaf, this was seized, being sometimes completely engulfed in their mouths, or a point very near the base was seized, and the leaf was then quickly dragged or rather jerked into their burrows. It appeared both to my son and myself as if the worms instantly perceived as soon as they had seized a leaf in the proper manner. Nine such cases were observed, but in one of them the worm failed to drag the leaf into its burrow, as it was entangled by other leaves lying near. In another case a leaf stood nearly upright with the points of the needles partly inserted into a burrow, but how placed there was not seen; and then the worm reared itself up and seized the base, which was dragged into the mouth of the burrow by bowing the whole leaf. On the other hand, after a worm had seized the base of a leaf, this was on two occasions relinquished from some unknown motive.

As already remarked, the habit of plugging up the mouths of the burrows with various objects, is no doubt instinctive in worms; and a very young one, born in one of my pots, dragged for some little distance a Scotch-fir leaf, one needle of which was as long and almost as thick as its own body. No species of pine is endemic in this part of England, it is therefore incredible that the proper manner of dragging pine-leaves into the burrows can be instinctive with our worms. But as the worms on which the above observations were made, were dug up beneath or near some pines, which had been planted there about forty years, it was desirable to prove that their actions were not instinctive. Accordingly, pine-leaves were scattered on the ground in places far removed from any pine-tree, and 90 of them were drawn into the burrows by their bases. Only two were drawn in by the tips of the needles, and these were not real exceptions, as one was drawn in for a very short distance, and the two needles of the other cohered. Other pine-leaves were given to worms kept in pots in a warm room, and here the result was different; for out of 42 leaves drawn into the burrows, no less than 16 were drawn in by the tips of the needles. These worms, however, worked in a careless or slovenly manner; for the leaves were often drawn in to only a small depth; sometimes they were merely heaped over the mouths of the burrows, and sometimes none were drawn in. I believe that this carelessness may be accounted for by the air of the room being warm, and the worms consequently not being anxious to plug up their holes effectually. Pots tenanted by worms and covered with a net which allowed the entrance of cold air, were left out of doors for several nights, and now 72 leaves were all properly drawn in by their bases.

• • •

If we consider these several cases, we can hardly escape from the conclusion that worms show some degree of intelligence in their manner of plugging up their burrows. Each particular object is seized in too uniform a manner, and from causes which we can generally understand, for the result to be attributed to mere chance. That every object has not been drawn in by its pointed end, may be accounted for by labour having been saved through some being inserted by their broader or thicker ends. No doubt worms are led by instinct to plug up their burrows; and it might have been expected that they would have been led by instinct how best to act in each particular case, independently of intelligence. We see how difficult it is to judge whether intelligence comes into play, for even plants might sometimes be thought to be thus directed; for instance when displaced leaves re-direct their upper surfaces towards the light by extremely complicated movements and by the shortest course. With animals, actions appearing due to intelligence may be performed through inherited habit without any intelligence, although aboriginally thus acquired. Or the habit may have been acquired through the preservation and inheritance of beneficial variations of some other habit; and in this case the new habit will have been acquired independently of intelligence throughout the whole course of its development. There is no *à priori* improbability in worms having acquired special instincts through either of these two latter means. Nevertheless it is incredible that instincts should have been developed in reference to objects, such as the leaves or petioles of foreign plants, wholly unknown to the progenitors of the worms which act in the described manner. Nor are their actions so unvarying or inevitable as are most true instincts.

As worms are not guided by special instincts in each particular case, though possessing a general instinct to plug up their burrows, and as chance is excluded, the next most probable conclusion seems to be that they try in many different ways to draw in objects, and at last succeed in some one way. But it is surprising that an animal so low in the scale as a worm should have the capacity for acting in this manner, as many higher animals have no such capacity. For instance, ants may be seen vainly trying to drag an object transversely to their course, which could be easily drawn longitudinally; though after a time they generally act in a wiser manner. M. Fabre states that a Sphex—an insect belonging to the same highly-endowed order with ants—stocks its nest with paralysed grasshoppers, which are invariably dragged into the burrow by their antennæ. When these were cut off close to the head, the Sphex seized the palpi; but when these were likewise cut off, the attempt to drag its prey into the burrow was given up in despair. The Sphex had not intelligence enough to seize one of the six legs or the ovipositor of the grasshopper, which, as M. Fabre remarks, would have served equally well. So again, if the paralysed prey with an egg attached to it be taken out of the cell, the Sphex after entering and finding the cell empty, nevertheless closes it up in the usual elaborate manner. Bees will try to escape and go on buzzing for hours on a window, one half of which has been left open. Even a pike continued during three months to dash and bruise itself against the glass sides of an aquarium, in the vain attempt to seize minnows on the opposite side. A cobra-snake was seen by Mr. Layard to act much more wisely than either the pike or the Sphex; it had swallowed a toad lying within a hole, and could not withdraw its head; the toad was disgorged, and began to crawl away; it was again swallowed and again disgorged; and now the snake had learnt by experience, for it seized the toad by one of its legs and drew it out of the hole. The instincts of even the higher animals are often followed in a senseless or purposeless manner: the weaver-bird will perseveringly wind threads through the bars of its cage, as if building a nest: a squirrel will pat nuts on a wooden floor, as if he had just buried them in the ground: a beaver will cut up logs of wood and drag them about, though there is no water to dam up; and so in many other cases.

Mr. Romanes who has specially studied the minds of animals, believes that we can safely infer intelligence, only when we see an individual profiting by its own experience. By this test the cobra showed some intelligence; but this would have been much plainer if on a second occasion he had drawn a toad out of a hole by its leg. The Sphex failed signally in this respect. Now if worms try to drag objects into their burrows first in one way and then in another, until they at last succeed, they profit, at least in each particular instance, by experience.

But evidence has been advanced showing that worms do not habitually try to draw objects into their burrows in many different ways. Thus half-decayed lime-leaves from their flexibility could have been drawn in by their middle or basal parts, and were thus drawn into the burrows in considerable numbers; yet a large majority were drawn in by or near the apex. The petioles of the Clematis could certainly have been drawn in with equal ease by the base and apex; yet three times and in certain cases five times as many were drawn in by the apex as by the base. It might have been thought that the foot-stalks of leaves would have tempted the worms as a convenient handle; yet they are not largely used, except when the base of the blade is narrower than the apex. A large number of the petioles of the ash are drawn in by the base; but this part serves the worms as food. In the case of pine-leaves worms plainly show that they at least do not seize the leaf by chance; but their choice does not appear to be determined by the divergence of the two needles, and the consequent advantage or necessity of drawing them into their burrows by the base. With respect to the triangles of paper, those which had been drawn in by the apex rarely had their bases creased or dirty; and this shows that the worms had not often first tried to drag them in by this end.

If worms are able to judge, either before drawing or after having drawn an object close to the mouths of their burrows, how best to drag it in, they must acquire some notion of its general shape. This they probably acquire by touching it in many places with the anterior extremity of their bodies, which serves as a tactile organ. It may be well to remember how perfect the sense of touch becomes in a man when born blind and deaf, as are worms. If worms have the power of acquiring some notion, however rude, of the shape of an object and of their burrows, as seems to be the case, they deserve to be called intelligent; for they then act in nearly the same manner as would a man under similar circumstances.

To sum up, as chance does not determine the manner in which objects are drawn into the burrows, and as the existence of specialized instincts for each particular case cannot be admitted, the first and most natural supposition is that worms try all methods until they at last succeed; but many appearances are opposed to such a supposition. One alternative alone is left, namely, that worms, although standing low in the scale of organization, possess some degree of intelligence. This will strike every one as very improbable; but it may be doubted whether we know enough about the nervous system of the lower animals to justify our natural distrust of such a conclusion. With respect to the small size of the cerebral ganglia, we should remember what a mass of inherited knowledge, with some power of adapting means to an end, is crowded into the minute brain of a worker-ant.

• • •

THE AMOUNT OF FINE EARTH BROUGHT UP BY WORMS TO THE SURFACE.

We now come to the more immediate subject of this volume, namely, the amount of earth which is brought up by worms from beneath the surface, and is afterwards spread out more or less completely by the rain and wind. The amount can be judged of by two methods,—by the rate at which objects left on the surface are buried, and more accurately by weighing the quantity brought up within a given time. We will begin with the first method, as it was first followed.

Near Maer Hall in Staffordshire, quick-lime had been spread about the year 1827 thickly over a field of good pasture-land, which had not since been ploughed. Some square holes were dug in this field in the beginning of October 1837; and the sections showed a layer of turf, formed by the matted roots of the grasses, ½ inch in thickness, beneath which, at a depth of 2½ inches (or 3 inches from the surface), a layer of the lime in powder or in small lumps could be distinctly seen running all round the vertical sides of the holes. The soil beneath the layer of lime was either gravelly or of a coarse sandy nature, and differed considerably in appearance from the overlying dark-coloured fine mould. Coal-cinders had been spread over a part of this same field either in the year 1833 or 1834; and when the above holes were dug, that is after an interval of 3 or 4 years, the cinders formed a line of black spots round the holes, at a depth of 1 inch beneath the surface, parallel to and above the white layer of lime. Over another part of this field cinders had been strewed, only about half-a-year before, and these either still lay on the surface or were entangled among the roots of the grasses; and I here saw the commencement of the burying process, for worm-castings had been heaped on several of the smaller fragments. After an interval of 4¾ years this field was re-examined, and now the two layers of lime and cinders were found almost everywhere at a greater depth than before by nearly 1 inch, we will say by ¾ of an inch. Therefore mould to an average thickness of .22 of an inch had been annually brought up by the worms, and had been spread over the surface of this field.

• • •

The Sinking of great Stones through the Action of Worms.—When a stone of large size and of irregular shape is left on the surface of the ground, it rests, of course, on the more protuberant parts; but worms soon fill up with their castings all the hollow spaces on the lower side; for, as Hensen remarks, they like the shelter of stones. As soon as the hollows are filled up, the worms eject the earth which they have swallowed beyond the circumference of the stones; and thus the surface of the ground is raised all round the stone. As the burrows excavated directly beneath the stone after a time collapse, the stone sinks a little.* Hence it is, that boulders which at some ancient period have rolled down from a rocky mountain or cliff on to a meadow at its base, are always somewhat embedded in the soil; and, when removed, leave an exact impression of their lower surfaces in the underlying fine mould. If, however, a boulder is of such huge dimensions, that the earth beneath is kept dry, such earth will not be inhabited by worms, and the boulder will not sink into the ground.

A lime-kiln formerly stood in a grass-field near Leith Hill Place in Surrey, and was pulled down 35 years before my visit; all the loose rubbish had been carted away, excepting three large stones of quartzose sandstone, which it was thought might hereafter be of some use. An old workman remembered that they had been left on a bare surface of broken bricks and mortar, close to the foundations of the kiln; but the whole surrounding surface is now covered with turf and mould. The two largest of these stones had never since been moved; nor could this easily have been done, as, when I had them removed, it was the work of two men with levers. One of these stones, and not the largest, was 64 inches long, 17 inches broad, and from 9 to 10 inches in thickness. Its lower surface was somewhat protuberant in the middle; and this part still rested on broken bricks and mortar, showing the truth of the old workman's account. Beneath the brick rubbish the natural sandy soil, full of fragments of sandstone was found; and this could have yielded very little, if at all, to the weight of the stone, as might have been expected if the sub-soil had been clay. The surface of the field, for a distance of about 9 inches round the stone, gradually

* This conclusion, which, as we shall immediately see, is fully justified, is of some little importance, as the so-called bench-stones, which surveyors fix in the ground as a record of their levels, may in time become false standards. My son Horace intends at some future period to ascertain how far this has occurred.

FIG. 6.

Transverse section across a large stone, which had lain on a grass-field for 35 years. A A, general level of the field. The underlying brick rubbish has not been represented. Scale ½ inch to one foot.

sloped up to it, and close to the stone stood in most places about 4 inches above the surrounding ground. The base of the stone was buried from 1 to 2 inches beneath the general level, and the upper surface projected about 8 inches above this level, or about 4 inches above the sloping border of turf. After the removal of the stone it became evident that one of its pointed ends must at first have stood clear above the ground by some inches, but its upper surface was now on a level with the surrounding turf. When the stone was removed, an exact cast of its lower side, forming a shallow crateriform hollow, was left, the inner surface of which consisted of fine black mould, excepting where the more protuberant parts rested on the brick-rubbish. A transverse section of this stone, together with its bed, drawn from measurements made after it had been displaced, is here given on a scale of ½ inch to a foot (Fig. 6). The turf-covered border which sloped up to the stone, consisted of fine vegetable mould, in one part 7 inches in thickness. This evidently consisted of worm-castings, several of which had been recently ejected. The whole stone had sunk in the thirty-five years, as far as I could judge, about 1½ inch; and this must have been due to the brick-rubbish beneath the more protuberant parts having been undermined by worms. At this rate the upper surface of the stone, if it had been left undisturbed, would have sunk to the general level of the field in 247 years; but before this could have occurred, some earth would have been washed down by heavy rain from the castings on the raised border of turf over the upper surface of the stone.

• • •

At Stonehenge, some of the outer Druidical stones are now prostrate, having fallen at a remote but unknown period; and these have become buried to a moderate depth in the ground. They are surrounded by sloping borders of turf, on which recent castings were seen. Close to one of these fallen stones, which was 17 ft. long, 6 ft. broad, and 28½ inches thick, a hole was dug; and here the vegetable mould was at least 9½ inches in thickness. At this depth a flint was found, and a little higher up on one side of the hole a fragment of glass. The base of the stone lay about 9½ inches beneath the level of the surrounding ground, and its upper surface 19 inches above the ground.

A hole was also dug close to a second huge stone, which in falling had broken into two pieces; and this must have happened long ago, judging from the weathered aspect of the fractured ends. The base was buried to a depth of 10 inches, as was ascertained by driving an iron skewer horizontally into the ground beneath it. The vegetable mould forming the turf-covered sloping border round the stone, on which many castings had recently been ejected, was 10 inches in thickness; and most of this mould must have been brought up by worms from beneath its base. At a distance of 8 yards from the stone, the mould was only 5½ inches in thickness (with a piece of tobacco pipe at a depth of 4 inches), and this rested on broken flint and chalk which could not have easily yielded to the pressure or weight of the stone.

A straight rod was fixed horizontally (by the aid of a spirit-level)

FIG. 7.

Section through one of the fallen Druidical stones at Stonehenge, showing how much it had sunk into the ground. Scale ½ inch to 1 foot.

across a third fallen stone, which was 7 feet 9 inches long; and the contour of the projecting parts and of the adjoining ground, which was not quite level, was thus ascertained, as shown in the accompanying diagram (Fig. 7) on a scale of ½ inch to a foot. The turf-covered border sloped up to the stone on one side to a height of 4 inches, and on the opposite side to only 2½ inches above the general level. A hole was dug on the eastern side, and the base of the stone was here found to lie at a depth of 4 inches beneath the general level of the ground, and of 8 inches beneath the top of the sloping turf-covered border.

Sufficient evidence has now been given showing that small objects left on the surface of the land where worms abound soon get buried, and that large stones sink slowly downwards through the same means. Every step of the process could be followed, from the accidental deposition of a single casting on a small object lying loose on the surface, to its being entangled amidst the matted roots of the turf, and lastly to its being embedded in the mould at various depths beneath the surface. When the same field was re-examined after the interval of a few years, such objects were found at a greater depth than before. The straightness and regularity of the lines formed by the embedded objects, and their parallelism with the surface of the land, are the most striking features of the case; for this parallelism shows how equably the worms must have worked; the result being, however, partly the effect of the washing down of the fresh castings by rain. The specific gravity of the objects does not affect their rate of sinking, as could be seen by porous cinders, burnt marl, chalk and quartz pebbles, having all sunk to the same depth within the same time. Considering the nature of the substratum, which at Leith Hill Place was sandy soil including many bits of rock, and at Stonehenge, chalk-rubble with broken flints; considering, also, the presence of the turf-covered sloping border of mould round the great fragments of stone at both these places, their sinking does not appear to have been sensibly aided by their weight, though this was considerable.

• • •

We have no means of judging how great a weight of earth a single full-sized worm ejects during a year. Hensen estimates that 53,767 worms exist in an acre of land; but this is founded on the number found in gardens, and he believes that only about half as many live in corn-fields. How many live in old pasture land is unknown; but if we assume that half the above number, or 26,886 worms live on such land, then taking from the previous summary 15 tons as the weight of the castings annually thrown up on an acre of land, each worm must annually eject 20 ounces. A full-sized casting at the mouth of a single burrow often exceeds, as we have seen, an ounce in weight; and it is probable that worms eject more than 20 full-sized castings during a year. If they eject annually more than 20 ounces, we may infer that the worms which live in an acre of pasture land must be less than 26,886 in number.

Worms live chiefly in the superficial mould, which is usually from 4 or 5 to 10 and even 12 inches in thickness; and it is this mould which passes over and over again through their bodies and is brought to the surface. But worms occasionally burrow into the sub-soil to a much greater depth, and on such occasions they bring up earth from this greater depth; and this process has gone on for countless ages. Therefore the superficial layer of mould would ultimately attain, though at a slower and slower rate, a thickness equal to the depth to which worms ever burrow, were there not other opposing agencies at work which carry away to a lower level some of the finest earth which is continually being brought to the surface by worms. How great a thickness vegetable mould ever attains, I have not had good opportunities for observing; but in the next chapter, when we consider the burial of ancient buildings, some facts will be given on this head. In the two last chapters we shall see that the soil is actually increased, though only to a small degree, through the agency of worms; but their chief work is to sift the finer from the coarser particles, to mingle the whole with vegetable débris, and to saturate it with their intestinal secretions.

Finally, no one who considers the facts given in this chapter—on the burying of small objects and on the sinking of great stones left on the surface—on the vast number of worms which live within a moderate extent of ground—on the weight of the castings ejected from the mouth of the same burrow—on the weight of all the castings ejected within a known time on a measured space—will hereafter, as I believe, doubt that worms play an important part in nature.

CONCLUSION.

Worms have played a more important part in the history of the world than most persons would at first suppose. In almost all humid countries they are extraordinarily numerous, and for their size possess great muscular power. In many parts of England a weight of more than ten tons (10,516 kilogrammes) of dry earth annually passes through their bodies and is brought to the surface on each acre of land; so that the whole superficial bed of vegetable mould passes through their bodies in the course of every few years. From the collapsing of the old burrows the mould is in constant though slow movement, and the particles composing it are thus rubbed together. By these means fresh surfaces are continually exposed to the action of the carbonic acid in the soil, and of the humus-acids which appear to be still more efficient in the decomposition of rocks. The generation of the humus-acids is probably hastened during the digestion of the many half-decayed leaves which worms consume. Thus the particles of earth, forming the superficial mould, are subjected to conditions eminently favourable for their decomposition and disintegration. Moreover, the particles of the softer rocks suffer some amount of mechanical trituration in the muscular gizzards of worms, in which small stones serve as millstones.

The finely levigated castings, when brought to the surface in a moist condition, flow during rainy weather down any moderate slope; and the smaller particles are washed far down even a gently inclined surface. Castings when dry often crumble into small pellets and these are apt to roll down any sloping surface. Where the land is quite level and is covered with herbage, and where the climate is humid so that much dust cannot be blown away, it appears at first sight impossible that there should be any appreciable amount of sub-aerial denudation; but worm-castings are blown, especially whilst moist and viscid, in one uniform direction by the prevalent winds which are accompanied by rain. By these several means the superficial mould is prevented from accumulating to a great thickness; and a thick bed of mould checks in many ways the disintegration of the underlying rocks and fragments of rock.

The removal of worm-castings by the above means leads to results which are far from insignificant. It has been shown that a layer of earth, .2 of an inch in thickness, is in many places annually brought to the surface per acre; and if a small part of this amount flows, or rolls, or is washed, even for a short distance down every inclined surface, or is repeatedly blown in one direction, a great effect will be produced in the course of ages. It was found by measurements and calculations that on a surface with a mean inclination of 9° 26', 2.4 cubic inches of earth which had been ejected by worms crossed, in the course of a year, a horizontal line one yard in length; so that 240 cubic inches would cross a line 100 yards in length. This latter amount in a damp state would weigh 11½ pounds. Thus a considerable weight of earth is continually moving down each side of every valley, and will in time reach its bed. Finally this earth will be transported by the streams flowing in the valleys into the ocean, the great receptacle for all matter denuded from the land. It is known from the amount of sediment annually delivered into the sea by the Mississippi, that its enormous drainage-area must on an average be lowered .00263 of an inch each year; and this would suffice in four and half million years to lower the whole drainage-area to the level of the sea-shore. So that, if a small fraction of the layer of fine earth, .2 of an inch in thickness, which is annually brought to the surface by worms, is carried away, a great result cannot fail to be produced within a period which no geologist considers extremely long.

Archæologists ought to be grateful to worms, as they protect and preserve for an indefinitely long period every object, not liable to decay, which is dropped on the surface of the land, by burying it beneath their castings. Thus, also, many elegant and curious tessellated pavements and other ancient remains have been preserved; though no doubt the worms have in these cases been largely aided by earth washed and blown from the adjoining land, especially when cultivated. The old tessellated pavements have, however, often suffered by having subsided unequally from being unequally undermined by the worms. Even old massive walls may be undermined and subside; and no building is in this respect safe, unless the foundations lie 6 or 7 feet beneath the surface, at a depth at which worms cannot work. It is probable that many monoliths and some old walls have fallen down from having been undermined by worms.

Worms prepare the ground in an excellent manner for the growth of fibrous-rooted plants and for seedlings of all kinds. They periodically expose the mould to the air, and sift it so that no stones larger than the particles which they can swallow are left in it. They mingle the whole intimately together, like a gardener who prepares fine soil for his choicest plants. In this state it is well fitted to retain moisture and to absorb all soluble substances, as well as for the process of nitrification. The bones of dead animals, the harder parts of insects, the shells of land-molluscs, leaves, twigs, &c., are before long all buried beneath the accumulated castings of worms, and are thus brought in a more or less decayed state within reach of the roots of plants. Worms likewise drag an infinite number of dead leaves and other parts of plants into their burrows, partly for the sake of plugging them up and partly as food.

The leaves which are dragged into the burrows as food, after being torn into the finest shreds, partially digested, and saturated with the intestinal and urinary secretions, are commingled with much earth. This earth forms the dark-coloured, rich humus which almost everywhere covers the surface of the land with a fairly well-defined layer or mantle. Von Hensen placed two worms in a vessel 18 inches in diameter, which was filled with sand, on which fallen leaves were strewed; and these were soon dragged into their burrows to a depth of 3 inches. After about 6 weeks an almost uniform layer of sand, a centimeter (.4 inch) in thickness, was converted into humus by having passed through the alimentary canals of these two worms. It is believed by some persons that worm-burrows, which often penetrate the ground almost perpendicularly to a depth of 5 or 6 feet, materially aid in its drainage; notwithstanding that the viscid castings piled over the mouths of the burrows prevent or check the rain-water directly entering them. They allow the air to penetrate deeply into the ground. They also greatly facilitate the downward passage of roots of moderate size; and these will be nourished by the humus with which the burrows are lined. Many seeds owe their germination to having been covered by castings; and others buried to a considerable depth beneath accumulated castings lie dormant, until at some future time they are accidentally uncovered and germinate.

Worms are poorly provided with sense-organs, for they cannot be said to see, although they can just distinguish between light and darkness; they are completely deaf, and have only a feeble power of smell; the sense of touch alone is well developed. They can therefore learn little about the outside world, and it is surprising that they should exhibit some skill in lining their burrows with their castings and with leaves, and in the case of some species in piling up their castings into tower-like constructions. But it is far more surprising that they should apparently exhibit some degree of intelligence instead of a mere blind instinctive impulse, in their manner of plugging up the mouths of their burrows. They act in nearly the same manner as would a man, who had to close a cylindrical tube with different kinds of leaves, petioles, triangles of paper, &c., for they commonly seize such objects by their pointed ends. But with thin objects a certain number are drawn in by their broader ends. They do not act in the same unvarying manner in all cases, as do most of the lower animals; for instance, they do not drag in leaves by their foot-stalks, unless the basal part of the blade is as narrow as the apex, or narrower than it.

When we behold a wide, turf-covered expanse, we should remember that its smoothness, on which so much of its beauty depends, is mainly due to all the inequalities having been slowly levelled by worms. It is a marvellous reflection that the whole of the superficial mould over any such expanse has passed, and will again pass, every few years through the bodies of worms. The plough is one of the most ancient and most valuable of man's inventions; but long before he existed the land was in fact regularly ploughed, and still continues to be thus ploughed by earth-worms. It may be doubted whether there are many other animals which have played so important a part in the history of the world, as have these lowly organised creatures. Some other animals, however, still more lowly organised, namely corals, have done far more conspicuous work in having constructed innumerable reefs and islands in the great oceans; but these are almost confined to the tropical zones.

EPILOGUE

If the verb "darwinise" preceded Charles Darwin's birth, the amorphous noun "Darwinism" has long survived him. This is not to say that the ideas he broached in the *Origin* and amplified in his subsequent works were immediately or eventually elevated to the status of paradigm and accepted as the point of departure for all scientific discussion since his day. During Darwin's lifetime, his work gained wide popular and scientific acceptance for evolution as a natural, rather than divinely directed, process. But comparatively few people accepted his claim that natural selection, which even Darwin saw as but the chief among several factors, is the principal explanation of evolutionary change. By the last decades of the nineteenth century, Darwin's theories were being assailed from a number of directions.

Although few scientists continued to support creationism or theistic evolution, religious opposition to Darwinism did not by any means become extinct; and a number of alternative scientific explanations rose up. Chief among these competing models were a resurgent Lamarckism, orthogenesis, and mutation theory. As the historian of science Peter J. Bowler has pointed out, the different evolutionary camps were not precisely defined entities. Overlapping territory existed between camps, and divisions were discernible within camps. For purposes of basic understanding, though, several key distinctions separate the competing theories—the main distinctions being whether evolution is an orderly process or an irregular diversification, whether conditions of the external environment or internal forces drive the process, and whether it is continuous, with cumulative change, or discontinuous, with evolutionary leaps.

Of these evolutionary camps, Lamarckism was perhaps the most comfortable for idealists. The evolutionary theory of the eighteenth-century French biologist Jean Baptiste de Lamarck was based on two tenets—the innate drive toward complexity that pushed all life-forms toward higher stages of development, and the ability of an organism to pass on characteristics acquired during its lifetime to its progeny. In asserting that new behavioral patterns, whether these involve conscious or unconscious responses to environmental pressures, produce inheritable bodily changes, Lamarckism allowed room for active purpose in evolution—for direction by some sort of "life force." Attractive to such speculators as Herbert Spencer, Samuel Butler, and George Bernard Shaw, it tended to have less appeal for naturalists because its workings could not be observed in nature or verified through experimentation: a blacksmith's children do not inherit the iron-band arms their father developed.

Like Lamarckism, orthogenesis retained direction as crucial to evolution. Orthogenesis literally means "linear development," but the German zoologist Theodor Eimer, its chief proponent up to the turn of the twentieth century, used the term more restrictively. For Eimer, orthogenesis had to do with nonadaptive linear development. Under orthogenesis, the stimulus to develop along a particular path comes from within and has no relation to environmental demands. Thus an evolving species blindly follows its own course down a road that may lead to extinction. The fossil record, with its numerous apparent examples of elaboration leading to dead ends, offered particularly compelling evidence to those asserting a nonadaptive, internally directed idea of evolution.

Meanwhile Darwinism, which had in Darwin's day been a broad tent under which diverse theories (such as Lamarckism) might shelter, had narrowed. Neo-Darwinism, which depended mainly upon the work of the German naturalist August Weismann, dismissed the inheritance of acquired characteristics as biologically impossible. Having hypothesized that the chromosomes in the nucleus of the cell were the source of genetic information, Weismann asserted that both parents make equal contributions to the hereditary character of their progeny and, further, that individual characteristics are transmitted independent of one another. Parting company with Darwin's old unsatisfactory hyopothesis of pangenesis, Weismann claimed that because the germ plasm, in sperm and egg, is isolated from other cells of the body, those other cells modified by a lifetime of adaptive change have no means of passing on their acquired characteristics.

Julian Huxley, the grandson of Darwin's "bulldog" T. H. Huxley and himself an eminent biologist, has observed that "With the rediscovery of Mendel's laws of heredity in 1900, one would have thought that natural selection would have returned as the explanation for evolutionary change." Nevertheless, the first result of appreciating Mendel's significance was not a strengthening of Darwinian theory but a further attack on it. Weismann and the other Darwinians, most of them naturalists, continued to affirm that natural selection operates through small, gradual changes that are perpetuated because they give their possessors a slight competitive advantage in a particular environment. In contrast the mutationist camp of Mendelian geneticists, mostly laboratory scientists, believed that genetic mutations express themselves as sudden phenotypic variations and thus that evolution proceeds discontinuously, by abrupt jumps rather than by steady creeping. Because Mendel's experiments with garden peas focused on pairs of physical characteristics that differed in absolute ways, they seemed to leave no room for gradualism.

Since the 1930s, however, mutation and variation have come together in what has been called the Modern Synthesis, which wedded Darwinian theory with the population genetics of R. A. Fisher and J. B. S. Haldane in Britain and Sewell Wright in the United States. The Modern Synthesis has established Darwinian theory on a firmer hereditary footing than was available in his day by grounding it in experimental genetics and population study. Key developments have included the recognition that radical mutations are exceptions rather than rules and that organisms cannot ordinarily survive them, the understanding that relationships between genetic factors and somatic characteristics are not straightforward one-to-one correspondences but that many genes may affect a character, and that evolution is most profitably studied through populations rather than individuals. The result: revision of both natural selection and genetic theory to remove the oversimplifications that seemed to thwart their reconciliation. This reconciliation, however, has not meant the end to all controversy. Experimental studies of variation and heredity continue to seek cases for modified developmental and progressionist theories. Relying largely on the fact that the fossil record does not demonstrate gradual evolution, Niles Eldredge and Stephen Jay Gould have argued against evolutionary continuity with their idea of "punctuated equilibrium"—that the rise of new species proceeds swiftly and that once established a species remains essentially unchanged. Creationism continues to offer a non-scientific alternative to evolution. Although there has been no last word in the Darwinian debates, Darwin's unshakable contribution to science was his demonstration that evolution (whatever the mechanisms) is the natural process by which life has changed through time.

FURTHER READING

The literature on Darwin is vast. A complete bibliography might be as long as this book. We include here some important recent works that will lead the interested reader to the secondary Darwin literature, plus a few classics. All of Darwin's books are currently in print, although not usually in their first editions. Books in this list that are available in paperback are indicated by an asterisk.

Allan, Mea. *Darwin and His Flowers. The Key to Natural Selection.* London: Faber & Faber; New York: Taplinger, 1977. The only work devoted to Darwin's botanical research.

*Barlow, Nora, ed. *The Autobiography of Charles Darwin 1809–1882. With Original Omissions Restored.* London: Collins, 1958; reprinted New York: Norton, 1969. Edited by Darwin's granddaughter.

*Barrett, Paul H., ed. *The Collected Papers of Charles Darwin.* 2 vols. Chicago: University of Chicago Press, 1977.

Barrett, Paul H., Peter J. Gautrey, Sandra Herbert, David Kohn, and Sydney Smith, transcribers and eds. *Charles Darwin's Notebooks, 1836–1844. Geology, Transmutation of Species, Metaphysical Enquiries.* London: British Museum (Natural History); Ithaca, NY: Cornell University Press, 1987.

*Bowlby, John. *Charles Darwin. A New Life.* New York and London: Norton, 1990. One of the best recent biographies.

*Bowler, Peter J. *The Eclipse of Darwinism: Anti-Darwinian Evolution Theories in the Decades around 1900.* Baltimore: Johns Hopkins University Press, 1983. What happened to Darwinism after Darwin.

*Bowler, Peter J. *Evolution: The History of an Idea.* Berkeley and Los Angeles: University of California Press, rev. ed., 1989. From Genesis to the present.

Burkhardt, Frederick, and Sydney Smith, eds. *The Correspondence of Charles Darwin. Volume 1 1821–1836.* Cambridge: Cambridge University Press, 1985. *Volume 2 1837–1843*, 1986. *Volume 3 1844–1846*, 1987. *Volume 4 1847–1850*, 1988. *Volume 5 1851–1855*, 1989. *Volume 6 1856–1857*, 1990. *Volume 7 1858–1859*, 1991. The definitive work.

Burkhardt, Frederick, Duncan M. Porter, Janet Browne, and Marsha Richmond, eds. *The Correspondence of Charles Darwin. Volume 8 1860.* Cambridge: Cambridge University Press, 1993.

*Darwin, Charles. *The Illustrated Origin.* Abridged and introduction by Richard Leakey. London: Faber & Faber, 1979.

Darwin, Francis, ed. *The Life and Letters of Charles Darwin, Including an Autobiographical Chapter.* 3 vols. London: John Murray, 1887; 2 vols. New York: D. Appleton, 1887.

Darwin, Francis, and A. C. Seward, eds. *More Letters of Charles Darwin. A Record of His Work in a Series of Hitherto Unpublished Letters.* London: John Murray, 1903.

Desmond, Adrian, and James R. Moore. *Darwin.* London: Michael Joseph, 1991; New York: Warner Books, 1992. Based on the most recent research; reads like a novel.

Freeman, R. B. *The Works of Charles Darwin. An Annotated Bibliographical Handlist.* Folkestone, Kent: Dawson; Hamden, CT: Archon Books, 2nd. ed., 1977.

———. *Charles Darwin: A Companion.* Folkestone, Kent: Dawson; Hamden, CT: Archon Books, 1978. A Darwin encyclopedia.

*Glick, Thomas F., ed. *The Comparative Reception of Darwinism.* Austin, TX: University of Texas Press, 1974; reprinted Chicago: University of Chicago Press, 1988. How they saw Darwin from England to the Islamic world.

*Gould, Stephen Jay. *Ever Since Darwin: Reflections in Natural History.* New York and London: Norton, 1974. The first in a series of collected essays, with Darwin being one of the main themes.

*Gruber, Howard E. *Darwin on Man: A Psychological Study of Scientific Creativity.* Chicago: University of Chicago Press, 2nd. ed., 1981.

*Hull, David L. *Darwin and His Critics.* Cambridge, MA: Harvard

University Press, 1973, reprinted Chicago: University of Chicago Press, 1983. Commentary on reviews of *The Origin of Species.*

Keynes, R. D., ed. *Charles Darwin's Beagle Diary.* Cambridge: Cambridge University Press, 1988.

*Kohn, David. *The Darwinian Heritage.* Princeton: Princeton University Press, 1985. Essays on Darwin and his influence.

*Mayr, Ernst. *The Growth of Biological Thought: Diversity, Evolution, and Inheritance.* Cambridge, MA: Harvard University Press, 1982. A history by one of the main participants in the forging of the Modern Synthesis.

*Miller, Jonathan, and Borin Van Loon. *Darwin for Beginners.* London: Writers and Readers, 1982. The best short introduction: lucid, entertaining, cleverly illustrated.

*Moore, James R. *The Post-Darwinian Controversies: A Study of the Protestant Struggle to Come to Terms with Darwin in Great Britain and America.* Cambridge: Cambridge University Press, 1979.

*Moorehead, Alan. *Darwin and the Beagle.* London: Penguin; New York: Viking Penguin, rev. ed., 1979. Pretty pictures, readable text.

*Ruse, Michael. *Darwinism Defended. A Guide to the Evolution Controversies.* Reading, MA: Addison Wesley, 1982.

*Stauffer, Robert C., ed. *Charles Darwin's Natural Selection, Being the Second Part of His Big Species Book Written from 1856 to 1858.* Cambridge: Cambridge University Press, 1975.

THE VIKING PORTABLE LIBRARY

The Portable Sherwood Anderson
Edited by Horace Gregory

The Portable Beat Reader
Edited by Ann Charters

The Portable Blake
Edited by Alfred Kazin

The Portable Cervantes
Edited by Samuel Putnam

The Portable Chaucer
Edited by Theodore Morrison

The Portable Chekhov
Edited by Avrahm Yarmolinsky

The Portable Conrad
Edited by Morton Dauwen Zabel

The Portable Conservative Reader
Edited by Russell Kirk

The Portable Malcolm Cowley
Edited by Donald W. Faulkner

The Portable Stephen Crane
Edited by Joseph Katz

The Portable Dante
Edited by Paolo Milano

The Portable Emerson
Edited by Carl Bode and Malcolm Cowley

The Portable Faulkner
Edited by Malcolm Cowley

The Portable Greek Historians
Edited by M. I. Finley

The Portable Greek Reader
Edited by W. H. Auden

The Portable Graham Greene
Edited by Philip Stratford

The Portable Hawthorne
Edited by Malcolm Cowley

The Portable Henry James
Edited by Morton Dauwen Zabel

The Portable Thomas Jefferson
Edited by Merrill D. Peterson

The Portable James Joyce
Edited by Harry Levin

The Portable Jung
Edited by Joseph Campbell

The Portable Kipling
Edited by Irving Howe

The Portable D. H. Lawrence
Edited by Diana Trilling

The Portable Abraham Lincoln
Edited by Andrew Delbanco

The Portable Machiavelli
Edited by Peter Bondanella and Mark Musa

The Portable Karl Marx
Edited by Eugene Kamenka

The Portable Medieval Reader
Edited by James Bruce Ross and Mary Martin McLaughlin

The Portable Arthur Miller
Edited by Harold Clurman

The Portable Milton
Edited by Douglas Bush

The Portable Nietzsche
Edited by Walter Kaufmann

The Portable North American Indian Reader
Edited by Frederick Turner

The Portable Dorothy Parker
Edited by Brendan Gill

The Portable Plato
Edited by Scott Buchanan

The Portable Poe
Edited by Philip Van Doren Stern

The Portable Romantic Poets
Edited by W. H. Auden and Norman Holmes Pearson

The Portable Renaissance Reader
Edited by James Bruce Ross and Mary Martin McLaughlin

The Portable Roman Reader
Edited by Basil Davenport

The Portable Shakespeare
Edited by Marshall Best

The Portable Bernard Shaw
Edited by Stanley Weintraub

The Portable Steinbeck
Edited by Pascal Covici, Jr.

The Portable Swift
Edited by Carl Van Doren

The Portable Thoreau
Edited by Carl Bode

The Portable Tolstoy
Edited by John Bayley

The Portable Mark Twain
Edited by Bernard De Voto

The Portable Twentieth-Century Russian Reader
Edited by Clarence Brown

The Portable Victorian Reader
Edited by Gordon S. Haight

The Portable Voltaire
Edited by Ben Ray Redman

The Portable Walt Whitman
Edited by Mark Van Doren

The Portable Oscar Wilde
Edited by Richard Aldington and Stanley Weintraub

The Portable World Bible
Edited by Robert O. Ballou

THE VIKING PORTABLE LIBRARY

Darwin

CHARLES DARWIN was born at Shrewsbury in 1809 and educated at Shrewsbury Grammar School, at Edinburgh University, and Christ's College, Cambridge. He took his degree in 1831 and in the same year embarked on a five-year voyage on HMS *Beagle* as a naturalist; the purpose of the voyage was to chart the coasts of Patagonia and Tierra del Fuego, and to carry a chain of chronometric readings round the world. While he was away some of his letters on scientific matters had been privately published, and on his return he at once took his place among the leading men of science. He became secretary of the Geological Society in 1838, and in 1839 he was elected a Fellow of the Royal Society. Most of the rest of his life was occupied in publishing the findings of the voyage and in documenting his theory of the transmutation of species. *The Origin of Species by Means of Natural Selection* appeared in 1859. It provoked strong responses, some with violent and malicious criticism. Darwin spent his last 40 years with his wife, his cousin Emma Wedgwood, whom he had married in 1839, and their children at Down House in Kent. He died in 1882 and was buried in Westminster Abbey.

DUNCAN M. PORTER is professor of botany at Virginia Polytechnic Institute and State University and senior editor of the Darwin Correspondence Project at the Cambridge University Library. He is coauthor, with Ira L. Wiggins, of the *Flora of the Galapagos Islands* and has published extensively on Darwin's plant, animal, and geological collections from the voyage of the *Beagle*.

PETER W. GRAHAM is professor of English at Virginia Polytechnic Institute and State University. His publications on nineteenth-century British literature and culture include *Byron's Bulldog*, *Don Juan and Regency England*, and *Articulating the Elephant Man: Joseph Merrick and His Interpreters*, written with Fritz Oehlschlaeger.